Protective Thin Coatings Technology

Advances in Materials Science and Engineering
Series Editor
Sam Zhang

Biological and Biomedical Coatings Handbook: Processing and Characterization
Sam Zhang

Aerospace Materials Handbook
Sam Zhang and Dongliang Zhao

Thin Films and Coatings: Toughening and Toughness Characterization
Sam Zhang

Semiconductor Nanocrystals and Metal Nanoparticles: Physical Properties and Device Applications
Tupei Chen and Yang Liu

Advances in Magnetic Materials: Processing, Properties, and Performance
Sam Zhang and Dongliang Zhao

Micro- and Macromechanical Properties of Materials
Yichun Zhou, Li Yang, and Yongli Huang

Nanobiomaterials: Development and Applications
Dong Kee Yi and Georgia C. Papaefthymiou

Biological and Biomedical Coatings Handbook: Applications
Sam Zhang

Hierarchical Micro/Nanostructured Materials: Fabrication, Properties, and Applications
Weiping Cai, Guotao Duan, and Yue Li

Biological and Biomedical Coatings Handbook, Two-Volume Set
Sam Zhang

Nanostructured and Advanced Materials for Fuel Cells
San Ping Jiang and Pei Kang Shen

Hydroxyapatite Coatings for Biomedical Applications
Sam Zhang

Carbon Nanomaterials: Modeling, Design, and Applications
Kun Zhou

Materials for Energy
Sam Zhang

Protective Thin Coatings and Functional Thin Films Technology, Two-Volume Set
Sam Zhang, Jyh-Ming Ting, Wan-Yu Wu

For more information about this series, please visit:
https://www.routledge.com/Advances-in-Materials-Science-and-Engineering/book-series/
CRCADVMATSCIENG

Series: Advances in Materials Science and Engineering
Series Editor: Sam Zhang

Advances in Materials Science and Engineering
Series Statement
Materials form the foundation of technologies that govern our everyday life, from housing and household appliances to handheld phones, drug delivery systems, airplanes, and satellites. Development of new and increasingly tailored materials is the key to advance important applications with the potential to dramatically enhance and enrich our experiences.

The *Advances in Materials Science and Engineering* series by CRC Press/Taylor & Francis is designed to help meet new and exciting challenges in Materials Science and Engineering disciplines. The books and monographs in the series are based on cutting-edge research and development, and thus are up-to-date with new discoveries, new understanding, and new insights in all aspects of materials development, including processing and characterization and applications in metallurgy, bulk or surface engineering, interfaces, thin films, coatings, and composites, just to name a few.

The series aims at delivering an authoritative information source to readers in academia, research institutes, and industry. The Publisher and its Series Editor are fully aware of the importance of Materials Science and Engineering as the foundation for many other disciplines of knowledge. As such, the team is committed to making this series the most comprehensive and accurate literary source to serve the whole materials world and the associated fields.

As Series Editor, I'd like to thank all authors and editors of the books in this series for their noble contributions to the advancement of Materials Science and Engineering and to the advancement of humankind.

Sam Zhang

Protective Thin Coatings Technology

Edited by Sam Zhang, Jyh-Ming Ting and Wan-Yu Wu

CRC Press
Taylor & Francis Group
Boca Raton London New York

CRC Press is an imprint of the
Taylor & Francis Group, an **informa** business

First edition published 2022
by CRC Press
6000 Broken Sound Parkway NW, Suite 300, Boca Raton, FL 33487-2742

and by CRC Press
2 Park Square, Milton Park, Abingdon, Oxon, OX14 4RN

© 2022 Taylor & Francis Group, LLC

CRC Press is an imprint of Taylor & Francis Group, LLC

ISBN: 978-0-367-54250-4 (hbk)
ISBN: 978-0-367-54248-1 (pbk)
ISBN: 978-1-003-08834-9 (ebk)

Typeset in Times
by SPi Global, India

Contents

Preface to Protective Thin Coatings and Functional Thin Films Technology – 2-Volume Set

Over the decades, films and coatings have been developed and applied in industries that spread all over people's life of the current society and the defense. Films and coatings also evolved from single compound to multicompound to multilayer and to nanostructures and nanocomposites. All the portable electronic devices, such as cell phone and iPad, and removable storage media, such as memory cards and USB flash drives, involve, heavily, the use of micro-/nanoscaled films and coatings. In all, films or coatings either provide protection of the surface they are attached to or provide certain functionality through the film itself. To capture the most recent advances in both aspects, we put together a 2-volume set: one on protection and other on functional applications.

Protective Thin Coatings Technology
Functional Thin Films Technology

Films and Coatings, in essence, are two different things. Coatings have to realize their usefulness through attaching on to the surface of a substrate, as in the case of hard coatings on a mode or drill to provide lubrication for antisticking or hardening of the drill surface to lengthen drill life. Films, on the other hand, provide their functionality by standing alone with or without substrate. Even in the case of having a substrate, the substrate is there to provide a backing support, not necessarily having too much to do with the functionality the films are there to provide. As such, for protective applications, we use "coatings" (thus "thin" coatings are used only to differentiate from "thick" coatings) and for functional applications, we use "films".

Protective Thin Coatings Technology focuses on deposition/processing technologies and property characterizations in protective applications. *Functional Thin Films Technology* deals with deposition/processing technologies and property characterizations in functional applications.

In some fields, deposition and processing carry somewhat different meanings. For instance, in device-making technologies, processing includes assembly. However, in other fields, it simply means making or fabrication of films or coatings. As here, we do not focus on assembly of devices, but give ourselves the freedom of using both deposition and processing to refer to fabrication.

To be specific, *Protective Thin Coatings Technology* covers technologies for Sputtering of Flexible Hard Nanocoatings, Deposition of Solid Lubricating Films, Multilayer Transition Metal Nitrides, Integrated Nanomechanical Characterisation of Hard Coatings, Corrosion and Tribo-Corrosion of Hard Coating, High-Entropy Alloy Films and Coatings, Thin Films and Coatings for High-Temperature Applications, Nanocomposite Coating on Magnesium Alloys, and the Correlation Between Coating Properties and Industrial Applications.

Functional Thin Films Technology deals with technologies aiming at functionality when used in Nanoelectronics, Solar Selective Absorbers, Solid Oxide Fuel Cells, Piezo Applications, Sensors, Absorbers, Catalysts, Anodic Aluminum Oxide, Superhydrophobics, Semiconductor Devices, etc. Also included is a chapter that deals with the Transport Phenomena in Nanostructured Coating.

In summary, these two books highlight the development and advance in the preparation, characterization, and applications of protective and functional micro-/nanoscaled films and coatings. People working in areas related to semiconductor, optoelectronics, plasma technology, solid-state energy storages, 5G, etc., and students studying electrical, mechanical, chemical, materials, etc., engineering will find these books useful. To be specific, these include Senior Undergraduate Students, Graduate Students, Industry Professionals, Researchers, and Academics.

The editors would like to thank all chapter authors for their active contribution and timely effort to ensure the smooth publication of these high-quality books to catch the new trend and development in the topical matters. The editors thank Allison and Gabrielle of the publisher for their support along the way. Sam would also like to acknowledge the Fundamental Research Funds for the Central Universities SWU118105.

Editors

Prof. Sam Zhang Shanyong (張善勇), academically better known as Sam Zhang, was born and brought up in the famous "City of Mountains" Chongqing, China. He received his Bachelor of Engineering in Materials in 1982 from Northeastern University (Shenyang, China), Master of Engineering in Materials in 1984 from Iron & Steel Research Institute (Beijing, China) and PhD Degree in Ceramics in 1991 from the University of Wisconsin-Madison, USA. He was a tenured full professor (since 2006) at the School of Mechanical and Aerospace Engineering, Nanyang Technological University. Since January 2018, he joined School of Materials and Energy, Southwest University, Chongqing, China, and assumed duty as Director of the Centre for Advanced Thin Film Materials and Devices of the university.

Prof. Zhang was the founding Editor-in-Chief for *Nanoscience and Nanotechnology Letters* (2008 to December 2015) and Principal Editor for the *Journal of Materials Research* (USA) responsible for thin films and coating field (since 2003). Prof. Zhang has been serving the world's first "Thin Films Society" (www.thinfilms.sg) as its founding and current president since 2009. Prof. Zhang has authored/edited 13 books, of which 12 were published with CRC Press/Taylor & Francis. Of these books, *Materials Characterization Techniques* has been adopted as core textbook by more than 30 American and European universities since October 2015. That book had also been translated into Chinese and published by China Sicence Publishing Co in October 2010 and distributed nationwide in China (available online at Amazon.cn). Very recently, Prof. Zhang has published his book *Materials for Energy* (Sam Zhang (ed.), October 2020: 6-1/8 x 9-1/4: 528pp Hb: 978-0-367-35021-5 eBook: 978-0-429-35140-2. URL:https://www.routledge.com/Materials-for-Energy/Zhang/p/book/9780367350215. Prof. Zhang's new book on "Materials for Devices", to be published by CRC Press, is also in the pipeline. Meanwhile, Prof. Zhang is the Series Editor for *Advances in Materials Science and Engineering* Book Series published by CRC Press/Taylor & Francis.

Prof. Zhang has been elected as Fellow of Royal Society of Chemistry (FRSC) and Fellow of Thin Films Society (FTFS) in 2018, and Fellow of Institute of Materials, Minerals and Mining (FIoMMM) in 2007. Prof. Zhang's current research centers on areas of Energy Films and Coatings for solar cells, hard yet tough nanocomposite coatings for tribological applications by physical vapor deposition; measurement of fracture toughness of ceramic films and coatings, and Electronic/Optical Thin Films. Over the years, he has authored/co-authored over 360 peer-reviewed international journal papers. As of December 8, 2020, as per Web of Science (https://publons.com/researcher/2817766/sam-zhang/), the sum of the times cited is 10,342, with average citations per article: 29.1, h-index: 54.

Prof. Zhang holds Guest Professorship at Shenyang University (2002), Institute of Solid State Physics, Chinese Academy of Sciences (2004), Zhejiang University (2006), Harbin Institute of Technology (2007), Xian Jiaotong University (2018), Kashi University (2018), and Xiamen University of Science and Technology (2020). Prof. Zhang also serves as an International Advisor to Shenzhen Association for Vacuum Technology Industries.

Dr. Jyh-Ming Ting is an Advanced Semiconductor Engineering Chair Professor of the Department of Materials Science and Engineering at National Cheng Kung University (NCKU) in Taiwan. He received a BS degree in Nuclear Engineering from National Tsing Hua University in Taiwan 1982, and completed MS and PhD works at the Department of Materials Science and Engineering in University of Cincinnati in 1987 and 1991, respectively. From 1990 to 1997, Dr. Ting worked at Applied Sciences, Inc. (ASI), as a Scientist and then the R&D Director. In August 1997, Dr. Ting became a faculty in NCKU, working on various carbon materials and low-dimensional materials, and their (nano)composites. At the early stage, nanoscaled (composite) thin films and CNT were of interest. In particular, Prof. Ting developed and patented several novel techniques that allow the growth of aligned CNTs on metallic substrates, also holding the record of the growth rate until now. In recent years, in response to the serious issues related to environment, his research focuses on applying low-dimensional materials, mainly nanoparticles and 2D layered structures, and their composites, to energy generation/storage and photodegradation. More recently, Prof. Ting is working in high-entropy oxides. Prof. Ting has authored over 170 journal articles, given more than 30 invited talks and a few keynote speeches, and holds more than 30 patents.

Dr. Wan-Yu Wu was born in Taipei, Taiwan. She obtained her Bachelor's degree in 2000 and PhD degree in 2008 from the Department of Materials Science and Engineering of National Cheng Kung University, Tainan, Taiwan. In 2010, Dr. Wu became an Assistant Professor in MingDao University in Taiwan where she was also a key member of the Surface & Engineering Research Center. In 2014, she joined the Department of Materials Science and Engineering in Da-Yeh University in Taiwan as an Associate Professor. Due to her well-recognized expertise in thin film and coating technology, Prof. Wu was commissioned to establish a Coatings Technology Research Center in Da-Yeh university and became a full professor in 2019. Over the years, she has focused her research on thin film and nanostructure materials synthesized using High-Power Impulse Magnetron Sputtering (HiPIMS) deposition and wet chemistry processes, respectively. Prof. Wu also actively participates in professional society. She is in the board of directors in Taiwan Association for Coating and Thin Film Technology (TACT). She also serves as a symposium chair in several international conferences including International Conference on Metallurgical Coatings and Thin Films (ICMCTF), Thin Films Conference, and TACT International Thin Films Conference.

Contributors

Ben D. Beake
Micro Materials Ltd.
Willow House, Ellice Way, Yale Business Village
Wrexham, UK

Materials Research Institute
University of Huddersfield
Huddersfield, UK

Faculty of Science and Engineering
Manchester Metropolitan University
Manchester, UK

Yin-Yu Chang
Department of Mechanical and Computer-
    Aided Engineering
National Formosa University
Taiwan

Dandan Chen
School of Mechanical, Electronic and Control
    Engineering
BeiJing Jiaotong University
BeiJing, China

Yujie Chen
Centre for Advanced Thin Films and Devices,
    School of Materials and Energy
Southwest University
Chongqing, China

and

School of Mechanical Engineering
The University of Adelaide
Adelaide, Australia

Yung-I Chen
Department of Optoelectronics and Materials
    Technology
National Taiwan Ocean University
Keelung, Taiwan

Jenq-Gong Duh
Department of Materials Science and
    Engineering
National Tsing Hua University
Hsinchu, Taiwan

Jui-Ting Hsu
School of Dentistry
China Medical University
Taiwan

Heng-Li Huang
School of Dentistry
China Medical University
Taiwan

Yuehua Huang
College of Engineering and Technology
Southwest University
Chongqing, China

Zhe Jiang
School of Mechanical, Electronic and Control
    Engineering
BeiJing Jiaotong University
BeiJing, China

JieJin
School of Mechanical, Electronic and Control
    Engineering
BeiJing Jiaotong University
BeiJing, China

Šimon Kos
Department of Physics, Faculty of Applied
    Sciences
University of West Bohemia
Czech Republic

Jyh-Wei Lee
Department of Materials Engineering
Ming Chi University of Technology
New-Taipei City, Taiwan

Bin Liao
School of Nuclear Science and Technology
Beijing Normal University
Beijing, China

Tomasz W. Liskiewicz
Faculty of Science and Engineering
Manchester Metropolitan University
Manchester, UK

Xingmin Liu
School of Materials Science and Engineering
Northwestern Polytechnical University
Xi'an, PR China

Xingang Luan
School of Materials Science and Engineering
Northwestern Polytechnical University
Xi'an, PR China

Paul Munroe
School of Materials Science and Engineering
The University of New South Wales
Sydney, Australia

Jindřich Musil
Department of Physics, Faculty of Applied
    Sciences
University of West Bohemia
Pilsen, Czech Republic

Yuchang Qing
School of Materials Science and Engineering
Northwestern Polytechnical University
Xi'an, PR China

Deen Sun
School of Materials and Energy
Southwest University
Chongqing, P.R. China

Ming-Tzu Tsai
Department of Biomedical Engineering
Hungkuang University
Taiwan

Vladimir M. Vishnyakov
Materials Research Institute
University of Huddersfield
Huddersfield, UK

Fan-Bean Wu
Department of Materials Science and
    Engineering
National United University
Miaoli, Taiwan

Wenling Xie
School of Materials and Energy
Southwest University
Chongqing, China

Zonghan Xie
School of Mechanical Engineering
The University of Adelaide
Adelaide, Australia

Sam Zhang
Centre for Advanced Thin Films and Devices,
    School of Materials and Energy
Southwest University
Chongqing, China

1 Advanced Sputtering Technologies of Flexible Hard Nanocoatings

Jindřich Musil and Šimon Kos
University of West Bohemia, Czech Republic

CONTENTS

1.1 INTRODUCTION

Advanced hard nanocoatings should exhibit new unique thermally stable properties and an enhanced resistance to cracking. Only such nanocoatings can be good protective and functional coatings due to avoiding cracking causing changes of their properties during the operation. The creation of the flexible nanocoatings resistant to cracking is, however, a difficult and important task. On the other hand, it is a very huge challenge for the development of new advanced nanocoatings. In recent years, great attention has been devoted to the solution of this problem [1–10].

It is well known that most of metallic, alloy, and compound coatings easily crack when loaded by an external force, for instance, by bending or pressing. Cracks of the protective coatings almost always finish their correct function and must be avoided. It requires to find correct deposition parameters under which the coatings resistant to cracking can be formed. Recently, it was found that the flexible hard coatings with an enhanced resistance to cracking must simultaneously fulfill four necessary conditions [2, 4]:

1. The material of the coating must have a low effective Young's modulus E^* ensuring a high ratio $H/E^* \geq 0.1$
2. The material of the coating should exhibit a high elastic recovery $W_e \geq 60\%$
3. The microstructure of the coating must be dense and void-free
4. The coating should exhibit a compressive macrostress ($\sigma < 0$)

Many experiments carried out so far clearly indicate that these conditions necessary for the formation of the flexible coatings with enhanced resistance to cracking are of a general validity [11–20]. The coatings fulfilling these requirements can be prepared by (i) the addition of a selected element in the base material of coating and/or (ii) the delivery of a sufficient energy ε by bombarding and condensing particles into the coating during its growth [2, 4].

The finding that the ratio $H/E^* \geq 0.1$ is of key importance for the development of advanced flexible hard nanocoatings because it is a necessary condition to form the coatings resistant to cracking. This chapter shows that new advanced nanocoatings with new unique properties and even with a high ratio $H/E^* \geq 0.1$ can be formed by technologies based on strongly non-equilibrium processes based at an atomic level. It is shown that these technologies are of great potential for many applications.

1.2 THE LOW-TEMPERATURE SPUTTERING OF NANOSTRUCTURED COATINGS

At present, there is an urgent need to master the formation of hard nanostructured coatings on heat-sensitive substrates such as papers, polymer foils, plastics, and polycarbonate. It is a very difficult task because many coatings exhibit only an amorphous structure when they are deposited on unheated substrates without the substrate heating. It means that instead of the substrate heating, another kind of energy ε must be delivered to the growing coating to stimulate its nanocrystallization or full crystallization [2, 21]. This energy can be delivered by bombarding and condensing particles ($\varepsilon_p = \varepsilon_{bi} + \varepsilon_{fn}$), that is, by bombarding ions (ε_{bi}) and/or condensing fast neutrals (ε_{fn}).

The principle of the low-temperature (low-T) sputtering of hard nanocoatings is based on replacement of the conventional equilibrium heating (T_s/T_m) of the growing film by the non-equilibrium atomic scale heating controlled by the sputtering gas pressure p and the bombardment of film by ions (ε_{bi}) and/or the fast neutrals (ε_{fn}); see Figure 1.1. This figure displays four regions of the film microstructure in the two-dimensional (2D) Thornton's structural zone model (SZM) [22, 23]: (1) zone 1 composed of tapered crystallites separated by voids, (2) zone T composed of fibrous grains embedded in a void-free amorphous matrix, (3) zone 2 composed of grains separated by dense intercrystalline boundaries [24] and (4) zone 3 composed of recrystallized grains. The hard nanocrystalline coatings with dense, void-free, non-columnar microstructure and enhanced resistance to cracking at low-T are formed in zone T of the Thornton's SZM. Such coatings, sputtered at low temperatures ($T_s/T_m \leq 0.1$), have to be formed at low sputtering gas pressures p ($\leq 10^{-1}$ Pa) only, as it is seen in the marked region in Figure 1.1. For example, the TiN coating with $T_m = 2950\ °C = 3223$ K [25, 26] can be sputtered at a very low substrate temperature $T_s = 322$ K $= 49\ °C$ (corresponding to $T_s/T_m = 0.1$) at $p < 10^{-1}$ Pa. In this case, the energy necessary for the nanocrystallization of the coating can be delivered only by fast neutrals (ε_{fn}) [27]. The maximum energy $\varepsilon_{p\,max}$ delivered to the growing coating is, however, limited to a value that still prevents a thermal destruction of the heat-sensitive substrate.

The possibility to sputter the hard nanocrystalline coatings with void-free non-columnar microstructure only by their bombardment by fast neutral particles at low pressures below 0.1 Pa is also a big challenge for the development of new advanced coatings. Further important advantage of this

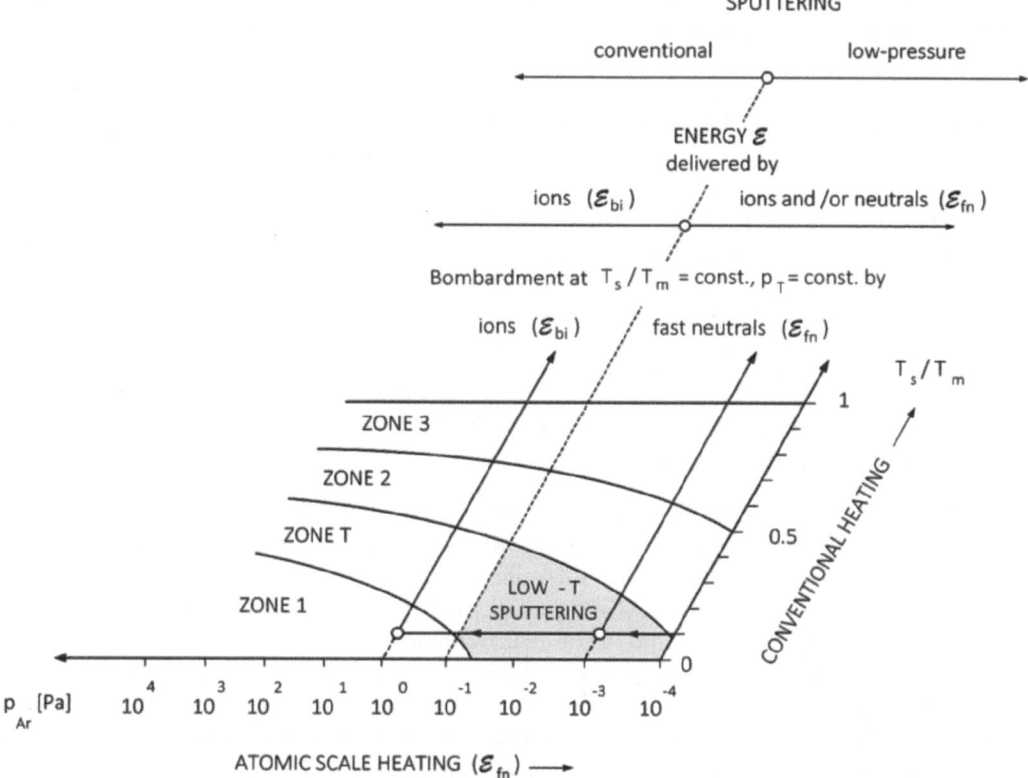

FIGURE 1.1 Schematic illustration of the two-dimensional (2D) Thornton's structural zone model (microstructure = f (p_{Ar}, T_s/T_m, ε_{bi}, ε_{fn})) with the marked region where hard, dense, void-free, non-columnar coatings can be created at low temperatures. Adapted with permission from Ref. [7]. J. Musil, J. Sklenka, R. Čerstvý: Protection of brittle film against cracking, Appl. Surf. Sci. 370C (2016), 306–311. Copyright © 2016, Elsevier.

process is the sputtering of coatings held at the floating potential U_{fl}, that is, at the substrate bias, $U_s = U_{fl}$. In this case, the same number of the electrons N_e and the ions N_i is incident on the surface of the growing coating. It eliminates the accumulation of the negative charge on the coating surface and thereby also the arcing on its surface and the generation of defects in the growing coating. For more details, see Ref. [28–31].

The possibility to form the hard films with the microstructure corresponding to zone T of the Thornton's SZM at low substrate temperatures ($T_s \leq 100$ °C) is of a great potential for many advanced applications, namely, for flexible electronics, flat panel displays, micro-electro-mechanical systems (MEMS), formation of functional films on polymer foils and fabrics, etc. The formation of nanocrystalline or even crystalline films with the microstructure corresponding to zone T of the SZM at low substrate temperatures $T_s \leq 100$ °C can be realized also by the addition of selected elements in the film. The melting temperature T_m of the material of such films can be decreased compared to that of pure films and thereby the T_s/T_m ratio can be increased and zone T can be reached without increasing of the particle bombardment ($\varepsilon_p = \varepsilon_{bi} + \varepsilon_{fn}$). At the end, it is worthwhile to note that the formation of hard nanocrystalline thin films and coatings with dense, void-free, non-columnar microstructure at low temperatures requires to master the sputtering at low pressures below 0.1 Pa. It means that new magnetrons operated at low pressures (≤ 0.1 Pa) need to be developed.

In the development of low-pressure magnetron sputtering, it is necessary to strongly increase the ionization of the sputtering gas at low magnetron voltages U_d, approximately 500 V. It can be achieved by (i) the additional ionization of sputtering gas, for instance, using a hot cathode electron emitter, hollow cathode, inductively coupled rf discharge and microwave discharge, and/or (ii) the magnetic confinement of an additional discharge in front of the target of the magnetron [32–37]. Additional ionization is crucial also for sputtering of superhard metallic coatings and of overstoichiometric nitride coatings where it is realized by the hybrid dual magnetron (HDM); see Sections 1.4 and 1.5.

In summary, it can be concluded that the technology of the low-T sputtering of thin films and coatings, based on the replacement of the conventional equilibrium heating (T_s/T_m) with the non-equilibrium atomic scale heating ($\varepsilon_p = \varepsilon_{bi} + \varepsilon_{fn}$) makes it possible (1) to move zone T of SZM into the region of low values of $T_s/T_m \leq 0.1$ and (2) to avoid an ion bombardment of the growing coating and this way to form the flexible hard nanocoatings at low pressures $p \leq 0.1$ Pa even on unheated substrates held at a floating potential.

1.3 THE SPUTTERING OF HIGH-TEMPERATURE, BETA-PHASE ALLOY COATINGS

In recent years, a great effort has been devoted to the development of new advanced films with new unique properties. The alloy films with high-temperature (high-T), beta (β-) phases, which differ from the films with low-temperature (low-T), α-phases with different crystal structures, is one group of such films. At present, there is little information on formation of the high-T, β-phase films and mainly on the mechanisms and conditions of their formation.

At present, there are several papers reporting on the formation of the β-phase (Me$_1$, Me$_2$) alloy films prepared by magnetron sputtering [38–50]. All these papers report on the so-called "stabilizing elements" which enable to form the high-T, β-phase films based on the phase diagrams of binary alloys formed under equilibrium conditions. The content of the stabilizing element can change from several at.% (e.g., TiFe, Figure 1.2a [52]) to 100 at.% (e.g., TiCr, Figure 1.2b [53]). For instance, the papers [44–50, 52, 53] show that Mo, V, Ta, Nb, Fe, Ni, and Cr are the β-stabilizing elements in titanium alloys. However, there is no information how (i) high temperatures necessary to reach the β-phase region in phase diagrams of the binary alloys can be achieved and (ii) the high-T, β-phase can be stabilized. It is the subject of this section.

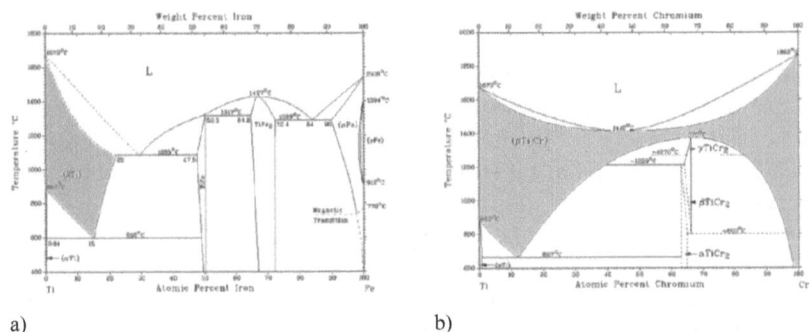

a)   b)

FIGURE 1.2 Phase diagram of (a) the Ti-Fe alloys and (b) the Ti-Cr alloys showing regions of temperatures and compositions (marked regions) in which high-T, β-phase alloys with the bcc structure, that is, the high-T, β-Ti(Fe) alloy (Figure 1.2a) and the high-T, β-(Ti,Cr) alloy (Figure 1.2b), can be formed. Adapted with permission from Ref. [51]. J. Musil, Š. Kos, S. Zenkin, Z. Čiperová, D. Javdošňák, R. Čerstvý: β-(Me$_1$, Me$_2$) and MeN$_x$ films deposited by magnetron sputtering: Novel heterostructural alloy and compound films, Surf. Coat. Technol. 337 (2018), 75–81. Copyright © 2018, Elsevier.

1.3.1 Principle of Formation of High-T, β-phase Coatings

The principle of the formation of "the high-T, β-phase coatings" by the magnetron sputtering is based on three non-equilibrium processes at an atomic level:

(1) The extremely fast heating of the coating material
(2) The extremely fast cooling of the created material down to RT ($t_0 \leq 10$ ps) [54, 55]
(3) The very short cooling time.

The extremely fast heating of the created coating material is achieved by the energy ε delivered into the surface of the growing coating by bombarding and condensing particles incident just like for the low-temperature sputtering discussed in Section 1.2. The energy ε of the bombarding ions ε_{bi} (tens to hundreds eV) and/or condensing fast neutrals ε_{fn} (several eV), delivered into very small areas of an atom size (approximately 0.04 nm^2), is very high (1 eV = 11,600 K). It is sufficient to heat the material of the created alloy coating to high temperatures T, corresponding to the β-phase of the alloys or compounds in the phase diagrams of binary alloys, at an atomic level. It was demonstrated in sputtering of the Ti(Fe) films; see Figure 1.3. This figure compares XRD patterns from (a) the pure, low-T, h-αTi film (Figure 1.3a) and (b) the high-T, c-βTi(Fe) alloy film (Figure 1.3b) sputtered on unheated glass substrates. The low-T, h-αTi film was sputtered from a pure Ti target and high-T, c-βTi(Fe) alloy film was sputtered from a Ti(14 wt.% Fe) alloy target, both films on unheated glass substrates; here h and c are the hexagonal and cubic phase, respectively. The perfect coincidence of the angular positions of reflections of the measured XRD pattern with those of the standard h-αTi clearly demonstrates that the pure-Ti film, sputtered onto an unheated glass substrate at $T_s < 100$ °C, forms a low-T hexagonal h-αTi structure, as predicted by the equilibrium phase diagram; see Figure 1.3a. On the other hand, Figure 1.3b clearly shows that the measured XRD pattern fits very well with the reflections of the standard high-T, c-βTi(Fe) alloy (calculated for a = 0.32482 nm corresponding to 14 wt.% of Fe in Ti).

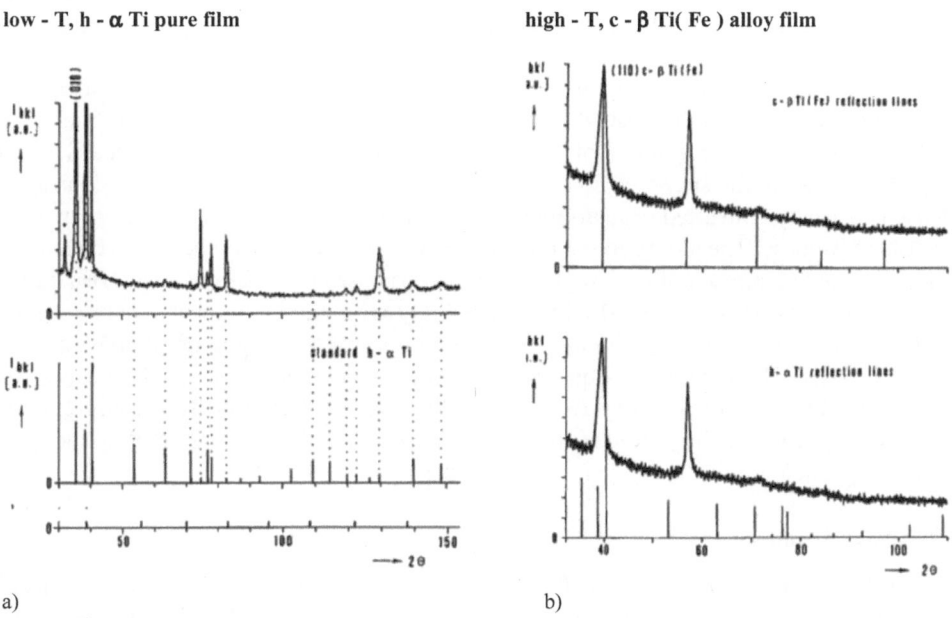

low - T, h - α Ti pure film   high - T, c - β Ti(Fe) alloy film

a)   b)

FIGURE 1.3 Comparison of XRD structure of (a) low-T, h-αTi pure film and (b) high-T, c-βTi(Fe) alloy film sputtered on unheated glass substrate. Adapted with permission after Ref. [38]. J. Musil, A.J. Bell, J. Vlček, T. Hurkmans, Formation of high-temperature phases in sputter deposited Ti-based films below 100 deg. C, J. Vac. Sci. Technol. A 14 (4)(1996) 2247–2250. Copyright © 1996, American Vacuum Society.

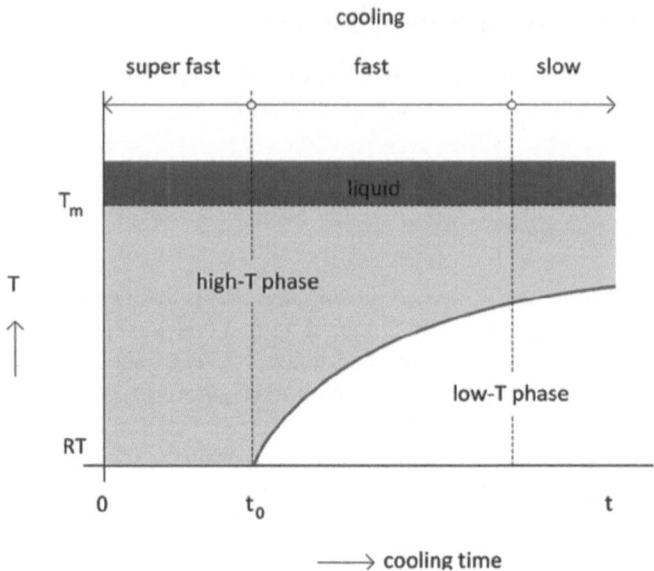

FIGURE 1.4 Schematic illustration of the structure of a solid material (coating) as a function of the final temperature T ranging from room temperature RT to the melting temperature of the material T_m and the cooling time t of the created material of coating. Adapted with permission from Ref. [51]. J. Musil, Š. Kos, S. Zenkin, Z. Čiperová, D. Javdošňák, R. Čerstvý: β-(Me$_1$, Me$_2$) and MeN$_x$ films deposited by magnetron sputtering: Novel heterostructural alloy and compound films, Surf. Coat. Technol. 337 (2018), 75–81. Copyright © 2018, Elsevier.

The extremely fast cooling of the created coating from T_m down to RT makes it possible to avoid its phase conversion due to freezing of its high-T, β-phase; see Figure 1.4. This figure schematically shows how the phase of the coating, after being cooled down from an initial very high temperature greater than the melting temperature T_m, depends on its final temperature T and the cooling time t. From this figure, it is seen that the β-phase coating stable down to RT can be formed only in the case when the cooling of a created material is very fast, that is, when the cooling time is very short $t \le t_0 \le 10$ ps [54, 55]; here t_0 is the longest time at which the β-phase coating exhibits no conversion to the low-T, α-phase. In the case when the cooling is slower, that is, the cooling time is longer $(t > t_0)$, a two-phase coating, composed of both the high-T and low-T phases, is formed. The content of the low-T, α-phase in the sputtered coating increases with increasing cooling time during which the β-phase is partially converted (transformed) into the α-phase with the crystal structure different from that of the β-phase. The slow cooling of a hot alloy is the reason why the alloys created from a (Me$_1$, Me$_2$) melt are composed of both high-T and low-T phases of different crystal structures, that is, the coating is a mixture of two α-Me$_1$ (Me$_2$) + β-Me$_1$ (Me$_2$) phases. It means that *the very short cooling time* smaller than t_0 is *the third necessary condition* for the formation of coatings with the pure high-T, β-phase. These three non-equilibrium atomic-level processes are the basis of the formation also of superhard metallic coatings described in Section 1.4. Additionally, the non-equilibrium analogue of high pressure is important in that case. Hence, a detailed discussion of all the four processes including a quantitative analysis is provided there.

1.3.2 Thermal Stability

The key problem in the formation of the β-phase film is, however, the temperature stability of the structure of the high-T, β-phase material during its cooling-down to RT. The high-T, β-phase is converted into low-T, α-phase material in two cases during the slow cooling:

1. During creation of the high-T, β-phase coating as shown above in Figure 1.4
2. During operation of the high-T, β-phase coating at temperature $T > T_{stab}$ and subsequent slow cooling to RT.

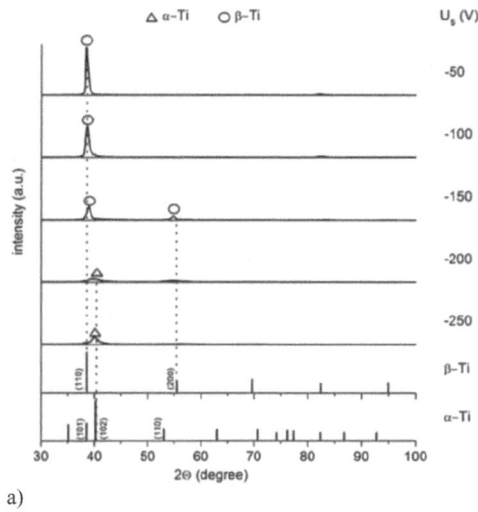
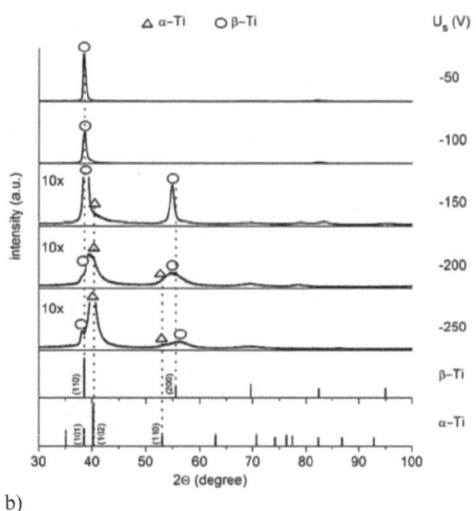

a) b)

FIGURE 1.5 Evolution of the structure of the Ti-Cr film DC sputtered by unbalanced magnetron equipped with TiCr (90/10 at.%) alloy target at $T_s = 450$ °C, $d_{s-t} = 60$ mm, $p_{Ar} = 1$ Pa with increasing negative substrate bias U_s from -50 V to -250 V. Figure 1.5a shows a strong nanocrystallization with increasing U_s. Figure 1.5b shows the conversion of single phase high-T, β-(c-Ti(Cr)) films into two phase β-(c-Ti(Cr)) and α-(h-Ti(Cr)) films with different crystal structures. Adapted with permission after Ref. [51]. J. Musil, Š. Kos, S. Zenkin, Z. Čiperová, D. Javdošňák, R. Čerstvý: β-(Me$_1$, Me$_2$) and MeN$_x$ films deposited by magnetron sputtering: Novel heterostructural alloy and compound films, Surf. Coat. Technol. 337 (2018), 75–81. Copyright (c) 2018, Elsevier.

The temperature T_{stab} is a maximum temperature at which the high-T, β-phase film can be operated without phase conversion. The conversion of the single-phase high-T, β-phase film into two-phase high-T, β-phase and low-T, α-phase films is illustrated in Figures 1.5 and 1.6. As examples, the Ti-Cr and Ti-W alloy films sputtered by an unbalanced magnetron from the TiCr (90/10 at. %) alloy target and the TiW target composed of a W round plate ($\varnothing = 100$ mm, 6 mm thick) fixed to the magnetron cathode body with a Ti fixing ring with the inner diameter of 30 mm, respectively, were used [51]. The structure of the sputtered Ti-Cr and Ti-W alloy films was characterized by X-ray diffraction.

Figure 1.5 shows the evolution of the structure of Ti-Cr alloy films sputtered by an unbalanced magnetron from the TiCr (90/10 at.%) alloy target and controlled by the substrate bias U_s, that is, by the energy ε_{bi} delivered into growing films by bombarding ions [51]. The energy ε_{bi} increases with increasing negative substrate bias. The increase of ε_{bi} above 0.4 MJ/cm^3 achieved at $U_s = -50$ V results in (a) a strong nanocrystallization of the Ti$_{90}$Cr$_{10}$ films; see Figure 1.5a, and (b) a partial conversion of the high-T, β-phase grains into the low-T, α-phase grains; see Figure 1.5b. It means that the homostructural β-(c-Ti(Cr)) films are converted into the heterostructural films composed of two β-(c-Ti(Cr)) and α-(h-Ti(Cr)) phases with different crystal structures. The conversion of β-phase grains into α-phase ones is caused by increase of the substrate temperature T_s above 450° during the film sputtering at high negative substrate biases, that is, by overstepping of the thermal stability of the high-T, β-phase ($T_s > T_{stab}$) and its subsequent slow cooling to RT after the film deposition.

Figure 1.6 shows the effect of a post-deposition thermal annealing on the structure of the as-deposited Ti$_{88}$W$_{12}$ film. From this figure, it is seen that the thermal annealing results in a partial conversion of the homostructural β-(c-Ti$_{88}$W$_{12}$) film into the heterostructural film composed of two high-T, β-(c-Ti$_{88}$W$_{12}$) and low-T, α-(c-Ti$_{88}$W$_{12}$) phases of different crystal structure. This fact shows that the high-T, β-(c-Ti$_{88}$W$_{12}$) is metastable. It means that its thermal stability is limited by the temperature T_{stab}. The values of the temperature stability T_{stab} of high-T, β-phase films need to be determined for every (Me$_1$, Me$_2$) alloy film.

In summary, it can be concluded that high-T, β-phase films represent a new class of "heterostructural films". These films are formed by mixing of elements with different crystal structure by

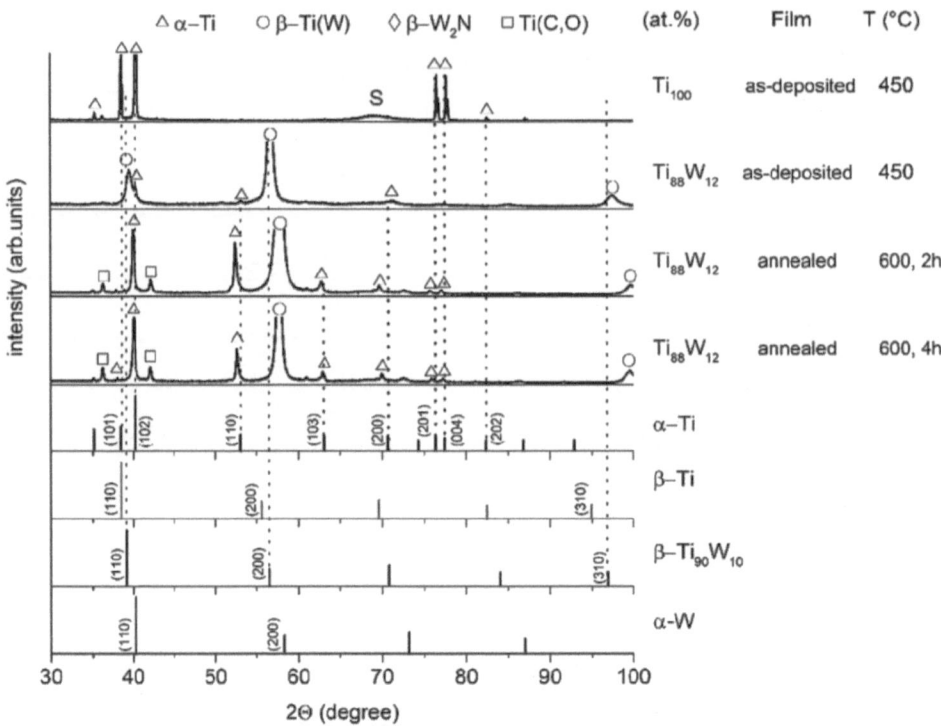

FIGURE 1.6 XRD patterns of the $Ti_{88}W_{12}$ alloy film DC sputtered by an unbalanced magnetron at $T_s = 450\ °C$, $U_s = -50\ V$, $d_{s\text{-}t} = 60\ mm$, $p_T = p_{Ar} + p_{N2} = 0.25 + 0.25 = 0.5\ Pa$ after their (i) deposition (as-deposited films) and (ii) thermal annealing in argon at the temperature $T_a = 600\ °C$ for 2 and 4 h at $p_{Ar} = 1\ Pa$. The XRD pattern of a pure Ti film is given for comparison. Adapted with permission after Ref. [51]. J. Musil, Š. Kos, S. Zenkin, Z. Čiperová, D. Javdošňák, R. Čerstvý: β-(Me$_1$, Me$_2$) and MeN$_x$ films deposited by magnetron sputtering: Novel heterostructural alloy and compound films, Surf. Coat. Technol. 337 (2018), 75–81. Copyright © 2018, Elsevier.

three strongly non-equilibrium processes at an atomic level: extremely fast heating of mixing atoms, extremely fast cooling of the created material and very short time of cooling avoiding the structure conversion.

1.4 THE SPUTTERING OF SUPERHARD METALLIC COATINGS

As stated in Section 1.1, one of the conditions necessary for a coating's enhanced resistance to cracking is the high ratio H/E* ≥ 0.1. A way to achieve this high ratio for the metallic coatings on substrates that might even be made of the same bulk metal, thereby avoiding issues due to the dissimilarity of the materials, is to increase the hardness of the coatings several times. Such superhard metallic coatings can be produced by a sputtering technology based on four simultaneous non-equilibrium processes at an atomic-level:

1. The heating of the coating material to high temperatures exceeding its melting temperature T_m by impacting particles.
2. The pressing of the coating material at high pressures of about 1000 GPa.
3. Extremely fast cooling of the coating material from the melting temperature down to RT.
4. Freezing of the melted coating material in atomic layers.

As these processes are realized at very small (the size of one ion) mutually separated areas from the melted material due to extremely fast cooling and at a very high flux of incident ions continuously covering the whole substrate surface, we call this new technology of the coating deposition – *the nanowelding* [56]. Unlike conventional welding or conventional formation of advanced superhard bulk materials at high temperatures (~2500 °C) and high pressures (50–100 GPa) [57–66] the nanowelding takes place in an inert atmosphere, thereby reducing material deficiencies. Continuous coatings can be deposited not only on large areas such as sheets but also on parts of complex shapes using this technology.

In this section, we shall discuss quantitatively these processes in turn, and estimate the values of parameters characterizing these processes.

1.4.1 Heating to High Temperatures

The energy of atoms and ions is given by the Thompson cascade [67] and the Bohm sheath criterion [68] to be the electronic energy given parametrically and by the order of magnitude by the Rydberg energy [69],

$$Ry = \hbar^2 / 2ma_B^2 \approx 14\,eV \tag{1.1}$$

where $\hbar$ is the reduced Planck constant, m is the mass of electron and

$$a_B = \hbar / \alpha mc = 4\pi\varepsilon_0\hbar^2 / me^2 \approx 0.5\,\text{Å} \tag{1.2}$$

is the Bohr radius that sets the scale of the interatomic distances, $\alpha \equiv e^2/4\pi\varepsilon_0\hbar c \approx 1/137$ is the fine-structure constant, and c is the speed of light in vacuum. This energy is a non-equilibrium analogue of a temperature Ry/k_B, where k_B is the Boltzmann constant, of tens of thousands Kelvin. This energy, and consequently its temperature analogue, can be further increased by an energy ε provided by an applied substrate bias, either directly in case of ions or by transfer in case of neutrals. This brief analysis clearly shows that the energy delivered into the growing coatings during strongly non-equilibrium processes is fully sufficient to exceed the melting temperature T_m of every material.

1.4.2 Pressing at High Pressures

The arriving particle, a neutral atom or an ion, of mass M exerts a high pressure that we can estimate from the momentum

$$P = Mv = M(2Ry / M)^{1/2} = (2MRy)^{1/2} \tag{1.3}$$

carried by the particle and the particle impact time τ giving the force

$$F = P / \tau = (2MRy)^{1/2} / \tau \tag{1.4}$$

and the pressure p

$$p = F / (10a_B)^2 \tag{1.5}$$

because the force F acts on an area of the size of the interatomic spacing estimated as $(10a_B)^2 \approx 0.28$ nm^2. We estimate the impact time τ by the vibrational time, that is, the inverse of the vibrational frequency ω on the order of terahertz (10^{12} Hz)

$$\tau \approx 1 / \omega = (M / k)^{1/2} \tag{1.6}$$

where k is the elastic constant obtained by considering that upon displacement of the order $10a_B$ the energy will increase by about Ry, so we may estimate the elastic constant as $k \approx Ry/(10a_B)^2$ giving [69].

$$\tau = \left(M/k\right)^{1/2} \approx 10a_B \left(M/Ry\right)^{1/2} = 10\hbar \left(2ma_B^2/\hbar^2 Ry\right)^{1/2} \left(M/2m\right)^{1/2} = 10\left(\hbar/Ry\right)\left(M/2m\right)^{1/2} \quad (1.7)$$

Thus, the vibrational time is longer by a factor $10(M/2\,m)^{1/2}\,M/m$ than the electronic time of the order

$$\hbar/Ry = 2ma_B^2/\hbar = 2a_B\left(m/\hbar\right)\left(\hbar/\alpha mc\right) = 2a_B/\alpha c \quad (1.8)$$

in agreement with the fact that the typical electronic speed ~ $(Ry/m)^{1/2}$ is ~ αc.

We see that both the momentum delivered upon the impact, Equation 1.3, and the impact time, Equation 1.7, are proportional to $(M)^{1/2}$, so in their ratio, the dependence on the particle mass cancels, and the force is given purely by electronic scales:

$$F = \left(2MRy\right)^{1/2}\left(1/\left(10a_B\right)\right)\left(Ry/M\right)^{1/2} = 2^{1/2}\,Ry/10a_B \quad (1.9)$$

Then, the resulting non-equilibrium pressure is

$$p = F/\left(10a_B\right)^2 = 2^{1/2}\,Ry/\left(10a_B\right)^3 \approx 2^{1/2}14 \times 1.6 \times 10^{-19}\,J/\left(5.3 \times 10^{-10}\,m\right)^3 = 2.1 \times 10^{10}\,Pa \quad (1.10)$$

The thermal pressure is given by the equation of state of the ideal gas

$$p_{th} = nk_B T \quad (1.11)$$

where n is the particle density in the deposition chamber. Hence

$$p_{th}/p = \left(n\left(10a_B\right)^3\right)\left(k_B T/\left(2^{1/2}Ry\right)\right) \quad (1.12)$$

Even if we take for temperature its non-equilibrium analogue discussed above, Ry/k_B or higher, making the second factor of order one, the first factor $n(10a_B)^3$ is smaller than one by orders of magnitude, and so is the thermal pressure compared to the non-equilibrium pressure.

With the additional energy ε given to the impacting particles by an applied bias, the momentum carried by the particles becomes $P = (2\,M(Ry + \varepsilon))^{1/2}$. If ε is significantly smaller than the atomic energy ~Ry, then the momentum carried by the particle becomes $(2\,M(Ry + \varepsilon))^{1/2} \approx (2M\,Ry)^{1/2}(1 + \varepsilon/2Ry)$ so the non-equilibrium pressure is enhanced by a factor $(1 + \varepsilon/2Ry)$. If the additional energy ε is significantly greater than the atomic energy ~Ry, then the momentum carried by the particle becomes $P \approx (2M\varepsilon)^{1/2}$ and so in the formula for non-equilibrium pressure, the energy of Ry in the numerator is replaced by the geometric mean of the energies ε and Ry, that is $p \approx (2\varepsilon Ry)^{1/2}/(10a_B)^3 = 2^{1/2}\,Ry/(10a_B)^3\,(\varepsilon/Ry)^{1/2} \approx 2.1 \times 10^{10}\,Pa(\varepsilon/Ry)^{1/2}$, so the non-equilibrium pressure is enhanced by a factor $(\varepsilon/Ry)^{1/2}$.

1.4.3 Fast Cooling

The Fourier law of heat conduction states that the heat flux **q** is proportional to the temperature gradient

$$\mathbf{q} = -\kappa\,grad\,T \quad (1.13)$$

where κ is the thermal conductivity. If we neglect the terminal ambient temperature compared to the temperature $T \approx Ry/k_B$ or higher after the impact, decreasing on the scale of interatomic distances we have been estimating as $10a_B$, then the power leaving the heated volume of the order $(10\,a_B)^3$ with a surface area of the order $(10\,a_B)^2$

$$\iint \mathbf{q}\,d\mathbf{S} \approx \kappa\left(T/\left(10a_B\right)\right)\left(10a_B\right)^2 \tag{1.14}$$

can be estimated by $C(10\,a_B)^3\,T/\tau_c$, where C is the specific heat per unit volume and τ_c is the cooling time which we can then estimate by approximately equating these two expressions

$$\tau_c \approx C\left(10a_B\right)^2 / \kappa \tag{1.15}$$

The thermal conductivity is dominated by phonons at high temperatures and is given by the specific heat C per unit volume, the sound velocity v, and the phonon mean free path l as $\kappa \approx Cvl$ so

$$\tau_c \approx \left(10a_B\right)^2 / vl \tag{1.16}$$

The sound velocity is given by the bulk modulus $B \approx Ry/(10\,a_B)^3$ and the material mass density $\rho \approx M/(10\,a_B)^3$ as $v = (B/\rho)^{1/2} \approx (Ry/M)^{1/2}$ and we may estimate the mean free path by the interatomic spacing of the order of $10\,a_B$ so

$$\tau_c \approx 10a_B\left(M / Ry\right)^{1/2} \tag{1.17}$$

Comparing with the above estimate of the vibrational time $1/\omega$, Equation 1.6, approximating the impact time τ, Equation 1.7, we see that the cooling time τ_c is of the same order of magnitude, the inverse terahertz, that is, picoseconds.

1.4.4 SPUTTERING OF SUPERHARD TI COATING

In this section, the sputtering of superhard Ti coatings using nanowelding technology is briefly described. The sputtering involving the four non-equilibrium atomic-level processes outlined at the beginning of this section may be realized by the HDM consisting of two magnetrons with a closed magnetic-field configuration [56]. The HDM is described in more detail in Section 1.5 on the over-stoichiometric nitrides for which it was first developed. The nanowelding technology process makes it possible to produce superhard flexible metallic coatings with the hardness several times greater than that of the soft coatings sputtered by a standard magnetron technology. To form the superhard metal coatings two necessary conditions must be fulfilled: (1) Sufficiently high energy ε must be delivered into the place of the incidence of bombarding ions ε_{bi} and (2) the number of bombarding ions must be very high to cover every point on the coated surface almost at the same time; see Figure 1.7. It can be achieved in the case when an extremely high flux of ions bombards the surface to be coated. The coating created under these conditions is very hard and its hardness H_c is several times greater than that of the soft bulk material $H_{bulk\ material}$. This fact is illustrated in Figure 1.8 for sputtered Ti coatings.

From Figure 1.8, it is seen that (1) the hardness H of the Ti coatings deposited by the HDM is more than 4 times greater than that of the Ti bulk material ($H \approx 5$ GPa) and 5 times greater than that of the Ti coatings deposited by the standard magnetron (SM) ($H \approx 4$ GPa) and (2) the Ti coatings deposited by the SM exhibit low values of hardness $H \approx 5$ GPa, ratio $H/E^* < 0.1$ and elastic recovery $W_e < 50\%$. On the other hand, the Ti coatings deposited by the HDM exhibit high values of $H \geq 13$ GPa, $H/E^* \geq 0.1$ and $W_e > 50\%$. A very important finding is the fact that the hardness of the Ti coatings deposited by the HDM increases with increasing ratio i_s/a_D and the energy ε_{bi} at $U_s = $ constant. It indicates that the ratio i_s/a_D is a more important parameter for the increase of the coating hardness than the substrate bias U_s. A great difference between the hardness of the soft Ti coating sputtered by the SM and the hard Ti coating sputtered by the HDM is illustrated in Figure 1.9.

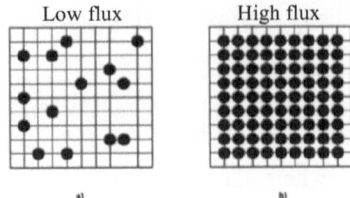

FIGURE 1.7 Schematic illustration of particles falling down on the unit coating surface and delivering the energy into separated points at (a) low flux and (b) high flux of particles and the same time $t_{aft\,inc}$ after their incidence.

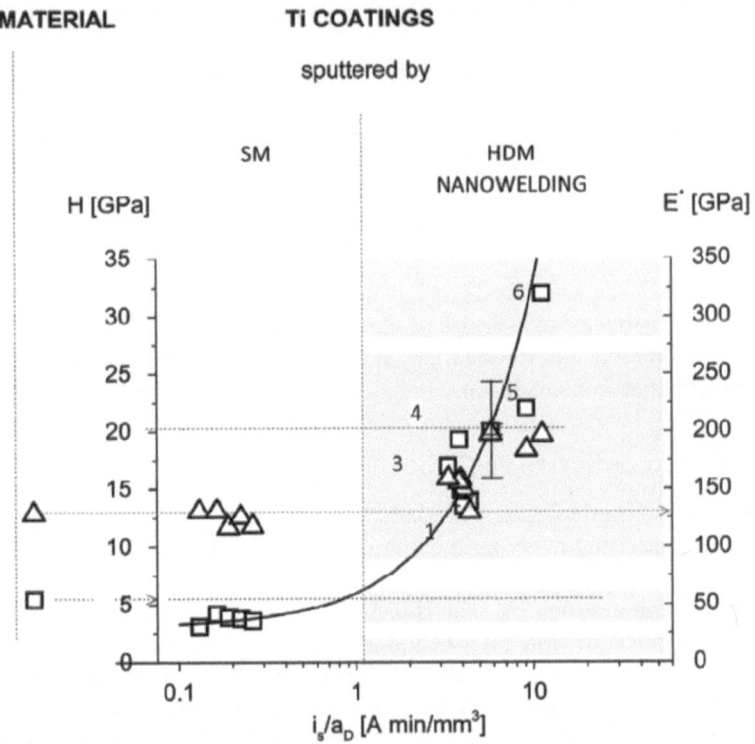

FIGURE 1.8 Comparison of the hardness H and the effective Young's modulus E^* of (i) the Ti bulk material, (ii) the soft Ti films sputtered by a DC SM, and (iii) the superhard Ti films sputtered by a pulsed HDM as a function of the ratio i_s/a_D on a soft Ti substrate. Squares denote the hardness H and triangles the effective Young's modulus $E^*$of both the Ti bulk material and the sputtered Ti films. Adapted with permission from Ref. [56]. J. Musil, M. Jaroš, Š. Kos: Superhard metallic coatings, Materials letters, 247 (2019), 32–35. Copyright © 2019, Elsevier.

The structure of the soft Ti coating and the hard Ti coating also strongly differ. It is illustrated by their XRD patterns; see Figure 1.10. The soft Ti coating with hardness H = 5.4 GPa is well crystalline with a grain size d ≈ 40 nm (Figure 1.10a). On the other hand, the hard Ti coating with H = 16.1 GPa, characterized with very low-intensity (13 times magnified) broad reflections, is a nanocrystalline Ti (002) coating (Figure 1.10b) with grain size d ≈ 10 nm (Figure 1.10c). In this case, the Ti (002) reflection is shifted to lower angles 2θ due to a high compressive macrostress (σ = −2.1 GPa). This comparison indicates that also a nanocrystallization of the Ti coating plays an important role in strong hardening of a soft Ti substrate covered by the flexible hard nanocrystalline

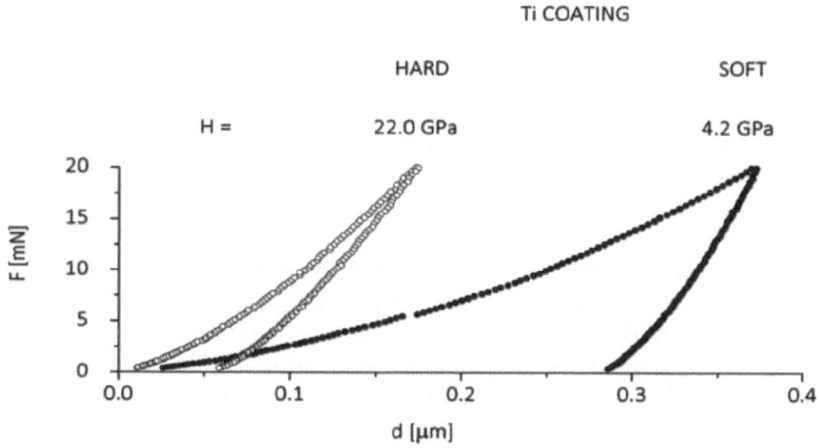

FIGURE 1.9 The comparison of loading/unloading curves of the soft Ti film sputtered by the SM and the hard Ti coating sputtered by the HDM on a soft Ti substrate.

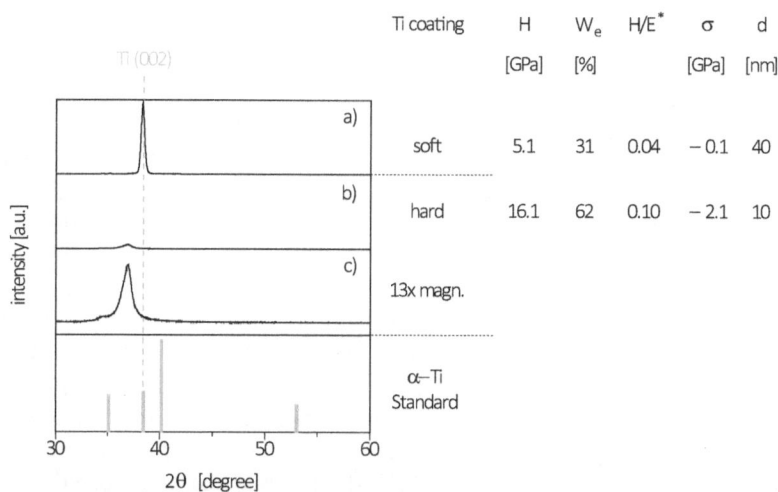

Ti coating	H	W_e	H/E^*	σ	d
	[GPa]	[%]		[GPa]	[nm]
soft	5.1	31	0.04	− 0.1	40
hard	16.1	62	0.10	− 2.1	10

FIGURE 1.10 Comparison of XRD patterns of the soft Ti coating deposited by the DC SM and the hard Ti coating deposited on Si substrates by the pulsed HDM. Adapted with permission from Ref. [56]. J. Musil, M. Jaroš, Š. Kos: Superhard metallic coatings, Materials letters, 247 (2019), 32–35. Copyright © 2019, Elsevier.

protective Ti coating with enhanced resistance to cracking. Hence, the increased ionization without an increase in the deposition rate in the HDM decreases the crystallinity of the coating unlike the atomic-scale heating that increases the crystallinity as described in Section 1.2. For a further discussion of the difference, see Section 1.6.2.

In summary, it can be concluded that the new sputtering technology of the thin films and coatings based on their formation from the melted material of sputtered coatings opens a new way for creation of the new generation of advanced hard coatings with unique properties. Next step in this direction should be the development of new advanced flexible hard alloy coatings with high ratio $H/E^* \geq 0.1$, high elastic recovery $W_e \geq 60\%$, and enhanced resistance to cracking. At present, almost all sputtered (Me_1, Me_2) alloy coatings are soft and brittle due to a low ratio $H/E^* < 0.1$; here Me_1 and Me_2 are metallic elements. Only recently, flexible hard alloy coatings were developed in the (Me, Met) system; here, Me is a metallic element and Met is a metalloid element [70]; see Section 1.6 for details. These results indicate that the brittleness of soft alloy coatings can be removed and flexible hard alloy coatings can be sputtered when their elemental composition and the energy ε delivered to them during their growth is correctly selected.

1.5 THE SPUTTERING OF TM OVERSTOICHIOMETRIC NITRIDE AND DINITRIDE COATINGS

Recently, theoretical studies based on ab initio calculations showed that overstoichiometric nitrides should exhibit extraordinary properties such as superhardness, electrical conductivity, and optical transparency that originate from metal–nitrogen charge transfer and the nature of N-N bonds [71–74]. TMN_2 materials such as TiN_2, MoN_2, TaN_2, WN_2, and ReN_2 are called dinitrides. The TMN_2 are declared as new materials exhibiting a unique crystal structure containing single-bonded dinitrogen units, that is, the pernitride ions, N_2^{4-} and TM^{4+}. The chemical bonding within the TMN_2 material is complex and clearly of mixed covalent/ionic character (covalent N-N bond and ionic bond between N-N ions and TM atoms). The N-N covalent bonds make the material more resistant to an external stress (ultraincompressible) [75]. It was also reported that bulk dinitrides were successfully synthesized under high pressures up to 73 GPa and at high temperatures up to 2400 K (high-pressure high-temperature (HPHT) synthesis, that is, using *an equilibrium process* [74–76]. For instance, the superhard TiN_2 dinitride with an ultraincompressible bulk modulus (360–385 GPa) was synthesized from the titanium nitride using a laser-heated diamond anvil cell surrounded by a dense N_2 pressure medium compressed to 73 GPa and heated to 2400 K. [75]. TiN_2 has also been synthesized in the form of thin film by an arc deposition with an enhanced ionization of the sputtering gas [32, 77].

Based on this information, it could be expected that the overstoichiometric $TMN_{x>1}$ nitrides and TMN_2 dinitrides could be formed in the form of thin films and coatings also using magnetron sputtering. The formation of the overstoichiometric nitride material could be realized utilizing a *non-equilibrium sputter deposition process* running at an atomic level. The condensing fast neutral atoms (the energy ε_{fn} of several eV controlled by sputtering gas pressure p) and/or bombarding ions with energy ε_{bi} controlled by the negative substrate bias U_s deliver to very small area of the size of atom (ca. 0.04 nm^2) of the growing coating sufficient energy $\varepsilon = \varepsilon_{bi} + \varepsilon_{fn}$ (i) to heat this area to very high temperatures, easily exceeding 2500 K, and (ii) to press this area at high pressure p of about 300 GPa [78]. Moreover, the synthesis of the material in *the non-equilibrium process* running at an atomic level (TM + 2 N → TMN_2) is more efficient compared with that of material in *the equilibrium process* running at a molecular level (TM + N_2 → TMN_2). Therefore, it can be expected that overstoichiometric $TMN_{x>1}$ nitrides and TMN_2 dinitrides could be prepared using magnetron sputtering.

1.5.1 PRINCIPLE OF SPUTTERING OF OVERSTOICHIOMETRIC $TMN_{x>1}$ NITRIDE COATINGS

The formation of TMN_x nitride films is carried out by a reactive sputtering in magnetron discharges generated in an ionized mixture of Ar + N_2 gases. A basic problem in reactive sputtering of overstoichiometric $TMN_{x>1}$ nitrides using a standard magnetron is a higher number of TM atoms n_{TM} compared with the number of N atoms n_N. It is due to a low ionization degree of the sputtering gas, n_i/n (p_T), of about 10^{-4} to 10^{-3} at operating pressures $p_T = p_{Ar} + p_{N2} \approx 0.133$ Pa (1 mTorr); here, n_i is the number of ionized gas atoms and $n(p_T)$ is the number of gas atoms in unionized gas at pressure $p_T = p_{N2}$; more details are given in [78]. Therefore, the deficiency of N atoms must be removed. In principle, there are three ways this could be achieved:

1. The reduction of the number of sputtered TM atoms to the value corresponding to the stoichiometry x of the $TMN_{x=2}$ dinitride film, that is, $n_N = 2n_{TM}$.
2. A strong increase in the ionization of sputtering gas, for instance by an additional ionization of sputtering gas [32, 77].

3. An additional nitriding of the growing TMN_x film using a substrate rotation (an interruptive deposition of the film) or a pulsed deposition and nitriding during pulse-off time.

These methods are described in more details in [78].

1.5.2 Hard Overstoichiometric $TMN_{x > 1}$ Nitride Films Prepared by Magnetron Sputtering

Very recently, magnetron sputtered overstoichiometric $TMN_{x > 1}$ nitride films were successfully developed. These $TiN_{x > 1}$ films were reactively sputtered by a pulsed HDM in a mixture $N_2 + Ar$. The HDM consists of two different magnetrons M1 and M2 with a closed magnetic field B. The magnetron M1 is a magnetron with a very low sputtering of its target and the magnetron M2 is a standard, well sputtering magnetron. The low sputtering of the magnetron M1 was achieved by extraction of the central magnet from the magnetron M1. By control of the powers P_{M1} and P_{M2} delivered into the magnetron M1 and the magnetron M2, respectively, it is possible to increase the ion bombardment of the growing film by an increase of the ion flux i_s on the substrate while keeping the film deposition rate a_D constant, and in this way to sputter overstoichiometric TiN_x films with the stoichiometry $x = N/Ti > 1$. The increase of i_s at constant a_D is achieved by increasing of the power P_{M1} delivered into the magnetron M1 and keeping the power P_{M2} delivered into the magnetron M2 constant. More details are given in [78]. Just like in case of the superhard metallic coatings discussed in Section 1.4, the HDM carries out the four non-equilibrium atomic-level processes outlined at the beginning of that section and subsequently analyzed quantitatively there.

The $TiN_{x > 1}$ films were reactively sputtered in an $N_2 + Ar$ mixture by pulsed HDM powered by a pulsed power supply AE Pinnacle Plus +5/5 kW (Advanced Energy, Inc., USA) and operated in a synchronous pulse mode at the repetition frequency $f_r = 1/T = 20$ kHz and duty cycle $\tau/T = 0.99$ onto Si (100) substrates in a deposition chamber evacuated to a base pressure $p_0 = 1 \times 10^{-3}$ Pa. Based on detailed investigations of the structure, elemental composition, mechanical properties, macrostress σ, and electrical resistivity of the $TiN_{x > 1}$ sputtered under different deposition parameters it was found that

1. The overstoichiometric $TiN_{x = 2.3}$ can be sputtered by pulsed HDM at a low total pressure $p_T = 0.17$ Pa. This strongly overstoichiometric $TiN_{x = 2.3}$ film is created thanks to a high energy $\varepsilon_{bi} = 8.2$ MJ/cm^3 delivered into it by bombarding ions and developing a very high pressure $p = 1340$ GPa in place of their incidence. The energy ε_{bi} was calculated from the formula $\varepsilon_{bi} = (U_s.i_s)/a_D$; here U_s and i_s is the substrate bias and the substrate ion current density, respectively [2]. This experiment clearly demonstrates that the formation of $TiN_{x = 2}$ dinitride films by magnetron sputtering is possible.

2. The decrease of the total gas pressure p_T from 0.30 to 0.17 Pa and mainly the increase of the powers P_{M1} and P_{M2} in the magnetrons M1 and M2, respectively, increasing the ionization of the N_2 gas results not only in an increase of the stoichiometry $x = N/Ti$ of the TiN_x film but also in a decrease of its hardness H from 25 to 16 GPa, elastic recovery W_e from 84 to 69% and H/E* ratio from 0.13 to 0.10. It indicates that the overstoichiometric $TMN_{x > 1}$ nitride films are highly flexible films with enhanced resistance to cracking.

3. Both overstoichiometric films, $TiN_{x = 1.4}$ and $TiN_{x = 2.3}$, are well conductive and exhibit a low electrical resistivity $\rho \approx 1.5 \times 10^{-4}$ Ωcm.

4. The stoichiometry x of the TiN_x nitride film strongly influences also its color; see Figure 1.11. The stoichiometric $TiN_{x = 1}$ nitride film is golden yellow, shown as light gray in Figure 1.11. On the other hand, a strongly overstoichiometric $TiN_{x = 2.3}$ film is brown, shown as dark gray in Figure 1.11. The real colors are shown in the source reference [79].

x = 1.0   x = 2.3

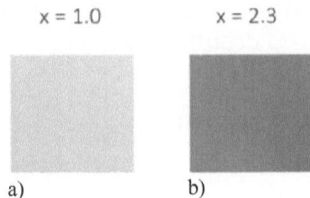

a)   b)

FIGURE 1.11 Color of TiN_x films with two different values of stoichiometry x. (a) The stoichiometric $TiN_{x=1}$ film with golden-like color (light gray) and (b) the strongly overstoichiometric $TiN_{x=2.3}$ film with brown color (dark gray). The real colors are in the source reference [79]. Adapted with permission from Ref. [79]. J. Musil et al.: Hard TiN2 dinitride films prepared by magnetron sputtering, J. Vac. Sci. Techno. A 36(4) (2018), 040602–1 to 040602–3. Copyright © 2018, American Vacuum Society.

1.5.3 General Properties of Sputtered Overstoichiometric $TMN_{x>1}$ Nitride Coatings

Besides the overstoichiometric $TiN_{x>1}$ nitride films also the overstoichiometric $ZrN_{x>1}$ nitride films were investigated in detail [78]. This study was of a key importance for understanding the formation of TMN_2 dinitride films. It was found that TMN_2 films can be formed only from TM elements which do not form TM_3N_4 nitrides because the formation enthalpy ΔHf_{TM3N4} of the TM_3N_4 is much higher than that of TMN_2. For instance, ΔHf_{ZrN2} = −37.35 kJ/mol and Δ = Hf_{Zr3N4} = − 817.05 kJ/mol. At present, it is not clear which TM elements do not form TM_3N_4 nitrides. Potential candidates for sputtering of TMN_2 dinitrides are Ti, Ta, Mo, etc. However, this statement requires to be experimentally verified.

In summary, it can be concluded that

1. Overstoichiometric $TMN_{x>1}$ films can be one-phase films with c-TiN structure (TiN_x) or two-phase films with c-ZrN and o-Zr_3N_4 structure (ZrN_x).
2. Two-phase overstoichiometric $TMN_{x>1}$ films cannot form dinitrides because the formation enthalpy of the TM_3N_4 phase is greater than that TMN_2 phase.
3. Overstoichiometric $TMN_{x>1}$ films with x = N/TM can be sputtered either by the HDM which allows to reach n_N/n_{TM} > 1 due to the reduction of the deposition rate a_D of the TM atoms [78, 79] or in the Magnetron/low-pressure arc evaporation system (magnetron/PINK) [77], or in the Magnetron/ECR discharge (magnetron/ECR), etc.

1.6 THE SPUTTERING OF HETEROSTRUCTURAL ALLOY COATINGS

In recent years, many papers reporting on magnetron sputtering of (Me_1, Me_2) alloy films containing two Me_1 and Me_2 elements were published [44, 80–88]. However, these (Me_1, Me_2) alloy films exhibit low values of the hardness H < 10 GPa and the resistance to cracking, and high values of the effective Young's modulus E^*. As examples (Ni,Ti) [80], (Zr,Ti) [81, 82], (Ti,Nb) [44], (Ti,Cr) [83], (Cu,Zr) [84], (Cu,Mo) [85], (Al,Mo) [86], (Ta,Ag) [87], and (Ni,Zr) [88] alloy films can be given. A low ratio $H/E^* \leq 0.1$ of these films is one of the key reasons why the (Me_1, Me_2) alloy films easily crack [2, 4]. Low H and low resistance to cracking of the (Me_1, Me_2) alloy films strongly limit their use in many practical applications. Therefore, the development of flexible hard alloy films with an enhanced resistance to cracking is a great challenge, limiting their near-term industrial applications.

In this section, a possibility of the formation of hard and simultaneously tough alloy films is demonstrated on metal + metalloid (Me, Met) alloy films. The metalloid is a chemical element with properties between those of typical metals and non-metals. The metalloid elements are found in the middle of the main group of the periodic table at the point where the metals and non-metals meet; see Figure 1.12. Typical metalloids have a metallic appearance, but they are brittle and only

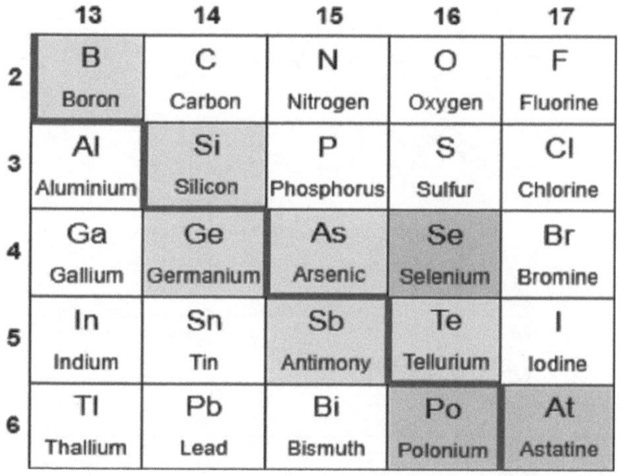

	13	14	15	16	17
2	B Boron	C Carbon	N Nitrogen	O Oxygen	F Fluorine
3	Al Aluminium	Si Silicon	P Phosphorus	S Sulfur	Cl Chlorine
4	Ga Gallium	Ge Germanium	As Arsenic	Se Selenium	Br Bromine
5	In Indium	Sn Tin	Sb Antimony	Te Tellurium	I Iodine
6	Tl Thallium	Pb Lead	Bi Bismuth	Po Polonium	At Astatine

FIGURE 1.12 Classification of metalloid elements: (1) Commonly recognized (93%): B, Si, Ge, As, Sb, Te, (2) irregularly recognized (44%): Po, At, (3) less commonly recognized (24%): Se, and (4) rarely recognized (7%): C, Al. Adapted with permission from Ref. [70]. J. Musil, Z. Čiperová, R. Čerstvý, S. Haviar: Flexible hard (Zr,Si) alloy films prepared by magnetron sputtering, Thin Solid Films 688 (2019), 137,216. Copyright © 2019, Elsevier.

fair conductors of electricity. Chemically, they behave mostly as non-metals. Despite the fact that the metalloid elements are brittle they can form flexible hard materials in a combination with TM elements, that is, in (TM, Met) alloys. It is demonstrated for the (Zr,Si) alloy films with a high Si content [70].

1.6.1 The Hardness of Sputtered of (Zr,Si) Alloy Films

The main task in the development of flexible, hard (Zr,Si) alloy films with an enhanced resistance to cracking is to increase their hardness H and to find conditions under which the ratio $H/E^* \geq 0.1$. The hardness H is strongly influenced mainly by the content of Si in Zr; see Figure 1.13. This figure gives a new insight into the correlations between the mechanical properties (H, σ) and the elemental composition (Si,Zr) of the $ZrSi_x$ films. From this figure the following important facts can be drawn:

1. The $ZrSi_x$ alloy films with the ratio Si/Zr slightly below 1 (≈ 0.9) exhibit maximum hardness up to $H \approx 20$ GPa.
2. The hardest $ZrSi_{x \approx 0.9}$ alloy films exhibit the highest compressive macrostress of about -2.5 GPa but still acceptable for protective films. These films do not follow the hardness calculated from the rule of mixture (denoted as H theory), that is, $H_{cal} = H_{Si}.V_{Si} + H_{Zr}.V_{Zr}$; here $H_{Si} = 13.2$ GPa and $H_{Zr} = 0.9$ GPa, and $V_{Si} = x/(1 + x)$ and $V_{Zr} = 1/(1 + x)$.
3. The measured hardness H approximately agrees with the hardness H_{cal} calculated from the rule of mixture only in case of the soft $ZrSi_x$ with a low hardness ($H \approx 10$ GPa) at $x \approx 0.5$ and $x > 1.5$ and tensile ($\sigma > 0$) or almost zero ($\sigma \approx 0$) stress.

Further, it was found that the $ZrSi_x$ films with $x = Si/Zr \approx 1$, which exhibit high values of $H \geq 18$ GPa, $H/E^* \geq 0.1$, $W_e \geq 60\%$, non-columnar, dense, void-free microstructure and compressive macrostress ($\sigma < 0$), are crystalline and resistant to cracking. The high hardness and enhanced resistance to cracking result from a combined action of their elemental composition and compressive macrostress. The crystallinity of alloy films strongly depends on the energy $\varepsilon_{bi} \approx (|U_s|(i_s/a_D)$ which is controlled by the negative substrate bias U_s and the argon pressure p; see Figure 1.14. The crystallinity of sputtered films is characterized by XRD patterns. Figure 1.14a and Figure 1.14b show the

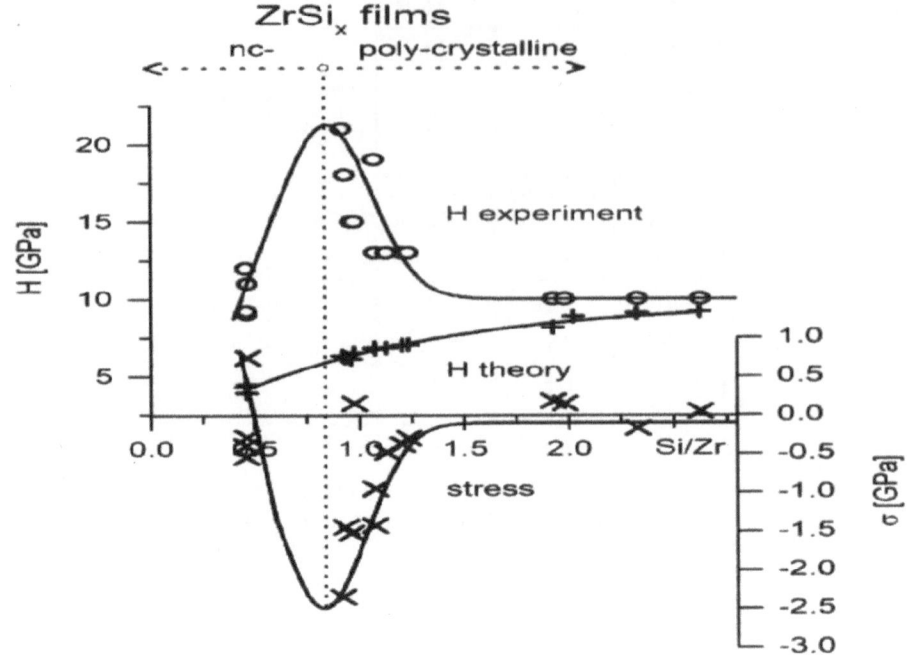

FIGURE 1.13 Hardness H and macrostress σ in (Zr,Si) alloy films sputtered at $T_s = 300$ °C, $U_s = -500$ V on Si (100) substrates as a function of the ratio Si/Zr. Adapted with permission from Ref. [70]. J. Musil, Z. Čiperová, R. Čerstvý, S. Haviar: Flexible hard (Zr,Si) alloy films prepared by magnetron sputtering, Thin Solid Films 688 (2019), 137,216. Copyright © 2019, Elsevier.

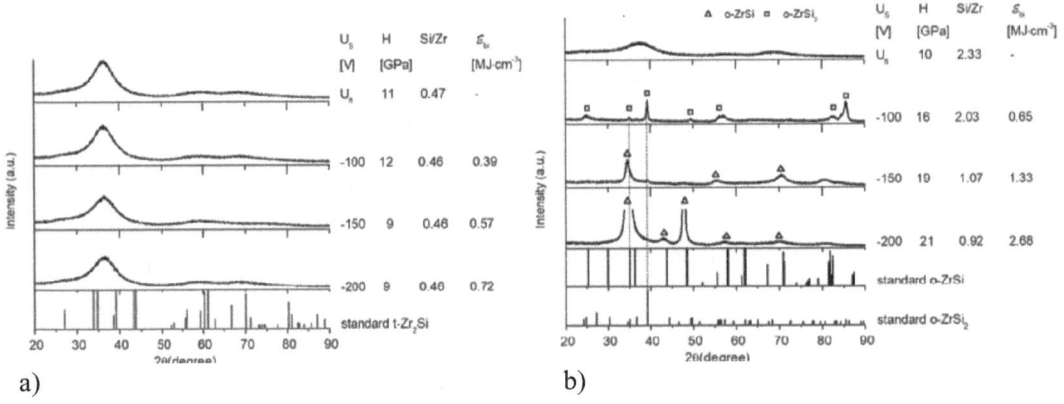

FIGURE 1.14 Evolution of XRD patterns of the ZrSi$_x$ films with (a) a low ratio Si/Zr ≈ 0.5 and at high p = 1.0 Pa and (b) a high ratio Si/Zr ≥ 1 and at low p = 0.3 Pa with increasing negative substrate bias U$_s$. Adapted with permission from Ref. [70]. J. Musil, Z. Čiperová, R. Čerstvý, S. Haviar: Flexible hard (Zr,Si) alloy films prepared by magnetron sputtering, Thin Solid Films 688 (2019), 137,216. Copyright © 2019, Elsevier.

evolution of the XRD patterns of the ZrSi$_x$ sputtered at a high (1 Pa) and a low (0.3 Pa) argon pressures. From these figures, three important results can be drawn:

1. The energy ε_{bi} strongly influences the crystallinity of ZrSi$_x$ films and their stoichiometry x = Si/Zr.
2. The ZrSi$_x$ alloy films sputtered at a low energy $\varepsilon_{bi} \leq 0.7$ MJ/cm^3 are X-ray amorphous and have either a low stoichiometry x = Si/Zr ≈ 0.46 at p = 1 Pa (Fig 1.14a) or a high stoichiometry x = Si/Zr ≈ 2 at p = 0.3 Pa (Figure 1.14b).

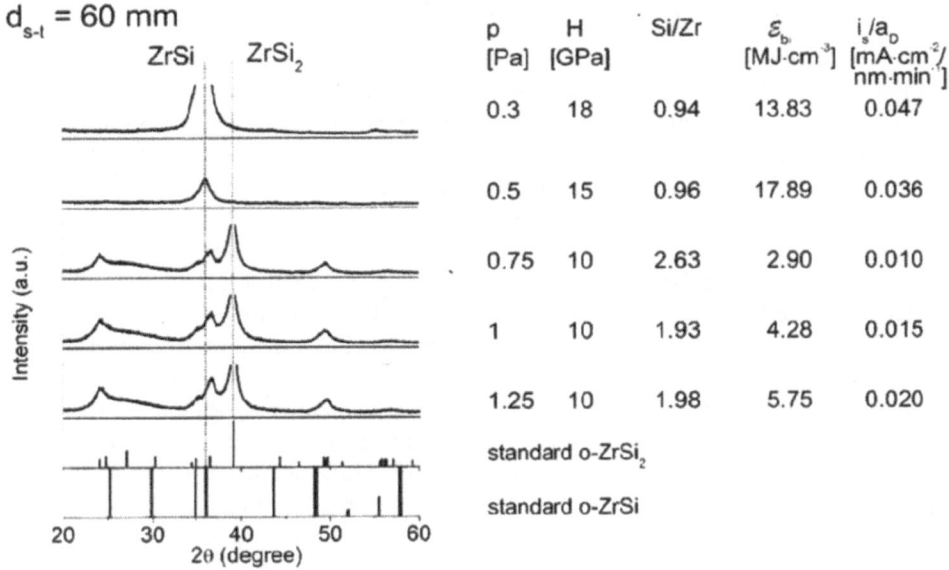

p [Pa]	H [GPa]	Si/Zr	ε_{bi} [MJ·cm^{-3}]	i_s/a_D [mA·cm^2/ nm·min]
0.3	18	0.94	13.83	0.047
0.5	15	0.96	17.89	0.036
0.75	10	2.63	2.90	0.010
1	10	1.93	4.28	0.015
1.25	10	1.98	5.75	0.020
standard o-ZrSi$_2$				
standard o-ZrSi				

FIGURE 1.15 XRD patterns of the ZrSi$_x$ alloy films sputtered at $T_S = 300$ °C, $U_s = -500$ V and different values of the energy ε_{bi} controlled by the argon pressure p. Adapted with permission from Ref. [70]. J. Musil, Z. Čiperová, R. Čerstvý, S. Haviar: Flexible hard (Zr,Si) alloy films prepared by magnetron sputtering, Thin Solid Films 688 (2019), 137,216. Copyright © 2019, Elsevier.

3. The ZrSi$_x$ alloy films sputtered at a high energy $\varepsilon_{bi} \geq 1$ MJ/cm^3 are hard and crystalline with the stoichiometry x ≈ 1 and a pure o-ZrSi crystal structure. The hardness of the ZrSi$_{x \approx 1}$ films increases with increasing negative substrate bias U_s.

The hardness H of the crystalline ZrSi$_{x \approx 1}$ films sputtered at a high negative bias U_s also strongly depends on their stoichiometry x = Si/Zr; see Figure 1.15. From this figure, it is seen that

1. The energy ε_{bi} delivered into the growing ZrSi$_x$ films increases with decreasing argon pressure p at constant bias $U_s = -500$ V.
2. The stoichiometry x of the ZrSi$_x$ films decreases from ~2 to ~0.95 with decreasing p.
3. The ZrSi$_x$ films with x ≈ 2 are polycrystalline composed of o-ZrSi$_2$ and o-ZrSi crystals and the ZrSi$_x$ with x ≈ 0.95 are single-phase polycrystalline o-ZrSi films.
4. The single-phase polycrystalline o-ZrSi films are the hardest alloy films.

1.6.2 Difference Between Standard Ion Bombardment and Nanowelding

Increase of hardness H of the sputtered film by increasing the negative substrate bias ($U_s < 0$) results in an increase of its crystallinity (the size of grains increases). In this process, the whole volume of the sputtered film is heated, enhancing its crystallization. This process is strongly different from the nanowelding. In the case of the sputtering of the film using the nanowelding process, the crystallinity of the created film is decreased (the size of grains decreases). It is due to the formation of the film from a melted coating material, its extremely fast cooling, freezing in very thin layers of about several monolayers, and the sputtering on a cold film material. This is a way to produce a hard nanocrystalline film. However, to master the nanowelding technology, a high ratio i_s/a_D has to be used; for more details, see Section 1.5.

In summary, it can be concluded the ZrSi$_x$ alloy films with x = Si/Zr ≈ 1 composed of one metallic and one metalloid element (Me, Met) exhibit high values of H ≥ 18 GPa, H/E* ≥ 0.1, $W_e \geq 60\%$, non-columnar, dense, void-free microstructure and compressive macrostress ($\sigma < 0$). However, both the stoichiometry x and the energy ε_{bi} must still be optimized.

1.7 THE HARDNESS OF HARD NANOCOATINGS

Up to recently, main attention was concentrated on the hardness H of nanomaterials (coatings), the ways of the hardness H enhancement and the achievement of the coating hardness H approaching or even exceeding that of diamond. Many new advanced single-phase and nanocomposite coatings based on (i) nitrides, particularly the composites of the type nc-MeN/a-Si$_3$N$_4$ with low ($\leq$ 10 at.%) Si content, where Me = Ti, Zr, Ta, V, Nb, Hf, Cr, Mo, W, Al, etc., (ii) oxides, (iii) oxynitrides, (iv) carbides, (v) carbonitrides, (vi) borides and other compounds were successfully developed. Some of these coatings exhibit an enhanced hardness H achieving up to 50–70 GPa but none of them exhibit the hardness H approaching that of diamond. It means that diamond still remains the hardest material. On the other hand, these new hard advanced coatings exhibit many new unique properties. This was a reason why many researchers consider the hardness H as a main parameter of the nanocoatings only.

Generally, the hardness H is measure of the resistance to localized plastic deformations induced by either mechanical indentation or abrasion. The hardness is not a real physical parameter. Usually, it is measured by the loading of the coating by the diamond indenter at a low external load L, ensuring that the depth d of its penetration into the coating is small (d/h $\leq$ 0.1), called the indentation test; here h is the thickness of the coating. The ratio d/h $\leq$ 0.1 is a necessary condition that the hardness H$_c$ of the measured coating is not influenced by the presence of the substrate. At high load, the hardness strongly depends on the hardness H$_s$ and the effective Young's modulus E*$_s$ of the substrate [6].

The coating hardness H$_c$ is strongly influenced by its macrostress σ generated inside the coating during its growth. The macrostress σ depends on the elemental and phase composition, structure and microstructure of the coating and mainly on the energy ε delivered into it during its growth. Very complex relations between σ and these parameters strongly complicate a correct explanation of the measured H$_c$ and the optimization of deposition parameters necessary to create the coatings with enhanced hardness. The stress generated in the hard coatings with columnar microstructure is tensile ($\sigma > 0$), in the hard coatings with dense, void-free, and non-columnar microstructure, is compressive ($\sigma < 0$). The low compressive stress has positive effect on the formation of coatings with enhanced resistance to cracking. On the other hand, high compressive stress results in easy cracking of coatings and also in their delamination from the substrates.

Very often, sputtered hard coatings exhibit a high compressive macrostress ($\sigma < 0$) up to several GPas. Such coatings delaminate from the substrate and easily crack. Therefore, it is necessary to form hard coatings with a low compressive macrostress of about –2 GPa and lower. In principle, the compressive macrostress can be reduced by an optimization of the sputtering process, that is, by the controlled stress healing during sputtering. The self-healing can be realized, for instance, in multilayer coatings composed of alternating amorphous and crystalline layers; see Figure 1.16. The compressive macrostress generated in a crystalline layer is released in a low-density amorphous layer. The controlled stress healing can be realized, for instance, in sputtering with a pulsed substrate bias with alternating negative and positive pulses. For illustration, two examples showing (i) the possibility of the macrostress σ reduction (healing) in hard coatings using a pulsed substrate bias U$_{sp}$ and (ii) the formation of hard coatings with zero macrostress σ are given below.

1.7.1 MACROSTRESS HEALING WITH PULSED BIPOLAR SUBSTRATE BIAS

The principle of a reduction (healing) of the macrostress σ in sputtered films at a pulsed substrate bias U$_{sp}$ is based on alternating of the ion and the electron bombardment of the film during its growth; see Figure 1.17. The alternating ion and electron bombardment of the growing film is realized by alternating negative and positive pulses at the substrate. The microstructure of the growing film is densified during the negative pulse by ion bombardment. Simultaneously, the compressive macrostress ($\sigma < 0$) is generated in the film and its magnitude increases with increasing voltage

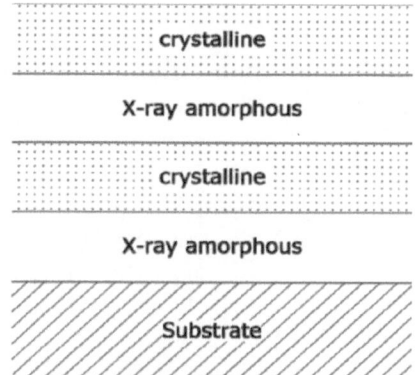

FIGURE 1.16 Multilayer coating composed of alternating X-ray amorphous and crystalline layers [6, 7]. Reprinted with permission from Ref. [6]. J. Musil: Advanced hard nanocomposite coatings with enhanced toughness and resistance to cracking, Chapter 7 in *Thin Films and Coatings: Toughening and Toughening Characterization*, S. Zhang (Editor) CRC Press, USA, 2015, pp. 377–463. Copyright © 2015, CRC Press.

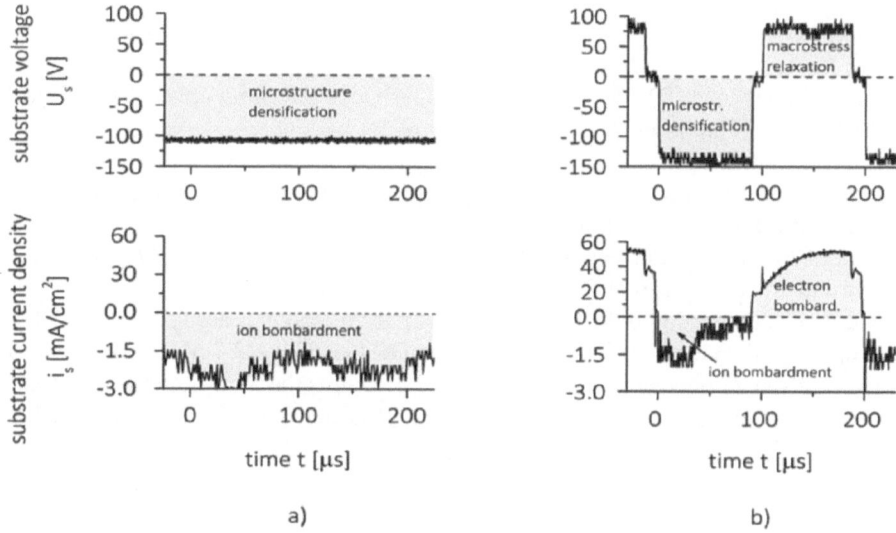

FIGURE 1.17 Comparison of DC and pulsed substrate bias used in sputtering of Ti(Al,V)N films by DC dual magnetron discharge generated at $I_d = 1$ A, $T_s = 500$ °C, $d_{s-t} = 60$ mm, $p_T = p_{Ar} + p_{N2} = 0.8 + 0.2 = 1$ Pa. (a) Continuous ion bombardment and (b) alternating ion/electron bombardment of the growing Ti(Al,V)N film by ions and electrons produced by DC bias ($U_s = -100$ V) and pulsed bias ($U_{sp} = -130/+70$ V, $f_r = 5$ kHz), respectively. Adapted with permission from Ref. [89]. M. Jaroš, J. Musil, R. Čerstvý, S. Haviar: Effect of energy on macrostress in Ti(Al,V)N films prepared by magnetron sputtering, Vacuum 158 (2018), 52–59. Copyright © 2018, Elsevier.

of the negative pulse. On the other hand, the macrostress σ, generated in the film during the ion bombardment, is relaxed by the electron current which thermally anneals the growing film during the positive pulse. It means that the films sputtered at a DC substrate bias $U_{s\,DC}$ will always exhibit a higher compressive macrostress ($\sigma < 0$) compared with that of the films sputtered at a pulsed bipolar substrate bias U_{sp}.

The relaxing of the compressive macrostress ($\sigma < 0$) in a sputtered film by the electron bombardment was confirmed by sputtering of the Ti(Al,V)N films under the same deposition conditions at

TABLE 1.1

Physical and mechanical properties and compressive macrostress ($\sigma < 0$) in the Ti(Al,V)N films sputtered by DC dual magnetron*

Bias	f_r	U_s	i_s	h	a_D	τ_e/τ_i	E_{bi}	σ	H	E^*	W_e	H/E^*	ε_{cr}	L_{cr}	Structure	Texture
voltage	[kHz]	[V]*	[mA/cm²]	[nm]	[nm/min]		[MJ/cm³]	[GPa]	[GPa]	[GPa]	[%]		[%]	[N]		
DC	0	−40	1.0	2100	36.0	0	1.6	−1.7	28.4	282	70	0.10	--	0.25	Crystalline	(200) + (220)
DC	0	−100	1.8	1100	37.5	0	3.7	−4.0	30.7	220	81	0.14	>2.0	>1.0	Crystalline	(220)
Pulsed	5	−100/+70	0.9	1000	33.0	1.3	1.6	−0.8	19.1	175	68	0.11	1.3	0.75	XRA	

ε_{bip} is the average energy of ios during the negative pulse of the pulsed substrate bias U_{sp}, the index „cr"denotes values at which the film cracks under a load L and XRA is the X-ray amorphous structure.

* Operated at Id = 1 A, Ts = 500 °C, ds-t = 60- mm, pT = pAr + pN2 = 0.2 + 0.8 = 1 Pa on the substrate held at (i) the DC substrate bias voltage Us DC and (ii) the pulsed substrate bias voltage Usp with repetition frequency of pulses fr = 5 kHz. The bending test was performed on the films sputtered on the Mo strip and the indentation test on the films sputtered on the Si substrates; for more details, see Ref. [6]. Adapted with permission from Ref. [89]. M. Jaroš, J. Musil, R. Čerstvý, S. Haviar: Effect of energy on macrostress in Ti(Al,V)N films prepared by magnetron sputtering, Vacuum 158 (2018), 52–59.

DC and pulsed bipolar bias [89]. Results of this experiment are summarized in Table 1.1. From Table 1.1, the following important issues can be drawn:

1. The film sputter deposited at a DC negative bias $U_s = -100$ V, that is, at the ion bombardment of the growing film only, exhibits a high compressive macrostress ($\sigma = -4$ GPa) compared with that of the film sputter deposited at the pulsed bias $U_{sp} = -100/+70$ V with alternating negative and positive pulses at $f_r = 5$ kHz ($\sigma = -0.8$ GPa).
2. The energy ε_{bi} delivered into the film growing at pulsed bipolar bias U_{sp} is lower (1.6 MJ/cm³) than the energy delivered to the film growing at DC bias $U_{s\,DC}$ (3.7 MJ/cm³). This is a reason why the film sputter deposited at a pulsed bias U_{sp} exhibits the X-ray amorphous structure and the film sputter deposited at DC bias $U_{s\,DC}$ is the crystalline with a dominant TiN (220) texture; see Figure 1.18.

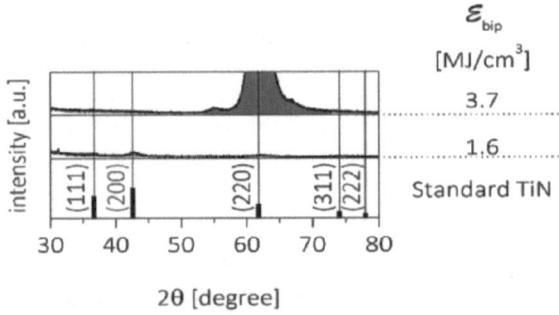

FIGURE 1.18 Comparison of the structure of the Ti(Al,V)N film sputter deposited at (i) DC substrate bias $U_{s\,DC} = -100$ V and a high energy $\varepsilon_{bi} = 3.7$ MJ/cm³ and (ii) the pulsed substrate bias U_{sp} and a low energy $\varepsilon_{bip} = 1.6$ MJ/cm³. Adapted with permission from Ref. [89]. M. Jaroš, J. Musil, R. Čerstvý, S. Haviar: Effect of energy on macrostress in Ti(Al,V)N films prepared by magnetron sputtering, Vacuum 158 (2018), 52–59. Copyright © 2018, Elsevier.

3. The electron bombardment of the growing film, however, results not only in a strong decrease of the compressive macrostress σ but also in decrease of its hardness H, elastic recovery W_e, H/E* ratio, and a lower resistance to cracking; see Table 1.1.

The electron bombardment of the growing film is the reason why the macrostress σ generated in the film sputter deposited at the pulsed substrate bias U_{sp} with alternating negative and positive pulses is considerably lower than that in the film sputter deposited at the DC negative substrate bias voltage U_s. The length of the negative pulse τ_i and the length of positive pulse τ_e can be different. It means that the efficiency of a relaxing of macrostress σ in the film can be controlled by the ratio τ_e/τ_i. More details are given in Ref. [89].

1.7.2 The Hard Coatings with Zero Macrostress

The hard coatings with zero macrostress ($\sigma = 0$) are important as much from the point of basic science as from the point of applications. In the first case, the hard coatings enable us to determine the intrinsic hardness of the coating material. In the second case, the knowledge of the deposition conditions necessary to form such coatings enables us to sputter free-standing coatings. At present, there are three methods allowing to produce hard coatings with $\sigma = 0$.

The first method is controlled by the film microstructure; see Figure 1.19. The boundary line between zone 1 and zone T in the SZM model discussed in Section 1.2 is claimed to correspond to the zero macrostress $\sigma = 0$ generated in sputtered films and it separates the films with microstructure composed of the columns separated by voids (zone 1) from the films with the dense, void-free microstructure composed of the fibrous grains embedded in an amorphous intergrain phase (zone T) [22, 23]. However, the boundary line $\sigma = 0$ is a curve depending not only on the film microstructure but also on the homologous temperature T_s/T_m, the sputtering gas pressure p_{Ar}, the energy of bombarding particles $\varepsilon_p = \varepsilon_{bi} + \varepsilon_{fn}$, that is, also on the negative substrate bias U_s. It is a reason why it is almost impossible to select correctly all deposition parameters necessary to form the film with $\sigma = 0$.

The second method is controlled by the energy of bombarding ions ε_{bi}; see Figure 1.20. This figure shows the evolution of the macrostress σ in sputtered δ-TiN and α-Ti(N) coatings with increasing energy ε_{bi}. The energy ε_{bi} was calculated from the formula $\varepsilon_{bi} = U_s i_s/a_D$ [91]. In this figure the photos of the SEM images of the cross sections of the α-Ti(N) and δ-TiN coatings with (i) the columnar microstructure and tensile stress ($\sigma > 0$) and (ii) dense, void-free microstructure and compressive stress ($\sigma < 0$), respectively, are also inserted. The measured dependence $\sigma = f(\varepsilon_{bi})$ and the finding

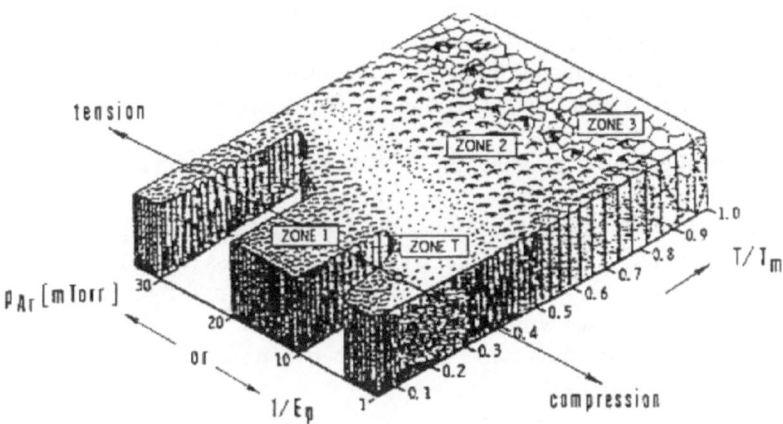

FIGURE 1.19 3D structural zone model (SZM) of sputtered films proposed by J.A. Thornton [22, 23]. Adapted with permission from reference [23]. J.A. Thornton: Recent developments in sputtering – Magnetron sputtering, Metal Finishing 77(5) (1979), 83–87. Copyright © 2017, Elsevier.

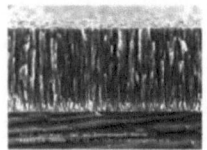

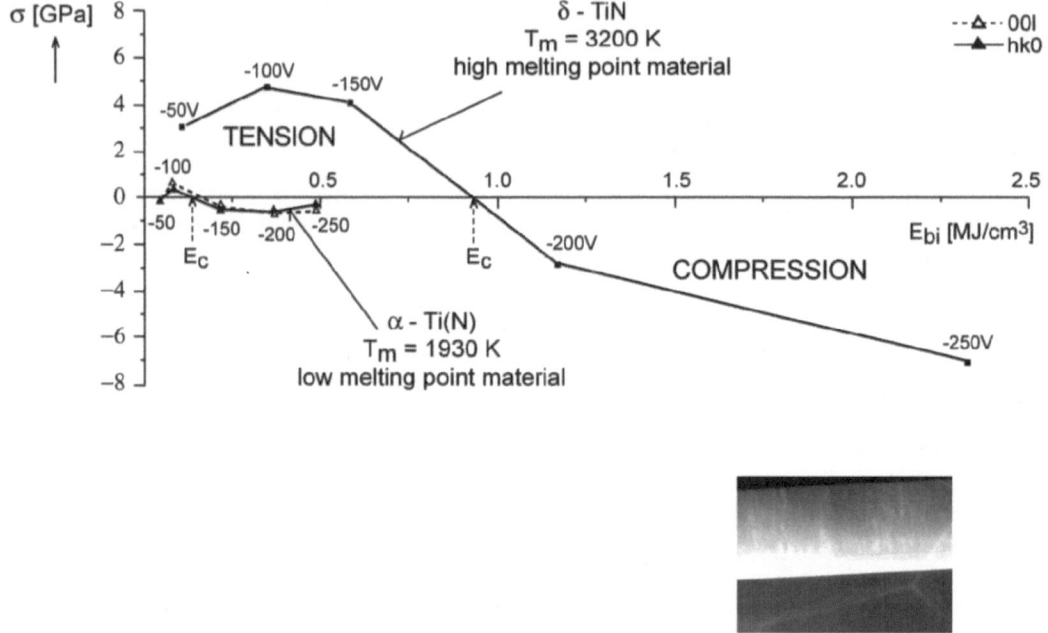

FIGURE 1.20 Macrostress σ in sputtered α-Ti(N) and δ-TiN$_{x \approx 1}$ films as a function of the energy $\varepsilon_{bi} = U_s i_s/a_D$ at $p_T = p_{Ar} + p_{N2} = 5$ Pa and $T_s = 350°C$, that is, at $T_s/T_m = 0.32$ and 0.19 for α-Ti(N) and δ-TiN$_{x \approx 1}$, respectively. Adapted with permission after reference [90]. J. Musil, V. Poulek, V. Valvoda, R. Kužel, Jr., H.A. Jehn, M.E. Baumgätner: Relation of deposition conditions of Ti-N films prepared by dc magnetron sputtering to their microstructure and macrostress, Surf.Coat.Technol. 60 (1993), 484–488. Copyright © 1993, Elsevier.

of a critical energy ε_c at which the macrostress σ in the sputtered coating is zero (σ = 0) show the direct connection of the energy ε_{bi} with the microstructure of the sputtered coating in the SZM via the borderline between zone 1 and zone T, the homologous temperature T_s/T_m and the sputtering gas pressure p; see Figure 1.19. From Figures 1.19 and 1.20, five important conclusions can be drawn.

1. There is a critical energy ε_c at which the sputtered coating exhibits zero macrostress, that is, σ = 0 at $\varepsilon_{bi} = \varepsilon_c$.
2. The value of ε_c decreases with decreasing ratio T_s/T_m and sputtering gas pressure p.
3. The coatings sputtered at $\varepsilon_{bi} < \varepsilon_c$ exhibit a tensile macrostress (σ > 0). On the other hand, the coatings sputtered at $\varepsilon_{bi} > \varepsilon_c$ exhibit a compressive macrostress (σ < 0).
4. The macrostress σ increases with increasing ratio T_s/T_m.
5. The energy ε_c exactly corresponds to a transition between zone 1 and zone T in the Thornton's SZM as presented in Ref. [92].

A similar dependence was reported by Shen and Mai in 2000 [93]; see Figure 1.21. This figure shows the evolution of the stress in the as-deposited WN$_x$ films on Si (100) substrates by DC reactive sputtering at a high pressure $p_{Ar + N2} = 25$ mTorr as a function of its stoichiometry x. The stress in the WN$_x$ films is converted from tension to compression when the amount of N$_2$ in the Ar + N$_2$

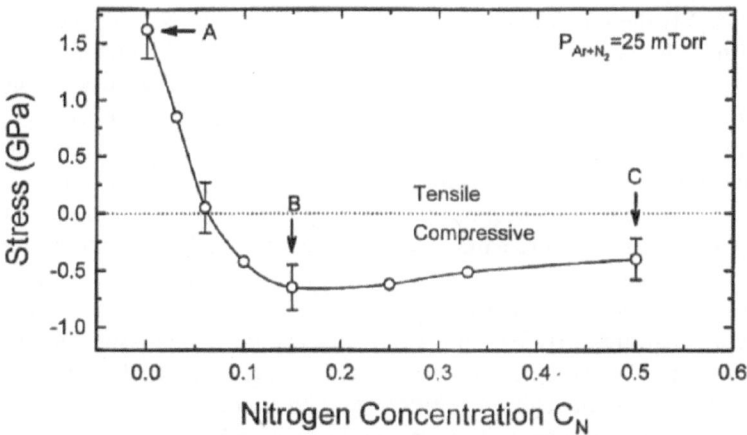

FIGURE 1.21 Macrostress σ of the WN_x film magnetron sputtered on Si (100) substrate at $T_s = 200$ °C and high pressure $p = p_{Ar} + p_{N2} = 25$ mTorr as a function of nitrogen concentration C_N in sputtering gas. Reprinted with permission from Ref. [93]. Y.G. Shen, Y.W. Mai: Effect of deposition conditions on internal stresses and microstructure of reactively sputtered tungsten nitride films, Surf. Coat. Technol. 127 (2000), 238–246. Copyright © 2000, Elsevier.

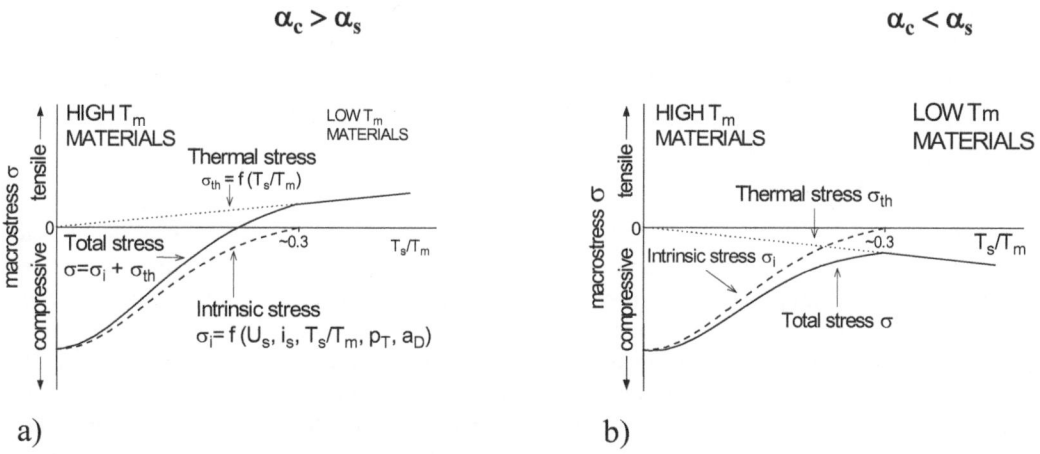

FIGURE 1.22 Schematic illustration of the macrostress σ as a function of T_s/T_m for coatings with compressive intrinsic stress ($\sigma_i < 0$) and for a) tensile thermal stress ($\sigma_{th} > 0$) and b) compressive thermal stress ($\sigma_{th} < 0$). Adapted with permission after references [92, 94]. J.A. Thornton, D.W. Hoffman: Stress-related effects in thin films. Thin Solid Films 171 (1989), 5–31. J. Soldán and J. Musil: The structure and mechanical properties of DC magnetron sputtered TiC/Cu films, Vacuum 81 (2006), 531–538. Copyright © 1989, 2006, Elsevier.

mixture increases. The stress $\sigma = 0$ at $C_N = 0.06$; here, C_N is the nitrogen concentration. The WN_x films sputtered at $C_N < 0.06$ are in tension, the films WN_x sputtered at $C_N > 0.06$ are in compression.

The third method allowing produce hard coatings with $\sigma = 0$ is controlled by the difference between the thermal contraction of the film and the substrate; see Figure 1.22. This difference in thermal expansion coefficients of the coating α_c and the substrate α_s results in the thermal macrostress σ_{th} and can be calculated from the following formula [92].

$$\sigma_{th} = \left[E_c / (1 - \nu_c) \right] (\alpha_c - \alpha_s)(T_s - T) \tag{1.18}$$

Here, $E_c/(1 - v_c)$ is the biaxial elastic modulus of the coating, and T_s and T are the substrate temperature during film deposition and the temperature at which the macrostress σ is measured, respectively. The thermal stress is controlled by the substrate temperature T_s and can be either compressive ($\sigma_{th} < 0$) or tensile ($\sigma_{th} > 0$) according to the ratio $\alpha_c/\alpha_s < 1$ or $\alpha_c/\alpha_s > 1$, respectively. The total macrostress σ_T generated in the coating during its deposition is composed of two components: thermal stress σ_{th} and intrinsic stress σ_i, that is, $\sigma_T = \sigma_{th} + \sigma_i$. The intrinsic stress σ_i occurs as a consequence of the accumulation of crystallographic defects the amount of which increases with increasing energy ε_p delivered into the growing coating by bombarding and condensing particles.

The dependence of the total macrostress σ on the ratio T_s/T_m is schematically displayed in Figure 1.22: the tensile thermal stress ($\sigma_{th} > 0$) in Figure 1.22a and the compressive thermal stress ($\sigma_{th} < 0$) in Figure 1.22b. From this figure, it is seen that the intrinsic stress σ_i dominates over the thermal stress σ_{th} at low values of ratio $T_s/T_m \leq 0.3$ [92]. On the contrary, σ_{th} dominates at high values of ratio $T_s/T_m \geq 0.3$, that is, at high deposition temperatures T_s and low melting temperatures T_m of the material of coating. Since the value of σ_{th} can be either tensile (> 0) or compressive (< 0) as Equation (1.18) clearly shows, the total macrostress σ_T in the coating produced at $T_s/T_m \geq 0.3$ may be also tensile or compressive if $\alpha_c > \alpha_s$ or $\alpha_c < \alpha_s$, respectively. In the first case, σ_i and σ_{th} are mutually compensated and it means that (i) the stress-free coating can be produced at a certain value of $T_s/T_m < 0.3$ when $\sigma_{th} = -\sigma_i$ and (ii) σ is tensile (> 0) in coatings produced at $T_s/T_m > T_s/T_m$ at $\sigma_{th} = -\sigma_i$. In the case when $\alpha_c < \alpha_s$, the total stress σ_T in coatings is compressive for all values T_s/T_m and reaches its minimum value at $T_s/T_m \approx 0.3$.

In summary, it can be concluded that (1) there is a critical energy $\varepsilon_c = \varepsilon_{bi} = f(T_s/T_m, p)$ at which sputtered film exhibits the zero macrostress ($\sigma = 0$), (2) the hardness of the coating sputtered at $\varepsilon_{bi} = \varepsilon_c$ determines the intrinsic hardness of the coating material, and (3) several micrometer thick, hard, and free-standing coatings with enhanced resistance to cracking could be sputtered at $\varepsilon_{bi} = \varepsilon_c$.

1.8 CONCLUDING REMARKS

At present, it is already well known that a key parameter for deciding on sputtering of the coating with a given elemental composition and required physical and functional properties is the energy ε delivered into the coating during its growth. This finding is of a general validity and will be also a basis for the development of new advanced nanocoatings with new unique properties. However, this development will require deeper knowledge in the field of thin films and coatings and particularly, a wide interdisciplinary insight. It will be important: (1) to develop new technological processes not only based on the knowledge of the material science and plasma physics but also on the electrical engineering, vacuum science and chemistry, and (2) to develop new systems for the deposition of thin films and coatings operating under new physical conditions.

Only correct interdisciplinary insights in this development will guarantee the success of this project in the future. New results will deepen our knowledge in the field and new advanced nanocoatings with a huge application potential will be developed. Main attention should be devoted to the development of

1. The heterostructural nanocoatings composed of elements with different crystal structure.
2. The nanocoatings with high thermal stability, enhanced wear, erosion, oxidation, corrosion, and radiation resistance for the operation in extreme conditions, for instance, at high temperatures above 1500 °C.
3. A new generation of the nanocoatings formed from the coating material melted in monolayers under high pressures (≥ 1000 GPa) and extremely fast cooling rate ($\geq 10^{10}$ K/s), that is, under strongly non-equilibrium deposition conditions at atomic level.

ACKNOWLEDGEMENTS

This work was supported in part by many projects of the Grant Agency of the Czech Republic and the Czech Ministry of Education, Youth and Sports in the period between 2010 and 2020.

REFERENCES

[1] S. Zhang, D. Sun, Y. Fu and H. Du: Toughening of hard nanostructured thin films: a critical review, *Surf. Coat. Technol.* 198 (2005), 2–8.

[2] J. Musil: Hard nanocomposite coatings: Thermal stability, oxidation resistance and toughness, *Surf. Coat. Technol.* 207 (2012), 50–65.

[3] S. Zhang, X.M. Zhang: Toughness evaluation of hard coatings and thin films, *Thin Solid Films* 520 (2012), 2375–2389.

[4] J. Musil: Flexible hard nanocomposie coatings, *RSC Adv.* 5 (2015), 60482–60495.

[5] Y.X. Wang, S. Zhang: Present status of hard-yet tough ceramic coatings, Chapter 1 in *Thin Films and Coatings: Toughening and Toughening Characterization*, S. Zhang (Ed.) CRC Press, USA, 2015, pp. 1–37.

[6] J. Musil: Advanced hard nanocomposite coatings with enhanced toughness and resistance to cracking, Chapter 7 in *Thin Films and Coatings: Toughening and Toughening Characterization*, S. Zhang (Editor) CRC Press, USA, 2015, pp. 377–463.

[7] J. Musil, J. Sklenka, R. Čerstvý: Protection of brittle film against cracking, *Appl. Surf. Sci.* 370C (2016), 306–311.

[8] J. Musil: Flexible antibacterial coatings, *Molecules* 22 (2017), 813 (10 pp).

[9] T. Jorg, D. Music, F. Hauser, M.J. Rodill, R. Franz, H. Kostenbauer, J. Winkler, J.M. Schneider, C. Mitterer: Deformation behavior of Re alloyed Mo thin films on flexible substrates: In situ fragmentation analysis supported by first-principles calculations, *Sci. Rep.* 7 (2017), 7374.

[10] H. Kindlund, D.G. Sangiovanni, I. Petrov, J.E. Greene, L. Hultman: A review of the *intrinsic* ductility and toughness of hard transition-metal nitride alloy thin films, *Thin Solid Films*, doi: 10.1016/j.tsf.2019.137479.

[11] J. Musil, J. Sklenka and R. Čerstvý: Transparent Zr-Al-O oxide coatings with enhanced resistance to cracking, *Surf. Coat. Technol.* 206 (2011), 2105–2109.

[12] J. Musil, J. Sklenka, R. Čerstvý, T. Suzuki, M. Takahashi, T. Mori: The effect of addition of Al in ZrO2 thin film on its resistance to cracking, *Surf. Coat. Technol.* 207 (2012), 355–360.

[13] J. Musil, M. Meissner, R. Jílek, T. Tolg, R. Čerstvý: Two-phase single layer Al-O-N nanocomposite films with enhanced resistance to cracking, *Surf. Coat. Technol.* 206 (2012), 4230–4234.

[14] J. Musil, J. Sklenka, J. Procházka: Flexible over-layer coating preventing cracking of thin films deposited on flexible substrates, *Surf. Coat. Technol.* 240 (2014), 275–280.

[15] J. Musil, R. Jílek, R. Čerstvý: Flexible Ti-Ni-N thin films prepared by magnetron sputtering, *J. Mater. Sci. Eng.* A4 (1) (2014), 27–33.

[16] J. Musil, J. Blažek, K. Fajfrlík, R. Čerstvý: Flexible antibacterial Al-Cu-N films, *Surf. Coat. Technol.* 264 (2015), 114–120.

[17] J. Procházka, R. Čerstvý, J. Musil: *Interrelationships between mechanical properties and resistance to cracking of magnetron sputtered (Ti,Al,V)N nitride films*, Abstract # 1396, *International Conference on metallurgical Coatings and Thin Films, ICMCTF 2015*, April 20, 2015, San Diego, USA

[18] J. Musil, M. Zítek, K. Fajfrlík, R. Čerstvý: Flexible antibacterial Zr-Cu-N thin films resistant to cracking, *J. Vac. Sci. Technol.* A 34 (2016), 021508.

[19] J. Musil, G. Remnev, V. Legostaev, V. Uglov, A. Lebedynskiy, A. Lauk, J. Procházka, S. Haviar, E. Smolyanskiy: Flexible hard Al-Si-N films for high temperature operation, *Surf. Coat. Technol.* 307 (2016), 1112–1118.

[20] J. Musil, S. Zenkin, R. Čerstvý: Flexible hydrophobic Zr-N coatings, *Vacuum* 131 (2016), 34–38.

[21] J. Musil, J. Šícha, D. Herman, R. Čerstvý: Role of energy in low-temperature, high-rate formation of hydrophilic TiO_2 thin films using pulsed magnetron sputtering, *J. Vac. Sci. Technol.* A 25(4) (2007), 666–694.

[22] J.A. Thornton: High rate thick films growth, *Annu. Rev. Mater. Sci.* 7 (1977), 239–260.

[23] J.A. Thornton: Recent developments in sputtering – Magnetron sputtering, *Met. Finish.* 77(5) (1979), 83–87.

[24] B.A. Movchan, A.V. Demchishin: Study of the structure and properties of thick vacuum condensates of nickel, titanium, tungsten, aluminum oxide and zirconium oxide, *Phys. Met. Mellogr.* 28 (1969), 83–90.

[25] H. Holleck: Materials selection for hard coatings, *J. Vac. Sci. Technol.* 4(6) (1986). 2661–2669.

[26] H.O. Pierson: *Handbook of refractory carbides and nitrides*, Noyes Publications, Westwood, NJ, USA, 1996, p. 193.

[27] J. Musil, P. Zeman, P. Baroch: Hard Nanocomposite Coatings, in *Comprehensive Materials Processing*, Vol. 4 Coatings and Films, Chapter 4.16, D. Cameron (Ed.), Elsevier, 2014, pp. 325–353.

[28] J. Musil, J. Šůna: The role of energy in formation of sputtered nanocomposite films, *Mater. Sci. Forum* Vol. 502 (2005), 291–296.

[29] M. Jaroš, J. Musil, R. Čerstvý, S. Haviar: Effect of energy on structure, microstructure and mechanical properties of hard Ti(Al,V)N$_x$ films prepared by magnetron sputtering, *Surf. Coat. Technol.* 332 (2017), 190–197.

[30] J. Musil, M. Jaroš, R. Čerstvý, S. Haviar: Evolution of microstructure and macrostress in sputtered hard Ti(Al,V)N films with increasing energy delivered during their growth by bombarding ions, *J. Vac. Sci. Technol.* A 35(2) (2017), 029601-1–020601-5.

[31] J. Musil: Advanced hard nanocoatings: Present State and trends, Chapter 5 in the book *Top 5 Contributions in Molecular Sciences: 5th Edition*, First Published, October 28, 2019 Avid Science, Telanga India, Berlin, Germany, (65 pages), published online and can be accessed at the URL: https://avidscience.com/book/top-5-contributions-in-molecular-sciences-6th-edition/.

[32] J. Musil, S. Kadlec, W.-D. Münz: Unbalanced magnetrons and new sputtering systems with enhanced plasma ionization, *J. Vac. Sci. Technol.* A 9(3) (1991), 1171–1177.

[33] J. Musil, K. Rusňák, V. Ježek, J. Vlček: Planar magnetron with additional plasma confinement, *Vacuum* 46(4) (1995), 341–347.

[34] S. Kadlec, J. Musil: Low pressure magnetron sputtering and selfsputtering discharges, *Vacuum* 47(3) (1996), 307–311.

[35] J. Musil: Low-pressure magnetron sputtering *Vacuum* 50(3–4) (1998), 363–372.

[36] M. Yamashita, Y. Setsuhra, S. Miyake, M. Kumagai, T. Shoji, J. Musil: Studies on magnetron sputtering assisted by inductively coupled rf plasma, *Jpn. J. Appl. Phys.* 38 (1999), 4291–4295.

[37] J. Vlček, P. Kudláček, K. Burcalová, J. Musil: High-power pulsed sputtering using a magnetron with enhanced plasma confinement, *J. Vac. Sci. Technol.* A 25(1) (2007), 42–47.

[38] J. Musil, A.J. Bell, J. Vlček, T. Hurkmans, Formation of high-temperature phases in sputter deposited Ti-based films below 100 deg C, *J. Vac. Scfi. Technol.* A 14 (4)(1996) 2247–2250.

[39] J. Musil, J. Vlček, Magnetron sputtering of alloy and alloy-based films, *Thin Solid Films* 343–344 (1999) 47–50.

[40] Y.M. Zhou, Z. Xie, Y.Z. Ma, F.J. Xia, S.L. Feng, Growth and characterization of Ta/Ti bi-layer films on glass and Si (111) substrates by direct current magnetron sputtering, *Appl. Surf. Sci.* 258 (2012) 7314–7321.

[41] S. Achache, S. Lamri, M. Arab Pour Yazdi, A. Billard, M. Francois, S. Sanchete: Ni-free superelastic binary Ti-Nb coatings obtained by DC magnetron co-sputtering, *Surf. Coat. Technol.* 275 (2015) 283–288.

[42] F.T.N. Wullers, R. Spolenak, Alpha- vs, beta-W nanocrystalline thin films: a comprehensive study of sputter parameters and resulting materials' properties, *Thin Solid Films* 577 (2015) 26–34.

[43] A. Gebert, D. Eigel, P.F. Gostin, V. Hoffman, M. Uhlemann, A. Helth, S. Pilz, R. Schmidt, M. Calin, M. Gottlicher, M. Rohnke, J. Janek, Oxidation treatments of beta-type Ti-40 Nb for biomedical use, *Surf. Coat. Technol.* 302 (2016) 88–99.

[44] D. Photiou, N.T. Panagiotopolous, L. Koutsokeras, G.A. Evangelakis, G. Constantinides, Microstructure and nanomechanical properties of magnetron sputtered Ti-Nb films, *Surf. Coat. Technol.* 302 (2016) 310–319.

[45] E.D. Gonzalez, T.C. Niemeyer, C.R.M. Afonso, P.A.P. Nascente, Ti-Nb thin films deposited by magnetron sputtering on stainless steel, *J. Vac. Sci. Technol.* A34 (2) (2016) (021511-1 to 021511-6).

[46] W.L. Wang, W.C. Chen, K.T. Peng, H.C. Kuo, M.H. Yeh, H.J. Chien, T.H. Ying, The influence of amorphous TaNx under-layer on the crystal growth of over-deposited Ta film, *Thin Solid Films* 603 (2016) 34–38.

[47] J.J. Colin, G. Abadias, A. Michel, C. Jaouen, On the origin of the metastable β-Ta phase stabilization in tantalum sputtered thin films, *Acta Mater.* 126 (2017) 481–493.

[48] J. Xu, J. Cheng, S. Jiang, P. Munroe, Z.H. Xie, Mechanical and electrochemical properties of a sputter-deposited β-Ta5Si3 nanocrystalline coating, *J. Alloys Compd.* 699 (2017) 1068–1083.

[49] E.D. Gonzales, C.R.M. Afonso, P.A.P. Nascente, Influence of Nb content on the structure, morphology, nanostructure, and properties of titanium-niobium magnetron sputter deposited coatings for biomedical applications, *Surf. Coat. Technol.* 326 (Part B) (2017) 424–428.

[50] F. Zhang, C. Li, M. Yan, J. He, Y. Yang, F. Yin, Microstructure and nanomechanical properties of as-deposited Ti-Cr films prepared by magnetron sputtering, *Surf. Coat. Technol.* 325 (2017) 636–642 on line 3 July 2017 doi:10.1016/j.surfcoat.2017.07.005.

[51] J. Musil, Š. Kos, S. Zenkin, Z. Čiperová, D. Javdošňák, R. Čerstvý: β-(Me$_1$, Me$_2$) and MeN$_x$ films deposited by magnetron sputtering: Novel heterostructural alloy and compound films, *Surf. Coat. Technol.* 337 (2018), 75–81.

[52] J.L. Murray, The Fe-Ti (Iron-Titanium) system, *Bull. Alloy Phase Diagr.* 2 (3) (1981) 320–334.

[53] J.L. Murray, The Cr-Ti (Chromnium-Titanium) system, *Bull. Alloy Phase Diagr.* 2 (2) (1981) 174–181.

[54] N.A. Marks: Evidence for subpicosecond thermal spikes in the formation of tetrahedral amorphous carbon, *Phys. Rev. A* 38 (1988), 3098.

[55] J. Houška, M.M.M. Bílek, D.R. McKenzie, J. Vlček: Ab initio simulation of nitrogen evolution in quenched CN$_x$ and SiBCN amorphous materials, *Phys. Rev. B* 72 (2017) e1700270 (7 June 20170, 7 pages).

[56] J. Musil, M. Jaroš, Š. Kos: Superhard metallic coatings, *Mater. Lett.*, 247 (2019), 32–35.

[57] K.-M. Chen, D.-A. Tsai, H.C. Liao, I.-G. Chen, W.-S. Hwang, Investigation of Al-Cr Alloy targets sintered by various powder metallurgy methods and their particle generation behaviors in sputtering process, *J. Alloys Compd.* 663 (2016) 52–59.

[58] A. Zeidler, W.A. Crichton, Materials under pressure, *MRS Bull.* 42 (10) (2017) 710–713.

[59] P. Carrez, P. Cordier, Plastic deformation of materials under pressure, *MRS Bull.* 42 (2017) 714–717.

[60] P. Postorino, L. Lalavasi, Chemistry at high pressure; Tuning functional materials properties, *MRS Bull.* 42 (10) (2017) 718–723.

[61] C.-S. Yoo, New states of matter and chemistry at extreme pressures: low-Z extended solid, *MRS Bull.* 42 (10) (2017) 724–728.

[62] H. Sumiya, Novel superhard nanopolycrystalline materials synthesized by direct conversion sintering under high pressure and high temperature, *MRS Bull.* 42 (2017) 729–733.

[63] X. Hong, Effect of 3 GPa pressure treatment on the solid-state phase transformation of Cu-11.76 Al alloy during heating process, J. *Mater. Sci. Eng. B* 8 (1–2) (2018) 8–12.

[64] M. Fan, Z. Luo, Z. Fu, X. Guo, J. Tao, Vacuum hot pressing and fatigue behaviors of Ti/Al laminate composites, *Vacuum* 154 (2018) 101–109.

[65] Q. Tang, M. Zhou, L. Fan, Y. Zhang, G. Quan, B. Liu, Constitutive behavior of AZ80 M magnesium alloy compressed at elevated temperature and containing a small fraction of liquid, *Vacuum* 155 (2018) 476–489.

[66] N. Nishiyama, J. Langer, T. Sakai, Y. Kojima, A. Holzheid, N.A. Gaida, E. Kulik, H. Hirao, S.I. Kawaguchi, T. Irifune, Y. Onishi, Phase relations in silicon and germanium nitrides up to 98 GPa and 2400 °C, *J. Am. Ceram. Soc.* (2018) 1–8.

[67] M.W. Thompson, Physical mechanisms of sputtering, *Phys. Rep.* 69 (1981), 335–371

[68] F.F. Chen, *Introduction to Plasma Physics and Controlled Fusion* (2nd Edition), (Plenum, New York, 984).

[69] A.B. Migdal and V. Krainov, *Approximation Methods in Quantum Mechanics* (W.A. Benjamin, New York, 1969)

[70] J. Musil, Z. Čiperová, R. Čerstvý, S. Haviar: Flexible hard (Zr,Si) alloy films prepared by magnetron sputtering, *Thin Solid Films* 688 (2019), 137216, 7 pages.

[71] A.F. Young, J.A. Montoya, C. Sanloup, M. Lazzeri, E. Gregoryanz, S. Scandolo: Interstitial dinitrogen makes PtN$_2$ an insulating hard solid, *Phys. Rev. B: Condens. Matter. Phys.* 73 (2006), 153102.

[72] Z.T.Y. Liu, D. Gall, S.V. Khare: Electronic and bonding analysis of hardness in pyrite-type transition-metal pernitrides, *Phys. Rev. B: Condens. Matter Mater. Phys.* 90 (2014), 134102.

[73] M. Wessel, R. Dronskowski: Nature of N-N bonding within high-pressure nobel-metal pernitrides and the prediction of lanthanum pernitride, *J.A. Chem. Soc.* 132 (2010), 2421–2429.

[74] E. Gregoryanz, C. Sanloup, M. Somayazulu, J. Bardo, G. Fiquet, H.K. Mao, R.J. Hemley: Synthesis and characterization of a binary noble metal nitride, *Nat. Mater.* 3 (2004), 294.

[75] V.S. Bhadram, D.Y. Kim, T.A. Strobel: High-pressure synthesis and characterization of incompressible titanium pernitride, *Chem. Mater.* 28 (2016), 1616–1620. doi: 10.1021/asc.chemmater.6b0042.

[76] J.C. Crowhurst, A.F. Goncharov, B. Sadigh, C.L. Evans, P.G. Morrall, J.L. Fereira, A.J. Nelson: Synthesis and characterization of the nitrides of platinum and iridium, *Science* 311 (2006), 1275.

[77] O.V. Krysina, N.N. Koval, I.V. Lopatin, V.V. Shugurov, S.S. Kowalsky: Generation of low-pressure arcs synthesis of nitride coatings, *J. Phys. Conf. Ser.* 669 (2016), 012032.

[78] J. Musil et al: Coating of overstoichiometric transition metal nitrides TMNx (x>1) by magnetron sputtering, *Jpn. J. Appl. Phys.* 58 SAAD10 (2019), 01–09.

[79] J. Musil et al.: Hard TiN2 dinitride films prepared by magnetron sputtering, *J. Vac. Sci. Technol.* A 36(4) (2018), 040602-1–040602-3.

[80] P.D. Tall, S. Ndiaye, A.C. Beye, Z. Zong, W.O. Soboyejo, J.-J. Lee, A.G. Ramirez, K. Rajan: Nanoindentation of Ni-Ti thin films, *Mater. Manuf. Process.* 22 (2007), 175–179.

[81] W.F. Cui, C.J. Shao: The improved corrosion resistance and anti-wear performance of Zr – xTi alloys by thermal oxidation treatment, *Surf. Coat. Technol.* 283 (2015), 101–107.

[82] J. Musil, S. Zenkin, R. Čerstvý, S. Haviar, Z. Čiperová: (Zr,Ti,O) alloy films with enhanced hardness and resistance to cracking prepared by magnetron sputtering, *Surf. Coat. Technol.* 322 (2017), 86–91.

[83] F. Zhang, C. Li, M. Yan, J. He, Y. Yang, F. Yin: Microstructure and nanomechanical properties of co-deposited Ti-Cr films prepared by magnetron sputtering, *Surf. Coat. Technol.* 325 (2017), 636–642.

[84] T. Oellers, R. Raghavan, J. Chakraborty, C. Kirchlechner, A. Kostka, C.H. Liebscher, G. Dehm, A. Ludwig: Microstructure and mechanical properties in the thin film system Cu-Zr, *Thin Solid Films* 645 (2018), 193–202.

[85] I. Souli, V. Terziyska, J. Zehner, C. Mitterer: Microstructure and physical properties of sputter-deposited Cu-Mo thin films, *Thin Solid Films* 653 (2018), 301–308.

[86] C. Wang, T. Wang, L. Cao, G. Wang, G. Zhang: Solid solution or amorphous phase formation in Al-Mo alloyed films and their mechanical properties, *J. Alloys Compd.* 746 (2018), 77–83.

[87] J. Li, J. Wang, A. Kumar, H. Li, D. Xiong: High temperatures tribological properties of Ta-Ag films deposited at various working pressures and sputtering powers, *Surf. Coat. Technol.* 349 (2018), 186–197.

[88] B.P. Sahu, Ch. Sarangi, R. Mitra: Effect of Zr content on structure and property relations of Ni-Zr alloy thin films with mixed nanocrystalline and amorphous structure, *Thin Solid Films* 660 (2018), 31–45.

[89] M. Jaroš, J. Musil, R. Čerstvý, S. Haviar: Effect of energy on macrostress in Ti(Al,V)N films prepared by magnetron sputtering, *Vacuum* 158 (2018), 52–59.

[90] J. Musil, V. Poulek, V. Valvoda, R. Kužel, Jr., H.A. Jehn, M.E. Baumgätner: Relation of deposition conditions of Ti-N films prepared by dc magnetron sputtering to their microstructure and macrostress, *Surf. Coat. Technol.* 60 (1993), 484–488.

[91] J. Musil: *Sputtering systems with enhanced ionization for ion plating of hard wear resistant coatings, Proc. of the 1 st Meeting on the Ion Engineering Society of Japan (IESJ-92)*, Tokyo, Japan 1992, 295–304.

[92] J.A. Thornton, D.W. Hoffman: Stress-related effects in thin films. *Thin Solid Films* 171 (1989), 5–31.

[93] Y.G. Shen, Y.W. Mai: Effect of deposition conditions on internal stresses and microstructure of reactively sputtered tungsten nitride films, *Surf. Coat. Technol.* 127 (2000), 238–246.

[94] J. Soldán and J. Musil: The structure and mechanical properties of DC magnetron sputtered TiC/Cu films, *Vacuum* 81 (2006), 531–538.

2 Solid Lubricating Films
Ion Beam Assisted Deposition

JieJin, Dandan Chen, and Zhe Jiang
University of Beijing Jiaotong University, China

CONTENTS

2.1 INTRODUCTION: SOLID LUBRICATION

Solid lubrication materials can be basically defined as materials which provide lubrication under essentially dry conditions [1, 2]. Maybe you have never heard of these "Solid Lubricants" before. "Liquid Lubricants" including oils or greases are what you usually notice. Liquid lubricants are usually what comes to mind when people think of lubrication used in an industrial setting or in your daily life [3]. For instance, car lubrication, mechanical drives, medical devices, skin protection, etc. When oils and greases are used or added in certain conditions, a layer of continuous adhesion of lubricant film forms in the tribological pair surfaces, thus reducing friction and wear and bringing you a pleasant experience.

With the development of science and technology, certain operating conditions strictly require the replacement of solid lubricants [4], such as vacuum, high voltage, high temperature, cryogenic, corrosion, overload, or radiation environments [5]. The most common examples involve aerospace/aviation/ocean applications, where low temperatures, vacuum, or high voltage preclude the use of liquid lubricants which simply become too viscous to effective lubrication or may even solidify [6, 7]. The most important thing is that solid lubricants motivated the advent of the space/ocean exploration in the second half of the 21th century.

A focused review of solid lubrication with molybdenum disulfide (MoS$_2$) and tungsten disulfide (WS$_2$) will be presented in this chapter. Section 2.1 makes an introduction about solid lubricants. Section 2.2 includes the structure and synthesis of MoS$_2$ and WS$_2$. Section 2.3 shows the deposited method: ion beam associated deposition (IBAD). Section 2.4 discusses the effects of doped silver

(Ag) elements on MoS_2 and WS_2, which have improved the long life of films. Finally, Section 2.5 introduces the applications of MoS_2 and WS_2 films.

2.2 SOLID LUBRICANTS WITH MOS₂ AND WS₂

As a typical two-dimensional (2D) material with unique solid lubrication properties, graphene has been widely used in various fields and been sought by the scientific and industrial circles. So, what exactly is a two-dimensional material? Two-dimensional materials are materials in which electrons can move freely (flat motion) on only two dimensions of non-nanoscale (1–100 nm), such as graphene, boron nitride, transition metal compounds (MoS_2, WS_2, tungsten disulfide, black phosphorus, etc.). Figure 2.1 is a list of annual publications of 2D materials that show the increasing trend of studying two-dimensional transition metal dichalcogenides.

2.2.1 STRUCTURE OF SOLID LUBRICATIONS

In the long process of natural evolution and selection, nature not only created a variety of material species, but also created a variety of material structures. Many of us know that water has different functions in different forms. Liquid water is drinkable, solid water becomes snowflakes with a subtle hexagon structure, and gaseous water present in every piece of air you breathe. Structure determines the nature and function in material world. Here, we're going to show you a secret about MoS_2/WS_2.

Two-dimensional transitional metal dichalcogenides(TMDCs) have the universal chemical formula MX_2, where M is a transition metal atom (groups 4–8, and 9–10 in the periodic table) and X is a chalcogen (group 16: S, Se, Te). These 2D nanomaterials consist of a monolayer of transition metal atoms sandwiched (X-M-X) between two layers of chalcogen atoms. Figure 2.2 displays various 2D TMDs possessing various physical properties.

MX_2 has several different typical structures: 1T, 2H, and 3R, decided by the bonding within the sheets and the stacks of the sheets (Figure 2.3). The two common structural phases are characterized by trigonal prismatic (2H structure with two sheets per cell in AB stacking) or octahedral (1T) coordination of Mo atoms. 1T means each single-sheet structure can be stacked into a crystal where the next sheet is exactly above the preceding one (AA stacking), and 2H structure has two sheets per cell in AB stacking. The less common 3R structure has three sheets per cell with ABC stacking [8–10].

MoS_2 an acknowledged two-dimensional material, is worthy of great concern to us. It consists of molybdenum and sulfur atoms, and is only three atoms thick, almost the same thinness as graphene. Most importantly, MoS_2 has an energy band gap of 1.8eV, while graphene does not. The team from

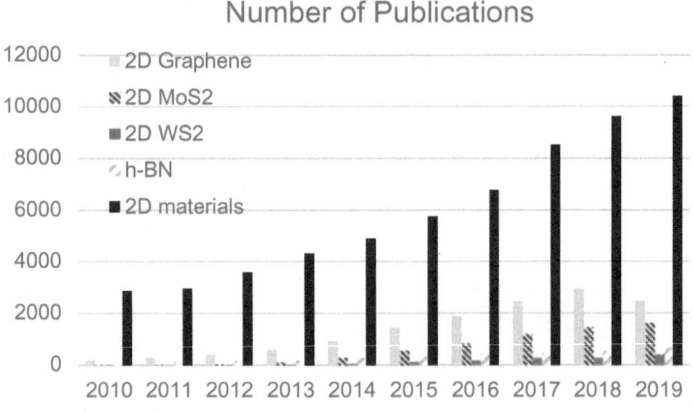

FIGURE 2.1 List of annual publications plots for 2D materials including graphene, MoS_2, WS_2, h-BN, and total 2D materials in the period of 2010–2019.

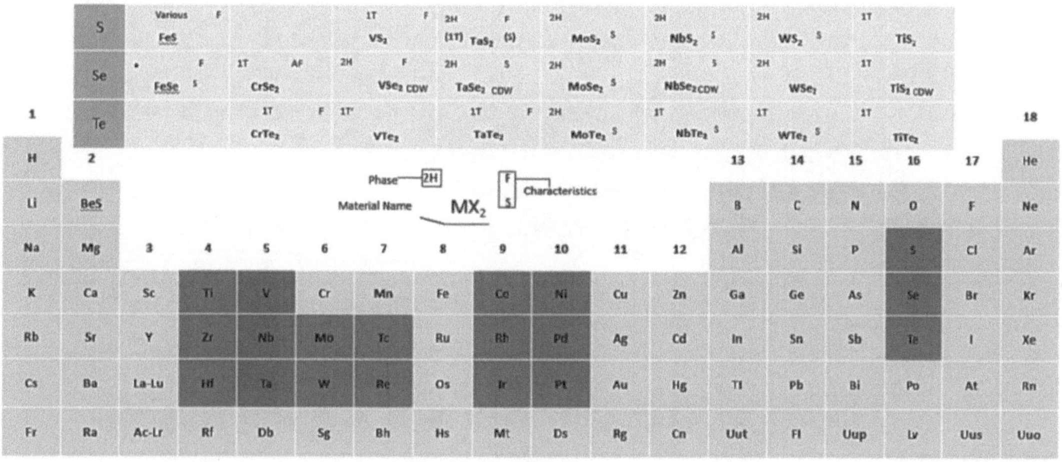

FIGURE 2.2 Table for various 2D TMDCs possessing various physical properties such as magnetism (ferromagnetic (F)/anti-ferromagnetic (AF)), superconductivity (s), and charge density wave (CDW) and crystal structures (2H, 1T, 1T').

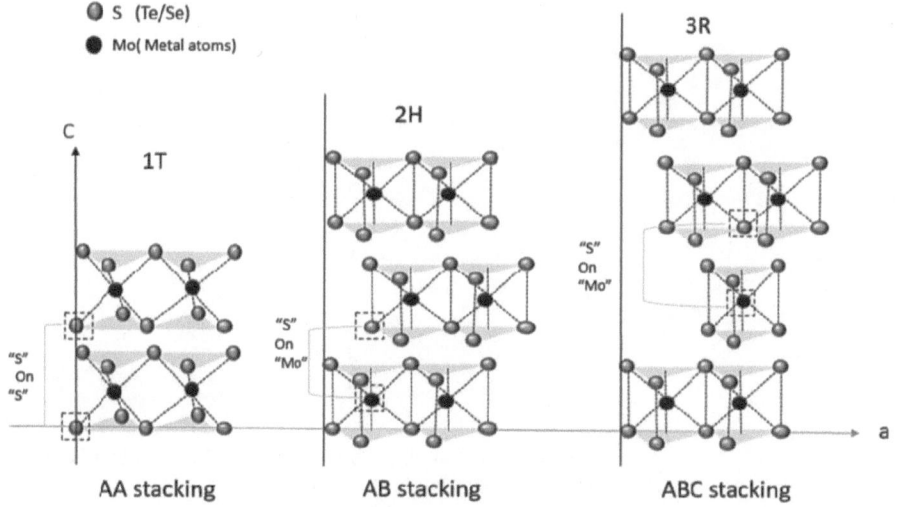

FIGURE 2.3 Structures of the polytypes of MX_2 (1T, 2H, and 3R with different stacking ways).

Lawrence Berkeley National Laboratory has accurately measured the energy band gap in MoS_2, a semiconductor two-dimensional material, and revealed a powerful tuning mechanism and a relationship between the electronic and optical properties of two-dimensional materials.

According to the description mentioned above, MoS_2 has a hexagonal layered structure with excellent solid lubricating properties. The lattice constants are: a = 0.32 nm and c = 1.23 nm. Similar to MoS_2, WS_2 has a close-packed hexagonal layered structure. The lattice constants are: a = 0.32 nm and c = 1.25 nm. Inside their unit layer, they are composed of stacks of atomically thin layers held by weak Van Der Waals Forces among them. How would such a similar structure be different in performance characteristics?

The crystal structure of MoS_2/WS_2 takes the form of a hexagonal plane of S atoms on either side of a hexagonal plane of Mo/W atoms. These triple planes stack on top of each other, with strong

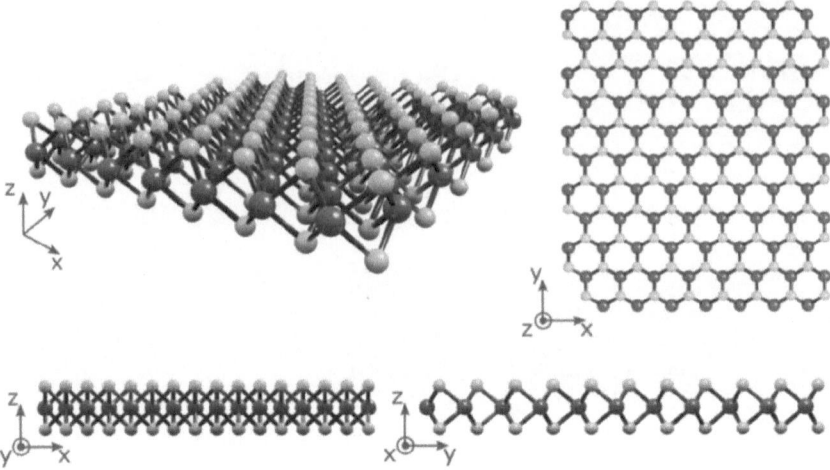

FIGURE 2.4 Crystal structure of TMDCs showing a layer of transition metal atoms (dark) sandwiched between two layers of chalcogen atoms (light).

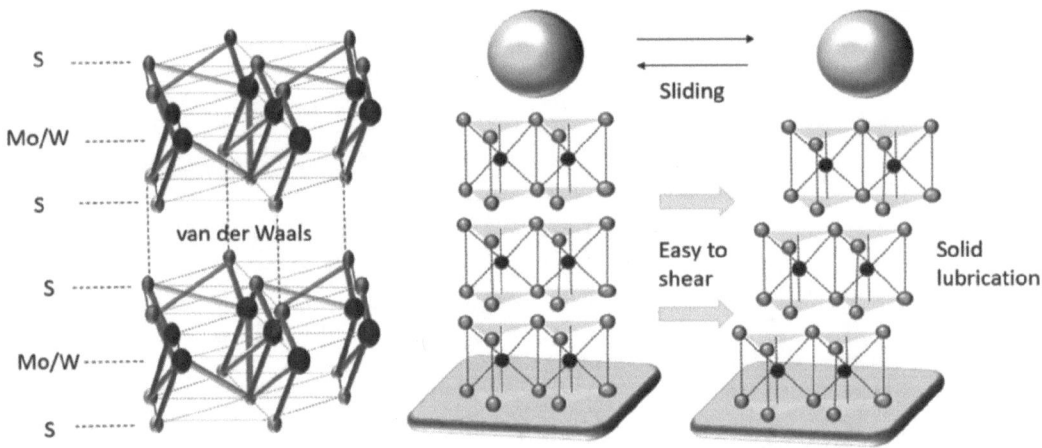

FIGURE 2.5 Schematic illustration of MoS_2/WS_2 lamellar structure (left). Schematic illustration of solid lubrication by lamellar compounds such as MoS_2/WS_2 (right) [79].

covalent bonds between the Mo/W and S atoms, but weak van der Waals fore holding layers together (S-S). This makes them separated mechanically to form 2D sheets of MoS_2/WS_2. The lubricant effect is based on this "Sandwiches" lamellar crystalline structure as shown in Figures 2.4 and 2.5, where the Sulfur lamellae are bonded only by weak interaction (van der Waals), which eases the shear [11].

In the process of sliding, Figure 2.5 shows that the crystalline layers of MoS_2/WS_2 will easily slide and relative motion direction is parallel to each other, which brings the lubricating effect [12, 13]. Undeniably speaking, the strong ionic bond between S and Mo/W provides the lamellae good resistance to asperity penetration [14]. Because of its special structure, the MoS_2/WS_2 nanocrystals can be used as nanobearings in the tribological contact, which greatly reduces the friction coefficient of the system [15, 16]. If the deposited films with WS_2 or MoS_2 could reduce friction, it will solve friction problems in a wide range of applications, such as gear, bearings, and shafts.

2.2.2 PROPERTIES OF MoS$_2$, WS$_2$, AND Ag

MoS$_2$ is a naturally forming layered transition metal dichalcogenide, that may be mined. It is a silvery-black crystal, which looks and feels similar to graphite crystal. Because its advantages over graphite, such as good semiconductor characteristics, small size, ultrathinness, and softness, it is particularly suitable for use in transistors, flexible electronics, LEDs, lasers, and solar cells. It is often used as a lubricant because the weak interlayer interactions makes it easy to slide against one another. It is also used as a substitute for graphite in high-vacuum applications, but its maximum operating temperature is really lower than that of graphite. Besides, the MoS$_2$ films have been successfully applied to aerospace devices since the 1950s by NASA, because of their excellent physical and chemical properties. MoS$_2$ possesses excellent thermal stability coupled with substantial lubricating properties that enables it to withstand high temperature and pressure; it keeps its lubricating property till 525°C though it begins to get oxidized gradually at about 400°C in air, and the oxidation speed sharply increases to form MoO$_3$ with a large friction coefficient after 540°C. Also, MoS$_2$ still possesses good lubricating property at a low temperature of −60°C. While the thermal stability of oil is poor, their applicable temperature range is far lower than that of MoS$_2$. The corrosion resistance of MoS$_2$ is very good, and normal acids except aqua regia, hydrogen nitrate, boiling hydrochloric acid, and concentrated sulfuric acid, do not have effect on it. However, the friction coefficient rises with the increase of humidity in air. The current solution is to improve deliquescence by doping other elements.

WS$_2$ in bulk has been used for many years as a dry lubricant due to it exhibiting properties similar to MoS$_2$, such as low coefficient of friction (COF), high chemical stability, thermal stability, and high extreme pressure resistance. WS$_2$ possesses better anti-oxidation performance and is adaptable to a wider range of temperature (begins to decompose at 450°C in air, and completely decomposes at 650°C; begins to decompose at 1000°C in vacuum, and completely decomposes at 2000°C). Meanwhile, the oxidation product WO$_3$ has a lower friction coefficient than MoO$_3$. WS$_2$ has an extremely low COF of 0.03 – lower than that of Teflon, graphite, and MoS$_2$. The film is remarkably durable compared to many other lubricant materials and can withstand tremendously high loads over 300,000 psi. It has unsurpassed performance properties for lubricity, non-stickiness, low drag, wear life, and load rating [17]. Meanwhile, WS$_2$ is considered to be a kind of valuable application material because of its crystal structure similar to MoS$_2$ and its better performance. Due to their direct band gap, they have great advantages over graphene for several applications, including solid lubrication, optical sensors, field-effect transistors, energy storages, gas sensors, photonic devices, and biosensors.

With the comprehensive properties of the MoS$_2$ and WS$_2$ described so far, the widely studied materials have been proved to be a well-endowed system for investigating the effect of surface treatments among the TMDCs owing to their stability in vacuum/air. The S atoms on the surface provide excellent potential for modification by way of doping. The most widely used doped element was Ag.

Silver, chemical element, is a white, soft, and lustrous metal valued for its comparative scarcity, malleability, ductility, and resistance to atmospheric oxidation. Being a soft metal, Ag is added as doped element to improve the tribological properties of self-lubricant films.

Beijing Jiao Tong University (BJTU) has spent a lot of time to study the doped Ag in MoS$_2$/WS$_2$ films all the time. Doped MoS$_2$-Ag and WS$_2$-Ag films have been shown to exhibit better tribological properties than pure MoS$_2$ and WS$_2$ films, as well as superior resistance to oxidation, which improves performance and storability in air. Basic functions of MoS$_2$/WS$_2$/Ag are shown in Table 2.1.

The evolution of nanostructured coatings in tribology is summarized in Figure 2.6. The researches began with basic and available materials and gradually became more complex (with different preparation methods), moving on to gradient structure, multicomponent coatings, and eventually to nanostructured smart surfaces. A good overview has been given by Donnet and Erdemir [4].

The impressive tribological properties of lamellar solids MoS$_2$/WS$_2$ under vacuum and in a wide range of temperature make it a natural choice as a solid lubricant for applications in space [18].

TABLE 2.1

Basic parameters of MoS$_2$/WS$_2$/Ag

Performance	MoS$_2$	WS$_2$	Ag
Relative atomic mass	160.08	248.02	107.87
Crystal type	Hexagonal lattice	Hexagonal lattice	Face-centered cubic (fcc)
Density/(g.cm^{-3})	4.5–4.8	7.4–7.5	10.49
Melting point/°C	1185	1250–1260	961.78
Friction coefficient	0.03–0.09	0.03–0.05	0.4–0.5
oxidation states	+4, +6 (normal)	+4, +6 (normal)	+1, +2, +3
Thermal stability	−0°C to 982°C	−188°C to 1316°C (in vacuum)	/
Chemical stability	Stable	Stable	Active

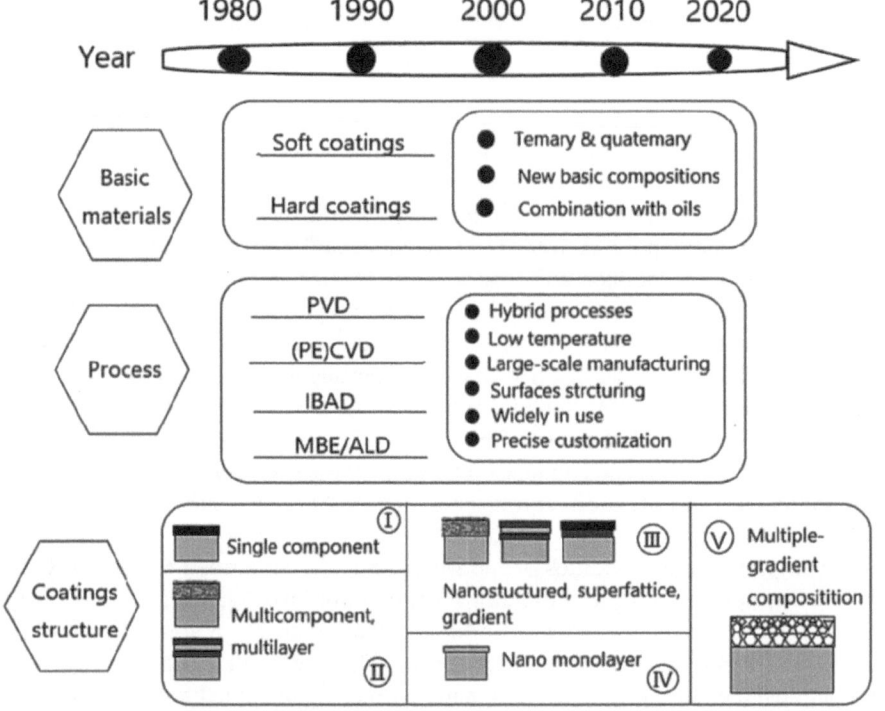

FIGURE 2.6 History trends in the development of tribological coatings [80].

However, the limitations of these conventional solid lubricants (including soft metals, such as lead, gold, and silver) can be removed by using nanoscale films. With the development of aerospace exploration, further research of MoS$_2$/WS$_2$ will include the use of atomic layer deposition (ALD) to coat particles [18] and self-lubricating films deposition [19, 20]. For example, WS$_2$ coatings can be sputtered with Ag or Ta to improve its stability or wear resistance. In any case, films containing various materials can be synthesized by the same or different preparation methods. They exist in different forms described as nanocomposite, gradient structure, and superlattice multilayer or monolayer.

2.2.3 Synthesis Methods of MoS₂ and WS₂ Films

At present, the preparation methods of MoS_2/WS_2 film include two main categories: the top-down approach and the bottom-up approach [21]. The former is normally called as physical preparation method, it is a strategy in which various forces break the weak interplanar Van Der Waals interactions in a layered material (MoS_2/WS_2 bulk) and no chemical reactions occur during this process. Whereas in the latter, the material is synthesized directly from various precursors through chemical reactions at certain experimental conditions [22]. Two commonly used techniques are physical vapor deposition (PVD) and chemical vapor deposition (CVD). The methods are given in Table 2.2.

1) Mechanical Exfoliation Method

The most basic and simplest method is mechanical exfoliation, through which Geim and Novoselov [23] exfoliated monolayer graphene. In this process, the adhesive (scotch) tape separates the bulk 3D stacked forms of 2D materials into atomically thin sheets. As no chemical reactions occurring during this process, the 2D materials present outstanding crystal quality with rarely defects. This method is suitable for research work and trials, but unsuitable for large-scale production in industrial application.

2) Liquid Exfoliation

Mechanical Exfoliation Method seems very easy in producing a nanosheet and can help us to analyze the structure of layer materials. However, it is difficult to control the position of flakes. There is a better way: Liquid exfoliation strategy produces dispersions of the nanoflakes in the solvent (Figure 2.7). Lee et al. used N-vinyl-pyrrolidone solvent to exfoliate MoS_2 to obtain MoS_2 thin films [24]. The TMD nanosheets can be obtained by dispersing and exfoliating sonication of bulk powders in organic solvents or aqueous solutions of surfactants, such as sodium dodecyl sulfate [24]. The liquid exfoliation method produces large volume of material and is scalable.

TABLE 2.2
Synthesis methods of MoS₂ and WS₂ films

Items	Methods	Advantages	Disadvantages
Top-Down Approaches	Mechanical exfoliation method	Outstanding crystal quality with few defects; High-quality flakes is suitable for research and lab uses.	Only layered materials can be stripped; Unsuitable for large-scale production for industrial uses.
	Liquid phase ultrasonic	Large volumes of material and scalable.	Very small sized flakes; Easy polluted by solutions
	Ion-intercalation	A very high yield of monolayers or single-layer nanosheet.	Very small sized flakes; Easy polluted by solutions
Bottom-Up Approaches	Chemical vapor deposition	Various crystalline and amorphous metals with high purity and required properties; More scalable; Cheap; Widely used.	Difficult to control the parameters of deposition; The deposition rate is low; Easily polluted by precursor gases.
	Physical vapor deposition	Controllable types of products; Low requirements for substrate; Outstanding crystal quality	Expensive preparation equipment.
	Molecular beam epitaxy	Large area growth of high-quality crystal films and heterostructures.	More suitable for research and lab uses.
	Atomic layer deposition	Produces a larger scale of several inorganic films.	Requires ultrahigh vacuum conditions; Not widely used.

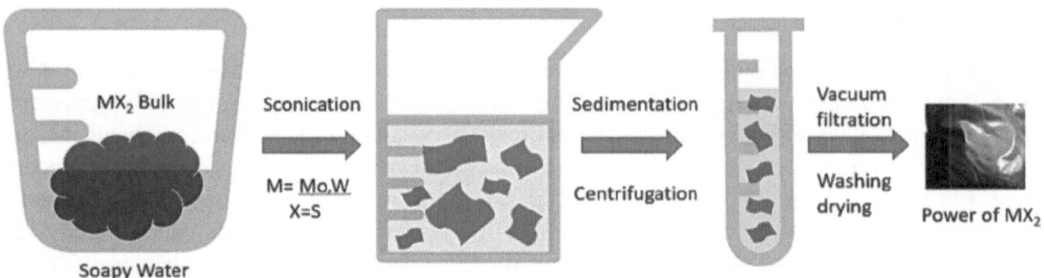

FIGURE 2.7 Schematic illustration of liquid exfoliation to produce dispersions.

3) Ion-Intercalation

An additional interesting process of exfoliating TMDCs is via alkali metal intercalation. The intercalated material is termed Lix MoS₂ and cleaned thoroughly with hexane to remove any organic residues. Johnson etal had exfoliated MoS₂ by performing lithium intercalation through n-butyl lithium dissolved in hexanes [25]. MoS₂ is promptly exfoliating when it is sonicated in water as the intercalated lithium reacts with water to generate hydrogen gas [26]. The resulting gas separates the flakes easily because weak attractive forces attach the sheets [27]. This separation is followed by centrifugation to further isolate the unexfoliated material that is forced down to the base of the vessel.

4) Chemical Vapor Deposition

CVD is a technique by using reactions between gas phases or gas and solid phases to generate solid sediment on the substrate surface. The chemical reactions of CVD include mainly two types: One is the sediment produced through the reactions between two gases or among several gases. Another is the reaction between a gas and solid substrate surface, depositing a thin film on the substrate. A typical CVD method for growing MoS₂ uses MoO₃ and S powders in an Ar environment, which can produce ultrathin samples of both single crystals up to hundreds of micrometers in size as well as large area films. Liu et al. described the CVD growth of large area MoS₂ with a few atomic layers based on the sulfurization of Mo metal films [28].

5) Physical Vapor Deposition

PVD covers a broad family of films processes, which the employed material is physically removed from a source or "target" by evaporation or sputtering. It is transported by the energy of the vapor particles and is condensed into a film on the surfaces of appropriately placed parts in vacuum. Chemical compounds are deposited either by using a similar source material or by introducing reactive gases containing the desired reactants, reacting with the metal vapor produced during the physical process. The major categories are evaporation, sputtering, and ion plating [29].

6) Molecular Beam Epitaxy

Molecular beam epitaxy (MBE) is a widely used commercial technique for the fabrication of specialized semiconductor devices, particularly those based on III-V materials and are used in optoelectronics and high-frequency applications. However, conceptually, it is a very simple method of crystal growth. Crystal growth by MBE requires ultrahigh vacuum (with pressure typically below 10^{-10} mbar) conditions and therefore, MBE is naturally compatible with many surface modified techniques. The group of Atsushi Koma published numerous reports on the MBE growth of TMDCs in the 1980s and 1990s [30–33].

7) Atomic Layer Deposition

ALD is a gas phase chemical process to deposit atomically thin films of various materials layer by layer, by using precursors to react them with substrate. The ALD technique is not widely used compared with other methods. However, it has the benefit of producing a larger scale of several inorganic films. The preparation of MoS$_2$ thin films by ALD is a relatively new technique by Tan et al. They used H$_2$S and MoCl$_5$ on a sapphire wafer at 300°C [34].

A simple and reliable fabrication method for the preparation of MoS$_2$/WS$_2$ film is very important for both research and application purposes. The most widely used methods include exfoliation, liquid phase ultrasonic and ion-intercalation. Two commonly used techniques are CVD and PVD. However, these methods more or less make the prepared film have different aspects of defects. Here, this paper mainly introduces a new device – ion beam associated deposition (IBAD) device, which was designed by the working group at BJTU. The design principle is based on PVD.

2.3 ION BEAM ASSOCIATED DEPOSITION

IBAD is one of the PVD methods, a surface-modified technique combines Ion Implantation with film deposition, which refers to the technique that a low-energy (several electron volts to thousands of electron volts) ion beam bombards the growing film during the deposition of the film, in order to improve properties or microstructures. The additional energy introduced into thin films by ion bombardment induces atomic displacements, surface migration, or transformation of growth mode. Ion irradiation during evaporation has been attracted many attentions in order to form nanostructures [35]. Also, low-energy ion beam irradiation during deposition can be used for biaxially aligned films [36].

IBAD technology can continuously grow film layer of any thickness under more stringent control conditions, significantly improve the crystallization and orientation of the film, increase the adhesion strength of the film, improve the denseness of the film, and synthesize the compound film with an ideal stoichiometric ratio at room temperature or near room temperature.

Within in a set of IBAD (Figure 2.8 shows the appearance of the IBAD devices), an ion beam from an assisted ion source and an ion source are directed simultaneously to a substrate, as shown in Figure 2.9. The steps of films deposited are as follows: first, low-energy ions bombard the substrate using Kaufman ion source for cleaning; second, magnetron sputtering is combined with high-energy ions produced by Kaufman ion source assisted deposition, which can gain solid lubrication films with the different thicknesses or graded structures. The advantages and disadvantages of magnetron sputtering and ion beam assisted deposition techniques are compared in Table 2.3. The typical energy of the assisted ions produced by these ion sources are several hundred eV. The metal or compound targets are placed on the wall of the vacuum chamber and used for thin film deposition. The ion beam from a magnetron sputtering source setup is generated by noble gases (e.g., argon – with physical ion influence by ion bombardment) or from gases like nitrogen or oxygen (with additional

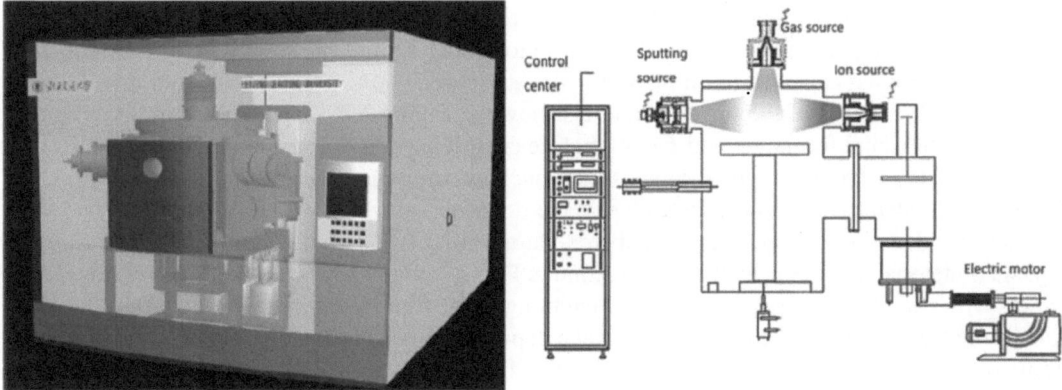

FIGURE 2.8 Ion beam associated deposition device.

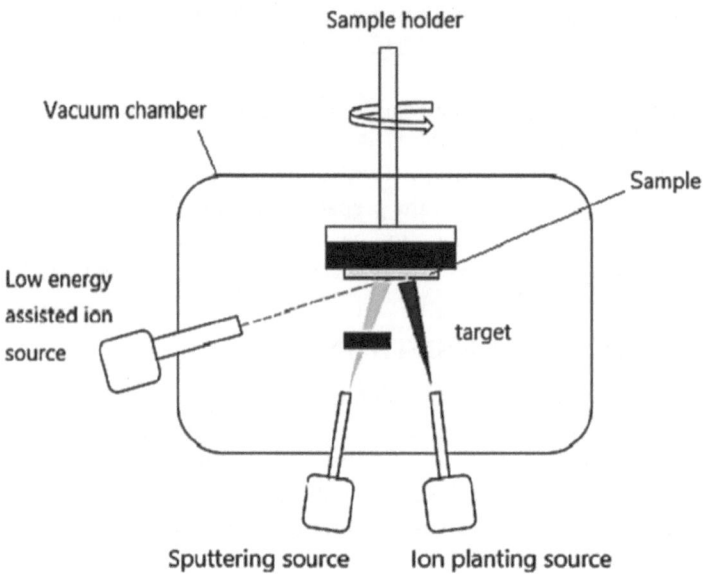

FIGURE 2.9 A diagram of an IBAD device equipped with a sputtering source, an ion source, and a low-energy ion beam assisted source.

TABLE 2.3
Method for magnetron sputtering and ion beam assisted deposition

Items	Magnetron Sputtering	Ion Beam Assisted Deposition (IBAD)
Binding force	Good	Better
Film density	Good	Better
Equipment costs	Expensive	More expensive
Growth process	Island to island	Layer by layer
Typical growth rates	10–300 nm/min	10–100 nm/min
Typical atom/ion ratio	10–1000	100–10^4
Typical applications	Metal, semiconductor, or isolating layers with defined compact structure and reduced inner stress	Metal, semiconductor, or isolating layers with defined compact structure and reduced inner stress

chemical influence on the layer deposition leading to changed stoichiometry of nitrides or oxides) [37, 38], which bombarded the surface of the target to sputter atoms. The source for ion implantation can introduce doping elements to improve the film structure.

There are three types of film growth, models shown in Figure 2.10 (left), known as: (1) Volmer–Weber growth (island formation and growth before coalescing). At the beginning of the film formation, the isolated islands in a three-dimensional nucleus form and then merge into a thin film; for this growth model, the atoms or molecules of the deposited material are more likely to bond with each other and avoid bonding with the substrate atoms, that is, the immersion is poor between the deposited material and the substrate. (2) Second is Frank–van der Merwe type (layer-by-layer film growth with complete coverage). From the beginning of the film formation, the film grows in a two-dimensional layer; when the immersion between deposited substance and the substrate is very good, atoms of the deposited substance are more inclined to bond with the substrate atom. (3) Third is Stranski–Krastanov growth (monolayer formation followed by island growth). In the early stage of film formation, the film grows in the two-dimensional layer, but after the formation of several layers,

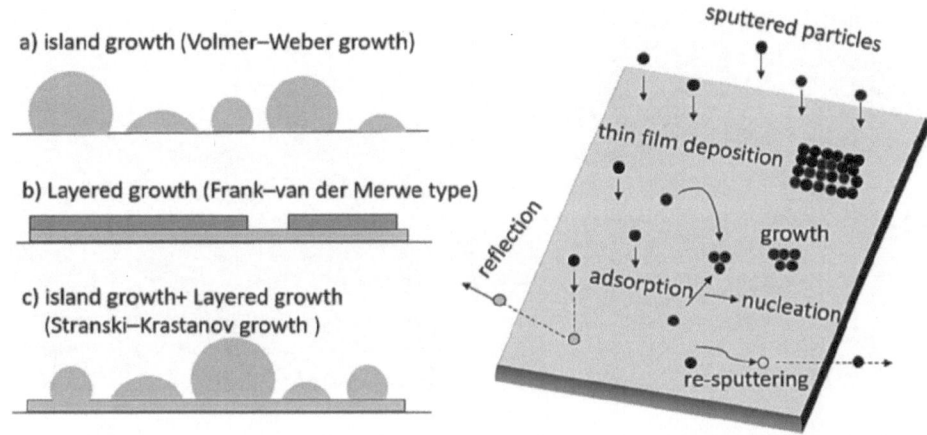

a) island growth (Volmer–Weber growth)

b) Layered growth (Frank–van der Merwe type)

c) island growth+ Layered growth
(Stranski–Krastanov growth)

sputtered particles

thin film deposition

reflection

adsorption

growth

nucleation

re-sputtering

FIGURE 2.10 Process of film growth.

the growth mode changes to an island pattern. When there are ion bombardments during thin film deposition, the growth mode changes dramatically. The hyperthermal ion bombardment increases the detachment of atoms from the tops of the growing islands. Therefore, the growth mode changes from 3D growth (island) to 2D (layer) growth with ion irradiation.

The physical process of nuclear formation and growth is described in Figure 2.10 (right). 1) The gas phase atoms from the evaporation source inject on the surface of the substrate. Part of the atoms possess high energy and are reflected back elastically, while the other part is adsorbed on the surface of the substrate. A small fraction of the adsorption gas phase atoms evaporates when the energy is slightly larger. 2) The absorbed gas phase atoms diffuse and migrate to the surface of the substrate, collide with each other into atomic pairs or small atom clusters, and condense on the surface of the substrate. 3) The atom clusters collide with other adsorption atoms, or release a single atom. Once the number of atoms in the cluster exceeds a certain threshold, the cluster collides further with other adsorption atoms, only to grow up to form a stable cluster. Atomic clusters containing critical numbers of atoms are called critical nucleus, and stable clusters are called stable nucleus.4) The stable nucleus can capture other adsorption atoms, or combine them with the incoming gas phase atoms to grow into an island.

At present, MoS_2 as a solid lubrication film has widely been used in the industry. There is a large amount of research published about pure MoS_2 films [38–46] or the films containing MoS_2 [47–55], produced by a variety of methods, such as PVD [39, 40, 42–44, 47, 51, 52, 54], electrodeposition [48, 49, 53], plating [38, 45], CVD + thermal diffusion [41], and others [46, 50, 55]. While there are still less researches and applications on WS_2 being reported, compared to MoS_2, WS_2 possesses better tribological properties; its oxidation product WO_3 has a lower friction coefficient than MoO_3. Therefore, it is worthy to pay more attention on WS_2 solid lubrication film. Now, there are many methods to prepare WS_2 film or coatings [56–61], but their microstructure and tribological properties are quite different. Meanwhile, to compare the properties of doped MoS_2-Ag with WS_2-Ag film, "pure" Ag films were also prepared. BJTU Working Group has been committed to the preparation of films deposited by IBAD technology like "pure" Ag film, MoS_2-Ag film, and WS_2-Ag film for a long time. Figures 2.11, 2.12, and 2.13 describe the above-mentioned analysis of the film surface morphologies, and elemental distribution maps can be observed. Figure 2.11 reveals the Scanning Electron Microscope (SEM) morphologies and Energy Dispersive Spectrometer (EDS) analysis of the two "pure" Ag films of different thickness deposited by IBAD on steel substrate. a), b), c), and f) shows a dense columnar microstructure with 500–600 nm thickness, d) and e) are layered films with 3.49 μm. Due to the different thickness of films deposited by IBAD (figure 2.9), it can be inferred that the growth of films begins with the columnar microstructure, followed by layer-by-layer growth

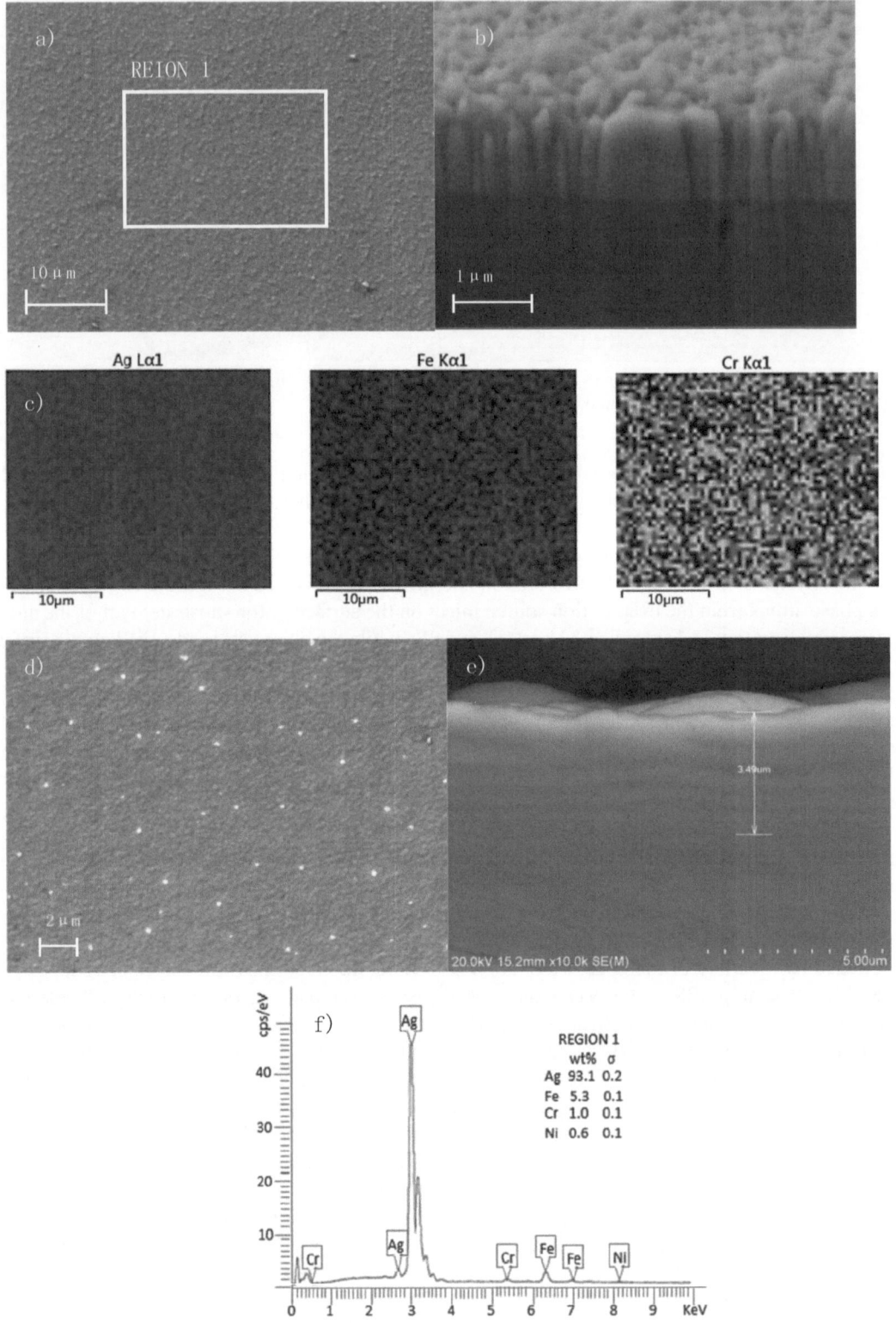

FIGURE 2.11 SEM morphologies and EDS analysis of the "pure" Ag films deposited by IBAD on steel substrate.

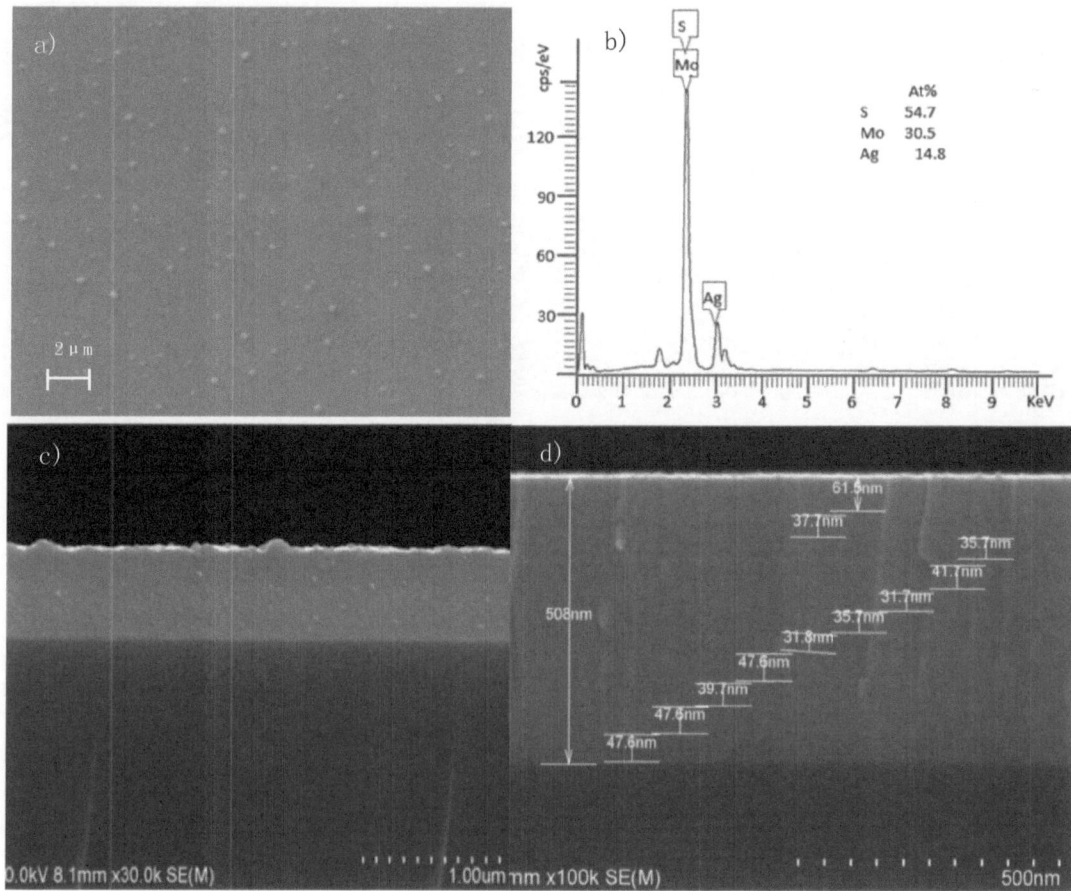

FIGURE 2.12 SEM surface, cross-section morphologies, and EDS analysis of MoS$_2$-Ag deposited by IBAD.

and then produces a well-defined multilayer Ag film. These results indicate that the Ag films fabricated by IBAD had a much denser microstructure than normal Ag films. Figure 2.12 presents the surface, cross-section SEM morphologies, and EDS analysis of layered MoS$_2$-Ag deposited by IBAD. The tiny Ag particles are distributed on the surface of the film and the thickness is 508 nm. The per layer film has a growth thickness of approximately 31–48nm. Figure 2.13 reveals the SEM surface, cross-section morphologies, and EDS analysis of WS$_2$-Ag deposited by IBAD; the WS$_2$-Ag films showed a dense columnar microstructure with a worm-like surface.

Figure 2.14 shows the tribology analysis results of Ag film. Compared with the friction coefficient curve of steel (0.8–0.9), the sputtered Ag films exhibited COF of 0.45–0.55, and the reports usually test the value of 0.4–0.5. From the line scanning analysis results and the film morphology photographed with a 3D white light interferometer, we can infer that maybe the deposited substance and the substrate between the immersion is not very good because of the delamination of Ag film.

Figure 2.15 displays the friction coefficient curve of the MoS$_2$-Ag film deposited on steel substrate at a load of 10g and 5g which indicates the heavier load is not appropriate for the MoS$_2$-Ag film with a thickness of 508 nm. Figures 2.15 (c) and (d) show the cross-sectional SEM microstructure of MoS$_2$-Ag film before the sliding trio-test; layer-by-layer film growth was observed (thickness: 508nm). The BJTU group designed special deposition processes of films to produce dense film microstructures. The Ag$^+$ ions were implanted on the steel substrate before deposited MoS$_2$ film to improve the surface binding energy of substrate and to form a gradient nanostructure. Due to the layer-to-layer deposited process, we can see the layer-by-layer deposition on the substrate.

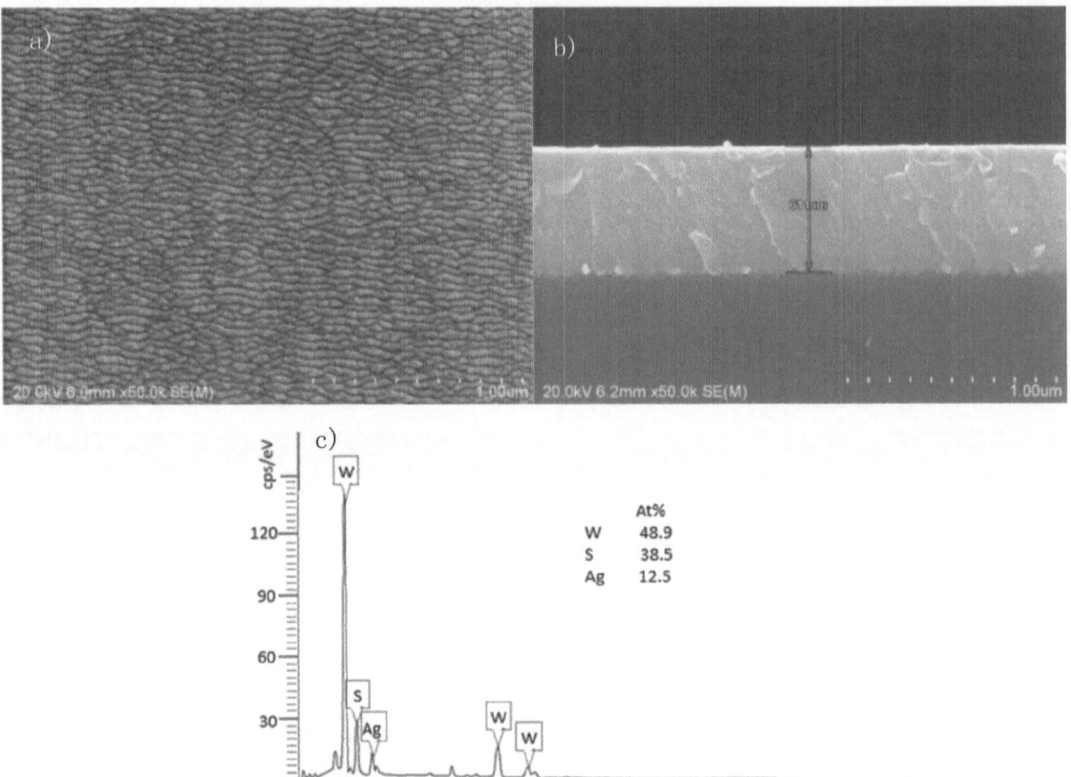

FIGURE 2.13 SEM surface, cross-section morphologies, and EDS analysis of WS$_2$-Ag deposited by IBAD.

Previous studies [62] have reported that this superior dense microstructure was expected to acquire better wear resistance of the TMD films. The value of friction coefficient reached an average 0.1 under 5g load.

Sputtered WS$_2$ films normally exhibited a porous columnar [63]. While the WS$_2$-Ag films deposited by the BJTU group possess a dense columnar microstructure and certain "worm" structure, and the right one showed a thickness of 611nm (Figure 2.13). The process carefully controlled the deposition processes of WS$_2$-Ag films to produce dense film microstructures. The tribological properties of the WS$_2$-Ag film are shown in Figure 2.16. Compared with a load of 10 g, the WS$_2$ film deposited on steel under load 5 g showed a stable, smooth line. Sputtered WS$_2$-Ag films exhibited well lubricant characteristics, a COF of 0.12–0.15. The morphology of the worn scar showed that there are clear lines of the plough grooves on the surface, and there are few debris distributed. It means good wear resistance.

2.4 EFFECTS OF DOPED ELEMENTS ON MOS$_2$ AND WS$_2$

2.4.1 STRUCTURE OF DOPED MX$_2$

Mohammad R. Vazirisereshk [18] provides a good idea on exploring the influence of doping elements on MX$_2$ structure. They focused on the structures of the doped MoS$_2$ monolayers from an atomistic characterization perspective, and DFT calculations played a key role in elucidating structure.

Dopants in MX$_2$ can be incorporated into several possible locations. a) "M" (Mo, W, Ta, etc.) or "X" (S, Te, Se) substitution, it can occur at either the "M" or "X" site. Atoms can be added (intercalated) between the MoS$_2$ layers or be adsorbed by the surface, as ad-atoms on an MX$_2$ sheet (basal

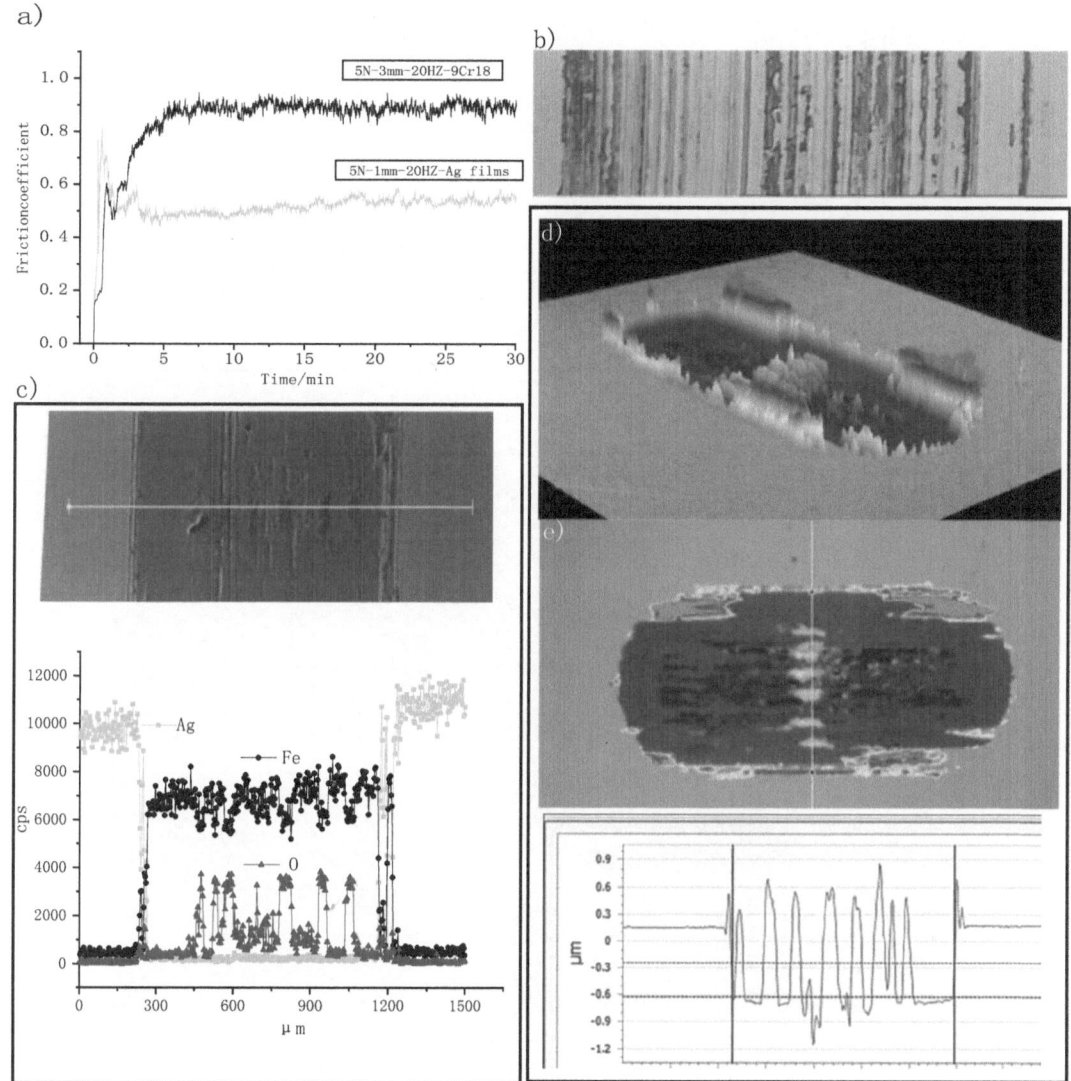

FIGURE 2.14 a) COF curve of "pure" Ag films. b) Worn morphology of steel observed using an optical microscope. c) Line scan analysis of "pure" Ag films. d) and e) "pure" Ag film morphology with a 3D white light interferometer.

plane) or on an edge, which is particularly likely for few-layer MX_2 or nanosheets; b) octahedral or tetrahedral intercalation, there could be interstitials within a MX_2 layer, in the sites within a "M" plane, in principle. While, there is very limited space and such a structure does not seem to be observed. Ad-atoms can occur in some high-symmetry sites such as atop "M", atop "X", bridging an "M-X" bond, or in the center of a hexagonal hollow [64]. If an atom intercalates between two layers in the 2H poly-type, two layers sites become just high-symmetry positions: tetrahedrally or octahedrally coordinated [65]. These possible dopant sites in 2H MX_2 are depicted in Figure 2.17.

2.4.2 Effect of Doped Elements on Properties of MoS_2 and WS_2

In order to improve the mechanical and tribological performance of MX_2 (M=Mo, W, X=S) films, most doping studies [18] found that dopants substituting for M the atom's ionic radius and its

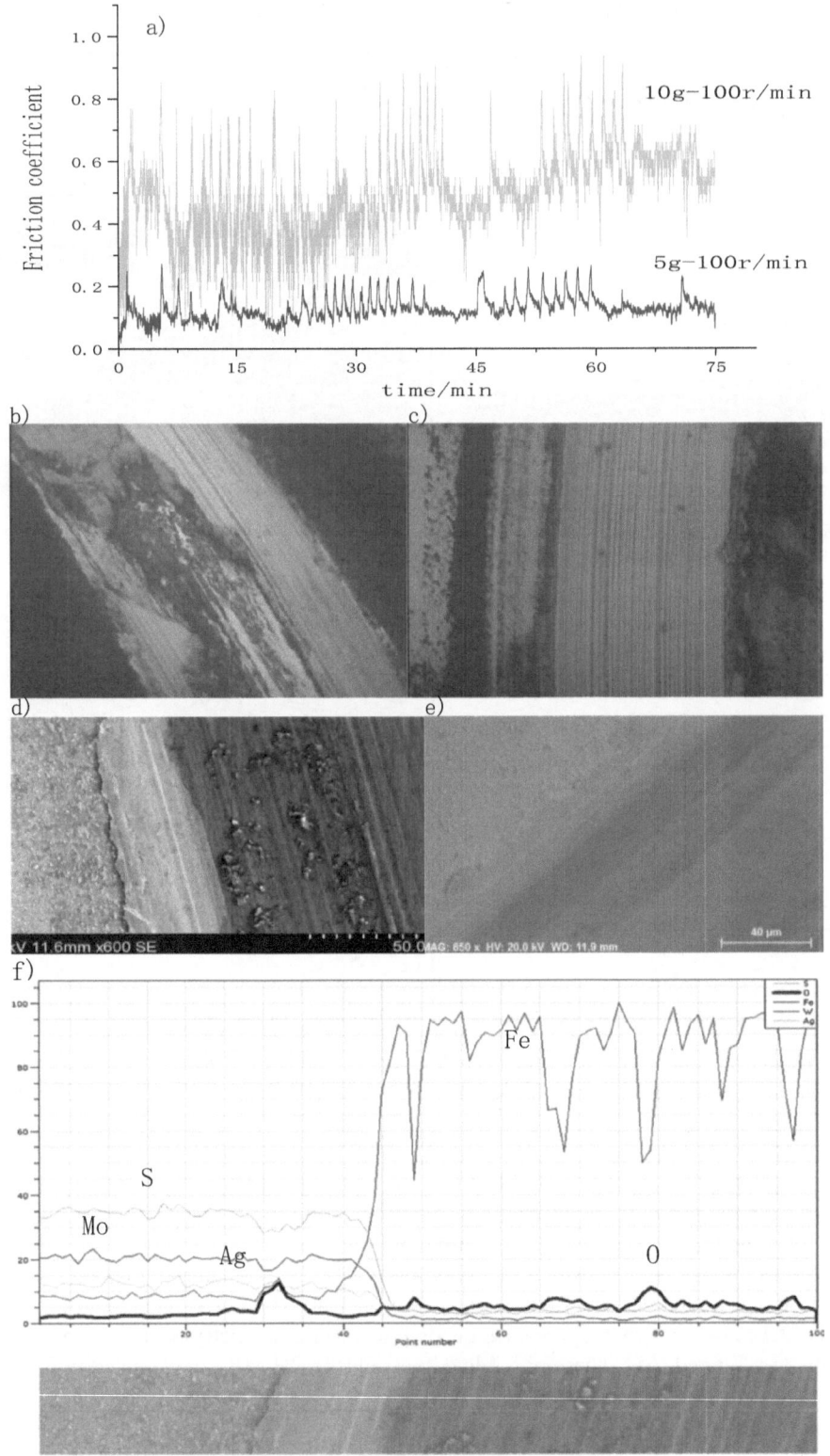

FIGURE 2.15 a) COF curve of MoS$_2$-Ag films deposited on steel substrate with loads 10g and 5g. b) and c) Worn morphology of Load-10g observed using an optical microscope. d) and e) Worn morphology of Load-5g observed by SEM. f) Line scan analysis of films.

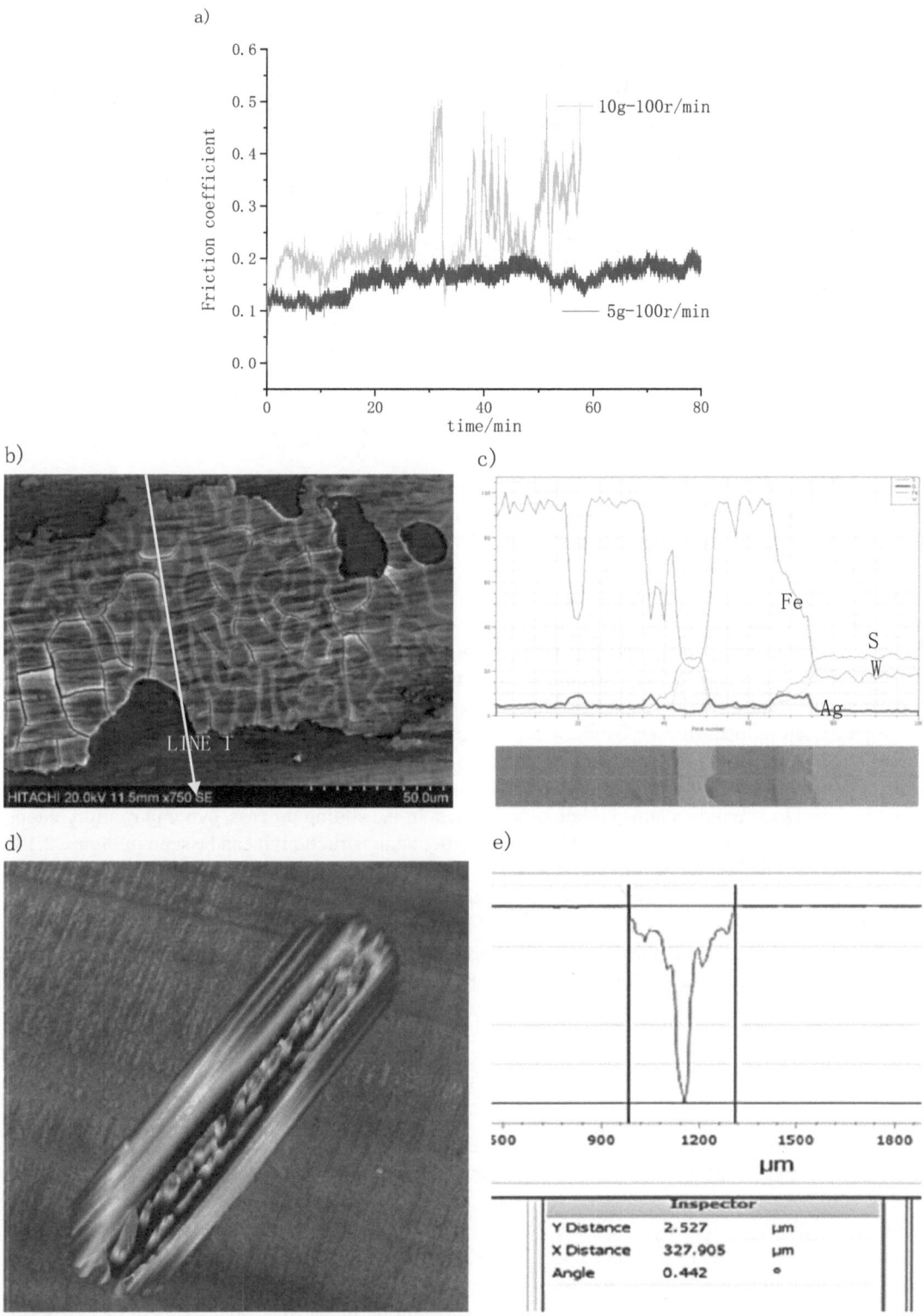

FIGURE 2.16 a) COF curve of WS$_2$-Ag films deposited on steel substrate with loads 10g and 5g. b) and c) Worn morphology and line scan analysis of WS$_2$-Ag deposited on steel. d) and e) WS$_2$-Ag film morphology with a 3D white light interferometer.

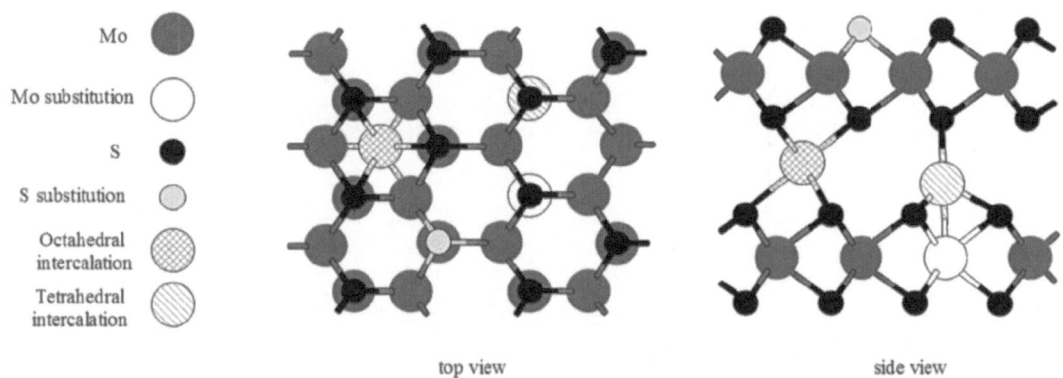

FIGURE 2.17 Possible doping sites in 2H MX_2: a) "M" (Mo, W, Ta, etc.) or "X" (S, Te, Se) substitution; b) octahedral or tetrahedral intercalation. Unsubstituted "M" and "X" are gray and black, respectively.

compatibility with "M" or "X" radii. Compared with those in MX_2, whether the atom has similar bond length with "M" or "X", and whether the atom forms a crystal with M or X of similar structure to MX_2 are key points. Dopant atoms can substitute in-situ atoms in layered structures through the model discussed in Section 2.4.1, and appear at interstitial sites between atoms in the lattice, intercalate between layers, or as a separate phase disperse in the material. Many doped MX_2 films are obtained by magnetron sputtering, IBAD, or ion implantation. The doped MoS_2/WS_2 films are shown to exhibit better tribological properties and superior resistance to oxidation than undoped MoS_2/WS_2 in vacuum or air. Understanding the mechanisms by which dopants improve the tribological and oxidation characteristics of MoS_2/WS_2 is still a hot topic of research.

Dopants can reduce MoS_2/WS_2 phase size in crystallite growth, and in turn increase density to improve oxidation resistance, as well as decrease the number of edge sites available to oxidize. More important is the structural design of the whole film. Multiple and gradient structures have been designed to strengthen anti-oxidation properties. In the sliding process, synergy reinforcement mechanism occurred between layered films and gradient nanostructure; it can be seen in Figure 2.18. The essence of gradient nanostructure is that the density of the crystal boundary (or other interface) changes spatially, which corresponds to many spatial gradient changes in physical and chemical properties. Multivariate means doped elements with high energy are deposited into the film to reach certain sites and strengthen the structure. Layered film lubricates and reduces abrasion, and the gradient structure can transfer the stress well to the substrate and enhance the life of the film. Here, the publications on self-lubricating composites with different doping elements in MoS_2 / WS_2 are summarized in Table 2.4.

2.4.3 MoS_2-Ag Films

In order to improve the tribological performance of MoS_2/WS_2 films, Ag, being a soft metal with good fluidity under 300–500°C, was added as the doped element. A perfect synergy effect is exhibited in MoS_2-Ag films, which showed lower wear rates and longer wear durability with low COF compared with pure MoS_2 films. Figure 2.19 presents the MoS_2-Ag films deposited by IBAD and ion implantation on steel substrate, which shows that globular Ag particles are distributed in the MoS_2 film, and the particles were uniformly covered with small white and bright particles. The particle size was less than 10 nm, and some white and bright particles were up to 200 nm in size. EDS surface scanning results showed that the particles were Ag as shown in Figure 2.19 (c), indicating that the MoS_2 film was doped with nanoscale Ag particles. Figure 2.19 (d) is the friction and wear curves of MoS_2-Ag solid lubricating film at room T = 25–28°C, load 10N, with CERT reciprocating

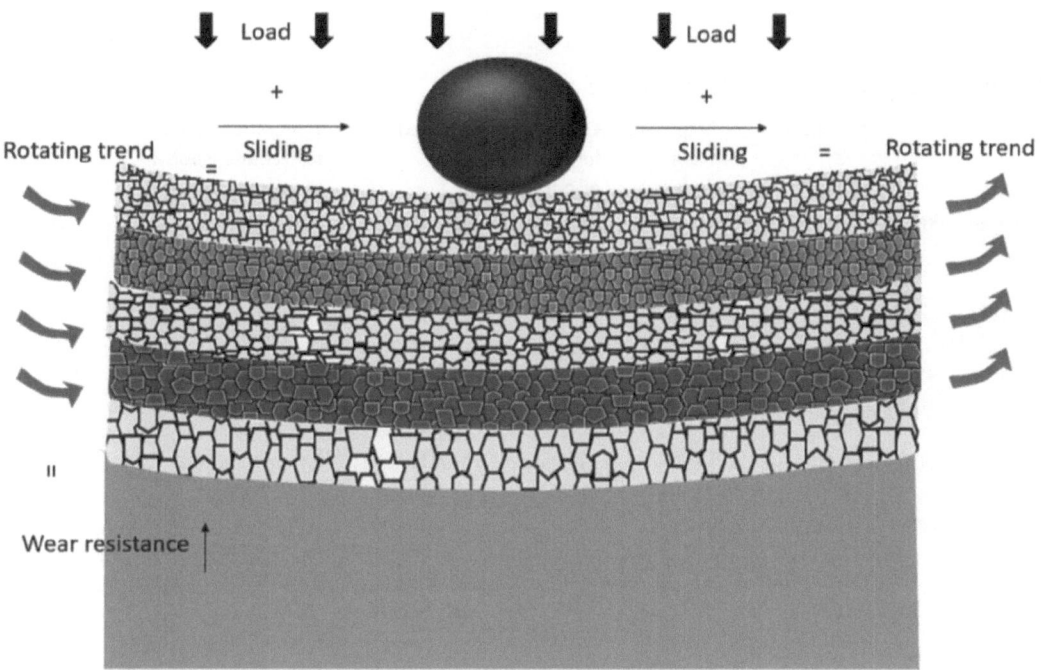

FIGURE 2.18 Reinforcement schematics of gradient and multistructure.

friction wear testing machine. It can be seen that the coefficient reached 0.08 in the beginning, and then reached stable values (COF ≈ 0.04). In the sliding process, synergy reinforcement mechanism occurred between layered films and gradient nanostructure. That's the reason for the good lubrication: COF ≈ 0.04. Figure 2.19 (h) exhibits the wear scar under a white light interferometer. The results show that the deepest position reached 2.5 microns and the average value is 0.9 microns, indicating good wear resistance.

2.4.4 WS$_2$-Ag Films

WS$_2$ films have many advantages over MoS$_2$ films, but the common fatal flaw is the moisture-prone solution in hygrothermal environments. This may be related to the conversion of S to H$_2$SO$_4$ [66] in WS$_2$ deliquescence and oxidation, and further forming sulfate. In order to inspect behaviors of solid lubricant under hygrothermal environment, WS$_2$-Ag composite films were prepared by IBAD technique, which combined a magnetron sputtering with a pair of Kaufman ion sources. In the test, the composite film was exposed to a humid environment (RH= 42%, T = 30°C). A series of tribo-test results can be found in Figure 2.20. The composite films with a COF of 0.1 was realized by the addition of Ag. It can be seen that the appearance of wear scar has smooth furrow, and a black solid transfer film forms locally on the smooth furrow. Based on this, the solid lubricating film brings obvious anti-friction effects. During the friction process, the formed transfer film makes the anti-friction effects more beneficial. WO$_3$ and AgO$_2$ are generated simultaneously on the WS$_2$-Ag film surface, while AgO$_2$ was generated under a certificated distance below the surface. Moreover, the amount of WO$_3$ decreases, or even disappears. However, the amorphous oxides created during the oxidation affect the formation of transfer film and behavior of abrasive wear. WS$_2$ in WS$_2$-Ag composite film has hexagonal crystal layered structure. Between the layered WS$_2$ crystals, each layer of WS$_2$ is composed of two sulfur atoms sandwiched with a layer of tungsten metal atoms. The two types atoms connect with strong covalent bonds, forming a pyramidal structure. Relatively

TABLE 2.4

Review of doped elements on MoS_2/WS_2

Ref.	Substrate	Deposited films	Other lubricants or additives	Method	Type of tribological test/counter-body/ atmosphere	Smaller average COF reported
Shu et al. (2009) [81]	Ag	MoS_2	/	Hot-pressed sintering	Pin-on-cylinder (brush-rotor)/Cu-5% Ag/air	≈0.20
Chen et al. (2013) [82]	Ag	MoS_2	Graphite	Hot-pressed sintering	Ring-on-disc/	≈0.07
Tsuya et al. (1972) [83]	Cu+Sn	MoS_2	WS_2	Hot-pressed sintering	Pin-on-disk, plane contact, pin-on-plate/-/ air and vacuum	0.15
Kato et al. (2003) [84]	Cu+Sn	MoS_2	Graphite	Hot-pressed sintering	Cylinder-on-plate/ AISI 1045/air	0.40
Deepthi et al. (2010) [85]	Si and mild steel	WS_2	Cr	Unbalanced magnetron sputtering	Ball-on-flat/WC ball/ air, 50–60% relative humidity	≈0.1
Zheng et al. (2008) [86]	Si and medium carbon steel	WS_2	Ag	RF magnetron sputtering	Ball-on-disk/GCr15/ vacuum and humid air	≈0.12
Xiaoming Gao et al. (2018) [87]	Si and AISI 440C	Pure MoS_2 target	Pure WS_2 target	Unbalanced magnetron sputtering	Ball-on-disk/AISI 440C/vacuum	≈0.05
Shusheng Xu et al. (2019) [88]	Si and 304 SS	WS_2	C	Magnetron sputtering	Ball-on disk/-/air	≈0.1
Shusheng Xu et al. (2018) [89]	Si	WS_2	/	Magnetron sputtering	Ball-on disk/GCr15/ rich oxygen vacuum	0.03–0.07
Jie Jin et al. (2016) [90]	9310 gear steel	WS_2	Ag	IBAD	Ball-on disk/Si_3N_4/air, 40% relative humidity	0.05–0.09
Jie Jin et al. (2010) [91]	Si and steel	Ag	/	IBAD and magnetron sputtering	-/-/air	0.4–0.7

weak Van der Waals forces combine the layers. The layered structure enables a small shear strength between WS_2 and Ag slip layers. During the friction process, WS_2 and Ag dissociate along the slip direction and some of them partially transfer to the corresponding surface, forming a transfer film. As a result, the friction process occurs between the transfer film and the lubrication film, showing a low COF. The radius of Ag atom is about 1.53Å (slightly higher than the radius of W atom, which is 1.46Å). Based on the 2H-WS_2 crystallographic parameters of Materials Project (https:// materialsproject.org/) database, the calculated W interlayer distance is about 6.16Å. In both directions of the a- and c-axes, the distance between S atoms is greater than 3Å, and the interstitial space is filled with Ag atoms. Therefore, Ag may exist in 2H-WS_2 in the following three situations: (1) Replace W; (2) filling the direction gap of an axis; (3) filling the gap in the direction of c-axis based on the anti-oxygen erosion protection ratio of doped elements to the suspended bonds in the layered structure of hexagonal crystal [67, 68].

In the test, the composite film was exposed to a humid environment and oxidized, forming amorphous WO_3 and AgO_2; the crystal structure of 2H-WS_2 /Ag was changed, leading to the increase of interlaminar shear strength, and the formation ability of the transfer film of the lubrication film was limited. Ag doped in WS_2 is more likely to be oxidized, which reduces oxygen's participation in the oxygenation process of WS_2 and maintains certain lubrication properties of the composite film.

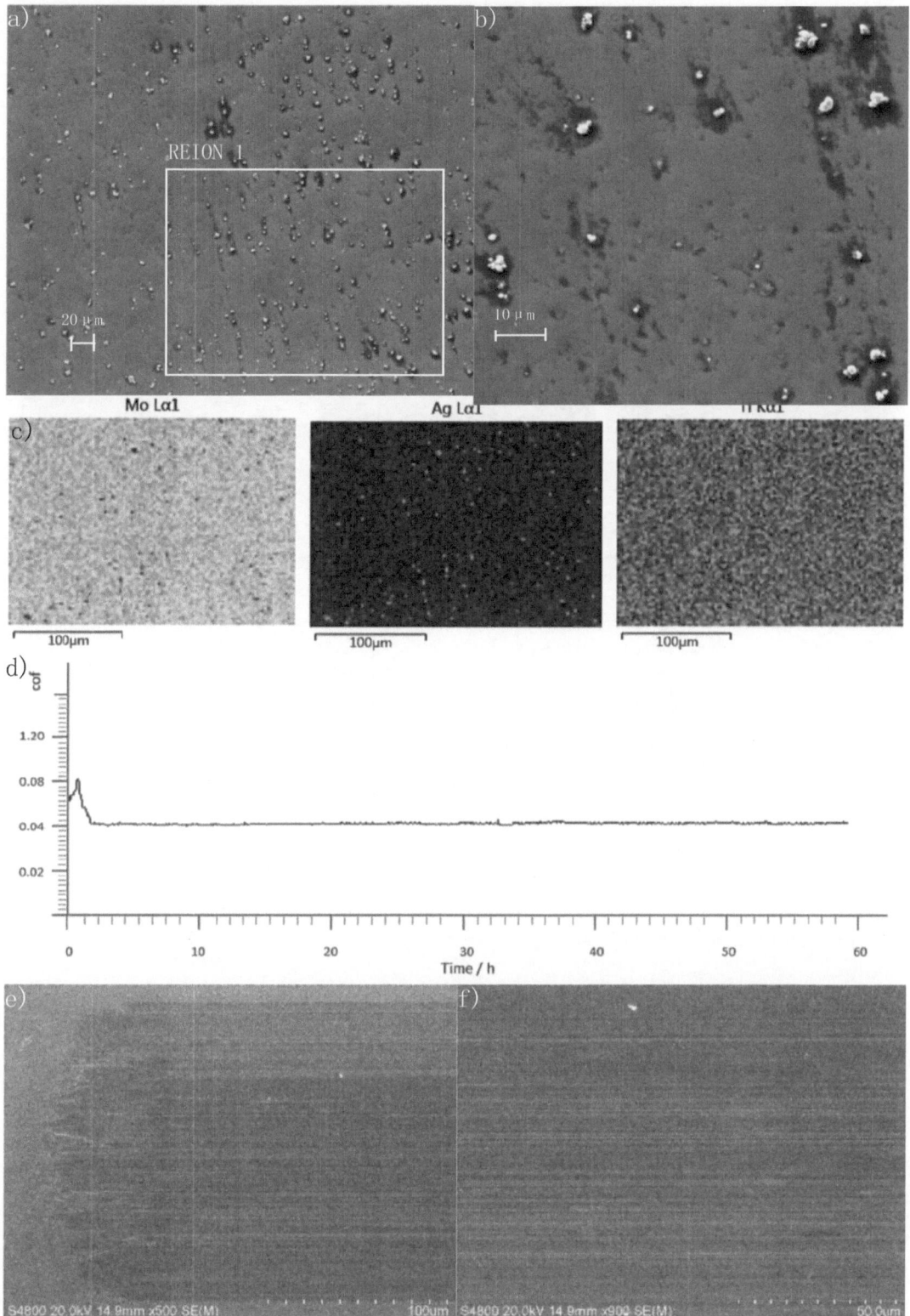

FIGURE 2.19 MoS$_2$-Ag films deposited by IBAD on steel substrate: (a) and (b) SEM surface morphologies; (c) EDS face distribution analysis of REION 1 in (a); (d) COF curve of MoS$_2$-Ag films; (e) and (f) worn morphology.

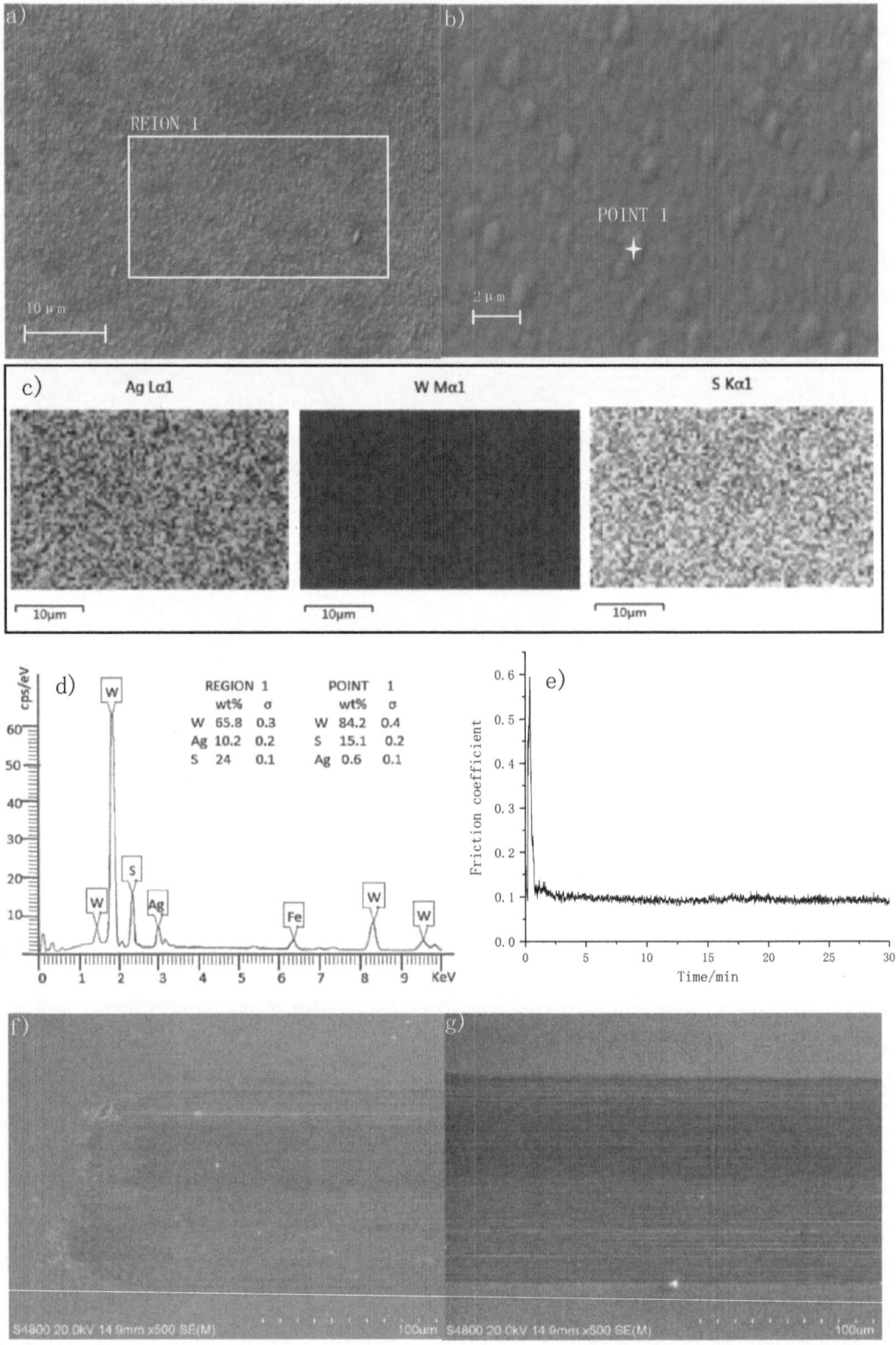

FIGURE 2.20 WS$_2$-Ag films deposited by IBAD on steel substrate: (a) and (b) SEM surface morphologies; (c) and (d) EDS face/point distribution analysis of REION 1/POINT 1 in (a) and (b); (e) COF curve; (f) and (g) worn morphology.

FIGURE 2.21 Applications of MoS_2/WS_2 films.

However, the oxide in the composite film and the oxide debris formed by the wear increased the wear of the composite film.

2.5 APPLICATIONS OF MOS_2 AND WS_2 FILMS

2.5.1 TRIBOLOGICAL APPLICATIONS OF MoS_2/WS_2 FILMS

Since most aerospace applications operated in vacuum environments, the typical solid lubricant is MoS_2/WS_2 [7, 69, 70]. Unlike graphite, which requires sufficient vapor pressure of water to provide lubrication, MoS_2/WS_2 performs best in vacuum environments [71, 72]. Representative examples of components in aerospace applications that already rely on MoS_2/WS_2 lubricants are ball bearings, pointing mechanisms, slip rings, gears, and release mechanisms [73, 74] (Figure 2.21). It has been suggested that most, if not all, of the American satellites and spacecraft contain MoS_2 in some applications [11].

2.5.2 OTHER APPLICATIONS OF MoS_2/WS_2 FILMS

It also needs to be mentioned that MoS_2/WS_2 has numerous other applications besides lubrication. One of the most important application is heterogeneous catalysis, in which active sites are edges, defects, or near "promoter" dopants (Ni, Co). MoS_2/WS_2 can catalyze reactions including hydrodesulfurization, hydrogen evolution reaction, oxygen reduction reaction, and photo-catalytic water splitting [1, 2]. The ability of MoS_2 to absorb Li between its layers (intercalation) has been utilized for batteries [75]. Electronic and optical applications have become a particular focus after the advent of few-layer MoS_2/WS_2, which have been reviewed recently [2–4]. Some unique features are strong excitonic effects, a direct-to-indirect band-gap transition with number of layers, tunability by strain, and ease of creating heterojunctions by stacking. The presence of multiple in equivalent low-energy valleys in the band structure might be used in "valley tronics", and the strong spin–orbit coupling suggests spintronic applications. These properties have been explored for using in transistors, sensors, photovoltaics, light-emitting diodes, and lasers [76–78].

REFERENCES

[1] Miyoshi K. *Solid lubrication fundamentals and applications*. UK: Taylor & Francis. 2001.

[2] Busch C., Solid Lubrication. *Lubricants and lubrication*, Wiley-VCH: Weinheim. 2007: 694–714.

[3] Mang, T., Bobzin, K., Bartels, T. *Industrial tribology: Tribosystems, friction, wear and surface engineering, lubrication*. Wiley-VCH: Weinheim, Germany. 2011.

[4] Donnet, C., Erdemir, A. Historical developments and new trends in tribological and solid lubricant coatings. *Surface and Coating Technology* 2004, 180, 76–84.

[5] Kostornov AG, Fushchich OI. Sintered antifriction materials. *Powder Metallurgy and Metal Ceramics* 2007, 46: 503–512.

[6] Roberts, E.W. Thin solid lubricant films in space. *Tribology International* 1990, 23, 95–104.

[7] Roberts, E.W. Space tribology: Its role in spacecraft mechanisms. *Journal of Physics D: Applied Physics* 2012, 45, 503001.

[8] Jain, A., Ong, S.P., Hautier, G., Chen, W., Richards, W.D., Dacek, S., Cholia, S., Gunter, D., Skinner, D., Ceder, G., et al. Commentary: The materials project: A materials genome approach to accelerating materials innovation. *APL Materials* 2013, 1, 011002.

[9] Materials Project, mp-2815. Available online: https://materialsproject.org/materials/mp-2815 (accessed on 13 June 2019).

[10] Materials Project, mp-1434. Available online: https://materialsproject.org/materials/mp-1434 (accessed on 13 June 2019).

[11] Lansdown AR. *Molybdenum disulphide lubrication*. UK: Elsevier; 1999.

[12] Drilube_corporation. DRILUBE® products containing molybdenum disulfide. http://www.drilube.co.jp/english/product/molybdenum.html. Accessed 8 January. 2016.

[13] Stachowiak GW, Batchelor AW. *Engineering tribology*. US: B.H. Elsevier; 2001: 411–442.

[14] I. Song, C. Park, H.C. Choi. Synthesis and properties of molybdenum disulphide: From bulk to atomic layers. *RSC Advances*, 2015, 5(10): 7495–8514.

[15] Rapoport L, Fleischer N, Tenne R. Fullerene-like WS_2 nanoparticles: Superior lubricants for harsh conditions. *Advanced Materials* 2003, 15: 651–655.

[16] Rapoport L, Leshchinsky V, Lvovsky M, Lapsker I, Volovik Y, Feldman Y, et al. Superior tribological properties of powder materials with solid lubricant nanoparticles. *Wear* 2003, 255: 794–800.

[17] Haidou Wang, Binshi Xu, Jiajun Liu. *Micro and nano sulfide solid lubrication [M]*. Beijing: Springer, 2012.

[18] Mohammad R. Vazirisereshk, Ashlie Martini. Solid Lubrication with MoS_2: A Review. *Lubricants* 2019, 7, 57. doi:10.3390/lubricants7070057.

[19] Scharf TW, Diercks DR, Gorman BP, Prasad SV, Dugger MT. Atomic layer deposition of tungsten disulphide solid lubricant nanocomposite coatings on rolling element bearings. *Tribology Transactions* 2009, 52: 284–292.

[20] Scharf TW, Prasad SV, Mayer TM, Goeke RS, Dugger MT. Atomic layer deposition of tungsten disulphide solid lubricant thin films. *Journal of Materials Research* 2004; 19: 3443–3446.

[21] Priyadarshi Kumar, Zibiao Li, Swee Liang Wong. Functionalized transition metal dichalcogenide-based nanomaterials for biomedical applications. In Bruno Sarmento, José das Neves, (Eds.), *Micro and Nano Technologies, Biomedical Applications of Functionalized Nanomaterials*, 2018: 289–314. doi: 10.1016/B978-0-323-50878-0.00010-0

[22] X. Q. Huang, S. H. Tang, X. L. Mu, et al. Freestanding palladium nanosheets with plasmonic and catalytic properties. *Nature Nanotechnology*, 2011, 6(1): 28.

[23] K.S. Novoselov, A.K. Geim, Morozov S V, et al. Electric fileld effect in atomically thin carbon films. *Science*, 2004, 306 (5696): 666–669.

[24] K. Lee, H. Y. Kim, M. Lotya, et al. Electrical characteristics of molybdenum disulfide flakes produced by liquid exfoliation. *Advanced Materials*, 2011, 23 (36): 4178–4182.

[25] M. A. Gee, R. F. Frindt, P. Joensen, et al. Inclusion compounds of MoS_2. *Materials Research Bulletin*, 1986, 21(5): 543–549.

[26] G. Eda, H. Yamaguchi, D. Voiry, et al. Photoluminescence from chemically exfoliated MoS2. *Nano Letters*, 2011, 11(12): 5111–5116.

[27] Yang J, Voiry D, Ahn S J, et al. Two-dimensional hybrid nanosheets of tungsten disulfide and reduced graphene oxide as catalysts for enhanced hydrogen evolution. *Angewandte Chemie. International Edition*: 2013, 52(51): 13751–13754.

[28] X. Liu, G. Zhang, Q. X. Pei, et al. Phonon thermal conductivity of monolayer MoS$_2$ sheet and nanorib-bons. *Applied Physics Letters*, 2013, 103(13): 133113.

[29] Bouzakis, K.-D., & Michailidis, N. Physical Vapor Deposition (PVD). *CIRP Encyclopedia of Production Engineering*, 2015: 1–8.

[30] Koma, A. & Yoshimura, K. Ultrasharp interfaces grown with van der waals epitaxy. *Surface Science* 174, 556–560 (1986).

[31] Ohuchi, F. S., Shimada, T., Parkinson, B. A., Ueno, K. & Koma, A. Growth of MoSe$_2$ thin-films with Van der Waals epitaxy. *Journal of Crystal Growth* 111, 1033–1037 (1991).

[32] Ohuchi, F. S., Parkinson, B. A., Ueno, K. & Koma, A. Van der Waals epitaxial growth and characterization of MoSe$_2$ thin films on SnS$_2$. *Journal of Applied Physics* 68, 2168–2175 (1990).

[33] Koma, A. Van der Waals epitaxy for highly latticemismatched systems. *Journal of Crystal Growth* 201–202, 236–241 (1999).

[34] L. K. Tan, B. Liu, J. H. Teng, et al. Atomic layer deposition of a MoS$_2$ film. *Nanoscale*, 2014, 6(18): 10584–10588.

[35] Paredez, P., Marchi, M. C., Maia da Costa, M. E. H., Figueroa, C. A., Kleinke, M. U., Ribeiro, C. T. M., Sanchez-Lopez, J. C., Rojas, T. C., Alvarez, F. Carbon nano-structures containing nitrogen and hydrogen prepared by ion beam assisted deposition. *Journal of Non-Crystalline Solids*. 2006, 352, 1303.

[36] Groves, J. R., Hammond, R. H., Matias, V., DePaula, R. F., Stan, L., Clemens, B. M. *Nuclear Instruments and Methods in Physics Research B* 2012, 272, 28.

[37] N. Toyoda, S. Matsui. Ion Beam Deposition: Recent Developments. *Comprehensive Materials Processing*. 2014,4,187-200..

[38] T. Itoh (ed.) *"Ion Beam Assisted Film Growth", Vol. 3 of "Beam Modification of Materials"*, Elsevier 1989.

[39] Shankara A, Menezes PL, Simha KRY, Kailas SV. Study of solid lubrication with MoS$_2$ coating in the presence of additives using reciprocating ball-on-flat scratch tester. *Sadhana - Acad Proc Eng Sci* 2008, 33: 207–220.

[40] Dunckle CG, Aggleton M, Glassman J, Taborek P. Friction of molybdenum disulfide-titanium films under cryogenic vacuum conditions. *Tribology International* 2011; 44: 1819–1826.

[41] Efeoglu I, Bulbul F. Effect of crystallographic orientation on the friction and wear properties of Mo$_x$S$_y$-Ti coatings by pulsed-dc in nitrogen and humid air. *Wear* 2005, 258: 852–860.

[42] Voronin S, Smorygo O, Bertrand P, Smurov I, Smirnov N, Makarov Y. Thermaldiffusion synthesis of thick molybdenum disulphide coatings on steel substrates. *Surface and Coating Technology* 2004, 180: 113–117.

[43] Wang DY, Chang CL, Chen ZY, Ho WY. Microstructural and tribological characterization of MoS$_2$-Ti composite solid lubricating films. *Surface and Coating Technology* 1999, 120: 629–635.

[44] Efeoglu I, Baran O, Yetim F, Altintas S. Tribological characteristics of MoS$_2$-Nb solid lubricant film in different tribo-test conditions. *Surface and Coating Technology* 2008; 203: 766–770.

[45] Teer DG, Hampshire J, Fox V, Bellido-Gonzalez V. The tribological properties of MoS$_2$/metal composite coatings deposited by closed field magnetron sputtering. *Surface and Coating Technology* 1997, 94–5: 572–577.

[46] Zhang XH, Liu JJ, Zhu BL. The tribological performance of Ni/MoS$_2$ composite brush-plating layer in vacuum. *Wear* 1992, 157: 381–387.

[47] Lee TS, Esposito B, Donley MS, Zabinski JS, Tatarchuk BJ. Surface and buriedinterfacial reactivity of iron and MoS$_2$: A study of laser-deposited materials. *Thin Solid Films* 1996, 286: 282–288.

[48] Ding XZ, Zeng XT, Goto T. Unbalanced magnetron sputtered Ti-Si-N: MoS$_x$ composite coatings for improvement of tribological properties. *Surface and Coating Technology* 2005, 198: 432–436.

[49] Fazel M, Jazi MRG, Bahramzadeh S, Bakhshi SR, Ramazani M. Effect of solid lubricant particles on room and elevated temperature tribological properties of NiSiC composite coating. *Surface and Coating Technology* 2014, 254: 252–259.

[50] Cardinal MF, Castro PA, Baxi J, Liang H, Williams FJ. Characterization and frictional behavior of nano-structured Ni-W-MoS$_2$ composite coatings. *Surface and Coating Technology* 2009, 204: 85–90.

[51] Steinmann M, Muller A, Meerkamm H. A new type of tribological coating for machine elements based on carbon, molybdenum disulphide and titanium diboride. *Tribology International* 2004, 37: 879–885.

[52] Wu Y, Li H, Ji L, Ye Y, Chen J, Zhou H. Preparation and properties of MoS$_2$/a-C films for space tribology. *Journal of Physics D: Applied Physics* 2013: 46.

[53] Li Z, Wang J, Lu J, Meng J. Tribological characteristics of electroless Ni-P-MoS$_2$ composite coatings at elevated temperatures. *Applied Surface Science* 2013, 264: 516–521.

[54] Moskalewicz T, Zimowski S, Wendler B, Nolbrzak P, Czyrska-Filemonowicz A. Microstructure and tribological properties of low-friction composite MoS$_2$(Ti,W) coating on the oxygen hardened Ti-6Al-4V alloy. *Metals and Materials International* 2014, 20: 269–276.

[55] Skarvelis P, Papadimitriou GD. Microstructural and tribological evaluation of potential self-lubricating coatings with MoS$_2$/MnS additions produced by the plasma transferred arc technique. *Tribology International* 2009, 42:1765–1770.

[56] Vijay K. Singh, Rahul Pendurthi, Joseph R. Nasr, Hitesh Mamgain, Radhey Shyam Tiwari, Saptarshi Das, Anchal Srivastava. Study on the growth parameters and the electrical and optical behaviors of 2D tungsten disulfide. *ACS Applied Materials & Interfaces* 2020, 12 (14), 16576–16583.

[57] Hang Liu, Guopeng Qi, Caisheng Tang, Maolin Chen, Yang Chen, Zhiwen Shu, Haiyan Xiang, Yuanyuan Jin, Shanshan Wang, Huimin Li, Miray Ouzounian, Travis Shihao Hu, Huigao Duan, Shisheng Li, Zheng Han, Song Liu. Growth of large-area homogeneous monolayer transition-metal disulfides via a molten liquid intermediate process. *ACS Applied Materials & Interfaces* 2020**,** 12 (11), 13174–13181.

[58] Yuchen Yue, JianCui Chen, Yu Zhang, ShuaiShuai Ding, Fulai Zhao, Yu Wang, Daihua Zhang, RongJin Li, Huanli Dong, Wenping Hu, Yiyu Feng, Wei Feng. Two-dimensional high-quality monolayered triangular ws$_2$ flakes for field-effect transistors. *ACS Applied Materials & Interfaces* 2018**,** 10 (26), 22435–22444.

[59] Mei Er Pam, Yumeng Shi, Junping Hu, Xiaoxu Zhao, Jiadong Dan, Xue Gong, Shaozhuan Huang, Dechao Geng, Stephen Pennycook, Lay Kee Ang, Hui Ying Yang. Effects of precursor pre-treatment on the vapor deposition of WS2 monolayers. *Nanoscale Advances* 2019, 1 (3), 953–960.

[60] Lianqing Dong, Yuyan Wang, Jiacheng Sun, Caofeng Pan, Qinghua Zhang, Lin Gu, Bensong Wan, Cheng Song, Feng Pan, Cong Wang, Zilong Tang, Junying Zhang. Facile access to shape-controlled growth of WS2 monolayer via environment-friendly method. *2D Materials* 2019, 6 (1), 015007.

[61] Soumyadip Majumder, Minhua Shao, Yuanfu Deng, Guohua Chen. Two dimensional WS$_2$/C nanosheets as a polysulfides immobilizer for high performance lithium-sulfur batteries. *Journal of The Electrochemical Society* 2019**,** 166 (3), A5386–A5395.

[62] M.C. Simmond, A. Savan, E. Pflüger, H. Van Swygenhoven, *Surface and Coating Technology* 126 (2000) 15–24.

[63] Shusheng Xu, Jiayi Sun, Lijun Weng, Yong Hua, et al. In-situ friction and wear responses of WS$_2$ films to space environment: Vacuum and atomic oxygen. *Applied Surface Science* 2018, 447: 368–373.

[64] Liu, G. L., Robertson, A.W., Li, M.M.J., Kuo, W.C.H., Darby, M.T., Muhieddine, M.H., Lin, Y.C., Suenaga, K, Stamatakis, M., Warner, J.H., et al. MoS$_2$ monolayer catalyst doped with isolated Co atoms for the hydrodeoxygenation reaction. *Nature Chemistry* 2017, 9, 810–816.

[65] Ivanovskaya, V.V., Zobelli, A., Gloter, A., Brun, N., Serin, V., Colliex, C. Ab initio study of bilateral doping within the MoS$_2$-NbS$_2$ system. *Physical Review B* 2008, 78, 134104.

[66] Xu S S, Gao X M, Sun J Y, et al. Comparative study of moisture corrosion to WS$_2$ and WS$_2$/Cu multilayer films. *Surface and Coating Technology*, 2014, 247(5): 30–38.

[67] Teer D G. New solid lubricant coatings. *Wear*, 2001, 251 (282): 1068–1074.

[68] Renevier N M, Fox V C, Teer D G, et al. Coating characteristics tribological properties of sputter-deposited MoS$_2$/metal composite coatings deposited by close filed unbalanced magnetron sputter ion plating. *Surface and Coating Technology*, 2000, 127(1): 24–37.

[69] Lince, J.R.; Fleischauer, P.D. Solid Lubricants. In *Space vehicle mechanisms: Elements of successful design*; Conley, P.L., Ed.; Wiley-Interscience: New York, NY, USA, 1998.

[70] S. Brown, J. L. Musfeldt, I. Mihut, J. B. Betts, A. Migliori, A. Zak, and R. Tenne. Bulk vs Nanoscale WS$_2$: Finite Size Effects and Solid-State Lubrication. *Nano Letters* 2007, 7: 2365–2369.

[71] Chen, Z.; He, X.; Xiao, C.; Kim, S.H. Effect of humidity on friction and wear: A critical review. *Lubricants*. 2018, 6, 74.

[72] L. Rapoport, O. Nepomnyashchy, I. Lapsker, A. Verdyan, Y. Soifer, R. Posovitz-Biro, R. Tenne. Friction and wear of fullerene-like WS$_2$ under severe contact conditions: Friction of ceramic materials. *Tribology Letters* 2005, 19: 143–149.

[73] Lince, J.R. Doped MoS$_2$ Coatings and Their Tribology. In *Encyclopedia of tribology*; Wang, Q.J., Chung, Y.W., Eds.; Springer: Boston, MA, USA, 2013.

[74] Sun, X. Solid Lubricants for Space Mechanisms. In *Encyclopedia of tribology*; Wang, Q.J., Chung, Y.W., Eds. Springer: Boston, MA, USA, 2013.

[75] Mao, J.; Wang, Y.; Zheng, Z.L.; Deng, D.H. The rise of two-dimensional MoS$_2$ for catalysis. *Frontiers of Physics* 2018, 13, 138118.

[76] S. X. Cao, T. M. Liu, S. Hussain, W. Zeng, X. H. Peng, F. S. Pan, Synthesis and characterization of novel chrysanthemum-like tungsten disulfide (WS$_2$) nanostructure: Structure, growth and optical absorption property. *Journal of Materials Science: Materials in Electronics* 2015, 26: 809–814.

[77] Ganatra, R.; Zhang, Q. Few-Layer MoS$_2$: A Promising Layered Semiconductor. *ACS Nano* 2014, 8, 4074–4099.

[78] Bernardi, M.; Ataca, C.; Palummo, M.; Grossman, J.C. Optical and electronic properties of two-dimensional layered materials. *Nano* 2017, 6, 479–493.

[79] Kaline Pagnan, Jose Daniel Biasoli de Mello, Aloisio Nelmo Klein. Self-lubricating composites containging MoS$_2$: A review. *Tribology International.* 2018 (120): 280–298. doi:10.1016/j.triboint.2017.12.033

[80] Kilbury OJ, Barrett KS, Fu XW, Yin J, Dinair DS, Gump CJ, et al. Atomic layer deposition of solid lubricating coatings on particles. *Powder Technology* 2012; 221: 26–35.

[81] Shu L, Yi F, Yang X, Shuxian C, Juan W. Structure and formation mechanism of surface film of Ag-MoS2 composite during electrical sliding wear. *Rare Metal Materials and Engineering* 2009, 38: 1881–1885.

[82] Chen FY, Feng Y, Shao H, Li B, Qian G, Liu YF, et al. Tribological behaviour of silver based self-lubricating composite. *Powder Metallurgy* 2013, 56: 397–404.

[83] Tsuya Y, Shimura H, Umeda K. A study of the properties of copper and copper-tin base self-lubricating composites. *Wear* 1972, 22: 143–162.

[84] Kato H, Takama M, Iwai Y, Washida K, Sasaki Y. Wear and mechanical properties of sintered copper-tin composites containing graphite or molybdenum disulfide. *Wear* 2003: 573–578.

[85] B. Deepthi, Harish C. Barshilia, K.S. Rajam, Manohar S. Konchady, Devdas M. Pai, Jagannathan Sankar. Mechanical and tribological properties of sputter deposited nanostructured Cr–WS2 solid lubricant coatings. *Surface & Coatings Technology.* 2010, 205: 1937–1946.

[86] X.H. Zheng, J.P. Tu, D.M. Lai, S.M. Peng, B. Gu, S.B. Hu. Microstructure and tribological behavior of WS$_2$–Ag composite films deposited by RF magnetron sputtering. *Thin Solid Films* 2008: 5404–5408.

[87] Xiaoming Gao, Yanlong Fu, Dong Jiang, Desheng Wang, Shusheng Xu, Weimin Liu, Lijun Weng, Jun Yang, Jiayi Sun, Ming Hu. Constructing WS$_2$/MoS$_2$ nano-scale multilayer film and understanding its positive response to space environment. *Surface and Coating Technology* 2018: 8–17.

[88] Shusheng Xu, Yuzhen Liua, Mingyu Gao, Kyeong-Hee Kang, Dong-Gap Shin, Dae-Eun Kim. Superior lubrication of dense/porous-coupled nanoscale C/WS$_2$ multilayer coating on ductile substrate. *Applied Surface Science* 2019: 724–732.

[89] Shusheng Xu, Jiayi Sun, Lijun Weng, Yong Hua, Weimin Liu, Anne Neville, Ming Hu, Xiaoming Gao. In-situ friction and wear responses of WS$_2$ films to space environment: Vacuum and atomic oxygen. *Applied Surface Science* 2018: 368–373.

[90] JieJin, Xiaolin Huang, Weiwei Qiu, Huawei Cui, Xiangyu Meng. Weatherable property of WS$_2$-Ag composite solid lubrication films prepared by IBAD under hygrothermal environment in Hainan. *China Surface Engineering.* 2016.

[91] JieJin, Weiwei Qiu, Jinhui Wang, Huibin Guo, Microstructure and tribological behavior of Ag films deposited by magnetron sputtering and IBAD. *Nuclear Techniques.* 2010.

3 Multilayer Transition Metal Nitride Protective Coatings

Fan-Bean Wu
National United University, Taiwan

Yung-I Chen
National Taiwan Ocean University, Taiwan

Jyh-Wei Lee
Ming Chi University of Technology, Taiwan

Jenq-Gong Duh
National Tsing Hua University, Taiwan

CONTENTS

3.1 INTRODUCTION

Transition metal nitride, TMN, coatings have been developed and applied as protective surface layers due to its excellent mechanical properties, such as high hardness and elastic modulus, good adhesion, sufficient resistance against scratch and indentation, outstanding tribology, and great chemical and thermal durability. Those combined benefits make TMN films good candidates to meet the requirements and to be applied in versatile fields, especially where mechanical, thermal, tribological, and chemical protection are needed. For a few decades, titanium nitride, TiN, surface layer, for example, has gathered tremendous attentions and efforts in protective coating aspects, such as ball bearing, vehicle transmission component, engine parts, drilling bits, cutting tools, decoration... etc. The use of TiN coating seems in a mature stage, yet plenty of studies to push its performance limits and to explore brand new applications are still ongoing [1–3]. As expected, in recent decade more researches have continuously been carried out to strengthen the conventional one-component TMN, such as TiN, CrN, TaN, MoN, WN, HfN...etc., by forming binary, ternary, multicomponent, multiphase, nanocomposite, and multilayer TMN to work against complicated and harsh environments [1, 3, 4]. In a wind power mining field, coatings on critical components need to withstand wear, heat, and cyclic stress, not to mention additional chemical attack when the wind farm allocates on shore or at sea shed. Coatings on turbine blades might encounter erosion by fine particles and thermal cycling at a large temperature difference simultaneously in desert territory for instance. One ought to consider high humidity and salinity and even sea water contact if a protection layer is utilized on an island or near coast region. These practical cases all imply that multiple functions are required for the coatings in field use.

TMN protective coatings can be manufactured with monolithic, dual-phase, graded, and multilayer structure according to compositional, phase, and layer distribution. The comparison between TMN films with various configurations is provided and an intense discussion on multilayer TMN protective coatings is focused in this chapter. Selection of coating material system, design and deposition methods, and microstructure are the keys toward high performance of the TMN multilayer protective coatings. For instance, the TMN multilayer coatings have been adopted for a variety of industrial fields, including rolling contact fatigue of steels, protection of titanium aluminides for aerospace usage, molding for high-temperature metal semi-solid treatments, advanced complex printing, medical material surfaces, textile field, cutlery, hard disk and tribological applications, and periodontal instruments [2, 3]. And especially for machining, such as stainless steel machining, dry high-speed machining, machining of Al alloys for aerospace and automotive components, Inconel superalloys, high-speed machining of stainless steel, dry machining of gray and ductile cast iron, and machining of difficult-to-cut materials... and so forth, the TMN multilayers enable the tool components to be pushed forward for better performance, elongated lifetime, and combined properties. The robustness of the multilayer TMN coatings comes from the layer configuration, material systems, integrity of interfaces, layer microstructure, controlled defects, and the intrinsic properties of the building layers. The selection of nitride building layers, thickness ratio and modulation characteristics, interface integrity, crystal growth, and microstructure evolution are elucidated through case studies of versatile TMN multilayer coatings. The indentation behavior, hardness, elastic modulus, scratch and tribological characteristics, failure modes, crack propagation, as well as thermal stability and corrosion resistance are mentioned and discussed. The relationships between microstructure features, design and deposition, and related mechanical characteristics are spotlighted to give ideas for the readers of wide-ranging interests.

The contents of this chapter include a brief introduction, the insight of multilayer transition metal nitride coatings in the field of protective application, the design of manufacture, detail deposition methods and control, the microstructure features and evolution, mechanical behavior, such as hardness, modulus, scratch and wear resistance, and strengthening mechanism, as well as chemical and thermal stabilities. Recent reports and literatures are reviewed and quoted to elucidate the advances in transition metal nitride multilayer protective coatings. Although there exist plenty of intensive reviews on hard and protective multilayer coatings [1, 4–22], the focus points and discussion of the so-called TMN multilayer hereafter are referred to as the coating assemblies containing at least one TMN building layer. And the other one can be metal or nitrides of frequently used metals that are not categorized as transition metals or even non-metals. That is to say the combination in the form of TMN/metal, TMN/SiN, TMN/AlN, for example, is within the scope of this chapter.

3.2 TRANSITION METAL NITRIDE COATINGS

3.2.1 Single-Element TMN Coatings

The TMN films have been intensively developed for several decades owing to their excellent performances in combined mechanical, thermal, chemical, and tribological characteristics. Titanium nitride film, TiN, for example, is the most frequently used one due to its high hardness and resistance to wear and corrosion, not to mention its golden color appearance, which is favored in decoration aspect. Chromium nitride, CrN, coating which shows superior hardness and thermal stability, as compared to TiN, has become another candidate and attracted attentions in recent decades. Another example is the persuit of materials with better performance for diffusion barrier in microelectronics manufacture. The TiN thin film diffusion barrier for Al patterns was replaced by TaN-based thin nitrides to better resist diffusion between Si wafer and Cu metallization for the nanometer semiconductor fabrication in 1990s. The MoN can be used in turbine engine parts surface because of its reliable thermal stability and the forming of surface lubricant layer of MoO_3 during tribological contact. These profound successes in single-element TMN films induce more efforts in finding new types of nitride coating with functional elements to form binary, ternary, and multinary element TMN films for further enhancement. Although the majority of this chapter focuses on multilayers, the background knowledge about TMN coatings is fundamental and useful before looking into the multilayer films. The typical and most widely adopted TMN protective layers, including single, binary, multi-elements films, especially the recently developed thin film metallic glass, TFMG, and nitrogen incorporated high entropy alloy, HEA, films, will be introduced and discussed in this section. The understanding of their microstructure, specialty, and performance in fields of use would be helpful for materials engineering and selection when one needs a TMN multilayer assembly for real case.

3.2.2 Binary TMN Coatings

In 2014, Mayrhofer and coworkers have reviewed thoroughly the TMN coatings as protective surface layers [1]. The article presents the fundamentals of TMN hard coatings from the viewpoints of bondings, structure, composition, material selection, and thermodynamics to elucidate the insights of the TMN films. One effective approach to promote the protective performance of the single-element TMN is to introduce a second element into TMN to form a binary TM-X-N system, where X represents functional elements not limited to transition metals. The strengthening by addition of the second element could be in various forms and/or their combination, including, solid-solutioning, grain refinery, dual-phase formation. Figure 3.1 shows typical chemical bonding and corresponding change in protective characteristics of hard ceramic materials [1, 12, 23, 24]. The corresponding change in mechanical properties, including strength, adhesion, hardness, ductility, brittleness, stability, and so on is marked to indicate the bonding characteristics and tendency. For instance, the CrN and HfN single-element nitrides exhibit stronger adhesion and ductility but otherwise show lower hardness, while higher hardness and strength can be found for TiB_2 and CN materials. The transition metal nitride materials are categorized as metallic hard materials while nitrides of the frequently

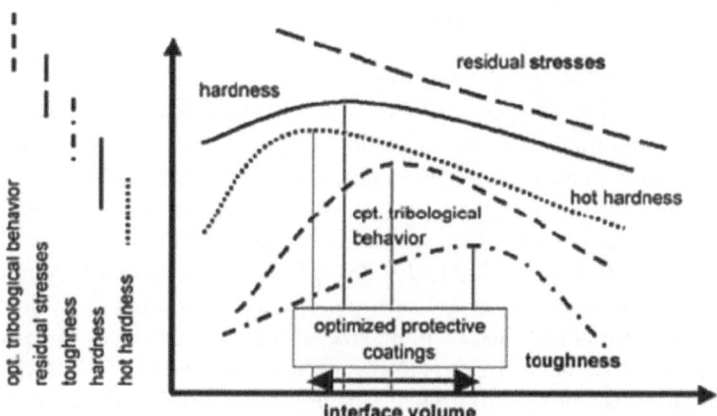

FIGURE 3.1 Chemical bonding and corresponding change in protective characteristics of hard ceramic materials [1, 12, 23, 24].

incorporated Al and Si lean to covalent hard materials side. Therefore, it is expected that the addition of Al or Si into hard TMN to generate the TM-X-N systems is an effective approach to promote the mechanical and tribological performances. Detailed description and intensive discussion on bonding characteristics and structure features of the Al-added TMN films are referred in the review chapter [1]. Several TM-X-N systems, which are often adopted in TMN multilayer assemblies, are reviewed in the following sections.

3.2.2.1 Al Incorporated Binary TMN Coatings

As compared to TMN coatings, the introduction of functional elements into TMN to generate the above-mentioned TM-X-N films is the game changer in the pursuit of advanced protective surface layers. Especially with Al or Si addition, TM-X-N coatings prevail in various properties due to solid solution, secondary phase formation, phase separation, by implementation of both composition and microstructure control during deposition and metallurgical evolution with heat treatments. Holleck checked the phase stability of the binary TM-Al-Ns, including TiAlN, ZrAlN, and HfAlN coatings [25]. Using low-temperature vapor deposition, the TM-Al-N films can be deposited with a meta-stable phase. Take TiAlN film for instance, the PVD-deposited TiAlN films show a supersaturated solid solution phase [26–31] of cubic or wurtzite structure depending on the fraction of Al [32, 33]. Generally the hardness rises with the Al content in TiAlN. However, beyond certain critical limits of Al incorporation, usually 30–35 at.%, the B1 structure of TM-Al-N evolves into a mixture of B1 and B4. This mixture triggers the degeneration of hardness, wear resistance, and oxidation resistance [34–39] simultaneously. Nevertheless, the TiAlN films with supersaturated feature usually exhibit an early stage age-hardening under heat treatment due to coherent phase separation [22, 40–42]. The TiAlN coatings fabricated with dedicated composition and phase control to figure out the feasibility in the fields of decoration [43] and dental prosthesis [44] can also be found.

Another typical TM-Al-N case attracting enormous attention is the CrAlN coating. The CrAlN hard coatings have been put into practice in versatile applications, including automobiles, aerospace components, cutting tools, due to its high hardness, excellent oxidation, and corrosion resistances [34, 45–52]. Though the CrAlN coating exhibits lower hardness as compared to TiAlN films, it accommodates larger amounts of AlN and shows a wider range of properties [34, 53, 54]. In CrAlN, the hardness increases with Al until 77%, at which a maximum solubility of cubic AlN in CrN is reached. Further increase in Al contents triggers the formation of wurzite AlN and the hardness descends accordingly. It should be noted that superior anti-oxidation and anti-corrosion can be obtained because of the formation of adherent and protective oxide scales during usage. The oxide scales of the CrAlN is often a dense layer of Al_2O_3, Cr_2O_3, and/or their mixture [34, 55–57], and thus retards further and severe oxidation once the dense oxide scale forms. Not only the above-mentioned typical binary TiAlN and CrAlN but also ZrAlN [58–62], NbAlN [63–66] attract research interests owing to their specific properties in recent decades.

3.2.2.2 Si-Incorporated Binary TMN Coatings

The introduction of Si into TMN to generate TM-Si-N initiates a new dimension in protective coatings. The merits of Si addition can be manifold, including strengthening by solid-solutioning and phase separation according to the Si amounts and the enhancement in mechanical, tribological, and anti-oxidation behavior [67–72]. The TiSiN film shows a nanocomposite microstructure feature in which the TiN forms nanograins and is embedded in amorphous SiN_x phases. Further studies suggest that the amorphous SiN_x phase is Si_3N_4 [68, 73–75]. Similarly, in CrSiN coatings, the increase in Si content up to approximately 10 at.% leads to a gradual promotion in hardness. It is demonstrated that the volume ratio of CrN and its grain size shrink as Si increases. As a consequence, the hardness rises by the grain refinery of CrN. However, the hardness of CrSiN drops drastically after 10 at.% Si incorporation because of too much soft amorphous Si_3N_4 phase. The research works from Kim et al. [76] and Lin et al. [77] both confirm that with smaller contents of Si, for example in Lin's study 4.8 at.% Si, the CrSiN exhibits a major CrN crystalline phase with Si solid solution. For further addition of 6.7 at.% Si in the CrSiN coatings, a nanocomposite feature shows up and remains until 10.2 at.% Si. Such microstructural evolution and corresponding mechanical behavior can also be found in MoSiN coatings [78, 79], Ta-Si-N, and Zr-Si-N [80], indicating a profound effect of Si in the TMN protective coatings.

3.2.3 MULTICOMPONENT TMN COATINGS

The demands for a further enhancement in TMN coating performance against harsh and critical environment have pushed the development of TMN systems with more functional elements. The concept is simply to integrate and to be benefited from the incorporated elements. Then the ternary, quaternary, or nitride with five or more elements are possible. For instance, the frequently used TiN and CrN films are translated into TiAlSiN [81–84] and CrAlSiN [85–88], respectively, owing to the merits of Al and Si incorporation as described in the last section. However, metallurgical effects, such as solid solution, segregation, precipitation, composition variation, defects ...and so forth, could occur in practice and resultant multiple phenomenon might hinder the thorough understanding of the multicomponent TMN coatings. The challenge then lies in material selection, composition homogeneity, microstructure or atomistic configuration manipulation, defect control, and strengthening mechanism. Thanks to the advances in multicomponent alloy systems, especially HEAs, and metallic glass, MG, the nitride films produced with multicomponent alloys gather tremendous attentions due to unprecedented characteristics.

3.2.3.1 High Entropy Alloy Nitride Coatings

HEA is defined as a metal material system composed of five or more elements with equal molar ratios [89]. The alloy systems are called with various names, like multiprinciple element alloy, complex concentrated alloy, or baseless alloy, based on the definitions of several interesting terms. As compared to amorphous structure of MG, the crystal structure of HEA presents randomly occupied lattice sites by various elements, through which configuration entropy increases from the viewpoint of thermodynamics. For composition aspect, the definition of HEAs has recently been modified with the concentration limitation in principal elements from equimolar to a region of 5 to 35 at.% [90]. Furthermore, the HEAs have widened their constituent principal by doping beneficial elements in order to improve the performance. The extended definition terms of HEAs and modified technique of minor addition impart a variety to the alloy systems. The nitriding of HEA films, namely HEAFs, could be classified into several categories according to the principal elements in HEAs base, including quaternary [91, 92], quinary [93–99], senary [100–104], and more-component systems [105–107]. By doping nitrogen, the HEAFs show a complex relationship of phase formation processes, where exists an intense competition among solid solution, segregation, precipitation, and amorphicity to reach the equilibrium of entire system under deposition. Therefore, a slight change in the composition could bring vast structural diversity of HEA nitride films including single-phase solid solution, intermetallic compound, amorphous, and multiphase coexisting features [92, 103, 108–113].

The large variety in phase and structures implies versatile and alternating properties to the nitride coatings. The HEA nitride films with a fair nitrogen doping concentration generally reveal higher hardness and elastic modulus than those of HEAs base [95, 96, 100, 106, 107, 112, 114–119]. The enhancement in hardness and modulus could be ascribable to the formation of nitride phases and the increase in packing density, which are correlated to high content of covalent bonds and low volume of voids in the coatings, respectively. For thermal properties, the effects of high entropy and lattice distortion invest the HEA nitride films with high thermal stability [101]. The HEA nitride films inhibit the interdiffusion between copper and silicon under elevated temperature because of the decrease in rapid diffusion paths and prolonging of diffusion paths for either side of the interlayer [102, 109, 120, 121]. The HEA nitride coatings are thus proposed as promising candidates for diffusion barriers and reveal a potential application to copper metallization in semiconductor industry. To further improve the mechanical properties, several research works tend toward process control to modify the phase, microstructure, and residual stress through nitrogen flow, deposition temperature, substrate bias, and post-treatment [93, 99, 122–125]. The HEA nitride films encompassing a wide range of available compositions and variety of process parameters are well worth pitching in with numerous studies to break new ground of material systems.

High compressive stress level of −3.5 GPa can be built into sputtered AlCrMoSiTiN HEAFs under −150 to −100 V substrate bias and 600 to 800°C deposition temperature [93]. With the fine tune of controlling parameters, the optimized mechanical properties, hardness, reduced Young's modulus, and H/E ratio are obtained as 34GPa, 325 GPa, and 0.11, respectively. As compared with dual-transition-metal nitride coatings, the enhanced mechanical performance for AlCrMoSiTiN HEAFs produced at elevated temperature is beneficial from bondings between multiple elements and nitrogen. Similar explanation for mechanical improvements could also be found for AlCrNbSiTiVN HEAFs [100, 101]. An even higher peak hardness value of 48 GPa is reached for TiVCrZrHfN HEA films treated at 450°C during manufacturing [122, 124]. Slight differences between indentation evaluation techniques might lead to discrepancy in hardness and modulus explanation. Nevertheless, the maximum hardness, ranging from 15 to 48 GPa, deduced from nanoindentation method for various HEAFs [95, 99, 100, 106, 114, 125–128], is promising for surface protective coating applications.

3.2.3.2 Thin Film Metallic Glass and Its Nitride Films

The TFMG has drawn a lot of attention due to its amorphous structural feature and unique properties in many applications. A detailed review work on the unique properties and potential applications has been accomplished by Chu et al. [129, 130]. The nitrogen incorporated TFMGs, like transition metal nitride systems, provided significant improvements to working components in hardness, adhesion strength, anti-scratch, anti-corrosion, thermal behavior, surface characteristic...and so forth. For instance, the hardness of 10.1 GPa for ZrCuNiAlN, which is superior as compared to Zr-based TFMGs, is resulted from a denser packing when N atoms are introduced [131]. In the ZrCuAlAgN film series with various N amount, the hardness increases from 7.8 to a maximum value of 11.1 GPa at 8at.%N codeposition [132]. The addition of N in Zr-based TFMG reached a hardness level of Fe-based ones. Moreover, less non-negative corrosion potential could be obtained as nitrogen cosputtered in TFMGs.

3.2.4 Multilayer Feature and Its Advantages

In order to enhance the mechanical properties of the protective surface layers, Musil et al. summarize four kinds of possible nanostructures, as shown in Figure 3.2 [133, 134]: (1) bilayers with period λ, (2) the columnar nanostructure, (3) nanograins surrounded by thin tissue phase, and (4) the mixture of nanograins with different crystallographic orientations and/or different phases. The above-mentioned single-layer TMN protective coatings fall in the cases of (2), (3), and (4). The columnar structure is frequently found for single-layer coating, where mechanical force and chemical attack are able to penetrate through column boundaries and to generate through-thickness failures. The multilayer coating concept today is well established and straightforward for the protective thin films for wear and tribological applications [21]. Multilayers can be interpreted as materials engineering on atomic scales, with structures made up of layers only a few atomic monolayers thick [3]. As schematically shown in Figure 3.2 [3], the choice of materials in a multilayer and the thicknesses of the individual layers play key roles in determining its properties. In particular, the bilayer thickness, commonly known as the modulation wavelength (Λ), critically influences the properties of the multilayer coatings. A variety of multilayers concerning protective coatings, including metal/metal, metal/ceramic, and ceramic/ceramic, have been investigated by many researchers. The metal or alloy layers introduced in multilayer system, in general, provide ductility, adhesion, stress, or other considerations, such as corrosion, in general owing to their metallic bonding nature. On the contrary, ceramic multilayers like oxide and carbides are also frequently produced for specific

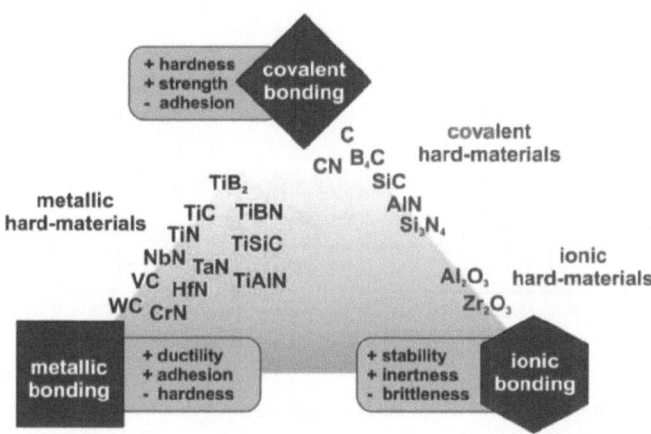

FIGURE 3.2 Schematic illustration of four nanostructures of coatings with enhanced hardness: (a) nanosize bilayers, (b) columnar nanostructure, (c) nanograins surrounded by a tissue phase, and (d) mixture of nanograins with different crystallographic orientation [133, 134].

applications including optics, thermal, and mechanical fields. The employment of nitride films relies on their high hardness, elastic modulus, chemical and thermal stabilities, as compared to metallic layers. In this part, nitride multilayers are focused as a main theme to meet the present topic of interest on transition metal nitride coatings.

The nitride/nitride multilayers have been proven to exhibit superior mechanical and wear-resistant properties among above-mentioned multilayer systems. The multilayer feature is constructed with building layers of distinguishable microstructures stacked alternatively, known as modulation. The bilayer module mostly comes from two distinct nitride films to bring sharp and clear interfaces and modulate crystalline structures in order to strengthen the film assemblies. The technologically important and frequently employed building nitride layers for multilayer systems include TiN [39, 135–145], CrN [135–138, 143, 146–152], VN [146, 153, 154], NbN [147], MoN [139, 141, 149, 150, 155], SiN [156, 157], AlN [144, 148, 155], WN [151], …etc. Not limited to single-element nitrides, dual element nitrides with promising performances are also frequently used, such as TiAlN [39, 150], CrAlN [140, 154, 157, 158], CrVN [153], and TiSiN [158, 159]. The design concept of the above-mentioned building layers is based on their microstructure difference, including composition, crystal structure, grain orientation, etc. One not only can take advantages of the single-layer TMN but also can be benefited from the multilayer configuration and effects of strengthening that single protection layer cannot provide. Detailed strategy, design, and fabrication methods of the multilayer TMN protective coatings will be elucidated in the following section.

3.3 MULTILAYER DESIGN AND DEPOSITION METHODS

Consider the design of multilayer protective coatings, Stueber and coworkers have reviewed the state-of-the-art PVD nanoscale multilayer thin films in 2009 and suggested the design concepts [21] which are now widely found in versatile fields of researches. Herein the microstructure control, layered configuration, and related interfacial characteristics are the key issues for a profound multilayer protective coatings. Basically the necessity of forming a multilayer coating is the difference between building layers. The discrepancies between adjacent layers could be material types, crystal structures, compositions, and intrinsic mechanical behavior, whenever interfaces are identified. For example, crystalline TMN1/TMN2, crystalline TMN/amorphous phase, superlattice with distinguishable lattice parameters of coherent layers, are typical cases having significant interfaces that are of great importance toward their mechanical behaviors. Through the suggestion from Holleck

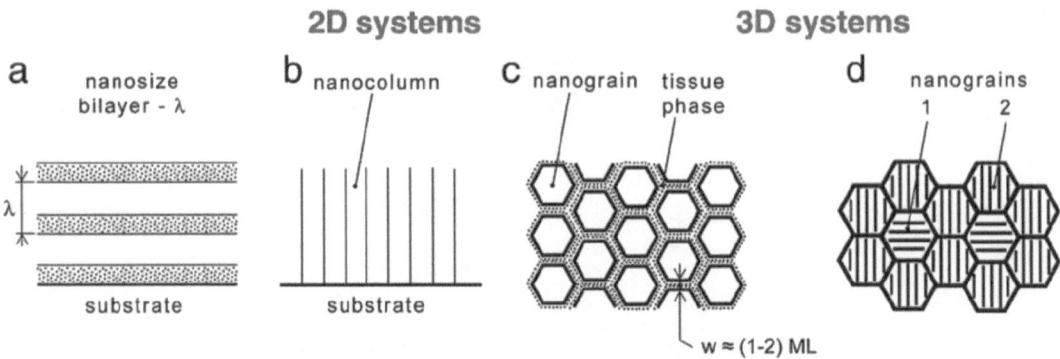

FIGURE 3.3 Optimization of mechanical behavior of multilayer protective coatings as a function of interface volume [12, 21].

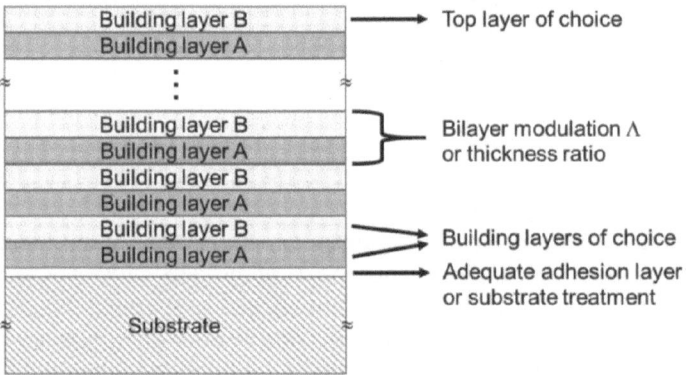

FIGURE 3.4 Schematic illustration of a multilayer configuration with adhesion layer (or interlayer), building layers, and top layer of choices.

et al. [12] on the mechanical behavior with respect to the interface volume, as shown in Figure 3.3, a surface coater would be able to engineer the way out and find approximately optimum of the properties needed. Generally speaking, the hierarchy of a multilayer coating assembly contains various building layers of at least two material systems, namely building layers A and B, as plotted in Figure 3.4. The building layers A and B are stacked sequentially with their thickness of coater's design. Often the adhesion layer, known as interlayer, substrate surface treatment, top layer, and the thickness ratio and bilayer modulation are critical and essential to overall performance of the protective multilayer.

Yashar and Sproul demonstrate the deposition of multilayer coatings using a dual-target vacuum coating system, as illustrated in Figure 3.5 [11]. The targets are arranged in opposite positions on the chamber wall facing the rotaing sample holder. With adequate rotation control the multilayer can be produced with designed thickness modulation or layer period. Several physical vapor deposition (PVD) methods equipped with various power sources have also been widely used [3]. These methods include ion beam deposition, sputtering, electron beam evaporation, arc evaporation, etc. Among them, the reactive magnetron sputtering process is the most frequently used one for multilayer nitride coating deposition. And there exists a variety of sputtering sources, such as radio frequency (RF), pulsed, direct current (DC), planar or cylindrical magnetron, triode or diode, or a combination thereof. Nevertheless, there are two main issues to be addressed cautiously during reactive sputtering. First, the control of the stoichiometry of each layer and second, the quality

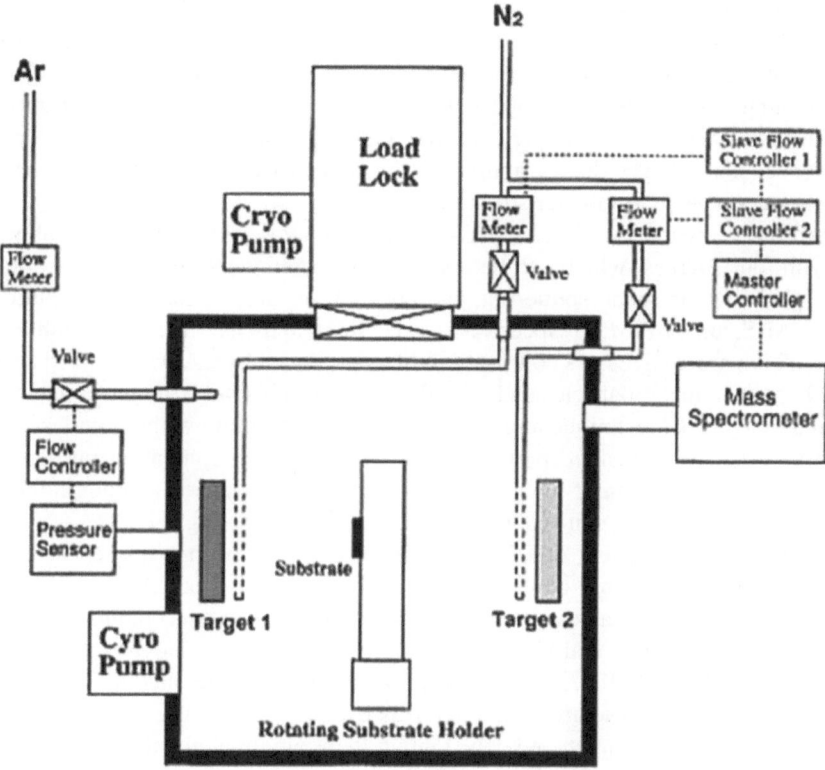

FIGURE 3.5 Schematic plot of the dual-gun vacuum deposition apparatus with rotating substrate holder for multilayer assembly [11].

of the interfaces [160]. With respect to the transition metal nitrides, specifically, the control of the stoichiometry becomes a critical issue because it is known that the heat of formation of the transition metal nitrides differs significantly [11]. Different partial pressures of nitrogen are thus required in the very same deposition chamber to adjust the stoichiometry of the metal nitride layers. In the case of obtaining a single-phase nitride, since the homogeneity range of single-phase nitrides is very narrow, the optimization of deposition parameters, such as modulation wavelength, nitrogen partial pressure, substrate bias, substrate temperature, and working pressure, are essential to achieve high-quality multilayer coatings.

Also, the control on the interface quality is very crucial as the properties of the multilayer coatings are highly dependent on the nature of the interface. A frequent approach to fabricate the multilayer coatings is to put the substrates in rotating holder in between the fixed sources. However, the interfaces of the multilayers are often graded since coating materials generated from sputtering sources are overlapped geometrically. Therefore, extreme care must be taken to achieve repeatability and uniformity in the multilayer coatings [3]. As a good example given by Qiu et al. [146] an In-Line PVD system (PSUVHL-200 C, HOPE Vacuum Technology Co., Ltd., Taiwan) is utilized to prepare CrN/VN multilayer coatings on SUS420 substrates. The CrN/VN multilayer coatings are prepared by two pulsed DC powers simultaneously, that is, the target shutter is set to block off the plasma from one of the targets while the other is working on deposition. The thickness of each bilayer is then controlled by the on–off of the shutters on Cr and V targets. In this way, three types of CrN/VN multilayers with period numbers of 24, 48, and 94 can be prepared.

In recent years, depositing transition metal nitride multilayer coatings at a relative high film growth rate to push up the throughput for industrial applications is in an urgent request. Cathodic arc evaporation technique is one of the solutions to meet the demand. Chen et al. [161] produce CrN/ZrN multilayer coatings with a cathodic arc deposition system. The Zr and Cr targets were allocated on two sides of the vacuum chamber facing each other with a barrel sample holder in between. The barrel mounted with substrates rotates against two sets of plasmas from Cr and Zr targets in sequence. In such a way the multilayer coatings could be generated with designed layer thickness by rotation speed control. The design minimizes the geometrical overlapping for a better chance to produce clear interfaces in multilayer structure.

Purandare and coworkers fabricate CrN/NbN multilayer films with a combinatorial setup equipped with four sources, which can be operated in steered cathodic arc deposition, unbalanced magnetron sputtering, arc bond sputtering, and high-power impulse magnetron sputtering, namely CAD, UBM, ABS, and HiPIMS, respectively [147]. Sharp and clear interfaces with a bilayer thickness of 2.5–4.2 nm are realized. As compared to films produced by CAD [162, 163] and UBM [164] with possible built-in macroparticles and granular voids, respectively, the intact and flat nanolayers are generated by HiPIMS technique and a dense film microstructure without undesired intercolumnar defects is revealed. Intensified power dissipation to increase ionization rate up to 40% [165, 166] and resulted densification of the growing film due to an enhanced adatom mobility have been demonstrated by using HiPIMS in producing TMN protective coatings [135].

An interesting single-element TMN multilayer coating has been proposed to produce a multilayer coating with nanolayered configuration using simplified deposition setup and recipe. The deposition is accomplished in an easy way with magnetron sputtering technique using just one transition metal target, such as Ta, and Hf [167, 168]. Input power and inlet gas mixture are modulated to produce TaN, MoN, and HfN, with composition and phase difference. According to the zone-structure model, the microstructure of the deposits evolve by changing the deposition conditions, such as temperature, bias, gas inlet, and power input. In the single-element TMN coating fabrication, the gas inlet ratio, Ar/N$_2$, is tuned to generate TMN films with different composition and microstructure feature for building layers. For example, TaN films can be deposited under Ar/N$_2$ ratios of 18/2 and 12/8 and show microstructure features of crystalline and amorphous, respectively. Those layers with phase discrepancy are taken as building layers for the stacking of the TaN multilayer systems. Similarly, the MoN films deposited under various Mo target input powers from 100 to 300 W on a Φ-2inch sputtering source and series of Ar/N$_2$ ratios from 18/2 to 10/10 exhibit microstructure features ranging from preferred orientation, columnar crystalline, to amorphous. The single-element MoN multilayers are then manufactured with building MoN layers featuring preferred/crystalline, preferred/amorphous, and crystalline/amorphous for further comparison. The layered configuration and detailed microstructure investigation can be seen in the following Section 3.4.

3.4 MICROSTRUCTURE FEATURES

The bilayer and multilayer coatings are stacked by sublayers with various structures and compositions. TiN and AlN crystallize into face-centered cubic and hexagonal phases, respectively. Both of the two structures are closed packed phases with N interstitial atoms. However, the Ti$_{1-x}$Al$_x$N composite coatings with $x \leq 0.66$ show a metastable cubic structure [169]. The hardness of Ti$_{0.34}$Al$_{0.66}$N coatings retain at 37 ± 2 GPa after 2 h-annealing at 600–950 °C in flowing Ar at atmospheric pressure. Moreover, the TiN and AlN films have the advantages of good adhesion properties and chemical stability, respectively. Therefore, TiN/AlN bilayers with high thickness values of 990 and 410 nm for TiN and AlN sublayers exhibit the merits of individual sublayers [144]. For multilayer films, because the various sublayers are built up sequentially, the crystalline structure of a sublayer is affected by its previous sublayer especially for the films with stacking periods in the scale of nanometer. Such multilayer films composed of alternative (semi-)coherent materials, namely superlattice, exhibit outstanding mechanical properties, such as TiN/AlN [170], CrN/AlN [171], TiN/NbN [172], and CrN/NbN [173]. The multilayer films could be consisted of alternative sublayers with various materials. The fabrication of two hard coatings, such as CrAlSiN/W$_2$N multilayer films, exhibits the function of crack deflection, energy dissipation, and stress relaxation [174]. Moreover, amorphous sublayer is one choice. Amorphous and cubic TaN [167] and MoN [175] are developed to improve the toughness and to suppress crack initiation and propagation, respectively.

Despite the above-mentioned multilayer with larger bilayer thickness, multilayer TMN coatings are usually produced with a building layer thickness down to several nanometers range for protective application. The ultra-thin layers of the multilayer coatings are frequently investigated through electron microscopy, EM, in cross-sectional view. The transmission electron microscopy, TEM, has been a useful technique owing to its analyzing powers on nanoscale resolution. Despite its difficulty in the sample preparation, cross-sectional TEM provides valuable information about uniformity, grain size, orientation, and texture and defects. More importantly, cross-sectional TEM can be used to reveal the microstructure features of the multilayer coatings, including layer thickness, interface, and interfacial roughness [176–180].

One quick example is the study by Wang et al. concerning the microstructure of CrAlN/VN multilayer coatings using cross-sectional TEM [177]. The bilayer periods were controlled with fine-tuned parameters during deposition to 10, 16, 20, 30, and 40 nm. The cross-sectional TEM images and corresponding selected area electron diffraction (SAED) patterns of multilayer coating with a bilayer period of 16 nm, denoted as CV16, in different regions could be revealed in Figure 3.6. While CrAlN and VN were visible as the light and dark layers, respectively, a flat and smooth multilayer architecture was clearly observed in Figure 3.6(a) and (b). Besides, the columnar grains along the growth direction were also discovered. These TEM images suggest that the bilayer period was precisely controlled and the sublayer thickness ratio of CrAlN to VN was about 1, which was a common design for multilayer systems. In the early growth of the multilayer, the interfaces between CrAlN and VN sublayers are sharp and flat, as shown in Figure 3.6(a). The SAED pattern shows an unapparent texture on this region and random orientations composed of (111), (200), and (220) are shown. For comparison, Figure 3.6(b) presents the microstructure in the middle region of the multilayer, showing that with the development of coherent structure, columnar grains can pass through several sublayers and the laminated structure becomes unobvious. Instead, a wave-like configuration is discovered, and the texture in the SAED is evident. The dark-field TEM image obtained from (220) diffraction ring of CV16 multilayer coating is presented in Figure 3.6(c). The columnar grains are found to range from 20 to 70 nm in width, and 250 to 350 nm in length.

Another interesting microstructural characterization in multilayer coatings is obtained by Chan et al. [178], wherein the cross-sectional TEM images directly confirm the microstructure evolution in TiAlN/SiN$_x$ multilayers. Figure 3.7(a) reveals that the structure of both SiN$_x$ and TiAlN is crystalline for the multilayer prepared with individual layer thickness of TiAlN and SiN$_x$ of 4 nm and 0.4 nm, respectively (l_{TiAlN}/l_{SiNx}=4/0.4), leading to a coherent growth. Columnar structure is developed by TiAlN and SiN$_x$ growing on each other layer by layer continuously across several modulation periods, indicating that SiN$_x$ strained into the cubic TiAlN lattice. It is suggested that to minimize the interfacial energy, nanoscale multilayers composed of materials having different crystal lattice structure often tend to form coherent interfaces between the layers through an epitaxial stabilization [181]. Consequently, both SiN$_x$ and TiAlN grow epitaxially to form a superlattice structure. Figure 3.7 (b) shows the amorphous SiN$_x$ of 1 nm and the crystalline TiAlN of 4 nm piled up in order. The light and dark fringes correspond to SiN$_x$ and TiAlN, respectively. The interfaces between TiAlN and SiN$_x$ are distinct, implying that the amorphous SiN$_x$ confines the growth of the crystalline TiAlN and interrupts the columnar growth. In other words, one might differentiate between the epitaxial and non-epitaxial growth by observing the microstructure evolution of multilayers.

On the other hand, in some cases, cross-sectional secondary electron microscopy may also provide sufficient microstructural analysis, as shown in Figure 3.8 [179]. Figure 3.8 shows a cross-sectional SEM image of CrAlSiN/W$_2$N multilayer coating with a bilayer thickness of 20 nm (labeled as M20) and a total thickness of 1 μm. The multilayer can be divided into two distinct regions. The first deposited ten layers are flat, while the latter grown ninety layers are wave-like region. The interfaces between CrAlSiN and W$_2$N sublayers are rather sharp. The wave- like layered configuration

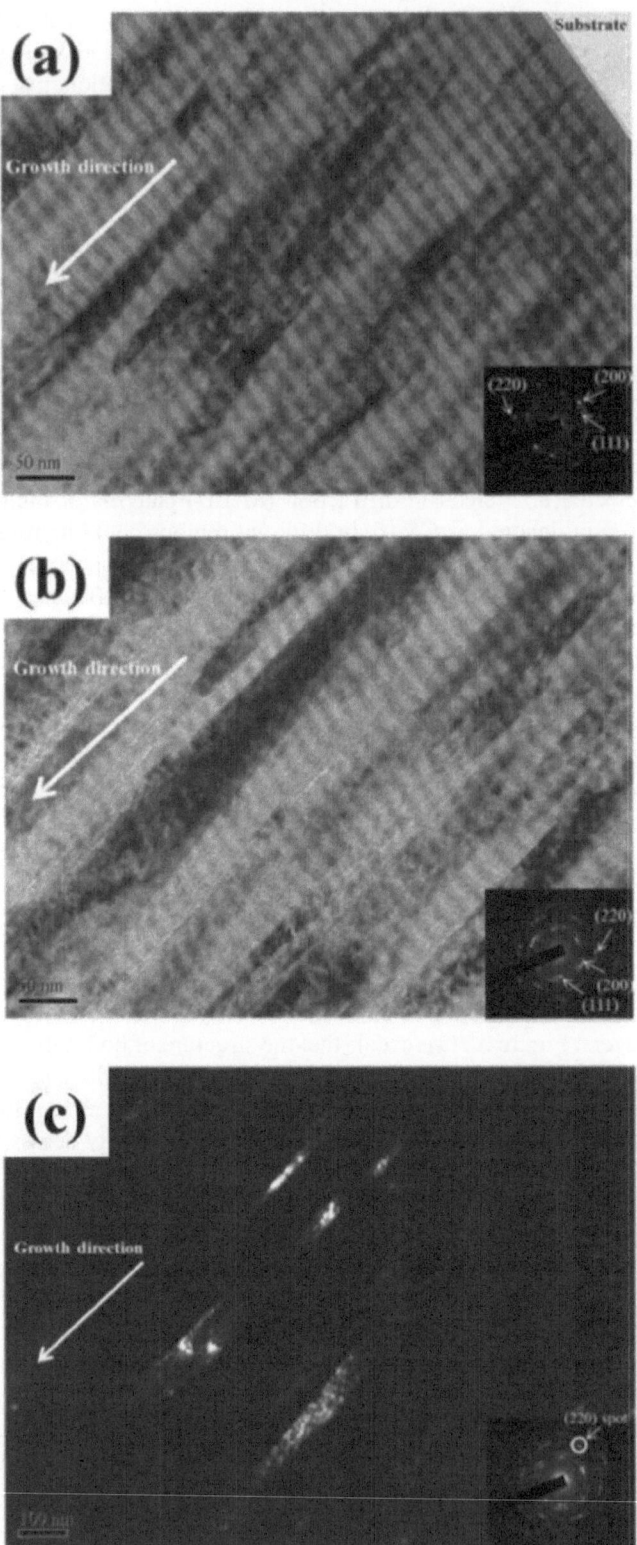

FIGURE 3.6 Cross-sectional TEM analysis on CrAlN/VN multilayer coating. (a) bright field image, (b) bright field image in coating center, and (c) dark field image. [177].

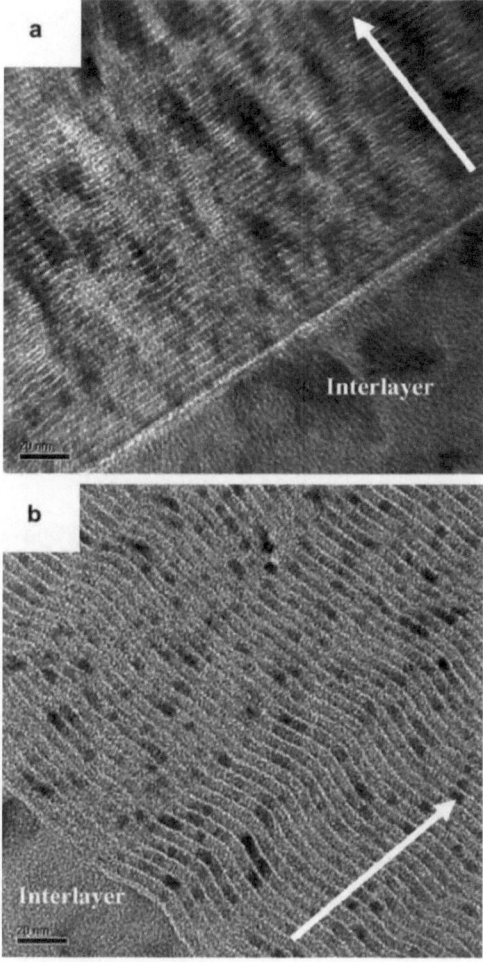

FIGURE 3.7 The cross-sectional TEM micrographs of the TiAlN/SiN$_x$ multilayer coatings with a fixed 4 nm TiAlN and (a) 0.4, and (b) 1.0 nm SiN$_x$ layer [178].

apparently results from the roughness accumulation of CrAlSiN and W$_2$N sublayers. During the initial growth of the multilayer, the grain size and surface roughness are quite small, thus the nano-layered configuration is flat. The abnormal grain growth is always anticipated with the growth of thin films. As multilayer gets thicker, it is revealed that the roughness of each sublayer starts to accumulate, and the configuration becomes more irregular.

Still related to CrAlSiN/W$_2$N multilayer coating [179], the cross-sectional TEM images and the corresponding SAED patterns of the multilayer with a bilayer period of 8 nm (labeled as M8) in different regions are shown in Figure 3.9. A nano-layered architecture with sharp interfaces is shown. The light and dark fringes correspond to CrAlSiN and W$_2$N, respectively. Similar to cross-sectional SEM image in Figure 3.8 (M20), Figure 3.9(a) shows that the nano-layered configuration of earlier deposited sublayers of M8 coating is smooth and flat, and the SAED pattern of this region shows a typical ring pattern, indicating that these sublayers exhibit nanocrystalline structure. The ring pattern is further identified as (111), (200), and (220) directions of B1-NaCl structure. Besides, it is noticed that the B1-NaCl grains grow randomly along various orientations, while the grain size is rather small. Figure 3.9(b) shows the microstructure of M8 coating near the film surface. The width of columnar grains is around 10–20 nm. Since the subsequently deposited sublayers grow along the direction of roughness accumulation, a wave-like configuration is thus revealed.

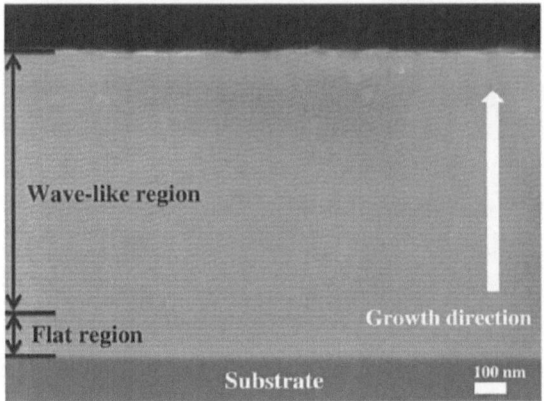

FIGURE 3.8 Cross-sectional SEM analysis on CrAlSiN/W$_2$N multilayer coating [179].

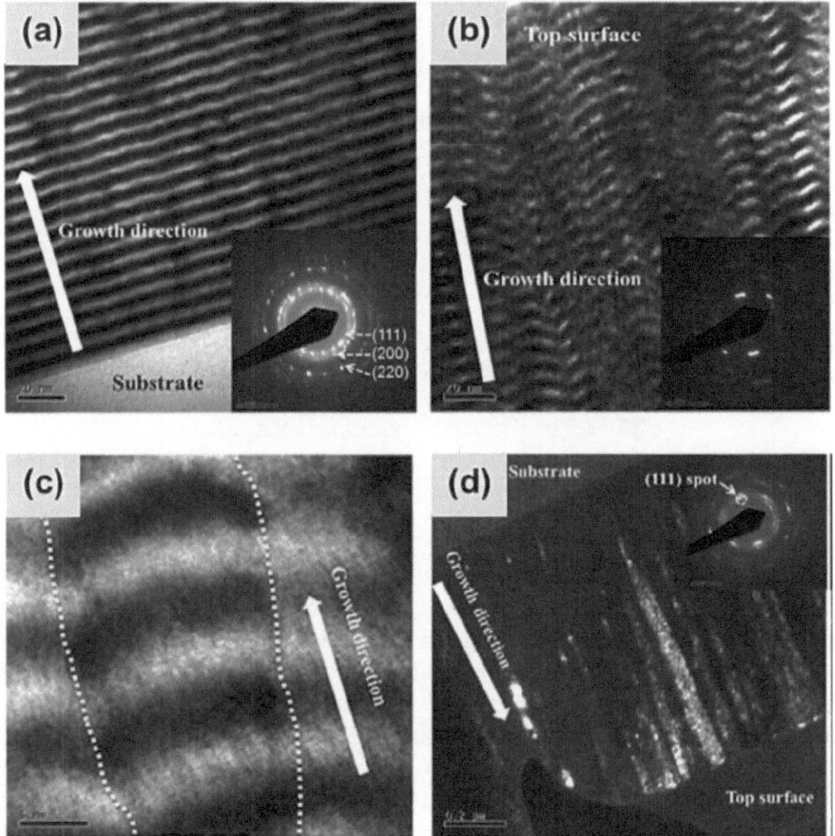

FIGURE 3.9 Cross-sectional TEM analysis on CrAlSiN/W$_2$N multilayer coating [179].

Instead of ring pattern obtained in Figure 3.9(a), the corresponding SAED of this region is spot pattern. It's demonstrated that the crystallinity near the surface in thin film is much higher than that near the substrate, which is in agreement with the result acquired from SEM. Figure 3.9(c) shows the high-resolution transmission electron microscopy, HRTEM, image of M8 multilayer coating [179]. One-dimensional lattice fringe and nano-layered structure are simultaneously observed in the columnar grain located at the figure center, which reveal a superlattice characteristic. CrAlSiN and

W_2N sublayers are sequentially stacked to form the columnar grains. The dark-field TEM image obtained by (111) diffraction ring of M8 multilayer is depicted in Figure 3.9(d). The grains range from 40 to 70 nm in width and 200 to 800 nm in length. SAED pattern of the whole M8 coating is shown for comparison, which demonstrates that (111) reflection is stronger than others. In the dark-field image, it is also revealed that the growth of columnar grains with a nano-layered configuration is predominant in (111) direction.

An interesting multilayer coating, heterostructural single-element transition metal nitride coating, which was assembled with stacking of only one transition metal nitride with alternating microstructure, was developed. The coating systems, including TaN, HfN, and MoN, are fabricated with crystalline, nanocrystalline, preferred orientation and/or even amorphous for building layers in the multilayer structures, as mentioned in Section 3.3. For instance, TaN is produced with amorphous and crystalline features through parameter control of Ar/N_2 inlet gas ratio and input power regulation during magnetron reactive sputtering technique. The amorphous and crystalline TaN layers are tailored into multilayer film with altering c-TaN/a-TaN microstructure [167], as shown in Figure 3.10.

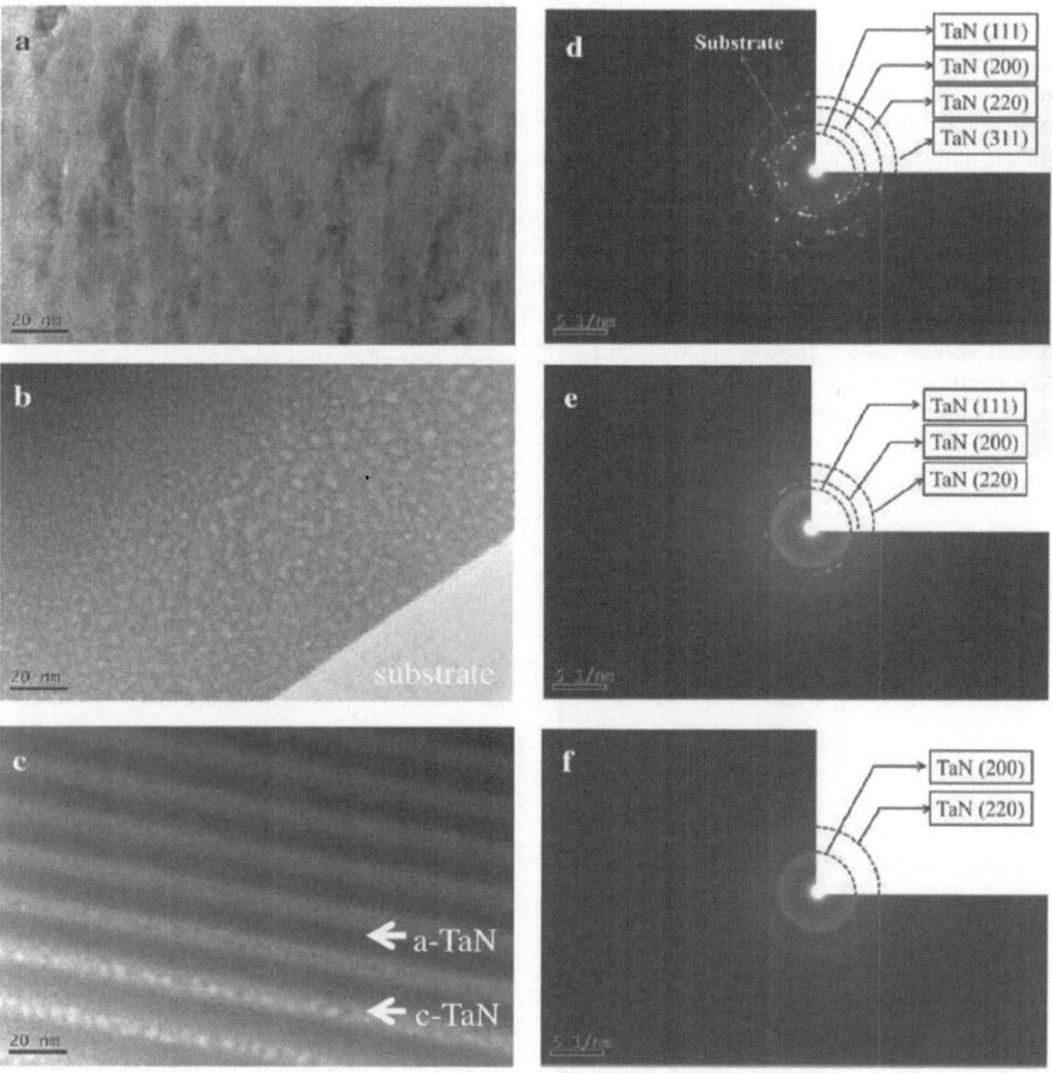

FIGURE 3.10 TEM images of (a) c-TaN, (b) a-TaN, (c) multilayer TaN, and SAED patterns of (d) c-TaN, (e) a-TaN, and (f) multilayer TaN coatings [167].

Though the hardness and Young's modulus of the multilayer c-TaN/a-TaN systems fall in between those of the single a-TaN and c-TaN films, the tailored multilayers show superior resistance to wear and scratch, a tougher mechanical behavior is observed as well. Series of studies of TaN, HfN, and MoN multilayers are investigated and discussed [168, 175, 182–184]. The alternatively stacked layers with distinguishable structure features, such as amorphous, crystalline, nanocrystalline, and prefer-oriented crystalline, generate intact interfaces between building layers and lead to superior tribological behavior for the multilayer coatings.

3.5 MECHANICAL BEHAVIOR AND STRENGTHENING MECHANISM

The mechanical properties of the metal nitrides based multilayer coatings are of great interest both to academia and industry owing to their very high strength and hardness at very low modulation wavelengths [3]. Thanks to a variety in mechanisms of hardness enhancement in multilayer coatings, such as the effect of elastic anomalies, coherency strains, elastic modulus differences, and Hall-Petch strengthening, the multilayer coatings are widely used as an effective approach to improve the fracture toughness of coatings, while maintaining hardness, wear resistance, and other mechanical properties [17, 21]. This is because the multilayer coatings have a unique feature in toughening. In multilayer structures, the mechanisms of toughening can be schematically illustrated in Figure 3.11 which includes crack splitting at the boundaries of small-sized grains, crack deflection at the interface between layers, reduction of stress concentration by interface opening, and plastic deformation at the interface for energy dissipation and stress relaxation [8, 14, 17, 21, 161]. It implies that internal interfaces in multilayer coating have a beneficial effect on the microscopic and macroscopic coating properties when their amount is carefully adjusted to the overall volume of the coating [161]. Therefore, the mechanical properties of multilayer coatings are expected to be boosted from the properties of single-layer coatings. Modulation of the bilayer thickness, often expressed as wavelength Λ or period, is one of the key factors. The grain size effect on hardness is a well-known approach for protective hard coatings. A peak hardness can be obtained with a grain size control around several to tens nanometers [133]. Likewise, a superior hardness can be deduced with a building layer thickness in the range for the multilayer coating.

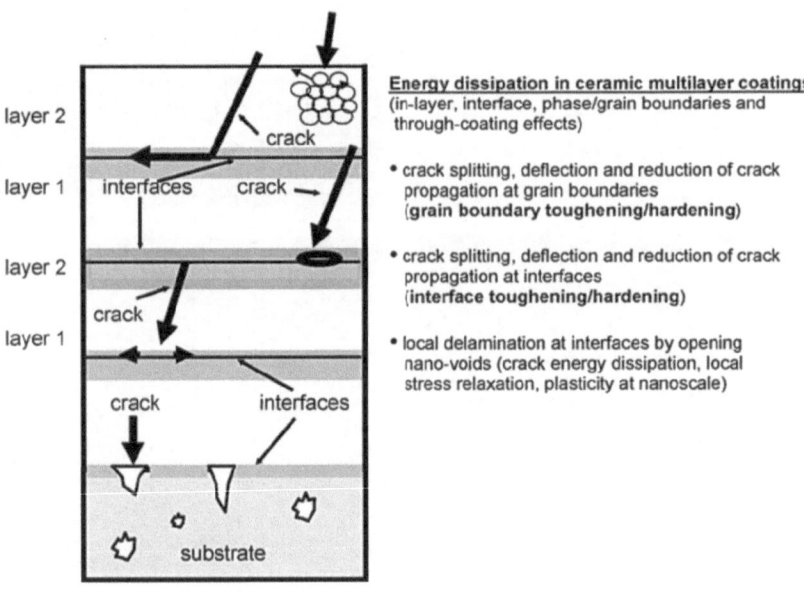

FIGURE 3.11 Schematic illustration of the toughening mechanisms in multilayer configuration [9].

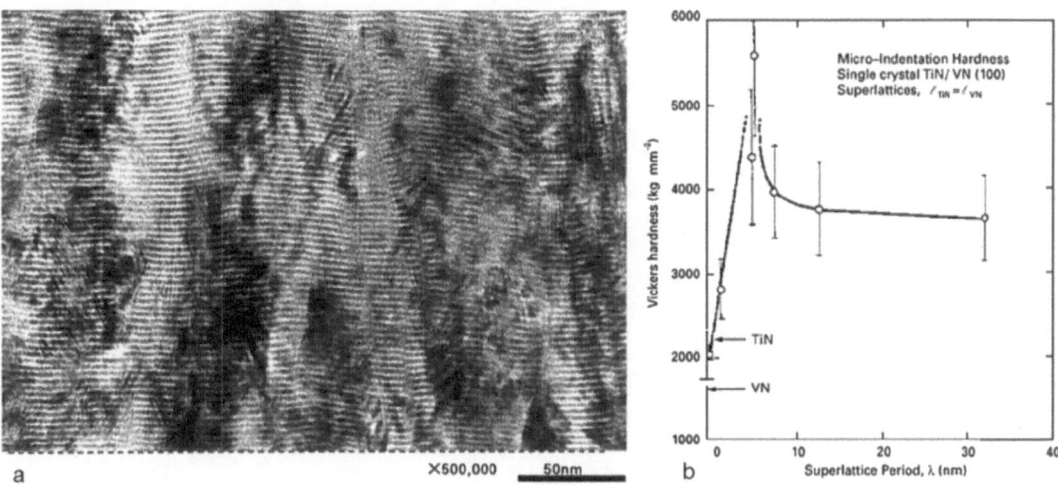

FIGURE 3.12 (a) Cross-sectional micrograph and (b) Vickers hardness of TiN/VN multilayer coatings [11].

Multilayer assembly using TiN layers has been one of the iconic coating systems for protective applications. Yashar and Sproul produce the TiN/VN multilayer coatings with superlattice periods of a few to 30 nm to correlate the building layer thickness and multilayer's mechanical properties [11]. The nanolayered feature can be observed in Figure 3.12(a), in which sequentially alternating thin TiN (bright) and VN (dark) building layers are represented. The resultant Vickers hardness as a function of superlattice period, i.e. bilayer period or modulation λ, is evaluated and plotted in Figure 3.12(b). The hardness values of the single layers of TiN and VN are also marked for comparison. A drastically increasing trend of hardness is shown, from approximately 2000 HV for TiN and VN layer alone to over 5000 HV for the TiN/VN multilayer with a modulation of 5 nm. With thicker period from 8 to 30 nm, the hardness of the TiN/VN multilayer descends and keeps steady at 3800 HV. The strengthening by several nanometer of the building layer thickness is thus manifested.

An improvement of mechanical properties and tribological performances of CrN coating via multilayer feature with VN has been shown by Qiu et al. [146]. In their study, the bilayer thickness Λ of the CrN/VN multilayer coatings is controlled at 7, 13 to 27 nm. The combination of CrN and VN coating gives rise to an increased hardness from 16.7 GPa for the CrN monolayer to 25.2 GPa for the multilayer and the corresponding coefficient of friction (COF) drops from 0.40 to 0.21. The increase of hardness is attributed to the elastic modulus mismatch and alternating-thermal stress strengthening. And the decrease of COF is achieved due to the formation of vanadium oxides in the tribological process [146]. Figure 3.13 plots the nanohardness (H) and elastic modulus (E) of the various coatings. The hardness of monolayer CrN and VN coatings is 16.7 and 23.7 GPa, while the elastic modulus E is 272 and 337 GPa, respectively. As compared with CrN or VN coatings, multilayer coatings exhibit an enhanced hardness of approximately 25.2 GPa. However, there is no obvious relationship between hardness and the CrN/VN bilayer thickness (from 7 to 27 nm). The elastic modulus of CrN/VN multilayer coating decreases with decreasing bilayer thickness owing to the decrease in VN layer thickness. Compared to the lower hardness of single-layer CrN (H=16.7 GPa), the maximum interfacial repulsive shear stress enhancement caused by modulus difference in multilayers can be calculated according to the difference in elastic modulus proposed by Koehler [185].

$$\Delta H_{\max} = 3RG_B \sin\theta / 8\pi m \tag{3.1}$$

where $R = (G_A - G_B/(G_A + G_B)$, G_A and G_B are the shear moduli of materials A and B, respectively, with G_A being larger than G_B. The θ is the smallest angle between the interface and the glide planes

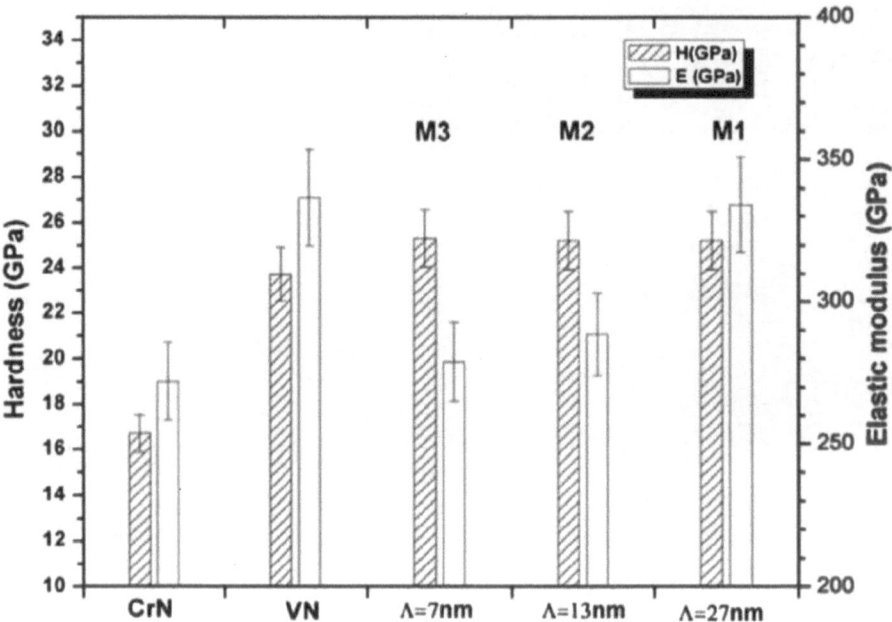

FIGURE 3.13 The hardness and elastic modulus of CrN and VN monolayers and CrN/VN multilayer with various bilayer wavelengths [146].

of crystal with the smaller elastic constants. The Taylor factor, m, is 0.3 for transition metal nitride [186], and $G = E/2(1 + v)$ (E is the elastic modulus and v is the Poisson's ratio). Using the data obtained from the single-layer films, E_{CrN}=272 GPa and E_{VN}=337 GPa, the Poisson's ratio is 0.28 for CrN and 0.279 for VN, and θ is 45° for multilayers, the shear modulus values of CrN and VN can be calculated to 106 and 131.6 GPa, respectively [146].

A hardness enhancement of 3.2 GPa, as compared with CrN coating, can be achieved, which is a little bit lower than the measured enhancement of 8.5 GPa (25.2 GPa-16.7 GPa). In addition to the Koehler's theory, the alternating-thermal stress strengthening mechanism [187] also influences the hardness of CrN/VN coatings. Since there is a heat process during coating deposition, the effect of heat-induced alternating stress fields on hardness cannot be ignored. The difference of thermal expansion coefficient of CrN (α_1=2.3 x 10^{-6} K^{-1} [3]) and VN (α_2=8.1 x 10^{-6} K^{-1} [188]) is large enough to cause thermal stress at the interfaces. The mechanical stresses of monolayer can be expressed as follows [189]:

$$\sigma_1 = E_1 E_2 d_2 \left(\alpha_2 - \alpha_1 \right) \Delta T \, / \left[\left(1 - v_1 \right) E_2 d_2 + \left(1 - v_2 \right) E_1 d_1 \right] \tag{3.2}$$

$$\sigma_2 = -E_1 E_2 d_1 \left(\alpha_2 - \alpha_1 \right) \Delta T \, / \left[\left(1 - v_1 \right) E_2 d_2 + \left(1 - v_2 \right) E_1 d_1 \right] \tag{3.3}$$

where σ_1 and σ_2 are the residual stresses in CrN layers and VN layers, respectively. E_1 and E_2 are Young's modulus, d_1 and d_2 are thickness, v_1 and v_2 are Poisson's ratio for the CrN layers and VN layers, respectively. And the calculated σ_1 is about 0.18 GPa, σ_2 is about −0.54 GPa (ΔT=277 K). Therefore, there are alternating stress fields in the interface, which will also induce the hardness enhancement in CrN/VN multilayer coating.

In addition, the enhanced wear resistance in multilayer coatings has been reported to be attributed to many factors such as high hardness, formation of tribo-oxidation on the worn surface, and decreases in the COF [3]. For instance, COF versus the sliding distance of the CrN/VN multilayer coatings is plotted in Figure 3.14 [146]. The values of COF for multilayer coatings are lower than that of coatings with monolayer structure. For multilayer coatings, the friction behavior can be divided into three distinguished stages. Initial "running in" stage (stage 1) with an increase of COF is resulted from the contact stress variations and the rough surface. Then COF becomes stable at sliding range from 70 to 370 m (stage 2). Compared with the friction coefficient of CrN coating (0.4) and VN coating (0.33), the CrN/VN multilayer coating revealed lower friction coefficient of about 0.21 in this stage. During sliding test, H_2O will react with coating debris and form metal oxides.

The combination of CrN and WN layers to form a CrN/WN multilayer by dual-gun magnetron sputtering has also been realized to enhance the hardness. Figure 3.15 shows the TEM micrographs of the CrN/WN multilayer coating with a bilayer period around 24 nm and a total thickness approximately 1 μm [190]. Clear interfaces between CrN and WN layers can be observed. The smooth interfaces also indicate strong coherency and limited flaws between layers. One can also find that the CrN crystal in CrN layer, as illustrated in the middle of Figure 3.15(b), is contained within adjacent WN layers. The CrWN composite coating is also deposited and investigated in this study for comparison. Using nanoindentation technique, the hardness for the CrN/WN and CrWN films are evaluated as 30.5 and 22.3 GPa, respectively. The confinement of grain growth and intact interfaces are argued to be the cause of hardness enhancement. The deformation mechanism involving disclination dipole motion along grain boundaries and grain boundary sliding and stress-induced mass transfer, causing significant plastic flow in nanocrystalline materials has been proposed by Ovid'ko in 2002 [191]. In contrast, the nanolayered structural feature and intact interface lower the feasibility of crystal grains to deform in the multilayer configuration. Further studies on CrN/WN multilayer coatings using ion beam assisted deposition and bilayer period design of 8 and 30 nm

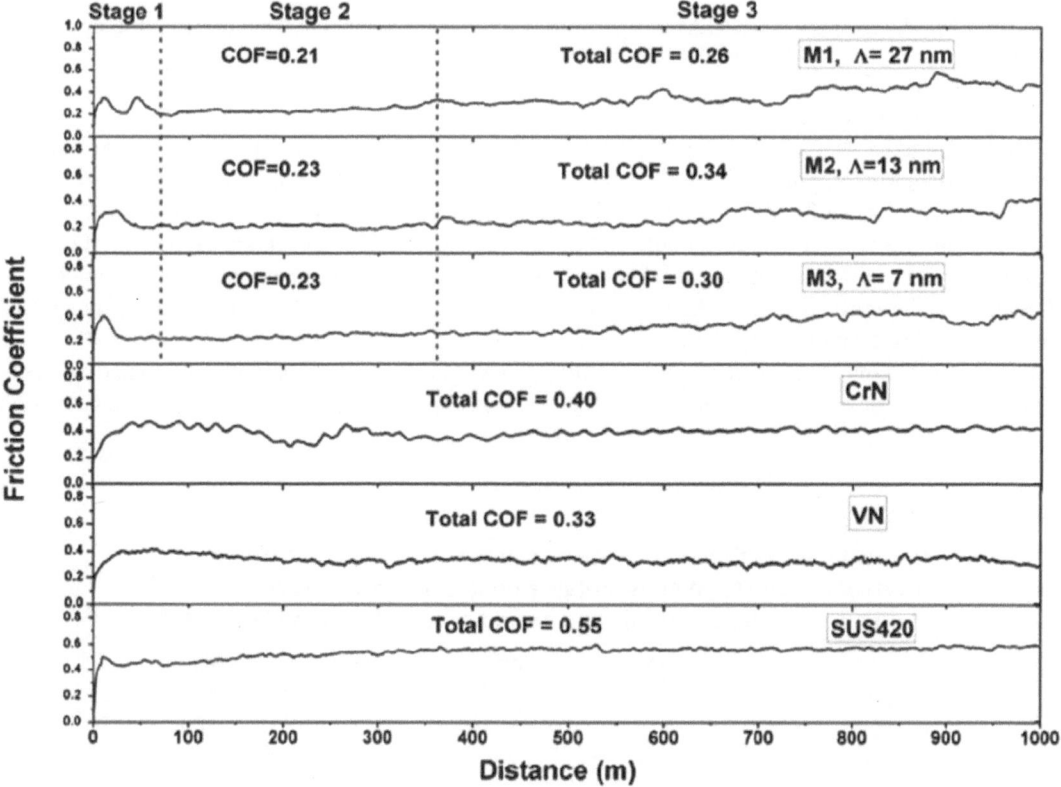

FIGURE 3.14 Coefficient of Friction, COF, of CrN and VN monolayers and CrN/VN multilayer with various bilayer wavelengths [146].

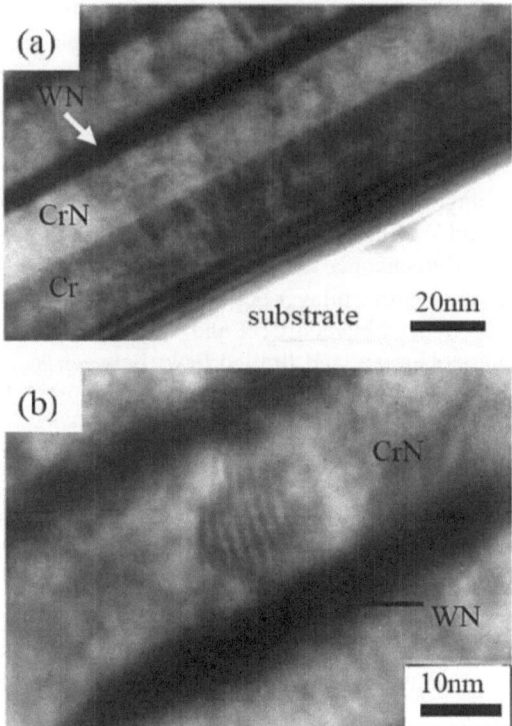

FIGURE 3.15 The cross-sectional TEM images of the CrN/WN multilayer coating (a) near substrate region, and (b) enlarged CrN crystal between WN layers [190].

are put into practice [192, 193]. The single CrN and WN single layers are fabricated and evaluated. The nanoindentation hardness of the CrN, WN single layers, and 8, and 30 nm bilayer period CrN/WN layers are 17.4, 24.3, 28.6, and 26.3 GPa. The hardness improvement by nanolayered configuration is again manifested. Furthermore, higher H/E ratios, which represents a higher resistance to wear, and lower surface roughness and friction coefficients, which are also index for as depicted in Table 3.1 are the reasons why multilayer coatings possess a superior mechanical performances as compared to single-layer TMN films.

The strengthening mechanism of the multilayer TMN coatings can be well elucidated by Tsai and coworkers' researches concerning CrAlSiN/W$_2$N multilayer systems [174, 194]. The CrAlSiN

TABLE 3.1
The surface roughness, hardness, Young's modulus, and friction coefficient of CrN, WN monolayer coatings and CrN/WN multilayer coatings with bilayer period of 8 and 30 nm [192, 193]

	CrN	WN	MX	M30
Ra (nm)	.68	5.88	1.90	2.22
H (GPa)	17.4	24.3	28.6	26.3
E (GPa)	237	274	300	247
H/E ratio	0.073	0.087	0.095	0.106
Friction coefficient (μ)	0.63	0.52	0.42	0.31

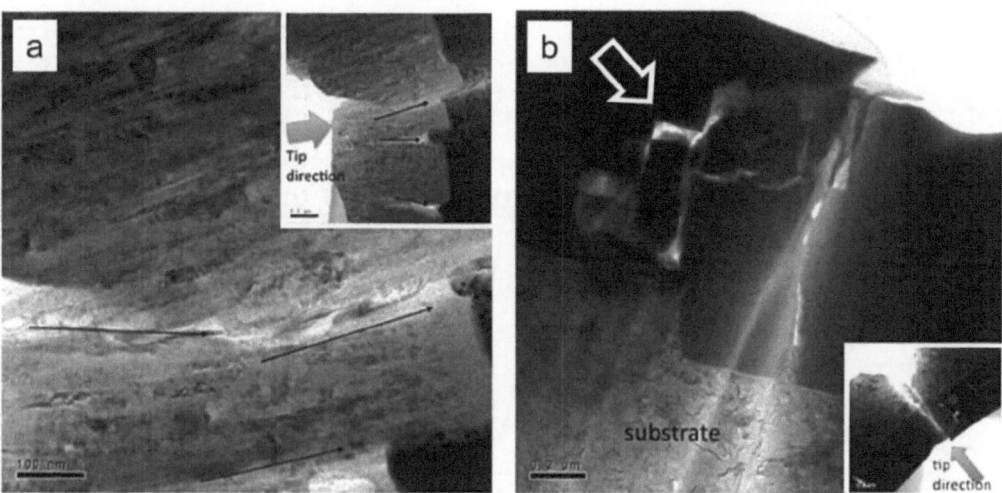

FIGURE 3.16 The cross-sectional TEM images of (a) CrAlSiN monolayer, and (b) CrAlSiN/W$_2$N multilayer coatings showing cracking behavior after wear test [194].

multicomponent coating exhibits a hardness of 25.4 GPa, while the CrAlSiN/W$_2$N multilayer films with modulation periods of 3, 8, and 20 nm show superior hardness values of 30.2, 40.3, and 29.9 GPa, respectively. The H/E values of the CrAlSiN/W$_2$N multilayer films are around 0.102 to 0.108, which are 20% higher than that of the CrAlSiN monolayer film. After wear test, the cross-sectional TEM images of the damaged CrAlSiN monolayer and CrAlSiN/W$_2$N multilayer films are analyzed, as illustrated in Figure 3.16. The monolayer shows a significant columnar structure and the easy path for crack propagation initiated from the contact point in surface along columnar grain boundaries is resulted. In the case of the CrAlSiN/W$_2$N multilayer, the wear cracks present a zigzag or stair-like pattern. The nanolayered configuration is thus active for crack deflection, energy dissipation, and stress relaxation, as predicted in Figure 3.11 [9] for toughening of protective coating using multilayer feature.

The thermal stability and oxidation resistance of the multilayer TMN coating are also focus points for protection applications. CrN/AlN and TiN/AlN multilayer coatings are investigated. Tien and coworkers fabricated CrN single and CrN/AlN multilayer coatings and looked into their thermal stability up to 900°C in ambient air for an hour [195]. Figure 3.17 shows the X-ray phase identification results of the annealed CrN single-layer film and CrN/AlN multilayer with modulation periods of 4 and 20 nm. The single-layer CrN film oxidizes at around 600°C, while the 20 nm-modulated CrN/AlN multilayer coating exhibits and onset oxidation temperature of 800°C. With a smaller modulation periods of 4 nm, the X-ray diffraction patterns do not show any evidence of oxidation for CrN/AlN multilayer even at 900°C. The microstructural study shows that there is a relatively thin, smooth, and continuous oxidation layer formed at the surface of the multilayer coatings, as illustrated in Figure 3.18. The dense oxide layer, which is identified as Al-rich oxide, i.e. $(Al_xCr_{1-x})_2O_3$ [196], on top of the CrN/AlN multilayer coatings prevents further oxidation into the films. The indentation hardness of the CrN single and CrN/AlN multilayer coatings are also evaluated to figure out the mechanical strength after high-temperature annealing. The hardness as a function of heat-treated temperature is plotted in Figure 3.19. The CrN single layer has a stable but relatively lower hardness value around 20 GPa till 800°C. After 850°C annealing, the hardness drops to 18 GPa. On the other hand, the CrN/AlN multilayer coatings present higher and stable hardness values from 25 to 28 GPa up to 850°C. And a highest hardness is observed for the CrN/AlN multilayer coating with a modulation period of 10 nm. This highest hardness is also an echo toward prediction in literatures [133].

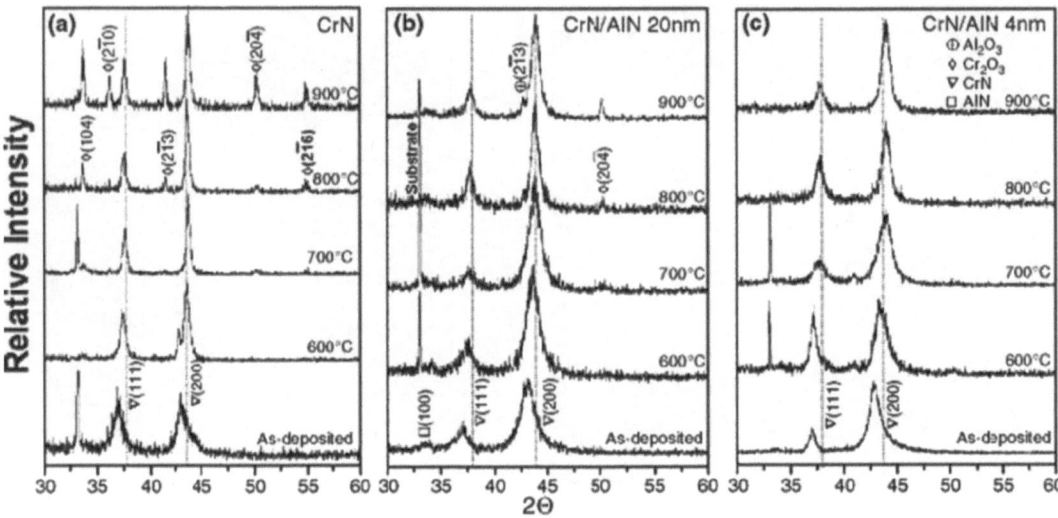

FIGURE 3.17 The X-ay diffraction patterns of (a) CrN single layer, (b) CrN/AlN 20 nm-, and (c) CrN/AlN/ 4 nm-modulated multilayer coatings in as-deposited and annealed states [195].

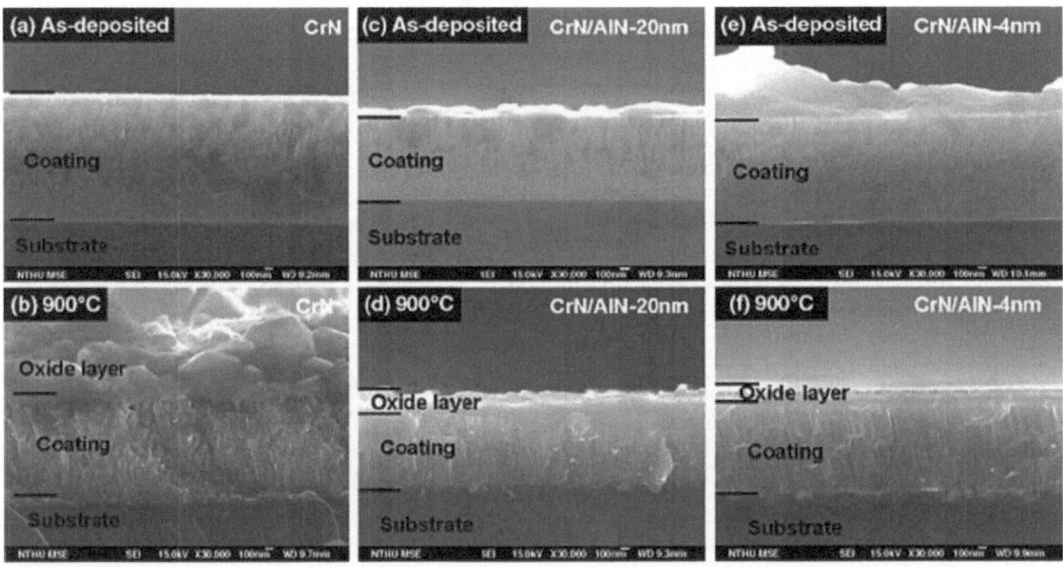

FIGURE 3.18 The cross-sectional SEM micrographs of (a) as-deposited and (b) 900°C annealed CrN single layer, and (c) as-deposited and (d) 900°C annealed CrN/AlN 20 nm-modulated, and (e) as-deposited and (f) 900°C annealed CrN/AlN 4 nm-modulated multilayer coatings [195].

Despite two kinds of TMNs combined to establish a nanolayered film, the silicon nitride films are frequently chosen as one of the building layers to form TMN/SiN multilayer protective coating systems. SiN is not only easy to fabricate but also structurally stable, meaning that intact interfaces with distinguishable microstructural difference can be produced for the multilayer systems. The thin SiN_x building layers have been integrated into TiAlN [178, 180], CrAlN [157, 197], CrMoN [198]…etc. to form multilayer assemblies. In Chan's reports, the designed bilayer period thickness of 4.4 and 5.0 nm in the TiAlN/SiN_x multilayer films with relatively thin SiN_x layer of 0.4 and 1.0 nm, respectively, are demonstrated. Figure 3.7 indicates cross-sectional images of the 4.4 and

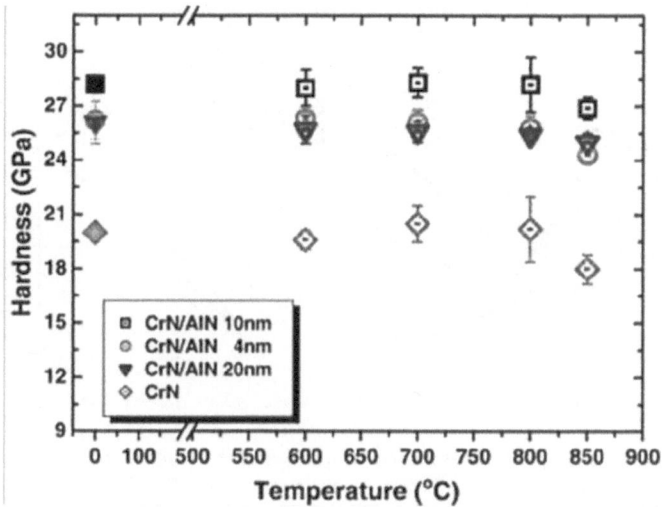

FIGURE 3.19 The hardness of CrN single layer and 4, 10, 20 nm-modulated CrN/AlN multilayer coatings after annealing [196].

5.0 nm-modulated TiAlN/SiN$_x$ multilayers. The 0.4 nm SiN building layers grow with a structure coherency with TiAlN layers. The 1.0 nm SiN$_x$ building layers turn into amorphous and interrupt the continuous crystalline growth of TiAlN layers, as shown in Figure 3.7(b). Through nanoindentation analysis, the hardness values for single-layer TiAlN and 4.4 and 5.0 nm-modulated TiAlN/SiN$_x$ multilayers are determined to be 34.4, 44.8, and 26.1 GPa, respectively [178, 180]. It is persuaded that the significant increase of coating hardness from 30.5 to 42.1 GPa by the introduction of 0.4 nm SiN$_x$ building layers is attributed to the structural coherency, i.e. the superlattice of TiAlN/SiN$_x$. Further increase of SiN thickness to 1.0 nm, the SiN$_x$ layers evolve into amorphous and the TiAlN/SiN$_x$ films lose coherent interfaces, leading to the appreciable decrease in hardness to 26.3 GPa. The elastic modulus, H/E ratio, and wear resistance of the TiAlN and TiAlN/SiN$_x$ films are investigated for comparison, as summarized in Table 3.2. The H/E ratios of the TiAlN/SiN$_x$ multilayer films are about 10–20% higher than that of TiAlN monolayer film, which is the indication of the enhanced wear resistance (approximately one third lower in wear rate). It should be noted that the elastic modulus of the 5 nm-modulated TiAlN/SiN$_x$ multilayer film has a relatively low value of 241.7 as compared to those of TiAlN monolayer and 4.4 nm-modulated TiAlN/SiN$_x$ multilayer film, 378 and 393, respectively. It is believed that the amorphous SiN$_x$ layer of 1.0 nm in the 5 nm-modulated TiAlN/SiN$_x$ multilayer film is soft and less stiff, causing a lower hardness and elastic moduli. Nevertheless, its H/E remains a higher value and it keeps a good performance in wear resistance.

TABLE 3.2

The hardness, elastic modulus, H/E ratio, and wear rate of single-layer TiAlN and TiAlN/SiN$_x$ multilayers [178, 180]

Sample	TiAlN	4.4 nm TiAlN/SiN$_x$	5.0 nm TiAlN/SiN$_x$
Hardness (H), GPa	34.4±1.1	44.8±0.7	26.1±1.2
Elastic Modulus (E), GPa	378.0±13.8	393.0±11.2	241.7±9.1
H/E ratio	0.091	0.114	0.108
Wear rate (W_R):(mm³/N/m)	$2.2*10^{-6}$	$9.1*10^{-7}$	$9.0*10^{-7}$

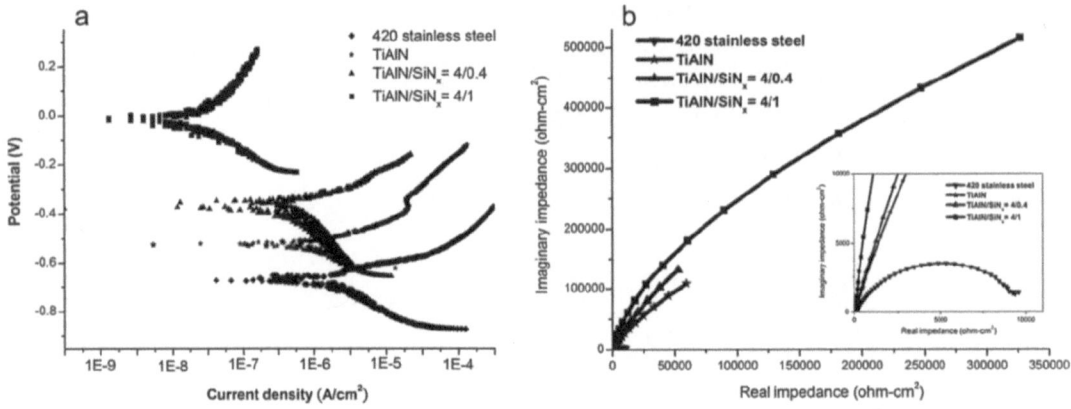

FIGURE 3.20 The (a) potentiodynamic polarization plots and (b) Nyquist plots, of substrate SS420, TiAlN monolayer, 4.4 and 5.0 nm-modulated TiAlN/SiN$_x$ multilayers [178, 180].

The corrosion characteristics of the TiAlN monolayer and TiAlN/SiN$_x$ multilayer coatings are also inspected in a 3.5 wt % NaCl aqueous solution under free air condition. Figure 3.20 depicts the corrosion potential-exchanging current and imaginary-real parts of impedance plots of the SS420 substrate, TiAlN monolayer, and 4.4 and 5.0 nm-modulated TiAlN/SiN$_x$ multilayer coatings. In Figure 3.20(a), the least non-negative potential and smallest exchanging current are evaluated for the 5.0 nm-modulated TiAlN/SiN$_x$ multilayer film, then the 4.4 nm-modulated TiAlN/SiN$_x$ multilayer, TiAlN monolayer, and SS420 substrate in sequence. The corrosion protection through TiAlN/SiN$_x$ multilayer assemblies is evident. It is worth noting that the 1.0 nm SiN$_x$ building layer with an amorphous microstructure retards chemical attack in potentiodynamic scanning more efficient than the 0.4 nm SiN$_x$ building layer does, as compared in Figure 3.20(a). It is believed that the amorphous SiN$_x$ layers in the 5.0 nm-modulated TiAlN/SiN$_x$ multilayer work against the through-thickness infiltration of the chemical solution better than the 4.4 nm-modulated TiAlN/SiN$_x$ multilayer with a superlattice configuration as a whole. This point is further demonstrated through Nyquist plot in Figure 3.20(b). The wider imaginary–real part impedance curve of the 5.0 nm-modulated TiAlN/SiN$_x$ multilayer with integrated amorphous SiN$_x$ layers implies a stronger resistance during corrosion. Serial studies concerning the effect of thin amorphous SiN$_x$ layers on the overall structural features of the TMN multilayer coatings, including CrAlN/SiN$_x$ [157, 197] and CrMoN/SiN$_x$ [198] films are conducted. With a single SiN$_x$ building layer thickness larger than 0.7 to 1.0 nm, the SiN$_x$ layer is amorphous and thus the CrAlN/SiN$_x$ and CrMoN/SiN$_x$ films exhibit refined grain size down to a few nanometers depending on the thickness of the accompanied single TMN building layer. On the contrary, the grain sizes of the superlattice TMN/SiN$_x$ multilayers can grow to an average size of 24–28 nm. To summarize, the thinner sub-nanometer SiN$_x$ building layer grows coherently with TMN layer into a superlattice TMN/SiN$_x$ multilayer and leads to a superior hardness, while the thicker amorphous SiN$_x$ building layer triggers a discontinuity in crystal growth and is helpful for the enhancement in corrosion resistance.

The insertion of SiN$_x$ layer into TMN to form multilayer protective coatings can also frequently show a Si$_3$N$_4$ phase and leads to hard and tough characteristics simultaneously. The CrN/Si$_3$N$_4$ multilayer coating has been recently reported [199]. The CrN/Si$_3$N$_4$ multilayer coating is bent to a significant deflection under mechanical forces without any delamination against substrate and cracking within coatings. The Si$_3$N$_4$ building layer is often in an amorphous state and does not present a high hardness like TMN film. Nevertheless, it can be easily codeposited and integrated in multicomponent and multilayer hard coating systems, and most importantly it shows an amorphous phase alone and works with the other TMN phase. Thus a multilayer composed by TMN/Si$_3$N$_4$

layered assemblies can be as hard to resist deformation and at the same time be as tough to avoid failure. The addition of amorphous Si_3N_4, known as a-Si_3N_4, layers with crystalline TMN layers is now attracting attentions and resultant superior characteristics for the TMN/a-Si_3N_4 can be realized [82, 187, 200–202].

The recent innovation in multilayer nitride coatings is mainly on the material choices, microstructure control, stacking configuration, including modulation and interface integration, and related characterization of properties for specific utilities. Multicomponent TMN layers is integrated [203–205] to provide thermal stability, mechanical strength, chemical inertness, …and so on, as described in the last part of Section 3.2. One of the building layer with multiple elements is utilized using the concept of nanocomposite, with which a complex nature of the layers consisting of a mixture of nitrides of constituting elements is expected. For example, the (TiZrNbTaHf)N/MoN multilayers are fabricated with vacuum-arc deposition technique with bias variation [203]. The highest values of 487 GPa in Young's modulus and 29 GPa in hardness could be obtained under variation bias from -100 to -300V during fabrication. The hardening mechanism comes from the solid-solutioning strengthening in the multicomponent nitride, (TiZrNbTaHf) N, layer and the modulation of multilayer at a 15nm/5nm:(TiZrNbTaHf)N/MoN thickness ratio. It is concluded that the multilayer system forms (TiZrNbTaHf) N and Mo_2N phases and their orientation changes from (111) and (311) to less packed (200) as substrate bias is more negative. The distinguishable and intact interface between (TiZrNbTaHf)N and Mo_2N building layers to hold up dislocation propagation and to confine the grain size of each nitride layer are one of the key reasons to mechanical enhancement.

The TMN multilayer protective coatings advance with the development of the single-layer TMN coating because of the benefits not only from the intrinsic properties of the selected building layers but also from the combined effects by the layered configuration. That is to say whenever a new type of nitride coating is generated, like HEA nitride, MG nitride, and so on, the employment of the new coating into a multilayer system becomes an almost must-try issue. However, the well-defined and widely used single-element TMN films, such as TiN, CrN, VN, WN, TaN, MoN… etc., are still attracting researchers for advanced configuration and different field of use. One would notice that there exists a large number of real recent studies working on such simple nitride multilayer systems, including Ti/TiN [206], V/VN [207], Hf/HfN [208], TiN/WN [209], CrN/NbN [210], CrN/AlN [211], and other dual or more elements nitride multilayer coatings [150, 212–220]. The design philosophy is simply to deduce film properties or performances that one single layer cannot provide. Hopefully the multilayer feature can bring even superior characteristics by the effects of sophisticated layered integration, like superlattice, crystalline/amorphous interfaces, layer coherency, stabilization of layer materials…etc. [21]. Nevertheless, those who may concern need to identify the environment the protective multilayer coatings will be facing before working on materials selection and layer deposition design.

3.6 CONCLUSIVE REMARKS AND OUTLOOK

The TMN coating has its own characteristics and advantages in fabrication and performance in plenty fields. With a layered microstructure feature, the multilayer TMN films provoke required properties which one single building layer cannot have. One of the top selling desserts in sweet house is the multilayer cake, as shown in Figure 3.21, which is always the excellent choice in your afternoon teatime. The compiled layer structure makes it possible for a gourmand not just to taste two ingredients at the same time but also a texture of crispy and soft when one takes his/her bites. The interacted flavors of the distinct building layers, for example, Mocha chocolate and strawberry, brownie and Toffee, and most of the times just secret recipes of the baker, would take you right to the paradise of fragrance as you enjoy it. The multilayer TMN coatings, likewise, integrate merits of the building layers and are benefited from the tailored microstructure and the resultant interfaces.

The challenges toward protective coatings are highly application-oriented, meaning field tests are required for a thorough evaluation, especially for the use in complicated environments. For

FIGURE 3.21 The multilayer cake. The sliced multilayer cake is compiled with sequentially altered layers made of egg, flour, unsalted butter, caster sugar, and secret ingredients and baked with process know-how. The mixture of brown thin sweet and golden egg-scent layers touch joyful sense as one takes his/her bites.

characterization, the research-oriented investigations rely on in-house setups and/or optimistically industrial standards. The real problems may not be reflected under such kinds of evaluation. In-situ monitoring of the coating performance during practice would be the best way to honestly interpret the real time conditions. For instance, the thinning of protective coatings can be detected by their changes in electric resistance and cracking/delamination of the tribolayers can be noticed through acoustic or frictional bursts. Smart coating systems, which exhibit multiple functional properties and performance, including but not limited to self-healing, self-cleaning, anticorrosion…etc. are nowadays recognized and attractive for protection applications. Owing to the tailored configuration, the multilayer coatings possess a great chance to integrate and to benefit from functional building layers with various characteristics at the same time. And to select suitable materials, to fabricate with proper microstructure, to control adequate properties, and to incorporate functionalities of building layers of the multilayer coatings are the key issues for developing the next generation protective coatings.

REFERENCES

[1] P.H. Mayrhofer, R. Rachbauer, D. Holec, F. Rovere, and J.M. Schneider, "Protective transition metal nitride coatings", *Comp. Mate. Proc.*, 4, (2014) 355–388.
[2] K. Holmberg, A. Matthews, "Coatings Tribology: Properties, Techniques and Applications in Surface Engineering", Ed. by K. Holmberg, A. Matthews, *Chapter 7 Applications, Tribology Series*, 28 (1994) 335–388.
[3] S. Zhang, *"Nanostructured Thin Films and Coatings"*, CRC Press, Boca Raton, (2010).
[4] L. Wu, X. Guo and J. Zhang, "Abrasive resistant coatings: A review", *Lubricants* 2 (2014) 66–89.
[5] H. Holleck, "Material selection for hard coatings", *J. Vac. Sci. Technol. A* 4 (6) (1986) 2661–2669.
[6] O. Knotek, F. Löffler, G. Krämer, "Multicomponent and multilayer physically vapour deposited coatings for cutting tools", *Surf. Coat. Technol.* 54–55 (1992) 241–248.
[7] C. Subramanian, K.N. Strafford, "Review of multicomponent and multilayer coatings for tribological applications", *Wear* 165 (1993) 85–95.
[8] S.A. Barnett, M. Shinn, "Plastic and elastic properties of compositionally modulated thin films", *Annu. Rev. Mater. Sci.* 24 (1994) 481–511.
[9] H. Holleck, V. Schier, "Multilayer PVD coatings for wear protection", *Surf. Coat. Technol.* 76–77 (1995) 328–336.
[10] S.J. Bull, A.M. Jones, "Multilayer coatings for improved performance", *Surf. Coat. Technol.* 78 (1996) 173–184.
[11] P.C. Yashar, W.D. Sproul, "Nanometer scale multilayered hard coatings", *Vacuum* 55 (1999) 179–190.

[12] H. Holleck, in: A. Kumar, Y.-W. Chung, J.J. Moore, J.E. Smugeresky (Eds.), "Surface Engineering: Science and Technology I", *The Minerals, Metals & Materials Society*, Warrendale, Pennsylvania, (1999) 207–218.

[13] A.K. Sikder, A. Kumar, "Superhard Cocatings in C-B-N Systems: Growth and Characterization", Chapter. 3 *Handbook of Thin Film Materials*, Ed. H.S. Nalwa, Volume 2: Characterization and Spectroscopy of Thin Films, San Diego: Academic Press (2002).

[14] S.J. Lloyd, J.M. Molina-Aldareguia, "Multilayered materials: A palette for the materials artist", *Philos. Trans. Royal. Soc. A* 361 (2003) 2931–2949.

[15] S.A. Barnett, A. Madan, I. Kim, K. Martin, "Stability of nanometer-thick layers in hard coatings", *MRS Bull.* 28 (2003) 169–172.

[16] W.-D. Münz, "Large-scale manufacturing of nanoscale multilayered hard coatings deposited by cathodic arc/unbalanced magnetron sputtering", *MRS Bull.* 28 (2003) 173–179.

[17] S.A. Barnett, A. Madan, "Hardness and stability of metal-nitride nanoscale multilayers", *Scr. Mater.* 50 (2004) 739–744.

[18] A.A. Voevodin, J.S. Zabinski, C. Muratore, "Recent advances in hard, tough, and low friction nanocomposite coatings", *Tsinghua Sci. Technol.*, 10 6 (2005) 665–679. ISSN 1007-0214 02/11.

[19] L. Hultman, in: A. Cavaleiro, J.Th.M. De Hosson (Eds.), *"Nanostructured Coatings"*, Springer, New York, 2006, pp. 539–554.

[20] G. Abadias, A. Michel, C. Tromas, C. Jaouen, S.N. Dub, "Stress, interfacial effects and mechanical properties of nanoscale multilayered coatings", *Surf. Coat. Technol.* 202 (2007) 844–853.

[21] M. Stüber, H. Holleck, H. Leiste, K. Seemann, S. Ulrich, C. Ziebert, "Concepts for the design of advanced nanoscale PVD multilayer protective thin films", *J. Alloys Compd.* 483 (2009) 321–333.

[22] B. Navinsek and Sudipta Seal, "Transition metal nitride functional coatings", *JOM* 53 (2001) 51–54.

[23] P.H. Mayrhofer, C. Mitterer, H. Clemens, "Self-organized nanostructures in hard ceramic coatings", *Adv. Eng. Mater.* 7 12 (2005) 1071.

[24] H. Holleck, "Basic principles of specific applications of ceramic materials as protective layers", *Surf. Coat. Technol.* 43–44 (1990) 245–258.

[25] H. Holleck, "Metastable coatings - Prediction of composition and structure", *Surf. Coat. Technol.* 36 1–2 (1988) 151–159.

[26] O. Knotek, M. Bohmer, T. Leyendecker, "On structure and properties of sputtered Ti and Al based hard compound films", *J. Vac. Sci. Technol. A*, 4 (1986) 2695.

[27] W.-D. Münz, "Titanium aluminum nitride films: A new alternative to TiN coatings", *J. Vac. Sci. Technol. A* 4 (1986) 2717.

[28] G. Beensh-Marchwick, L. Kròl-Stępniewsk, W. Posadowski, "Structure of thin films prepared by the cosputtering of titanium and aluminium or titanium and silicon", *Thin Solid Films* 82 (1981) 313–320.

[29] H.A. Jehn, S. Hofmann, V.E. Rückborn, W.-D. Münz, "Morphology and properties of sputtered (Ti, Al) N layers on high speed steel substrates as a function of deposition temperature and sputtering atmosphere", *J. Vac. Sci. Technol. A*, 4 6 (1986) 2701.

[30] H. A. Jehn, S. Hofmann, W.-D. Münz, "Surface and interface characterization of heat- treated (Ti, Al)N coatings on high speed steel substrates", *Thin Solid Films* 153 1–3 (1987) 45–53.

[31] D. McIntyre, J. E. Greene, G. Håkansson, J.-E. Sundgren, W.-D. Münz, "Oxidation of metastable single-phase polycrystalline Ti0.5Al0.5N films: Kinetics and mechanisms", *J. Appl. Phys.* 67 3 (1990) 1542.

[32] P. H. Mayrhofer, D. Music, J. M. Schneider, "Influence of the Al distribution on the structure, elastic properties, and phase stability of supersaturated $Ti_{1-x}Al_xN$", *J. Appl. Phys.* 100 9 (2006) 094906.

[33] R. Rachbauer, S. Massl, E. Stergar, D. Holec, D. Kiener, J. Keckes, J. Patscheider, M. Stiefel, H. Leitner, and P. H. Mayrhofer, "Decomposition pathways in age hardening of Ti-Al-N films", *J. Appl. Phys.* 110 (2011) 023515.

[34] A. E. Reiter, V. H. Derflinger, B. Hanselmann, T. Bachmann, B. Sartory, "Investigation of the properties of Al1-xCrxN coatings prepared by cathodic arc evaporation", *Surf. Coat. Technol.* 200 (2005) 2114–2122.

[35] H. Hasegawa, T. Suzuki, "Effects of second metal contents on microstructure and micro-hardness of ternary nitride films synthesized by cathodic arc method", *Surf. Coat. Technol.* 188–189 (2004) 234–240.

[36] A. Kimura, H. Hasegawa, K. Yamada, T. Suzuki, "Effects of Al content on hardness, lattice parameter and microstructure of $Ti_{1-x}Al_xN$ films", *Surf. Coat. Technol.* 120–121 (1999) 438–s441.

[37] B. Y. Man, L. Guzman, A. Miotello, M. Adami, "Microstructure, oxidation and H2-permeation resistance of TiAlN films deposited by DC magnetron sputtering technique", *Surf. Coat. Technol.* 180–181 (2004) 9–14.

[38] L. Chen, J. Paulitsch, Y. Du, P. H. Mayrhofer, "Thermal stability and oxidation resistance of Ti-Al-N coatings", *Surf. Coat. Technol.* 206 (2012) 2954–2960.

[39] S. PalDay, S.C. Deevi, "Single layer and multilayer wear resistant coatings of (Ti,Al)N: A review", *Mater. Sci. Eng. A* 342 (2003) 58–79.

[40] P.H. Mayrhofer, A. Hörling, L. Karlsson, J. Sjölén, T. Larsson, C. Mitterer, L. Hultman, "Self-organized nanostructures in the Ti-Al-N system", *Appl. Phys. Lett.* 83 10 (2003) 2049.

[41] A. Hörling, L. Hultman, M. Oden, J. Sjölén, L. Karlsson, "Thermal stability of arc evaporated high aluminum-content $Ti_{1-x}Al_xN$ thin films", *J. Vac. Sci. Technol. A* 20 (2002) 1815.

[42] P.H. Mayrhofer, L. Hultman, J.M. Schneider, P. Staron, H. Clemens, "Spinodal decomposition of cubic $Ti_{1-x}Al_xN$: Comparison between experiments and modeling", *Int. J. Mater. Res.* 11 (2007) 1054.

[43] C.T. Huang, J.G. Duh, "Deposition of (Ti,Al)N films on A2 tool steel by reactive r.f. magnetron sputtering", *Surf. Coat. Technol.* 71 (1995) 259–266.

[44] G.T. Liu, J.G. Duh, K.H. Chung, and J.H. Wang, "Mechanical characteristics and corrosion behavior of (Ti,Al)N coatings on dental alloys", *Surf. Coat. Technol.* 200 (2005) 2100–2105.

[45] O. Knotek, F. Loffler, H.J. Scholl, "Properties of arc-evaporated CrN and (Cr, Al)N coatings", *Surf. Coat. Technol.* 45 (1991) 53–58.

[46] A. Sugishima, H. Kajioka, Y. Makino, "Phase transition of pseudobinary Cr-Al-N films deposited by magnetron sputtering method", *Surf. Coat. Technol.* 97 (1997) 590–594.

[47] Y. Makino, K. Nogi, "Synthesis of pseudobinary Cr-Al-N films with B1 structure by rf-assisted magnetron sputtering method", *Surf. Coat. Technol.* 98 (1998) 1008–1012.

[48] M. Kawate, A. Kimura, T. Suzuki, "Microhardness and lattice parameter of $Cr_{1-x}Al_xN$ films", *J. Vac. Sci. Technol. A* 20 (2002) 569–571.

[49] R. Kaindl, R. Franz, J. Soldan, A. Reiter, P. Polcik, C. Mitterer, B. Sartory, R. Tessadri, M. O'Sullivan, "Structural investigations of aluminum-chromium-nitride hard coatings by Raman micro-spectroscopy", *Thin Solid Films* 515 (2006) 2197–2202.

[50] R.F. Zhang, S. Veprek, "Phase stabilities and spinodal decomposition in the $Cr_{1-x}Al_xN$ system studied by ab initio LDA and thermodynamic modeling: Comparison with the $Ti_{1-x}Al_xN$ and TiN/Si_3N_4 systems", *Acta Mater.* 55 (2007) 4615–4624.

[51] B. Alling, T. Marten, I.A. Abrikosov, A. Karimi, "Comparison of thermodynamic properties of cubic $Cr_{1-x}Al_xN$ and $Ti_{1-x}Al_xN$ from first-principles calculations", *J. Appl. Phys.* 102 (2007) 044314.

[52] P.H. Mayrhofer, D. Music, T. Reeswinkel, H.G. Fuß, J.M. Schneider, "Structure, elastic properties and phase stability of $Cr_{1-x}Al_xN$", *Acta Mater.* 56 (2008) 2469–2475.

[53] K. Bobzin, E. Lugscheider, R. Nickel, N. Bagcivan, A. Kramer, "Wear behavior of $Cr_{1-x}Al_xN$ PVD-coatings in dry running conditions", *Wear* 263 (2007) 1274–1280.

[54] J. Lin, B Mishra, J.J. Moore, W.D. Sproul, "A study of the oxidation behavior of CrN and CrAlN thin films in air using DSC and TGA analyses", *Surf. Coat. Technol.* 202 (2008) 3272–3283.

[55] S. Hofmann, "Formation and diffusion properties of oxide films on metals and on nitride coatings studied with Auger electron spectroscopy and X-ray photoelectron spectroscopy", *Thin Solid Films* 193–194 (1990) 648–664.

[56] O. Banakh, P.E. Schmid, R. Sanjines, F. Levy, "High-temperature oxidation resistance of $Cr_{1-x}Al_xN$ thin films deposited by reactive magnetron sputtering", *Surf. Coat. Technol.* 2003, 163–164, 57–61.

[57] F. Rovere, P.H. Mayrhofer, A. Reinholdt, J. Mayer, J.M. Schneider, "The effect of yttrium incorporation on the oxidation resistance of Cr-Al-N coatings ", *Surf. Coat. Technol.* 202 (2008) 5870–5875.

[58] H. Hasegawa, M. Kawate, T. Suzuki, "Effects of Al contents on microstructures of $Cr_{1-x}Al_xN$ and $Zr_{1-x}Al_xN$ films synthesized by cathodic arc method", *Surf. Coat. Technol.* 200 (2005) 2409–2413.

[59] Y. Makino, M. Mori, S. Miyake, K. Saito, K. Asami, "Characterization of Zr-Al-N films synthesized by a magnetron sputtering method", *Surf. Coat. Technol.* 193 1–3 (2005) 219–222.

[60] R. Sanjinés, C.S. Sandu, R. Lamni, F. Lévy, "Thermal decomposition of $Zr_{1-x}Al_xN$ thin films deposited by magnetron sputtering", *Surf. Coat. Technol.* 200 22–23 (2006) 6308–6312.

[61] S.H. Sheng, R.F. Zhang, S. Veprek, "Phase stabilities and thermal decomposition in the $Zr_{1-x}Al_xN$ system studied by ab initio calculation and thermodynamic modeling", *Acta Mater.* 56 5 (2008) 968–976.

[62] D. Holec, R. Rachbauer, L. Chen, L. Wang, D. Luef, P.H. Mayrhofer, "Phase stability and alloy-related trends in Ti-Al-N, Zr-Al-N and Hf-Al-N systems from first principles", *Surf. Coat. Technol.* 206 (2011) 1698-1704.

[63] T.I. Selinder, D.J. Miller, K.E. Gray, M.R. Sardela, L. Hultman, "Phase formation and microstructure of $Nb_{1-x}Al_xN$ alloy films grown on MgO (001) by reactive sputtering: a new ternary phase", *Vacuum* 46 12 (1995) 1401–1406.

[64] D. Holec, R. Franz, P.H. Mayrhofer, C. Mitterer, "Structure and stability of phases within the NbN-AlN system", *J. Phys. D. Appl. Phys.* 43 14 (2010) 341.

[65] R. Franz, M. Lechthaler, P. Polcik, M. Rebelo de Figueiredo, C. Mitterer, "Tribological properties of arc-evaporated NbAlN hard coatings", *Tribol. Lett.*, 45 (2012) 143–152.

[66] M. Benkahoul, M.K. Zayed, C.S. Sandu, L. Martinu, J.E. Klemberg-Sapieha, "Structural, tribo-mechanical, and thermal properties of NbAlN coatings with various Al contents deposited by DC reactive magnetron sputtering", *Surf. Coat. Technol.* 331 (2017) 172–178.

[67] M. Arab Pour Yazdi, F. Lomello, J. Wang, F. Sanchette, Z. Dong, T. White, Y. Wouters, F. Schuster, A. Billard, "Properties of TiSiN coatings deposited by hybrid HiPIMS and pulsed-DC magnetron co-sputtering", *Vacuum*, 109 (2014) 43–51.

[68] L. Qiu, Y. Du, S. Wang, K. Li, L. Yin, L. Wu, Z. Zhong, and L. Albir, "Mechanical properties and oxidation resistance of chemically vapordeposited TiSiN nanocomposite coating with thermodynamically designed compositions", *Int. J. Refract. Met. Hard Mater.*, 80 (2019) 30–39.

[69] K.H. Kim, S.R. Choi, S.Y. Yoon, "Superhard Ti-Si-N coatings by a hybrid system of arc ion plating and sputtering techniques", *Surf. Coat. Technol.* 298 (2002) 243–248.

[70] J.H. Park, W.S. Chung, Y.R. Cho, K.H. Kim, "Synthesis and mechanical properties of Cr-Si-N ocatings deposited by a hybrid system of arc ion plating and sputtering techniques", *Surf. Coat. Technol.* 188-189 (2004) 425–430.

[71] E. Martines, R. Sanjine, A. Karimi, "Mechanical properties of nanocomposite and multilayered Cr-Si-N sputtered thin films", *Surf. Coat. Technol.* 180–181 (2004) 570–574.

[72] J.W. Lee, Y.C. Chang, "A study on the microstructures and mechanical properties of pulsed DC reactive magnetron sputtered Cr-Si-N nanocomposite coatings", *Surf. Coat. Technol.* 202 (2007) 831–836.

[73] S. Vepřek, S. Reiprich, "A concept for the design of novel superhard coatings", *Thin Solid Films* 268 (1995) 64–71.

[74] A. Flink, M. Beckers, J. Sjolen, T. Larsson, S. Braun, L. Karlsson, L. Hultman, "The location and effects of Si in $(Ti_{1-x}Si_x)N_y$ thin films", *J. Mater. Res.* 24 (2009) 2483–2498.

[75] G. Greczynski, J. Patscheider, J. Lu, B. Alling, A. Ektarawong, J. Jensen, I. Petrov, J.E. Greene, L. Hultman, "Control of $Ti_{1-x}Si_xN$ nanostructure via tunable metal-ion momentum transfer during HIPIMS/DCMS co-deposition", *Surf. Coat. Technol.* 280 (2015) 174–184.

[76] J.W. Kim, K.H. Kim, D.B. Lee, J.J. Moore, "Study on high-temperature oxidation behaviors of Cr-Si-N films", *Surf. Coat. Technol.* 200 (2006) 6702–6705.

[77] J. Lin, B. Wang, Y. Ou, W.D. Sproul, I. Dahan. J.J. Moore, "Structure and properties of Cr-Si-N nanocomposite coatings deposited by hybrid modulated pulse power and pulse dc magnetron sputtering", *Surf. Coat. Technol.* 216 (2013) 251–258.

[78] Z.X. Lin, Y.C. Liu, C.R. Huang, M. Guillon, F.B. Wu, "Input power effect on microstructure and mechanical properties of Mo-Si-N multilayer coatings", *Surf. Coat. Technol.*, 383 (2020) 125222.

[79] Y.C. Liu, B.H. Liang, C.R. Huang, F.B. Wu, "Microstructure evolution and mechanical behavior of Mo-Si-N films", *CoatingsTech* 10 (2020) 987.

[80] Y.I. Chen, Y.E. Ke, M.C. Sung, L.C. Chang, "Rapid thermal annealing of Cr-Si-N, Ta-Si-N, and Zr-Si-N coatings in glass molding atmospheres", *Surf. Coat. Technol.* 389 (2020) 125662..

[81] S.K. Kim, P.V. Vinh, J.H. Kim, T. Ngoc, "Deposition of superhard TiAlSiN thin films by cathodic arc plasma deposition", *Surf. Coat. Technol.* 200 (2005) 1391-1394.

[82] C.L. Chang, J.W. Lee, M.D. Tseng, "Microstructure, corrosion and tribological behaviors of TiAlSiN coatings deposited by cathodic arc plasma deposition", *Thin Solid Films*, 517 17 (2009) 5231–5236.

[83] Y.Y. Chang, S.M. Yang, "High temperature oxidation behavior of multicomponent TiAlSiN coatings", *Thin Solid Films* 518 21 (2010) S34–S37.

[84] Q. Ma, L. Li, Y. Xu, J. Gu, L. Wang, Y. Xu, "Effect of bias voltage on TiAlSiN nanocomposite coatings deposited by HiPIMS", *Appl. Surf. Sci.* 392 15 (2017) 826–833.

[85] S.K. Tien, C.H. Lin, Y.Z. Tsai, J.G. Duh, "Effect of nitrogen flow on the properties of quaternary CrAlSiN coatings at elevated temperatures", *Surf. Coat. Technol.* 202 (2007) 735–739.

[86] T. Polcar, A. Cavaleiro, "High-temperature tribological properties of CrAlN, CrAlSiN and AlCrSiN coatings", *Surf. Coat. Technol.* 206 (2011) 1244–1251.

[87] H.W. Chen, Y.C. Chan, J.W. Lee, J.G. Duh, "Oxidation behavior of Si-doped nanocomposite CrAlSiN coatings", *Surf. Coat. Technol.* 205 (2010) 1189–1194.

[88] Y.Y. Chang, C.P. Chang, D.Y. Wang, S.M. Yang, W. Wu, "High temperature oxidation resistance of CrAlSiN coatings synthesized by a cathodic arc deposition process", *J. Alloys Compd.*, 461 (2008) 336–341.

[89] J.W. Yeh, S.K. Chen, S.J. Lin, J.Y. Gan, T.S. Chin, T.T. Shun, C.H. Tsau, S.Y. Chang, "Nanostructured high-entropy alloys with multiple principal elements: Novel alloy design concepts and outcomes", *Adv. Eng. Mater.* 6 (2004) 299–303.

[90] J.W. Yeh, "Recent progress in high entropy alloys", *Ann. Chim. Sci. Mater.* 31 (2006) 633–648.

[91] X. Feng, G. Tang, M. Sun, X. Ma, L. Wang, "Chemical state and phase structure of (TaNbTiW)N films prepared by combined magnetron sputtering and PBII", *Appl. Surf. Sci.* 280 (2013) 388–393.

[92] C.H. Tsau, Y.H. Chang, "Microstructures and mechanical properties of $TiCrZrNbN_x$ alloy nitride thin films", *Entropy*, 15 (2013) 5012–5021.

[93] H.W. Chang, P.K. Huang, J.W. Yeh, A. Davison, C.H. Tsau, C.C. Yang, "Influence of substrate bias, deposition temperature and post-deposition annealing on the structure and properties of multi-principal-component (AlCrMoSiTi)N coatings", *Surf. Coat. Technol.* 202 (2008) 3360–3366.

[94] C.H. Lin, J.G. Duh, "Corrosion behavior of $(TiAlCrSiV)_xN_y$ coatings on mild steels derived from RF magnetron sputtering", *Surf. Coat. Technol.* 203 (2008) 558–561.

[95] D.C. Tsai, Y.L. Huang, S.R. Lin, S.C. Liang, F.S. Shieu, "Effect of nitrogen flow ratios on the structure and mechanical properties of (TiVCrZrY)N coatings prepared by reactive magnetron sputtering", *Appl. Surf. Sci.* 257 (2010) 1361–1367.

[96] S.Y. Chang, S.Y. Lin, Y.C. Huang, C.L. Wu, "Mechanical properties, deformation behaviors and interface adhesion of $(AlCrTaTiZr)N_x$ multi-component coatings", *Surf. Coat. Technol.* 204 (2010) 3307–3314.

[97] H.T. Hsueh, W.J. Shen, M.H. Tsai, J.W. Yeh, "Effect of nitrogen content and substrate bias on mechanical and corrosion properties of high-entropy films $(AlCrSiTiZr)100_{1-x}N_x$", *Surf. Coat. Technol.* 206 (2012) 4106–4112.

[98] V. Braic, Alina Vladescu, M. Balaceanu, C.R. Luculescu, M. Braic, "Nanostructured multi-element (TiZrNbHfTa)N and (TiZrNbHfTa)C hard coatings", *Surf. Coat. Technol.* 211 (2012) 117–121.

[99] W.J. Shen, M.H. Tsai, J.W. Yeh, "Machining performance of sputter-deposited $(Al_{0.34}Cr_{0.22}Nb_{0.11}Si_{0.11}Ti_{0.22})_{50}N_{50}$ high-entropy nitride coatings", *CoatingsTech* 5 (2015) 312–325.

[100] P.K. Huang, J.W. Yeh, "Effects of nitrogen content on structure and mechanical properties of multi-element (AlCrNbSiTiV)N coating", *Surf. Coat. Technol.* 203 (2009) 1891–1896.

[101] P.K. Huang, J.W. Yeh, "Inhibition of grain coarsening up to 1000°C in (AlCrNbSiTiV)N superhard coating", *Scr. Mater.* 62 (2010) 105–108.

[102] S.Y. Chang, C.E. Li, S.C. Chiang, Y.C. Huang, "4-nm thick multilayer structure of multi-component (AlCrRuTaTiZr)Nx as robust diffusion barrier for Cu interconnects", *J. Alloys Compd.* 515 (2012) 4–7.

[103] A.D. Pogrebnjak, I.V. Yakushchenko, O.V. Bondar, V.M. Beresnev, K. Oyoshi, O.M. Ivasishin, H. Amekura, Y. Takeda, M. Opielak, C. Kozak, "Irradiation resistance, microstructure and mechanical properties of nanostructured (TiZrHfVNbTa)N coatings", *J. Alloys Compd.* 679 (2016) 155–163.

[104] K.S. Chang, K.T. Chen, C.Y. Hsu, P.D. Hong, "Growth (AlCrNbSiTiV)N thin films on the interrupted turning and properties using DCMS and HIPIMS system", *Appl. Surf. Sci.* 440 (2018) 1–7.

[105] M.H. Tsai, C.W. Wang, C.H. Lai, J.W. Yeh, J.Y. Gan, "Thermally stable amorphous (AlMoNbSiTaTiVZr)50N50 nitride film as diffusion barrier in copper metallization", *Appl. Phys. Lett.* 92 (2008) 052109.

[106] M.H. Tsai, C.H. Lai, J.W. Yeh, J.Y. Gan, "Effects of nitrogen flow ratio on the structure and properties of reactively sputtered $(AlMoNbSiTaTiVZr)N_x$ coatings", *J. Phys. D. Appl. Phys.* 41 (2008) 235402.

[107] B. Ren, Z.X. Liu, L. Shi, B. Cai, M.X. Wang, "Structure and properties of (AlCrMnMoNiZrB0.1)Nx coatings prepared by reactive DC sputtering", *Appl. Surf. Sci.* 257 (2011) 7172–7178.

[108] A.D. Pogrebnjak, A.A. Bagdasaryan, V.M. Beresnev, U.S. Nyemchenko, V.I. Ivashchenko, O. Ya O. Kravchenko, Zh.K. Shaimardanov, S.V. Plotnikov, O. Maksakova, "The effects of Cr and Si additions and deposition conditions on the structure and properties of the (Zr-Ti-Nb)N coatings", *Ceram. Int.* 43 (2017) 771–782.

[109] S.Y. Chang, M.K. Chen, "High thermal stability of AlCrTaTiZr nitride film as diffusion barrier for copper metallization", *Thin Solid Films* 517 (2009) 4961–4965.

[110] K.H. Cheng, C.H. Weng, C.H. Lai, S.J. Lin, "Study on adhesion and wear resistance of multi-element (AlCrTaTiZr)N coatings", *Thin Solid Films* 517 (2009) 4989–4993.

[111] V. Braic, M. Balaceanu, M. Braic, A. Vladescu, S. Panseri, A. Russo, "Characterization of multi-principal-element (TiZrNbHfTa)N and (TiZrNbHfTa)C coatings for biomedical applications", *J. Mech. Behav. Biomed. Mater* 10 (2012) 197–205.

[112] C.W. Tsai, S.W. Lai, K.H. Cheng, M.H. Tsai, A. Davison, C.H. Tsau, J.W. Yeh, "Strong amorphization of high-entropy AlBCrSiTi nitride film", *Thin Solid Films* 520 (2012) 2613–2618.

[113] X. Feng, G. Tang, X. Ma, M. Sun, L. Wang, "Characteristics of multi-element (ZrTaNbTiW)N films prepared by magnetron sputtering and plasma based ion implantation", *Nucl. Instrum. Meth. B* 301 (2013) 29–35.

[114] H.W. Chang, P.K. Huang, A. Davison, J.W. Yeh, C.H. Tsau, C.C. Yang, "Nitride films deposited from an equimolar Al-Cr-Mo-Si-Ti alloy target by reactive direct current magnetron sputtering", *Thin Solid Films* 516 (2008) 6402–6408.

[115] Z.C. Chang, S.C. Liang, S. Han, Y.K. Chen, F.S. Shieu, "Characteristics of TiVCrAlZr multi-element nitride films prepared by reactive sputtering", *Nucl. Instrum. Meth. B* 268 (2010) 2504–2509.

[116] D.C. Tsai, Z.C. Chang, B.H. Kuo, M.H. Shiao, S.Y. Chang, F.S. Shieu, "Structural morphology and characterization of (AlCrMoTaTi)N coating deposited via magnetron sputtering", *Appl. Surf. Sci.* 282 (2013) 789–797.

[117] B. Ren, S.Q. Yan, R.F. Zhao, Z.X. Liu, "Structure and properties of (AlCrMoNiTi)Nx and (AlCrMoZrTi) Nx films by reactive RF sputtering", *Surf. Coat. Technol.* 235 (2013) 764–772.

[118] K.H. Cheng, C.H. Lai, S.J. Lin, J.W. Yeh, "Structural and mechanical properties of multi-element (AlCrMoTaTiZr)Nx coatings by reactive magnetron sputtering", *Thin Solid Films* 519 (2011) 3185–3190.

[119] B. Ren, Z. Shen, Z. Liu, "Structure and mechanical properties of multi-element (AlCrMnMoNiZr)Nx coatings by reactive magnetron sputtering", *J. Alloys Compd.* 560 (2013) 171–176.

[120] S.Y. Chang, D.S. Chen, "(AlCrTaTiZr)N/(AlCrTaTiZr)N$_{0.7}$ bilayer structure of high resistance to the interdiffusion of Cu and Si at 900°C", *Mater. Chem. Phys.* 125 (2011) 5–8.

[121] D.C. Tsai, Z.C. Chang, B.H. Kuo, T.N. Lin, M.H. Shiao, F.S. Shieu, "Interfacial reactions and characterization of (TiVCrZrHf)N thin films during thermal treatment", *Surf. Coat. Technol.* 240 (2014) 160–166.

[122] S.C. Liang, Z.C. Chang, D.C. Tsai, Y.C. Lin, H.S. Sung, M.J. Deng, F.S. Shieu, "Effects of substrate temperature on the structure and mechanical properties of (TiVCrZrHf)N coatings", *Appl. Surf. Sci.* 257 (2011) 7709–7713.

[123] C.T. Lee, W.H. Cho, M.H. Shiao, C.N. Hsiao, K.S. Tang, C.C. Jaing, "Effects of DC bias on the microstructure, residual stress and hardness properties of TiVCrZrTaN films by reactive RF magnetron sputtering", *Procedia Engineer.* 36 (2012) 316–321.

[124] D.C. Tsai, S.C. Liang, Z.C. Chang, T.N. Lin, M.H. Shiao, F.S. Shieu, "Effects of substrate bias on structure and mechanical properties of (TiVCrZrHf)N coatings", *Surf. Coat. Technol.* 207 (2012) 293–299.

[125] P.K. Huang, J.W. Yeh, "Effects of substrate temperature and post-annealing on microstructure and properties of (AlCrNbSiTiV)N coatings", *Thin Solid Films* 518 (2009) 180–184.

[126] C.H. Lai, M.H. Tsai, S.J. Lin, J.W. Yeh, "Influence of substrate temperature on structure and mechanical, properties of multi-element (AlCrTaTiZr) N coatings", *Surf. Coat. Technol.* 201 (2007) 6993–6998.

[127] T.K. Chen, M.S. Wong, T.T. Shun, J.W. Yeh, "Nanostructured nitride films of multi-element high-entropy alloys by reactive DC sputtering", *Surf. Coat. Technol.* 200 (2005) 1361–1365.

[128] C.H. Lin, J.G. Duh, J.W. Yeh, "Multi-component nitride coatings derived from Ti-Al-Cr-Si-V target in RF magnetron sputter", *Surf. Coat. Technol.* 201 (2007) 6304–6308.

[129] J.P. Chu, J.C. Huang, J.S.C. Jang, Y.C. Wang, P.K. Liaw, "Thin film metallic glasses: Preparations, properties, and applications", *JOM* 62 4 (2010) 19–24.

[130] J.P. Chu, J.S.C. Jang, J.C. Huang, H.S. Chou, Y. Yang, J.C. Ye, Y.C. Wang, J.W. Lee, F.X. Liu, P.K. Liaw, Y.C. Chen, C.M. Lee, C.L. Li, Cut Rullyani, "Thin film metallic glass: Unique properties and potential applications", *Thin Solid Films* 520 (2012) 5097–5122.

[131] J. Lee, H.C. Tung, J.G. Duh, "Enhancement of mechanical and thermal properties in Zr-Cu-Ni-Al-N thin film metallic glass by compositional control of nitrogen", *Mater. Lett.* 159 (2015) 369–372.

[132] J. Lee, M.L. Liou, J.G. Duh, "The development of a Zr-Cu-Al-Ag-N thin film mssetallic glass coating in pursuit of improved mechanical, corrosion, and antimicrobial property for bio-medical sapplication", *Surf. Coat. Technol.* 310 (2017) 214–222.

[133] J. Musil, "Hard nanocomposite coatings: Thermal stability, oxidation resistance and toughness", *Surf. Coat. Technol.* 207 (2012) 50–65.

[134] J. Musil, P. Baroch, P. Zeman, in: R. Wei (Ed.), "*Plasma Surface Engineering Research and its Practical Applications*", Research Signpost, Kerala, (2008) 1.

[135] J. Paulitsch, P.H. Mayrhofer, W.-D. Münz, M. Schenkel, "Structure and mechanical properties of CrN/TiN multilayer coatings prepared by a combined HIPIMS/UBMS deposition technique", *Thin Solid Films* 517 (2008) 1239–1244..

[136] Y. Zhou, R. Asaki, W.H. Soe, R. Yamamoto, R. Chen, A. Iwabuchi, "Hardness anomaly, plastic deformation work and fretting wear properties of polycrystalline TiN/CrN multilayers", *Wear* 236 (1999) 159–164.

[137] C.Y. Su, C.T. Pan, T.P. Liou, P.T. Chen, C.K. Lin, "Investigation of the microstructure and characterizations of TiN/CrN nanomultilayer deposited by unbalanced magnetron sputter process", *Surf. Coat. Technol.* 203 (2008) 657–660.

[138] Y.X. Ou, J. Lin, H.L. Che, W.D. Sproul, J.J. Moore, M.K. Lei, "Mechanical and tribological properties of CrN/TiN multilayer coatings deposited by pulsed dc magnetron sputtering", *Surf. Coat. Technol.* 276 (2015) 152–159.

[139] G. Zhang, T. Wang, H. Chen, "Microstructure, mechanical and tribological properties of TiN/Mo$_2$N nano-multilayer films deposited by magnetron sputtering", *Surf. Coat. Technol.* 261 (2015) 156–160.

[140] X. Chen, H. Gao, Y. Bai, H. Yang, "Thermal failure mechanism of multilayer brittle TiN/CrAlN films", *Ceram. Int.* 44 (2018) 8138–8144.

[141] A.D. Pogrebnjak, G. Abadias, O.V. Bondar, B.O. Postolnyi, M.O. Lisovenko, O.V. Kyrychenko, A.A. Andreev, V.M. Beresnev, D.A. Kolesnikov, M. Opielak, "Structure and properties of multilayer nanostructured coatings tin/mon depending on deposition conditions", *Acta Phys. Pol. A* 125 (2014) 1280–1283.

[142] L. Chen, S.Q. Wang, Y. Dua, J. Li, "Microstructure and mechanical properties of gradient Ti(C, N) and TiN/Ti(C, N) multilayer PVD coatings", *Mater. Sci. Eng. A* 478 (2008) 336–339.

[143] M. Nordin, M. Larsson, S. Hogmark, "Mechanical and tribological properties of multilayered PVD TiN/CrN", *Wear* 232 (1999) 221–225.

[144] M.A. Auger, R. Gago, M. Fernández, O. Sánchez, J.M. Albella, "Deposition of TiN/AlN bilayers on a rotating substrate by reactive sputtering", *Surf. Coat. Technol.* 157 (2002) 26–33.

[145] A. Feuerstein, A. Kleyman, "Ti-N multilayer systems for compressor airfoil sand erosion protection", *Surf. Coat. Technol.* 204 (2009) 1092–1096.

[146] Y. Qiu, S. Zhang, B. Li, Y. Wang, J.-W. Lee, F. Li, D. Zhao, "Improvement of tribological performance of CrN coating via multilayering with VN", *Surf. Coat. Technol.* 231 (2013) 357–363.

[147] Y.P. Purandare, A.P. Ehiasarian, P. Eh. Hovsepian, "Deposition of nanoscale CrN/NbN physical vapor deposition coatings by high power impulse magnetron sputtering", *J. Vac. Sci. Technol. A* 26 2 (2008) 288–296.

[148] Y.S. Yang, T.P. Cho, Y.C. Lin, "Effect of coating architectures on the wear and hydrophobic properties of Al-N/Cr-N multilayer coatings", *Surf. Coat. Technol.* 259 (2014) 172–177.

[149] B.O. Postolnyi, V.M. Beresnev, G. Abadias, O.V. Bondar, L. Rebouta, J.P. Araujo, A.D. Pogrebnjak, "Multilayer design of CrN/MoN protective coatings for enhanced hardness and toughness", *J. Alloys Compd.* 725 (2017) 1188–1198.

[150] B. Han, Z. Wang, N. Devi, K.K. Kondamareddy, Z. Wang, N. Li, W. Zuo, D. Fu, C. Liu, "RBS depth profiling analysis of (Ti,Al)N/MoN and CrN/MoN multilayers", *Nanoscale Res. Lett.*, 12 (2017) 161–168.

[151] F.B. Wu, S.K. Tien, J.W. Lee, J.G. Duh, "Comparison in microstructure and mechanical properties of nanocomposite CrWN and nanolayered CrN/WN coatings", *Surf. Coat. Technol.* 200 (2006) 3194–3198.

[152] B. Warcholinski, A. Gilewicz, Z. Kuklinski, P. Myslinski, "Hard CrCN/CrN multilayer coatings for tribological applications", *Surf. Coat. Technol.* 204 (2010) 2289–2293.

[153] Y.X. Qiu, B. Li, J.W. Lee, D.L. Zhao, "Self-lubricating CrVN coating strengthened via multilayering with VN", *J. Iron Steel Res. Int.*, 21 (5) (2014) 545–550.

[154] Y. Qiu, S. Zhang, J.W. Lee, B. Li, Y. Wang, "Self-lubricating CrAlN/VN multilayer coatings at room temperature", *Appl. Surf. Sci.* 279 (2013) 189–196.

[155] T. Wang, G. Zhang, Z. Liu, B. Jiang, "Oxidation behavior of magnetron sputtered Mo2N/AlN multilayercoatings during heattreatment", *Ceram. Int.* 41 (2015) 7028–7035.

[156] D. Zhang, Z. Qi, B. Wei, Z. Wu, Z. Wang, "Anticorrosive yet conductive Hf/Si3N4 multilayer coatings on AZ91D magnesium alloy by magnetron sputtering", *Surf. Coat. Technol.* 309 (2017) 12–20.

[157] S.H. Tsai, J.G. Duh, "Microstructure and mechanical properties of CrAlN/SiNx nanostructure multilayered coatings", *Thin Solid Films*, 518 (2009) 1480–1483.

[158] H.T. Wang, Y.X. Xu, L. Chen, "Optimization of Cr-Al-N coating by multilayer architecture with TiSiN insertion layer", *J. Alloys Compd.* 728 (2017) 952–958.

[159] Y. Zhang, Y. Yang, H. Ding, Y. Peng, S. Zhang, L. Yu, P. Zhang, "Combining magnetic filtered cathodic arc deposition with ion beam sputtering to afford superhard TiSiN multilayer composite films with tunable microstructure and mechanical properties", *Vacuum*, 125 (2016) 6–12.

[160] I. Safi, "Recent aspects concerning DC reactive magnetron sputtering of thin films: a review", *Surf. Coat. Technol.* 127 (2000) 203–218.

[161] S.F. Chen, Y.C. Kuo, C.J. Wang, S.H. Huang, J.W. Lee, Y.C. Chan, H.W. Chen, J.G. Duh, T.E. Hsieh, "The effect of Cr/Zr chemical composition ratios on the mechanical properties of CrN/ZrN multilayered coatings deposited by cathodic arc deposition system", *Surf. Coat. Technol.* 231 (2013) 247–252.

[162] E. Bemporad, C. Pecchio, S. De Rossi, F. Carassiti, "Tribological characterisation and wear properties of industrially produced nanoscaled CrN/NbN multilayer coatingbehaviour of hard coatings deposited by arc-evaporation PVD", *Surf. Coat. Technol.* 188–189 (2004) 319–330.

[163] R.J. Rodríguez, J.A. García, A. Medrano, M. Rico, R. Sánchez, R. Martínez, C. Labrugère, M. Lahaye, A. Guette, "Tribological behaviour of hard coatings deposited by arc-evaporation PVD", *Vacuum*, 67 (2002) 559–566.

[164] A.P. Ehiasarian, P. Eh. Hovsepian, L. Hultman, U. Helmersson, "Comparison of microstructure and mechanical properties of chromium nitride-based coatings deposited by high power impulse magnetron sputtering and by the combined steered cathodic arc/unbalanced magnetron technique", *Thin Solid Films*, 457 (2004) 270–277.

[165] A.P. Ehiasarian, R. New, W.-D. Münz, L. Hultman, U. Helmersson, V. Kouznetsov, "Influence of high power densities on the composition of pulsed magnetron plasmas", *Vacuum* 65 (2002) 147–154.

[166] V. Kouznetsov, K. Macák, J.M. Schneider, U. Helmersson, I. Petrov, "A novel pulsed magnetron sputter technique utilizing very high target power densities", *Surf. Coat. Technol.* 122 (1999) 290–293.

[167] K.Y. Liu, J.W. Lee, F.B. Wu, "Fabrication and tribological behavior of sputtering TaN coatings", *Surf. Coat. Technol.* 259 B (2014) 123–128.

[168] Y.H. Yang, K.Y. Liu, Y.X. Qiu, C.H. Wu, F.B. Wu, "Fabrication and characterization of nanolayered single element nitride coating", *Surf. Coat. Technol.* 284 (2015) 112–117.

[169] A. Hörling, L. Hultman, M. Odén, J. Sjölén, L. Karlsson, "Mechanical properties and machining performance of Ti$_{1-x}$Al$_x$N-coated cutting tools", *Surf. Coat. Technol.* 191 (2005) 384–392.

[170] M. Fallmann, Z. Chen, Z.L. Zhang, P.H. Mayrhofer, M. Bartosik, "Mechanical properties and epitaxial growth of TiN/AlN superlattices", *Surf. Coat. Technol.* 375 (2019) 1–7.

[171] G.S. Kim, S.Y. Lee, J.H. Hahn, S.Y. Lee, "Synthesis of CrN/AlN superlattice coatings using closed-field unbalanced magnetron sputtering process", *Surf. Coat. Technol.* 171 (2003) 91–95.

[172] X. Chu, M.S. Wong, W.D. Sproul, S.L. Rohde, S.A. Barnett, "Deposition and properties of polycrystalline TiN/NbN superlattice coatings", *J. Vac. Sci. Technol.* A 10 (1992) 1604.

[173] D.C. Cameron, R. Aimo, Z.H. Wang, K.A. Pischow, "Structural variations in CrN/NbN superlattices", *Surf. Coat. Technol.* 142–144 (2001) 567–572.

[174] Y.Z. Tsai, J.G. Duh, "Tribological behavior of CrAlSiN/W$_2$N multilayer coatings deposited by DC magnetron sputtering", *Thin Solid Films* 518 (2010) 7523–7526.

[175] J.Y. Xiang, F.B. Wu, "Gas inlet and input power modulated sputtering molybdenum nitride thin films", *Surf. Coat. Technol.* 332 (2017) 161–167.

[176] M. Shinn, L. Hultman, S.A. Barnett, Growth, structure, and microhardness of epitaxial TiN/NbN superlattices, *J. Mater. Res.* 7 (2011) 901–911.

[177] Y. Wang, J.W. Lee, J.G. Duh, "Mechanical strengthening in self-lubricating CrAlN/VN multilayer coatings for improved high-temperature tribological characteristics", *Surf. Coat. Technol.* 303 A (2016) 12–17.

[178] Y.C. Chan, H.W. Chen, P.S. Chao, J.G. Duh, J.W. Lee, "Microstructure control in TiAlN/SiN$_x$ multilayers with appropriate thickness ratios for improvement of hardness and anti-corrosion characteristics", *Vacuum*, 87 (2013) 195–199.

[179] Y.C. Chan, H.W. Chen, Y.Z. Tsai, J.G. Duh, J.W. Lee, "Texture, microstructure and anti-wear characteristics in isostructural CrAlSiN/W$_2$N multilayer coatings", *Thin Solid Films* 544 (2013) 265–269.

[180] Y.C. Chan, H.W. Chen, J.G. Duh, J.W. Lee, Texture, "microstructure, and tribological behavior in TiAlN/SiN$_x$ multilayers", *Int. J. Appl. Ceram. Technol.* 11 (2014) 611–617.

[181] L. Hultman, J. Bareño, A. Flink, H. Söderberg, K. Larsson, V. Petrova, M. Odén, J.E. Greene, I. Petrov, "Interface structure in superhard TiN-SiN nanolaminates and nanocomposites: Film growth experiments and ab initio calculations", *Phys. Rev. B* 75 (2007) 155437.

[182] Y.H. Yang, D.J. Chen, F.B. Wu, "Microstructure, hardness, and wear resistance of sputtering TaN coating by controlling RF input power", *Surf. Coat. Technol.* 303 A (2016) 32–40.

[183] Y.H. Yang, F.B. Wu, "Microstructure evolution and protective properties of TaN multilayer coatings", *Surf. Coat. Technol.* 308 (2016) 108–114.

[184] J.Y. Xiang Z.X. Lin, E. Renoux, F.B. Wu, "Microstructure evolution and indentation cracking behavior of MoN multilayer films", *Surf. Coat. Technol.* 350 (2018) 1020–1027.

[185] J.S. Koehler, "Attempt to design a strong solid", *Phys. Rev. B* 2 (1970) 547–551.

[186] S.H. Kim, Y.J. Baik, D. Kwon, "Analysis of interfacial strengthening from composite hardness of TiN/VN and TiN/NbN multilayer hard coatings", *Surf. Coat. Technol.* 187 (2004) 47–53.

[187] J. Xu, K. Hattori, Y. Seino, I. Kojima, "Microstructure and properties of CrN/Si$_3$N$_4$ nano-structured multilayer films", *Thin Solid Films* 414 (2002) 239–245.

[188] G. Farges, E. Beauprez, M.C.S. Catherine, "Crystallographic structure of sputtered cubic δ-VN$_x$ films: influence of basic deposition parameters", *Surf. Coat. Technol.* 61 (1993) 238–244.

[189] J. Xu, L. Yu, Y. Azuma, T. Fujimoto, I. Umehara, I. Kojima, "Thermal stress hardening of a-Si3N4/nc-TiN nanostructured multilayers", *Appl. Phys. Lett.* 81 (2002) 4139–4141.

[190] F.B. Wu, S.K. Tien, J.G. Duh, "Manufacture, microstructure and mechanical properties of CrWN and CrN/WN nanolayered coatings", *Surf. Coat. Technol.* 200 (2005) 1514–1518.

[191] I.A. Ovid'ko, "Deformation of nanostructures", *Science* 295 (2002) 2386.

[192] Y.Z. Tsai, J.G. Duh, "Thermal stability and microstructure characterization of CrN/WN multilayer coatings fabricated by ion-beam assisted deposition", *Surf. Coat. Technol.* 200 (2005) 1683–1689.

[193] Y.Z. Tsai, J.G. Duh, "Tribological behavior of CrN/WN multilayer coatings grown by ion-beam assisted deposition", *Surf. Coat. Technol.*, 201 (2006) 4266–4272.

[194] Y.Z. Tsai, J.G. Duh, "Enhanced hardness of CrAlSiN/W$_2$N superlattice coatings deposited by direct current magnetron sputtering", *J. Mater. Res.* 25 (2010) 2325–2329.

[195] S.K. Tien, J.G. Duh, J.W. Lee, "Oxidation behavior of sputtered CrN/AlN multilayer coatings during heat treatment", *Surf. Coat. Technol.* 201 (2007) 5138–5142.

[196] S.K. Tien, J.G. Duh, "Effect of heat treatment on mechanical properties and microstructure of CrN/AlN multilayer coatings", *Thin Solid Films* 494 (2006) 173–178.

[197] C.H. Lin, Y.Z. Tsai, and J.G. Duh, "Effect of grain size on mechanical properties in CrAlN/SiNx multilayer coatings", *Thin Solid Films* 518 (2010) 7312–7315.

[198] L.K. Yeh-Liu, S.Y. Hsu, P.Y. Chen, J.W. Lee, J.G. Duh, "Improvement of CrMoN/SiNx coatings on mechanical and high temperature Tribological properties through biomimetic laminated structure design" Surf. *Coat. Technol.* 393 (2020) 125724.

[199] Y.X. Ou, X.P. Ouyang, B. Liao, X. Zhang, and S. Zhang, "Hard yet tough CrN/Si$_3$N$_4$ multilayer coatings deposited by the combined deep oscillation magnetron sputtering and pulsed dc magnetron sputtering", *Appl. Surf. Sci.* 502 (2020) 144168.

[200] N. Ravi, R. Markandeya, S. V. Joshi, "Fracture behaviour of nc-TiAlN/a-Si$_3$N$_4$ nanocomposite coating during nanoimpact test", *Surf. Eng.* 33 (2017) 282–291.

[201] X. Bai, W. Zheng, T. An, Q. Jiang, "Effects of deposition parameters on microstructure of CrN/Si$_3$N$_4$ nanolayered coatings and their thermal stability", *J. Phys. Condens. Matter* 17 (2005) 6405–6413.

[202] H.C. Barshilia, B. Deepthi, K.S. Rajam, Deposition and characterization of CrN/Si3N4 and CrAlN/Si3N4 nanocomposite coatings prepared using reactive DC unbalanced magnetron sputtering, *Surf. Coat. Technol.* 201 (2007) 9468–9475.

[203] A.A. Bagdasaryan, A.V. Pshyk, L.E. Coy, P. Konarski, M. Misnik, V.I. Ivashchenko, M. Kempiński, N.R. Mediukh, A.D. Pogrebnjak, V.M. Beresnev, S. Jurga, "A new type of (TiZrNbTaHf)N/MoN nanocomposite coating: Microstructure and properties depending on energy of incident ions", *Compos. Part B*, 146 (2018) 132–144.

[204] A.A. Bagdasaryan, A.V. Pshyk, L.E. Coy, M. Kempiński, A.D. Pogrebnjak, V.M. Beresnev, S. Jurga, "Structural and mechanical characterization of (TiZrNbHfTa)N/WN multilayered nitride coatings", *Mater. Lett.* 229 (2018) 364–367

[205] Z. Liu, M. Shen, S. Zhu, Y. Huang, W. Ma, Y. Jia, and F. Wang, "Effect of nitrogen content on the phase transformation of alumina scale on a nanocrystalline NiCrAlYSiHfN/AlN multilayer coating", *Corros. Sci.*, 165 (2020) 108396.

[206] W. Yang, G. Ayoub, I. Salehinia, B. Mansoor, H. Zbib, "The effect of layer thickness ratio on the plastic deformation mechanisms of nanoindented Ti/TiN nanolayered composite", *Comput. Mater. Sci.*, 154 (2018) 488–498.

[207] T. Fu, Z. Zhang, X. Peng, S. Weng, Y. Miao, Y. Zhao, S. Fu, N. Hu, "Effects of modulation periods on mechanical properties of V/VN nanomultilayers", *Ceram. Int.*, 45 (2019) 10295–10303.

[208] D.F. Zhang, Z.B. Qi, B.B. Wei, H. Shen, Z.C. Wang, "Microstructure and corrosion behaviors of conductive Hf/HfN multilayer coatings on magnesium alloys", *Ceram. Int.*, 44 (2018) 9958–9966.

[209] H.J. Zhao, P.B. Mi, F.X. Ye, "Compared the oxidation behavior of TiN and TiN/W_2N ceramic coatings during heat treatment", *Mater. Chem. Phys.* 217 (2018) 445–450.

[210] P. Eh Hovsepian, A.P. Ehiasarian, Y.P. Purandare, P. Mayr, K.G. Abstoss, M. Mosquera Feijoo, W. Schulz, A. Kranzmann, M.I. Lasanta, J.P. Trujillo, "Novel HIPIMS deposited nanostructured CrN/NbN coatings for environmental protection of steam turbine components", *J. Alloys Compd.* 746 (2018) 583–593.

[211] Z. Chen, Q. Shao, M. Bartosik, P. H. Mayrhofer, H. Chen, Z. Zhang, "Growth-twins in CrN/AlN multilayers induced by hetero-phase interfaces", *Acta Mater.* 185 (2020) 157–170.

[212] A. Illana, E. Almandoz, G.G. Fuentes, F.J. Perez, S. Mato "Comparative study of CrAlSiN monolayer and CrN/AlSiN superlattice multilayer coatings: Behavior at high temperature in steam atmosphere", *J. Alloys Compd.* 778 (2019) 652–s.

[213] Z. Chen, D. Holec, M. Bartosik, P.H. Mayrhofer, Z. Zhang, "Crystallographic orientation dependent maximum layer thickness of cubic AlN in CrN/AlN multilayers", *Acta Mater.*, 168 (2019) 190–202.

[214] Y.Y. Chang, Y.J. Yang, S.Y. Weng, "Effect of interlayer design on the mechanical properties of AlTiCrN and multilayered AlTiCrN/TiSiN hard coatings", *Surf. Coat. Technol.* 389 (2020) 125637.

[215] F. Pei, H.J. Liu, L. Chen, Y.X. Xu, Y. Du, "Improved properties of TiAlN coating by combined Si-addition and multilayer architecture", *J. Alloys Compd.* 790 (2019) 909–916.

[216] W.M. Seidl, M. Bartosik, S. Kolozsvári, H. Bolvardi, P.H. Mayrhofer, "sInfluence of coating thickness and substrate on stresses and mechanical properties of (Ti,Al,Ta)N/(Al,Cr)N multilayers", *Surf. Coat. Technol.* 347 (2018) 92–98.

[217] A. Vereschaka, S. Grigorie, V. Tabakov, M. Migranov, N. Sitnikov, F. Milovich, N. Andreev, "Influence of the nanostructure of Ti-TiN-(Ti,Al,Cr)N multilayer composite coating on tribological properties and cutting tool life", *Tribol. Int.*, 150 (2020) 106388.

[218] M. Falsafein, F. Ashrafizadeh, A. Kheirandish, "Influence of thickness on adhesion of nanostructured multilayer CrN/CrAlN coatings to stainless steel substrate" *Surf. Interfaces* 13 (2018) 178–185.

[219] W.M. Seidl, M. Bartosik, S. Kolozsvári, H. Bolvardi, P.H. Mayrhofer, "Mechanical properties and oxidation resistance of Al-Cr-N/Ti-Al-Ta-N multilayer coatings", *Surf. Coat. Technol.* 347 (2018) 427–433.

[220] A.R. Shugurov, M.S. Kazachenok, "Mechanical properties and tribological behavior of magnetron sputtered TiAlN/TiAl multilayer coatings", *Surf. Coat. Technol.* 353 (2018) 254–262.

4 Integrated Nanomechanical Characterisation of Hard Coatings

Ben D. Beake
Micro Materials Ltd., U.K.
University of Huddersfield, U.K.
Manchester Metropolitan University, U.K.

Vladimir M. Vishnyakov
University of Huddersfield, U.K.

Tomasz W. Liskiewicz
Manchester Metropolitan University, U.K.

CONTENTS

4.1 INTRODUCTION

Mechanical properties have an important influence on tribological performance but hardness alone can be a poor predictor of wear performance. Instead insights from advanced nanomechanical and nano/microtribological tests combined with simulated stress distributions help to better understand performance in tribological applications and to design coating systems for enhanced performance. Detailed simulated stress distributions enable data to be interpreted more effectively. They can provide mechanistic information which can be the key to unlocking exactly where and why coating systems fail in scratch and fretting tests, and then to designing coatings with improved performance.

In this chapter the combined experimental-modelling approach is illustrated with data from a wide range of coating systems including (i) ion beam assisted $Ti(Fe,N)_x$ on silicon, (ii) multilayered hard carbon coatings on steel and (iii) nitride-based coatings on cemented carbide. While it is desirable to design coatings to avoid wear completely in highly loaded mechanical contact, this is not always possible and design rules for enhanced damage tolerance are suggested. There is an inherent competition between plastic deformation and fracture in highly loaded contact which should be considered in optimising coating systems for wear resistance.

In applications of hard coatings involving highly loaded mechanical contact, such as turning, milling and forming, high temperatures are generated in contact and mechanical properties determined at room temperature may be less relevant than those measured in high-temperature tests. In developing advanced wear-resistant coatings for high-speed metal cutting a critical requirement is to study their behaviour at elevated temperature since the cutting process generates frictional heat which can raise the temperature in the cutting zone to 700–900 °C or more. Nanomechanical characterisation at high temperature is experimentally challenging and requires more careful instrumental and experimental design than at room temperature. High temperature nano-tribological tests provide severe tests for coatings that can simulate high contact pressure sliding/abrasive contacts at elevated temperature. Analytical modelling shows that the behaviour in the high-temperature test can be explained by temperature-dependent changes to the stress distribution in the highly loaded sliding contact which influence the critical load. The high strain rate contact in the nano-impact test can simulate the performance of coating systems under highly loaded intermittent contact and the evolution of wear under these conditions.

4.2 MECHANICAL PROPERTIES AND WEAR

An important goal of mechanical property measurements is to predict wear resistance in mechanical contact. Classical theories of wear, for example as proposed by Archard [1], have emphasized the importance of hardness. However, it was realised that hardness itself is often not a particularly effective predictor of wear resistance [2–4]. For example, Leyland and Matthews studied the correlation between hardness and sand-slurry rubber-wheel abrasive wear resistance of 5–19 at % C-doped W coatings [3, 4]. They noted that the best wear resistance was shown by coatings with 15–20 GPa hardness with much inferior wear resistance shown by all C-doped W coatings that were harder than this.

Parameters combining hardness (H) and elastic modulus (E), such as H/E and H^3/E^2, may be equally important predictors of wear in many practical applications and, in some cases, have been used as proxies for fracture toughness. The dimensionless ratio H/E is a measure of the elastic strain to break and is strongly correlated with energy dissipation in mechanical contact. Greenwood and Williamson originally proposed a plasticity index that related the deformation in rough contacts to E_r/H multiplied by a geometric factor [5] as shown in Equation 4.1.

$$\psi = (E_r / H)\sqrt{(\sigma / \beta)} \tag{4.1}$$

where E_r is the reduced modulus, ψ = a plasticity index, σ = standard deviation of the height of the contacting asperities (i.e. surface roughness) and β = their average radius. They noted that if $\psi \gg 1$ plastic deformation of asperities occurs even at minimal contact pressure, while if $\psi < 0.6$ deformation is largely elastic and plastic contact can only be caused if surfaces are forced together under very large nominal pressure. At constant surface roughness and asperity radius H/E is inversely proportional to this plasticity index [6].

In a nanoindentation test H/E is related to the elastic and plastic work done in the indentation through Equation 4.2

$$W_p / (W_p + W_e) \approx 1 - x(H / E_r) \approx h_r / h_{max} \tag{4.2}$$

where W_p is the plastic or irreversible work done during indentation, W_e is elastic deformation, h_r is residual indentation depth and h_{max} is maximum indentation depth [7, 8]. $W_p/(W_p + W_e)$ is a convenient dimensionless plasticity index which is inversely related to the elastic strain to break (H/E). The reduced indentation modulus E_r appears in Equation 4.2 rather than the elastic modulus since the shape of the indentation curve is subtly influenced by the stiffness of the indenter. This effect is small but can be seen when indenting very stiff materials with indenters of lower stiffness than diamond (e.g. cBN or sapphire) [9].

H^3/E^2 is a measure of the resistance to plastic deformation. Tsui and Pharr [10] noted that from the contact mechanics analysis developed by Johnson [11], the yield pressure (P_y) in a rigid ball of radius R in elastic/plastic contact was given by:

$$P_y = 0.78R^2(H^3 / E^2) \tag{4.3}$$

It follows that for a given contact pressure the contact is more likely to be elastic if H^3/E^2 is increased. Coatings with high H^3/E^2 can display improved resistance to fracture initiation due to enhanced load support. Pei and co-workers studied the influence of thin film deposition substrate bias (e.g. energy introduced from the plasma into the growing film) on the properties of nc-TiC/a-C:H nanocomposite coatings [12]. They reported a correlation between the critical load at which radial cracks formed when the coatings were indented with a Berkovich indenter and the H/E ratio of the coatings. As Figure 4.1 shows, the correlation with H^3/E^2 is also striking.

Coatings with very high H^3/E^2 have shown enhanced erosion resistance, for example when eroded by 50 μm Al_2O_3 particles [13, 14], or 40–100 μm SiC [15]. H^3/E^2 has also been taken as a direct measure of fracture resistance [16]. In annealed AlTiN coatings Bartosik and co-workers reported strong correlation between H^3/E^2 and the fracture toughness determined from cantilever bending experiments [17]. Musil and co-workers have reported strong correlation between H/E and surface cracking in indentation and bending of various oxide and nitride coatings [18–21]. However, regarding the correlation with fracture toughness, Chen et al. noted that neither H/E nor H^3/E^2 address the contribution of plasticity to fracture toughness and there are several examples where no or inverse correlation between H^3/E^2 and toughness has been reported [6]. For example, Tiegel and co-workers reported that fracture toughness did not correlate with either H/E or H^3/E^2 for a series of aluminosilicate glasses varied network modifier ions [22]. Shi and co-workers reported an inverse correlation between fracture toughness and H/E or H^3/E^2 for VC, W and W/VC multilayer coatings [23].

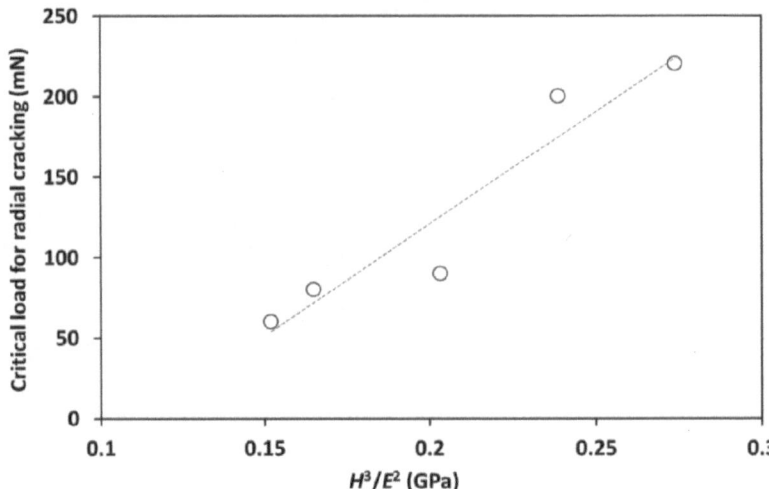

FIGURE 4.1 Relationship between the critical load at which radial cracks formed on indentation of nc-TiC/a-C:H nanocomposite coatings with a Berkovich indenter and the coating H^3/E^2 ratio.

There is an inherent competition between brittle fracture and plastic deformation (or crack resistance and strength). Damage tolerance in a given mechanical contact can require resistance to both. Bouzakis and co-workers noted that, while many studies of coating impact resistance have focussed on fracture initiation, the subsequent coating damage propagation is also important, as it relates to the ability of the coating to withstand loads after fatigue damage initiation [24]. For coating system design another complicating factor is that deformation mechanisms are also scale dependent [25, 26]. Due to the requirement to show high resistance to cracking and plastic flow in contact, hard coatings can show optimum wear resistance at intermediate values of hardness and other mechanical property indices such as H^3/E^2 [3, 4, 27]. For example, in a study of 6 μm TiAlN coatings on Ti6Al4V Yang and co-workers reported maximum resistance to erosion by alumina particles for coatings with intermediate H^3/E^2 [27].

The wear process itself can also modify the worn surface by smoothening it (altering its friction) and/or creating a nanocrystalline layer of different hardness. Liskiewicz et al. investigated tribologically transformed structures (TTS) formed on steel, titanium and copper using nanoindentation mapping approach [28]. High-resolution micro-mechanical measurements within the wear scars revealed modified mechanical properties of TTS layers. The approach was developed further with a modified Archard wear equation that accounted for changes in mechanical properties induced by friction [29]. It was shown that the modified Archard model, which accounted for TTS layer mechanical properties, could more reliably predict wear of steel and copper samples in long-cycle tests. In the following section strategies for accurate mechanical property measurement of thin hard coatings are discussed.

4.3 NANOMECHANICAL AND NANO/MICROTRIBOLOGICAL TEST METHODS

Nanomechanical testing has proved a revolutionary technique in improving our fundamental understanding of the basis of mechanical properties of materials and the importance of nanoscale behaviour on their performance [30]. Nanomechanical and nano/microtribological test methods can be broadly divided into two main categories – (i) those primarily designed for characterisation of mechanical properties and (ii) those primarily involved with simulating contact conditions. Examples of the first type include nanoindentation, micro-pillar compression and micro-cantilever bending. The stress state in micro-pillar and micro-cantilever testing is simpler than in nanoindentation testing, but these tests require extensive sample preparation by focussed ion milling. Sebastiani and co-workers have highlighted some of the challenges in determining K_{1C} fracture toughness [31]. Examples of the second type include nano- and micro-scratch testing, nano-/micro-impact and reciprocating nano-wear tests. The experimental conditions in both these types of test are aided by correct test dimensioning and modelling.

4.3.1 CHARACTERISATION OF MECHANICAL PROPERTIES

Due to their low thickness combined with the realisation that hardness alone does not control wear resistance, hard PVD coatings are increasingly being characterised by nanoindentation (also called depth-sensing indentation, DSI, or instrumented indentation testing, IIT) rather than more traditional microhardness tests. The nanoindentation technique has developed to have a well-established standard for best practice, ISO14577, part 4 of which ("Test method for metallic and non-metallic coatings" [32]) is aimed at strategies for testing coatings and determining true "coating-only" mechanical properties independent of substrate influence. Assuming the instrumentation has been correctly calibrated according to ISO 14577 methods or equivalent, there are several factors that influence whether comparable hardness and modulus measurements can be made, with particular challenges in reliably measuring the properties of ultra-thin coatings.

4.3.1.1 Hardness Determination

When a hard coating is deposited on a softer substrate its hardness should be determined with a sharp indenter that causes sufficient plastic yielding within the coating before the substrate yields (i.e. the maximum shear stress occurs within the coating causing plastic deformation while the stress in the substrate is lower than the substrate yield stress) [32]. A plateau in the measured hardness vs. the dimensionless relative indentation depth, h_c/t_c, (h_c = contact depth, t_c = film thickness) is usually taken as the coating hardness. However, the ISO indentation standard cautions that, for a given indenter sharpness and coating thickness, it may not be possible to reach the true hardness of the coating before plastic yield of the substrate occurs. In this case, for a hard coating on a soft substrate, the measured hardness will be less than the true hardness. The ISO standard gives an example with three DLC coatings on M2 steel of nominally the same composition but different thickness (0.46, 1.5 and 2.5 μm). For the thinnest coating substrate yield occurred before fully developed plasticity could be reached so that its measured hardness was only 15 GPa, which was 3 GPa lower than the value measured on both the thicker coatings, which was assumed to be the true coating hardness. In practice the plateau corresponding to the coating hardness is typically from a depth of ~0.2 R (where R is the indenter radius) to a relative indentation depth of 0.1. For hardness measurements on coatings of different thicknesses to be directly comparable the minimum thickness should therefore be at least $2R$. Since the tip end radius of a typical pyramidal Berkovich indenter is 100–200 nm it follows that the minimum thickness is of the order of 200 nm. On ultra-thin (≤100 nm) hard carbon films the hardness measured by nanoindentation may be less than the true film hardness, either at very shallow penetration depths where the indenter is more spherical in shape and the mean pressure is less than the hardness, or at greater depths where plastic yield in the substrate occurs before the plastic zone is fully developed within the coating. To measure the true film hardness of ultra-thin coatings it is recommended to use sharper indenters, for example cube corner (but note that the likelihood of cracking may increase) and/or deposit on a substrate with higher yield stress.

4.3.1.2 Elastic Modulus Determination

The hardness of a hard coating on a soft substrate can usually be taken as the value at a relative indentation depth of 0.1 (the Bückle rule, indentation to less than 1/10 of the total film thickness to determine the true coating hardness independent of any substrate influence). Using finite element analysis (FEA), Veprek-Heijman and Veprek have noted that for superhard coatings plastic substrate deformation was found even at relative indentation depth of 0.03 [33]. It has been erroneously assumed that the elastic modulus measurements at a relative indentation depth of 0.1 can be taken as being representative of the coating. The rule does not work as the elastic stress field extends much further into the substrate than the plastic stress field so that the value at a relative indentation depth of = 0.1 will contain appreciable contribution from the substrate.

Improved strategies for elastic modulus determination are (i) performing indentations to a range of indentation depths and extrapolating the results to zero penetration depth to obtain a result free of substrate influence or (ii) analytical treatments requiring knowledge of the substrate properties. Although in general the modulus–depth relationship is not linear, nevertheless the ISO approach can provide a more accurate measurement of the coating modulus than relying on the value at a relative indentation depth of 0.1. For hard/brittle coatings the data taken for the extrapolation should be in the range $a/t_c < 2$. For non-spherical indenters the effective contact radius (a) is approximated by a circle with the same area as the projected area of contact of the indenter (i.e. a = $\sqrt{(A_p/\pi)}$). Jennett and co-workers reported that when testing 1–3 µm TiN coatings on stainless steel the elastic modulus measurements for coatings with different thicknesses were almost indistinguishable provided they were compared at the same values of a/t_c [34], and the coating-only modulus from extrapolation to zero a/t_c varied by only 1% between them. While there is no safe depth range at which the elastic response of a coated system to indentation is completely free of substrate influence, its contribution is minimized by testing to 3–5% relative indentation depth [32], particularly when coating and substrate have similar moduli. The ISO-extrapolation method can also be used accurately with data from multi-cycle ("load-partial unload") tests when the coating does not exhibit a very strong time-dependent response. Table 4.1 shows the results of elastic modulus measurements at a relative indentation depth of 0.1 for 40 and 80 nm ta-C films on glass indented with a sharp cube corner indenter. At that depth the measurements contain a contribution from the lower stiffness glass substrate which is more noticeable for the stiffer 80 nm film. The ISO-extrapolation approach and an analytical treatment based on the effective indenter concept of Pharr [35, 36] agree to within a few %.

4.3.1.3 Indentation Energy and H/E

Curve fitting to FEA by Cheng and Cheng gives the value of the proportionality constant x in equation 2 as ~5 [7, 8]. For bulk materials experimental evidence suggests that x ~ 5 for glasses and x ~6–7 for metals [37, 38]. For hard coating systems x has been reported to be generally between these values [39, 40]. Although the relationship in Equation 2 implies a linear relationship between H/E and the dimensionless indentation energy plasticity index, $W_p/(W_p+W_e)$, closer examination by Chen and Bull has shown that it is not exactly linear, with the divergence from linearity becoming more marked as H/E_r increases beyond 0.1 [41]. Higher values of x are found for materials with higher H/E (or Y/E). Tabor [42] showed that H and Y are linked through the constraint factor, which was ~3 for metallic materials although much lower values have been proposed for high Y/E materials. In hard coating systems, x may additionally depend on (i) the peak indentation load (higher loads result in greater substrate contribution to elastic modulus and may result in cracking), (ii) whether the coating exhibits full plasticity at this load (otherwise the measured hardness will be lower), (iii) indenter rounding at the low indentation depths required to measure coating properties

TABLE 4.1

Elastic modulus measurements on ultra-thin films

ta-C thickness	E at $h_c/t_c = 0.1$	E (extrapolation)/GPa	E (analytical)/GPa
40 nm	105	115	111
80 nm	138	175	179

Measurements made with a cube corner indenter. Coatings deposited on glass substrate.

and (iv) elastic accommodation of deformed material under the indent. In very high H/E materials more displaced material is accommodated elastically under the indent and some of this is may not be recoverable due to the plastic deformation around the indentation. In this situation the irreversible term, W_p, may contain a contribution from locked-in elastic deformation in addition to normal plastic deformation.

4.4 SCRATCH AND WEAR TESTS

Instrumented wear test methods that simulate contact conditions so that the major deformation mechanism can be reproduced provide fundamental understanding and can be effective screening coating tools. Examples include the progressive load scratch test, repetitive scratch test, reciprocating tests and impact tests. A major advantage of this approach compared to bulk-scale testing is the simplified geometry in the nano-/micro-scale tests which allows modelling of the stresses involved and determination of the deformation mechanisms. The possibility to acquire multiple signals during the test (e.g. wear depth, friction, acoustic emission) and post-test microscopic analysis can provide further support. Predicting coating performance from small-scale wear tests is a more direct approach than relying solely on mechanical properties from indentation. Indentation testing has also been used to simulate the deformation mechanisms in single particle impact and Bull has noted that low-speed erosion damage of architectural and decorative components in the built environment, and the scale of damage produced, is comparable to that generated in indentation tests [25]. Modelling of indentation fracture has been used to simulate ductile machining in hard and brittle materials [43].

4.4.1 MACRO-SCALE SCRATCH TEST

The conventional (macro-scale) scratch test, where a 200 μm diamond probe is slid across the surface under a gradually increasing load, was initially primarily developed as a test of coating adhesion with higher critical loads directly corresponding to more adherent films [44, 45]. It is also a convenient method for producing damage to simulate abrasive wear [46, 47]. The major deformation mechanisms in abrasive wear including plastic deformation, cutting, micro-fracture and delamination can be replicated in the scratch test [47]. However, it has since been shown that it is the mechanical properties of the film and substrate that exert a significant influence on the deformation behaviour rather than the adhesion strength alone [48–64]. When the same probe is used for indentation and scratch testing, the deformation in both tests can be directly compared. In their study of TiN-coated high-speed steel, von Stebut and co-workers reported that the critical loads for spallation were significantly lower in scratch than in indentation testing (45 N vs. 195 N) [50]. The scratch test method does not measure the fundamental adhesion strength of the bond between the coating and the substrate. ASTM C1624 states that instead it "provides a quantitative engineering measurement of the practical (extrinsic) adhesion strength and damage resistance of the coating–substrate system as a function of applied normal force" [45]. The scratch test critical load is dependent on a range of extrinsic and intrinsic factors in addition to the interfacial strength. Intrinsic factors include scratching speed, loading rate, tip radius and extrinsic factors include substrate hardness, and the roughness, thickness and friction of the coating. Steinmann and co-workers studied how the critical load of a 3.5 μm thick CVD TiC coating on steel varied with scan speed and loading rate [44]. They reported that the critical load decreased at lower loading rate and increased at lower scratching speed for this system. However, when tests were performed at constant dL/dx (the increase in load per unit scratch distance where L = normal load and x = scratch distance), the critical load was approximately constant within the precision of their measurements. As lower critical loads were recorded when dL/dx decreased, it was recommended a fixed dL/dx of 10 N/mm be used.

Using a probe radius that is very large in comparison to the thickness of the coating, as is the case with the 200 μm probe radius in the conventional scratch test, produces maximum von Mises stresses that are typically located deep within the substrate at the critical load and substrate yield occurs before the coatings fail. The consequence of this is that the critical load increases with substrate hardness. For example, Ichimura and Ishii showed that for TiN and CrN films on steel substrates with varying hardness, the critical load varied linearly with substrate hardness and the failure behaviour was dominated by plasticity and the cohesive strength of the films rather than their adhesion [59]. Similarly, Wang and co-workers noted that the critical load for DLC and TiN coatings on titanium alloys did not depend on coating composition, but correlated directly with the hardness of the substrate [60]. Extensive substrate yield was revealed by cross-sectional profiles of the scratch track at failure which showed the scratch depth at failure was much greater than the coating thickness.

Three-dimensional FEA has shown that evolution of the stress field during progressive load scratch testing is strongly dependent on the mechanical properties of coating and substrate [61–63]. Bull and Rickerby reported that the changes in critical load with substrate bias for 2–3 μm TiN coatings on M2 tool steel did not reflect a change in adhesion but merely a change in the make-up of the coating removal stresses [54]. Heinke and co-workers reported that L_{c2} critical loads for CrN coatings on steel increased dramatically with their thickness, increasing from 21 N at 5 μm to 55 N for 20 μm coatings [57]. The sensitivity of the progressive load scratch test to adhesion may be low. For example, Ollendorf and Schneider purposely varied pre-sputtering time in TiN coatings on steel to control their adhesion but, contrary to their expectations, found an inverse correlation to scratch test critical load with more adherent films showing lower L_{c1} and L_{c2} critical loads [53].

4.4.2 MICRO- AND NANO-SCRATCH TESTS

The sensitivity of the scratch test to coating and interfacial properties can be increased by decreasing the probe radius and performing nano- or micro-scale scratch tests with instrumentation with greater sensitivity at lower load. By "dimensioning" the test in this way the influence of substrate deformation is reduced [64]. The influence of changing probe radius is illustrated schematically in Figure 4.2. Depending on the choice of indenter radius and load, the maximum stress can be positioned at the interface, or at interfaces between different layers in a multilayer coating system, to investigate potential deficiencies in adhesion.

Micro- and nano-scratch testing has become sufficiently popular that best practice standards are currently being developed [65, 66]. Nano-scratch test data on a wide range of carbon coatings have been critically analysed to determine the sensitivity to intrinsic and extrinsic factors, impact of scan speed and loading rate, influence of probe radius and geometry, estimation of tip contact pressure, film stress and thickness, and the role of ploughing on the load dependence of friction [67–69].

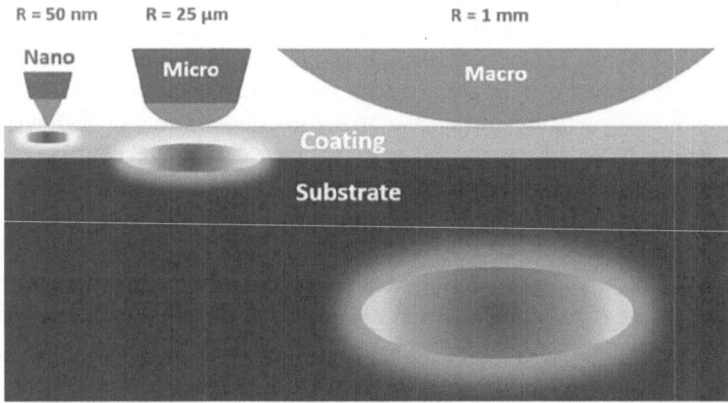

FIGURE 4.2 A schematic illustration of influence of changing probe radius on the depth of peak von Mises stresses in the spherical indentation of a coated substrate.

The influence of loading rate and scan speed on the critical load in the nano-scratch test has been investigated for DLC and Si-doped DLC coatings on glass, DLC and ta-C coatings on Si [67–69]. For these model brittle coating-brittle substrate systems, there was considerably less sensitivity to the scan parameters than has been reported in macro-scale scratch testing of hard PVD coatings [44]. In this study either (i) the scratch speed and load rate were both varied to probe a wide range of dL/dx or (ii) the loading rate was varied with a concurrent increase in scan speed so that the tests were done at the same dL/dx but over a wide range of loading rate. No clear dependence on critical load was found for changes in (1) scratching speed, (2) loading rate or (3) dL/dx, when dL/dx is <<1 N/mm, although critical loads were found to be slightly greater at higher dL/dx (1–5 N/mm). Minimal rate dependence of the critical load was found on a 80 nm ta-C film on silicon over a wider range of dL/dx, with the critical load for film failure being almost constant (113 ± 15 mN) over a 100-fold variation in dL/dx [68].

Within the Nanoindent-plus project [65] nano-scratch tests with a 6.5 μm end radius diamond probe on 450 nm and 962 nm ECR-CVD DLC films on Si were carried out at 0.167/1.67/16.7 mN/s with concurrent increase in scan speed at 0.5/5/50 μm/s so that the tests were all at the same dL/dx. The results are shown in Table 4.2. The films were deposited without an adhesion-promoting interlayer. They provided a suitable model system for studying the parameter sensitivity since they delaminated easily exhibiting small test-to-test variation in critical loads. The thicker film failed at around 60% of the critical load of the thinner film due to high stress. It can be seen from Table 4.2 that the 100-fold increase in loading rate with concurrent increase in scan speed so that the tests were done at the same dL/dx resulted in only a 10% increase in the L_{c1} and L_{c2} critical loads.

For a bulk material displaying no indentation size effects, Equation 4.3 implies a power law dependence with an exponent of two for the variation of the critical load with probe radius. For coatings the relationship will necessarily be more complex when failures occur not solely at a critical pressure but can also be associated with exceeding a critical bending strain in the coating that will be dependent on size of the contact (and hence the radius of the probe). Coating thickness can strongly influence the critical load in the nano-scratch test. Changes in thickness can have opposing effects on the critical load: 1) thicker films that are harder than the underlying substrate provide more load support delaying the onset of the substrate deformation that is often the precursor of film failure (higher critical load); 2) thicker films can be more highly stressed and more easily through-thickness crack and delaminate when deformed (lower critical load) since the driving force for spallation to reduce stored elastic energy is greater. In practice, several nano-scratch studies have reported that the critical load approximately scales with film thickness, i.e. L_c/t_f is a constant [69]. When coatings are stressed then the critical load for failure in front of the probe may still scale with thickness, possibly reflecting delayed substrate yielding for thicker coatings. In Table 4.3 a 5-fold increase in coating thickness resulted in large increases in critical load.

TABLE 4.2

Influence of loading rate and scan speed on the critical loads for CVD DLC films on Si

Film thickness (nm)	Loading rate (mN/s)	Scan speed (μm/s)	L_{c1} (mN)	L_{c2} (mN)
450	0.167	0.5	170 ± 14	172 ± 14
450	1.67	5	182 ± 5	184 ± 5
450	16.7	50	192 ± 4	193 ± 3
962	0.167	0.5	94 ± 1	102 ± 1
962	1.67	5	98 ± 1	107 ± 1
962	16.7	50	105 ± 0	113 ± 2

TABLE 4.3

Influence of thickness and substrate bias for a-C films on Si

Thickness (nm)	Bias (V)	H (GPa)	E (GPa)	H/E	L_c (mN)	L_c/t_f (mN/nm)
200	-40	11.0	189	0.058	48	0.24
200	-80	15.5	219	0.071	62	0.31
1000	-40	12.1	164	0.074	280	0.28
1000	-80	17.2	216	0.080	290	0.29

Note: The critical load stated is for film failure occurring in front of the scratch probe. A failure behind the probe was also observed on the thicker films at a lower load than this.

4.4.3 Modelling the Nano-Scratch Test

Moving from measurements of critical loads to understanding the coating failure mechanisms requires modelling (e.g. Hertzian, finite element or analytical) and/or in combination with microscopic analysis (e.g. SEM/FIB), in some cases with additional support from additional in-test signals (e.g. friction and AE monitoring).

An important refinement to the progressive load scratch is the 3-scan procedure involving (i) pre-scratch surface topography scan, (ii) progressive scratch and (iii) post-scratch surface topography scan. The first reported test of this type was by Wu in 1989 [70, 71]. It was later shown that removal of the instrument compliance contribution to the measured deformation allowed true nano-scratch depth data to be displayed after a levelling (surface slope removal) procedure [72]. It is possible to determine the critical load for the onset of non-elastic deformation (P_y) since this is the load at which the residual scratch depth is no longer zero.

A simple Hertzian analytical treatment of the scratch depth data enables the yield stresses and the pressure required for coating failure to be estimated from contact mechanics, assuming the geometry of indentation, provided spherical indenters are used and elastic recovery is assumed [73]. In spherical indentation the contact depth (h_p) is

$$h_p = \left(h_t + h_r \right) / 2 \tag{4.4}$$

where h_p is the contact depth, h_t is the on-load scratch depth and h_r is the residual depth from the final post-scratch topography scan. The contact radius (a) is determined from Equation 4.5, where R is the indenter radius. The contact pressure (P_m) at any point along the scratch track is given by Equation 4.6, where L is the applied load.

$$a = \sqrt{\left(2Rh_p - h_p^2 \right)} \tag{4.5}$$

$$P_m = L / \pi a^2 \tag{4.6}$$

For the treatment in Equations 4.4–4.6 to produce reasonable estimates of the contact pressure during the nano-scratch test, it is necessary to assume that (i) the presence of a tangential load does not influence the pressure distribution too greatly so that the measured friction coefficient is well below 0.3, (ii) the radius of the indenter is constant, (iii) the sliding speed is sufficiently slow and contact sufficiently close to elastic that the load is supported on the rear of the indenter and (iv) the indenter can reach the bottom of the scratch track in the final topographic scan. The approximation becomes increasingly less accurate as the load increases as the contact geometry moves away from Hertzian conditions due to increasing friction and/or plasticity [74 Hamilton/Goodman]. Comparison of indentation and scratch loading curves with the same spherical indenter [75] shows that at low load there is good agreement between them but they diverge (with more deformation in a scratch test) as the load increases.

The Hertzian treatment is useful, but more information can be gained from full (i.e. 3D) simulated stress distributions which can be evaluated with commercial software packages. One such tool is the Scratch Stress Analyzer (produced by the SIO company, Rugen, Germany [76]) which uses a physical-based analytical methodology to determine simulated stress distributions of the von Mises, tensile and shear stresses during the micro- or nano-scratch test data. Input parameters to the physical-based analytical model are experimental data – the mechanical properties of the coating and substrate, i.e., H, E, H/Y, their Poisson ratios, together with the applied load, scratch depth data, friction coefficient and probe radius used in the nano- and micro-scratch tests. Accurate experimental data is key to the accuracy of the simulations. Examples are shown in later sections. Tabor determined a constraint factor C, connecting hardness and yield stress according to $H = CY$ [42]. Based on experiments on metals a value of 2.8 was found to be a good fit experimentally.

Chudoba and co-workers used the Elastica software to study the dependence of the critical load for plastic deformation in an indentation contact on the yield stress of 2 and 3 μm TiN layers on a steel substrate with yield stress of 3.2 GPa as the radius of the counterpart [77]. The critical load vs. probe radius curves showed a sharp discontinuity at the transition between yield in coating (at low radius) and substrate (at higher radius). They noted a potential weakening effect. For very large radii the coating no longer had a protective effect due to the much higher stiffness of the coating than the substrate resulting in a stress concentration at the interface. This resulted in yield at lower load than the uncoated steel and disproves the idea that a hard stiff coating on a softer substrate will always mechanically protect. Schwarzer has studied the influence of intrinsic stresses on the critical load for yielding in normal (indentation) and mixed (scratch) loading [78]. Holmberg, Laukkanen and co-workers at VTT have used 3D FEM methods to study stress distributions and conditions for fracture in scratch tests of TiN- and DLC-coated hardened steels [61–63, 79, 80] and more recently have used their developed software (ProperTune) to study more complex systems such as thermal spray WC-CoCr coating and laser cladded WC-NiCrBSi thick composite coatings [81, 82].

4.4.4 MULTI-PASS SCRATCH TESTS

Constant load, unidirectional multi-pass scratch testing was described by Bull and Rickerby, and von Stebut and co-workers in 1989 [50, 83] and has been shown to be a reliable low cycle fatigue test. The same approach has proved effective in micro- and nano-scratch testing. The micro-scale scratch test can be considered as a model single asperity contact where the major abrasive wear mechanisms can be well reproduced [46, 47]. Studying single asperity contact at the nano- [84, 85] or micro-scale [47] has distinct advantages compared to standard tribological tests with much larger contact sizes. In standard tribological tests contact occurs only at the peaks of the asperities and the actual contact pressures responsible for the observed behaviour are not known. With the simplified single asperity geometry in the micro- or nano-scratch test the contact pressure can be estimated. A measure of control over the contact pressure and the location of maximum stress can be achieved by performing repetitive (sub-critical load) constant load nano- and micro-scratch tests. By suitable choice of applied load and probe radius either coating-dominated wear properties or interfacial behaviour can be studied [86]. When compared to progressive load nano-scratch testing, repetitive load nano-scratch testing has the advantage that the load can be varied to tune the maximum von Mises stress to be close to the coating–substrate interface and minimize (and potentially even eliminate in some cases) substrate deformation as a precursor of coating debonding. The experiments can be more informative regarding the influence of thin film stress leading to poor adhesion than single scratch tests. An example of this is repetitive constant load nano-scratch testing of 1 μm a-C films deposited at -20 to -120 V substrate bias [86]. The coatings fail in the progressive load nano-scratch test at a critical load of ~200 mN with a 4 μm diamond probe. Repetitive tests were performed at 50 and 150 mN. The load chosen for the low load test ensured that the maximum stresses were within

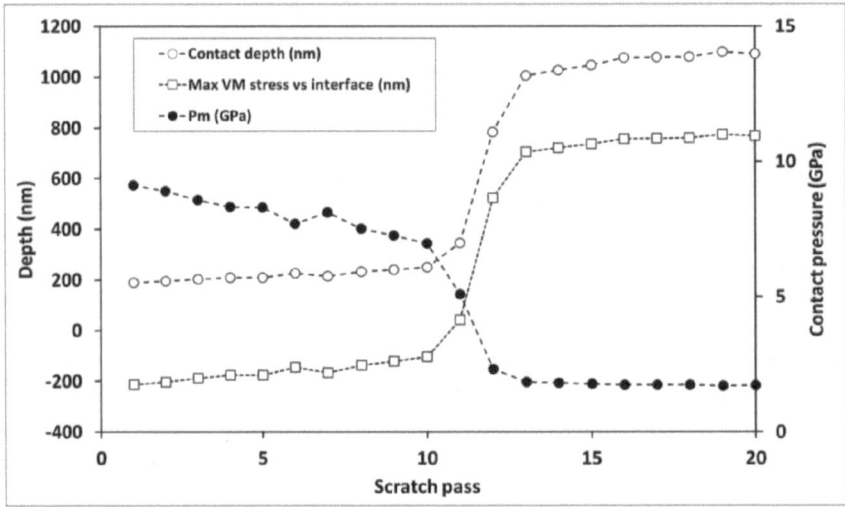

FIGURE 4.3 The variation in contact depth, mean pressure and position of maximum von Mises stress relative to the interface with number of scratch passes in a repetitive nano-scratch test at 70 mN on a 962 nm ECR-CVD DLC film deposited on Si with a $R = 6.5$ μm diamond probe.

the coating, so coating mechanical properties dominate and interfacial strength is less important. At 50 mN contact is almost completely elastic with residual wear depths under 100 nm. At 150 mN Hertzian analysis shows that the maximum von Mises stress is very close to the interface when using the 4 micron probe. Under these conditions plasticity and micro-fracture dominate and the harder films deposited under high substrate bias performed poorly with extensive delamination outside of the scratch track. The combination of repetitive scratch tests at different loading levels provides information regarding the suitability of the a-C films for contact applications. With their intrinsic low friction and high H/E, so that contact remains elastic or close to it over a wide load range, the Hertzian treatment (Equations 4.4–4.6) is particularly suitable to the repetitive nano-scratch testing of DLC films when spherical probes of ~5 μm end radius are used [75]. Alternating constant load scratches and surface topography scans enable the on-load and residual depth for each scratch to be calculated and the mean pressure determined. Figure 4.3 shows a repetitive nano-scratch test on a 962 nm ECR-CVD DLC film deposited on Si at 70 mN with a $R = 6.5$ μm diamond probe. The film failed at around 100 mN in a single progressive load scratch. Assuming the compressive strain in the coating is minimal enables the position of the maximum Von Mises stress to be tracked with respect to the coating–substrate interface for each scratch. In this example the calculations suggest the abrupt delamination failure occurred when the maximum stress was in the vicinity of the interface.

4.5 ILLUSTRATIVE CASE STUDIES

Studies on various thin film systems have revealed the importance of microstructural factors (e.g. grain size, multilayers), coating mechanical properties (e.g. H^3/E^2) and substrate (system response) factors in determining the deformation behaviour. Three illustrative coating systems – (i) ion beam assisted nitride coatings on silicon, (ii) multilayered hard carbon coatings on steel and (iii) nitride-based coatings on cemented carbide – are described in greater detail in the sections below with insights from modelling.

4.5.1 ION BEAM ASSISTED DEPOSITION OF THIN NITRIDE FILMS ON SILICON

Properties of Physical Vapour Deposition (PVD) films in a significant part are defined by the film growth processes. At what energy the atoms arrive to the thin film surface and how they can arrange afterward depend on the overall process conditions. Our ability to understand the effects of growing film surface processes and influence them are vital for attaining the desired thin film properties. Many PVD techniques involve sputtering. It is well known that the peak in energy distribution of the sputtered atoms lies at around 10 eV. The sputtering ions by themselves have not negligible probability to be backscattered as neutral particles from the surface they bombard. As the growing film substrate is in direct field of view then the backscattered atoms can bombard growing film too. This additional bombardment has relatively low (at best <500 eV) energy for magnetron sputtering but during ion sputter deposition the backscatter atoms energy is higher (up to the energy of sputtering ion beam, ~1.5 keV). It is probable that the backscattered atoms can lose significant part or even almost all of this primary energy before arriving to the substrate by collisions with the gas in the deposition chamber. The pressure in the deposition system during DC magnetron sputtering is higher than during ion sputtering and "unintentional" deposition of energy from the arriving to the growing surface atoms is much lower. However, use of reactive sputtering with oxygen is but another story.

In classical DC magnetron sputtering this additional film growth bombardment and energy are very small and cannot lead to film densification and additional film restructuring. The effects in principle are observable but usually not so significant and are ignored. In many cases, however, there is a requirement to add more energy to the growing film in order to enhance surface atom mobility and maybe even to help to achieve desired chemical composition without affecting the substrate. It is technologically easier, if possible, to get the desired effects by raising the substrate temperature. But the substrate temperature by itself only can do only that so much and the substrate itself might limit the possible temperature range.

Intentionally stimulated ion bombardment of the film growing surface is a powerful tool and allows us to significantly affect, and hopefully improve, the film properties. Here we will briefly discuss the model case techniques – ion sputter deposition and Ion Beam Assisted Deposition (IBAD). These deposition techniques have been reviewed by Colligon, Vishnyakov, Mattox, Smidt and others [87–91]. IBAD in some cases is implemented by use of two ion beams – one ion beam is used for material sputtering, while another ion beam is used to bombard the growing thin film. In this case the IBAD system is referred to as dual ion beam deposition. In contrast with complex magnetron sputtering arrangements, ion sputtering and IBAD allow more straightforward separation of deposition parameters and influences of energetic particles on growing film. This provides easier understanding and allows for high flexibility in tweaking of deposition parameters and film compositions. On the downside the deposition rates are relatively low and the techniques either have niche applications or mostly used as a first step in new material development and study of basic film growth processes. When the new material with desired properties is found by using ion sputtering and IBAD, and deposition parameters are settled it is possible to map them into magnetron sputtering for industrial production.

A possible configuration of the system for ion sputtering with IBAD (dual ion beam system) is shown in Figure 4.4(a). The ion sputtering source provides intensive low energy (usually below 1.5 keV and optimised for high sputtering yield) ion beam which bombards the sputtering target. A sputtering target can be geometrically composed from any conductive and vacuum compatible materials. The target materials can be pure metals, alloys and compounds. Geometrical adjustments of target material pieces so that the ion beam sputtered area changes allow the composition of the deposited film to be regulated. The amount of deposited energy on to the target is considerable and free standing without cooling target can reach temperatures in excess of 600 K.

It is very common to discuss how angle of incidence effects on the sputtering yield, for example how many target atoms are sputtered per an incident ion depends on the local angle of incidence. The yield is maximised for the angle somewhere between 60 and 70 degrees for a flat surface. Ion bombarded surfaces very fast develop significant surface roughness and the sputtering yield dependence on incidence to the macroscopic target angle averages over rough surface. The tilted targets shown in Figure 4.4 are only used to shorten distance between targets and the substrate. The working pressure in the chamber ($\sim 10^{-2}$ Pa) is controlled by the amount of gas flowing through the ion source

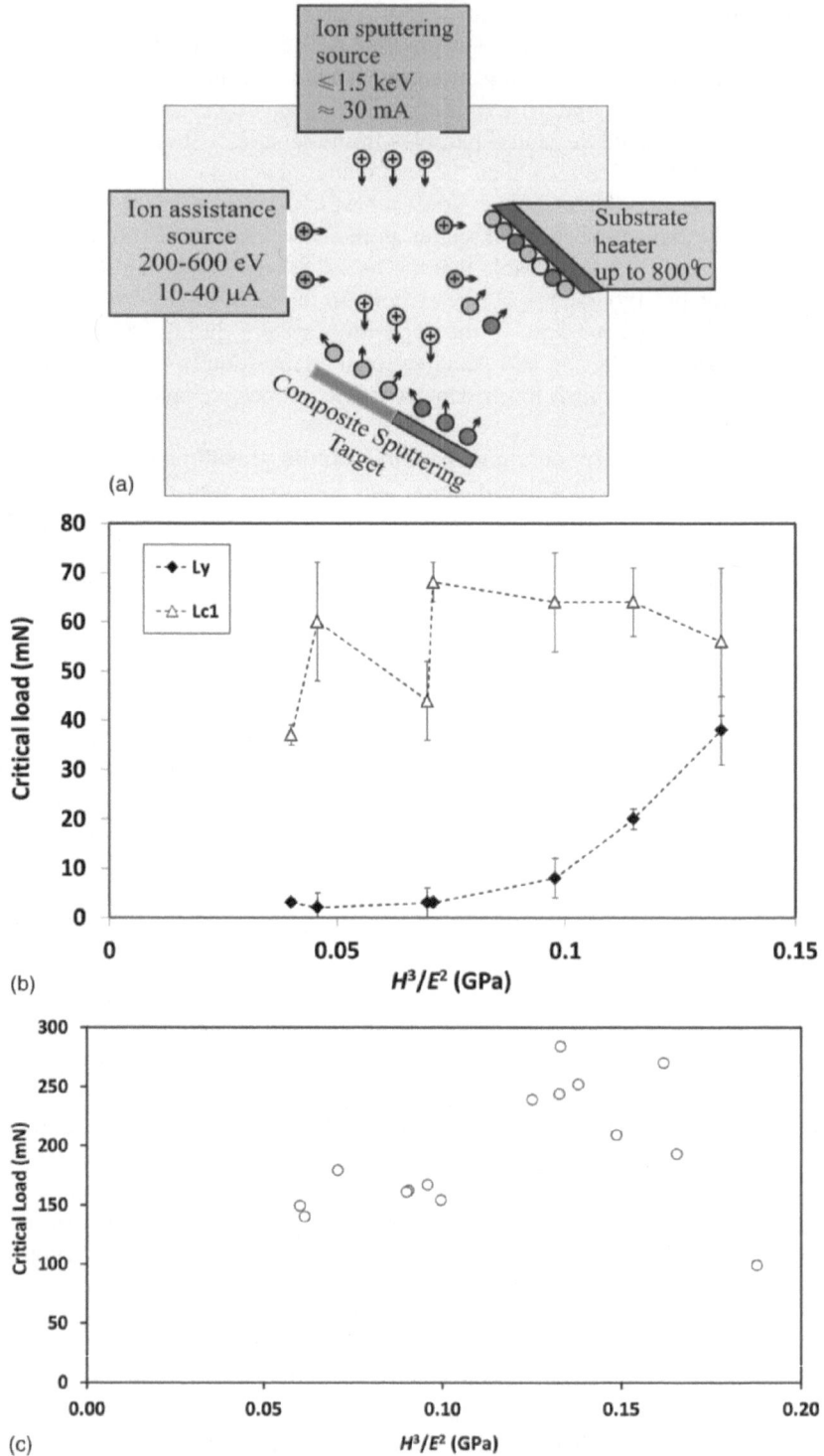

FIGURE 4.4 (a) Schematic representation of IBAD. (b) Variation in the critical load for yield and cracking with H^3/E^2 in nano-scratch tests of TiN/Si$_3$N$_4$ films on Si with a $R = 3$ μm diamond probe. (c) Variation in critical load for film failure for TiFeN and TiN coatings with H^3/E^2.

(Continued)

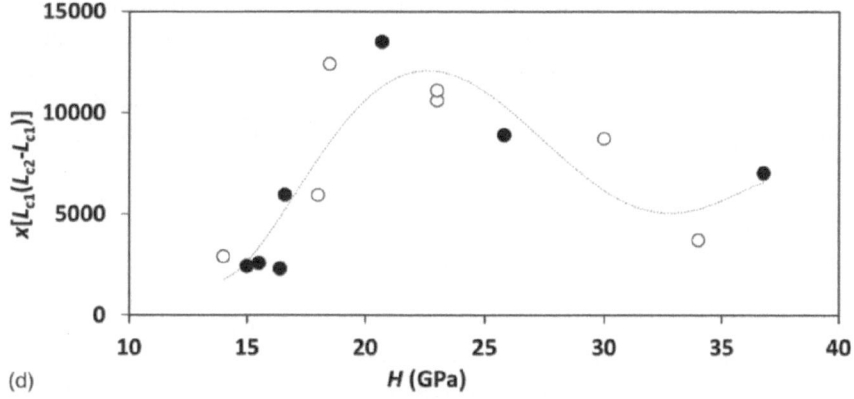

(d)

FIGURE 4.4 (CONTINUED) (d) Relationship between scratch toughness and hardness of nc-TiN/a-Si$_3$N$_4$ films produced by different deposition methods (filled circles = magnetron sputtering; open circles = ion beam assisted deposition).

and chamber pumping speed. Mean free path in this case allows atoms to fly almost without atomic collisions between hard surfaces in the deposition chamber. Sputtered atoms and backscattered atoms can reach the substrate almost without the loss of the initial energy. This atom backscattered atom bombardment leads to compressive stress and slightly higher density of thin film produced by ion sputtering when compared to classical DC MS. The grown film structure depends on the film composition, substrate temperature and gas pressure. Many thin films grown by ion sputtering will have columnar grain structure with long grain boundaries parallel to the deposited particles flypass. Dominance of vertical boundaries (from substrate toward the film surface) makes mechanical properties anisotropic and films tend more easily to develop cracks.

The desire to have the highest possible sputter rates and to produce ceramic films has led to the use of chemically active ambient in the chamber. In some cases, like growth if titanium containing films, it is possible to keep overall chamber pressure relatively low in order to achieve desired stoichiometry. Less chemically active compounds can present a challenge and the required reactive gas pressure can be too high for the effective ion sputtering or will need reactive gas activation. This is where IBAD can play the important supportive role. In some cases assisting ion beam is neutralised by the electron source but in the future discussion we will ignore this as apart from processes in gas phase the neutralisation changes close to nothing in thin film growth.

Ions from ion assistance source bring significant energy which mobilises surface atom migration and assists atom near-surface mobility in the growing film. However, the bombarding beam influence goes further than just to supply additional energy. One needs to remember that at the deposition vacuum conditions many potential atomic layers of gas in the chamber arrive per second to the freshly created film surface. Sticking and reaction probability for this continuous layering is not that high. Energetic ion bombardment increases rate of surface reactions and would even implant atoms from the surface into growing film by recoil collisions.

For many materials it is reasonable to assume that an atom displacement energy is ~20–40 eV. It means that each incoming ion can create a dozen vacancies, some of which can survive recombination and even can be further stabilised by implanted gas atoms. Depending on the ion assistance rate it is not unusual to also have clusters of vacancies and gas bubbles. While the first are relatively difficult to detect and generally almost completely ignored, the second usually manifest themselves as an inert gas signal (argon for instance) in EDX thin film analysis. Although TEM microscopy shows gas bubbles quite well under the right imaging conditions, usually there has not been enough interest shown by industry.

From the film structure point of view, the ion bombardment interrupts columnar structure and makes crystallites much smaller. The films attain almost maximum density. High intensity IBAD even allows amorphous ceramic films to be grown. All the above-mentioned processes modify stress and ultimate residual stress in the thin films. The stress behaviour depends on the thin film nature, deposition conditions, energy and flux of ion bombardment. Stress can change from compressive to tensile depending on the mentioned conditions. Using chemically active species as a gas in the ion assistance source will change film chemical composition. It means that in cases where chemical reactions are relatively slow, but it is desired to have to shift chemical composition to higher stoichiometric compositions, bombardment with a desired reactive gas can be employed. It is well known that ion bombardment leads to preferential sputtering. Sputtering yield for a given atomic bombarding species is complex function which depends on the target atomic species, binding energies, bombardment energy and incident angles (which are not just a fixed value for a corrugated surface). There is some evidence that growing films are a bit more complex than just a mixture of the composing atoms in a way that the atomic concentrations on the surface and in the bulk can be different. This in turn leads to unproportionally high sputtering of surface enriching atomic species.

In short – a film grown under IBAD conditions will change the chemical composition as compared to just deposition from the ion sputtered targets. Macroscopically the IBAD will modify all observed film properties and one can hope that this change at least in significant part is in the desired direction. Vishnyakov and co-workers used a dual ion beam system to deposit PVD nitride coatings on silicon, including TiN/Si_3N_4, TiFeMoN, TiFeN, with TiN and FeN also deposited for comparison [72, 92–95]. The dual ion beam approach provided the flexibility and control of nitrogen content in the deposited films to obtain a wide range of mechanical properties. Ion assistance during the deposition of TiN/Si_3N_4 and TiFeMoN produced harder coatings but ion assistance during the deposition of TiFeN resulted in compositional changes that created a larger fraction of softer FeN.

In one study nanocomposite TiN/Si_3N_4 coatings were produced with a wide range of mechanical properties (H = 13–30 GPa; E = 226–485 GPa; H/E = 0.050–0.068; H^3/E^2 = 0.040–0.133 GPa) on Si [72]. 3-scan ((i) pre-scratch surface topographic profile, (ii) ramped scratch, (iii) post-scratch surface topographic profile) nano-scratch tests were performed with a spheroconical diamond probe with a 3 μm end radius. The initial contact in the nano-scratch test was fully elastic. The following transitions were observed as the load increased: (i) elastic-to-plastic transition (at L_y), (ii) cohesive failure (edge or parallel cracking from L_{c1} critical load), (iii) external transverse cracks, (vi) "unloading failure" on the hardest films and (v) total film failure occurring in front of the probe. There was a clear correlation between the mechanical properties of the coating and the onset of plasticity as shown in Figure 4.4(b). However, this did not translate to a higher critical load for the onset of cracking. The critical load for compressive failure in front of the sliding probe was generally higher for harder films although the films with highest H/E and H^3/E^2 showed pronounced failure behind the probe at lower load. This failure resulted in large area delamination outside the scratch track. Tensile stresses are implicated in this failure behind the probe.

Subsequently TiFeMoN, TiN and N-rich TiFeN were deposited on Si(100) [94]. Delamination occurred in nano-scratch tests with a blunt Berkovich for all the coatings except a TiFeMoN film with moderate hardness (18 GPa) which showed intergranular cracking. It was suggested that formation of non-propagating cracks in this coating relieved the accumulated strain in the scratch test so that large-area delamination was avoided [94]. The dual ion beam system was then used to deposit a larger sample set of 18 $(Ti,Fe)N_x$, 2 FeN and 2 TiN ~1.3 μm thick films on Si(100) substrate with a ~20 nm Ti interlayer to improve their adhesion [95]. Sputtering of the target was by a 1.25 keV Ar^+ ion beam with a 280–600 eV N_2^+ ion beam also used for ion-assistance of the deposited film in some cases. Increasing ion bombardment typically lowered the coating hardness. The large sample set enabled the relationship between mechanical properties and the scratch test critical load and deformation over a wide range of coating hardness and H^3/E^2 to be determined.

The mechanical properties of the coatings strongly influenced the deformation in nano-scratch tests using a well-worn Berkovich indenter with end radius ~1000 nm scratching edge-forward with four main deformation behaviours observed for coatings with different mechanical properties. These were:-

(i) Fe(Ti)N and FeN coatings with low hardness and very low H^3/E^2 displayed predominantly ductile behaviour in the nano-scratch test. Failure was cohesive with ductile chipping at the edges of the scratch track without dramatic coating failures. For some of the coatings with hardness ~15 GPa the ductile deformation was accompanied by delamination from the side of the scratch track.

(ii) The presence of more Ti in the coating increased hardness to ~20 GPa and produced delamination events in the scratch test that were accompanied by fluctuations in the friction force.

(iii) For TiFeN with higher hardness periodic fine cracks were observed in the scratch track. At high load a more localised brittle machining mode was accompanied by a delamination failure behind the contact zone. The friction force almost completely insensitive to the coating failure when this occurred by delamination or brittle machining starting behind the probe.

(iv) Coatings with very high H^3/E^2 showed either brittle machining failure or adhesion failure at low load.

The variation in the L_{c2} critical load with H^3/E^2 for all the (non-ductile) coatings that exhibited a clear L_{c2} failure is shown in Figure 4.4(c). The highest L_{c2} critical loads were found for TiFeN coatings with intermediate mechanical properties (H ~ 25 GPa, H/E ~ 0.075 and H^3/E^2 ~ 0.13 GPa). The three hardest coatings tested exhibited a dramatic unloading failure in the vicinity of L_{c1} (50–100 mN).

Friction data at failure and stress modelling help interpret these results. The frictional response at failure shows the location of the first failure relative to the sliding probe. Marked oscillations in the friction were observed when failure occurs either in front of, or at the front/side of the moving scratch probe, but when failure begins behind the contact zone it was barely noticeable in the friction force despite leading to large-area delamination in many cases. Finite element modelling of the first principal stress in scratch testing of TiN on high-speed steel has shown that there is a tensile stress maximum at the surface at a distance of around half-contact width behind the sliding contact [61–63, 79], in agreement with previous analytical results by Hamilton and Goodman [74] and Johnson [11]. The position of failure onset behind the probe (rather than at the trailing edge of the contact) was consistent with the cracking not being observed in the friction force. The size of the tensile stress at the coating side of the interface was estimated by an analytical method. At 50 mN (~L_{c1}) plastic flow was constrained within the coating but substrate yield did not occur. Tensile stresses were present on the coating side of the interface from low load. With increasing load the peak von Mises stress on the substrate side rapidly approached the substrate yield strength. Three of the coatings (the TiN coatings and the hardest TiFeN coating) exhibited a dramatic unloading failure at a relatively low load (50–100 mN). The stress analysis revealed these coatings had the highest tensile stresses which also increased more rapidly with load than any other coatings studied. The behaviour on these coatings is consistent with coating failure due to a combination of high tensile stress and substrate yield. In contrast, for the softer coatings a different failure mechanism occurred with yielding starting within the coating at lower tensile stress, so that the interface was weakened from the coating side which did not produce the same dramatic unloading failures as when the interface was weakened first by substrate yield [95].

Several authors have investigated the correlation between the critical load in a scratch test and toughness [72, 96–102]. Zabinski and Voevodin proposed that the lower critical load in a scratch test could provide a measure of the fracture toughness [103]. This concept was later extended by Zhang and co-workers [96–102] who equated L_{c1} with the resistance to the initiation of cracks and $(L_{c2} - L_{c1})$ as a measure of crack propagation (L_{c2} = load for total coating failure). They proposed a parameter (later termed scratch crack propagation resistance [CPRs] or "scratch toughness") representing a combined resistance to crack initiation and propagation:

$$\text{Scratch toughness} = L_{c1}\left(L_{c2} - L_{c1}\right) \qquad\qquad 4.7$$

Values of the scratch toughness are inevitably a function of the geometry of the scratch probe (increasing with probe radius) making them difficult to compare between studies when different probe radii have been used. Nevertheless, they provide a useful qualitative assessment of coating behaviour and have been used in several studies aimed at developing "hard yet tough" coatings [96, 98, 100, 102]. The relationship between scratch toughness and hardness of nc-TiN/a-Si$_3$N$_4$ produced by magnetron sputtering [99] or IBAD [72] has been reported with maximum scratch toughness found for coatings with H ~20 GPa (shown graphically in Figure 4.4(d)). The scratch toughness definition accounted for the possibility of unloading failure. The critical load for coating failure (L_{tf}) was taken as the lower value of either L_{c2} or the unloading failure (L_u) so that scratch toughness was defined as:

$$\text{Scratch toughness} = L_{c1}\left(L_{tf} - L_{c1}\right) \qquad\qquad 4.8$$

Similarly the (Ti,Fe)N$_x$ films on Si discussed above exhibit a maximum in scratch toughness at intermediate hardness (~25 GPa). Despite differences in film thickness and scratch probe geometry in these studies, the thin films deposited on silicon show broadly similar trends in their scratch behaviour with mechanical properties. The enhanced toughening in the nano-scratch tests of TiFeN at intermediate hardness appears to be due to a combination of factors: (i) dense nanocomposite microstructure removing the weak columnar boundaries present in TiN, (ii) avoidance of the high tensile stresses that develop in the nano-scratch testing of the hardest films, (iii) sufficient hardness and (iv) localised energy dissipative mechanisms.

4.5.2 DLC Coatings on Hardened Steel

Friction reduction in automotive engines is important to reduce fuel and thus meet environmental and legislative requirements [104–106]. Diamond-like carbon (DLC) coatings are being increasingly used to coat many components in the power train (e.g. tappets, pistons, piston rings, fuel injectors) [107–109]. Although they work well in fully lubricated conditions it is important to design for components operating under mixed or boundary lubrication. Under boundary conditions the surface is critical to achieve low friction and wear. With current power train trends to downsizing, turbocharging, low lubricant viscosity and start-stop the number of components operating in boundary/mixed regime will increase. However, DLC coatings commonly display poor durability under severe loading conditions. Under these conditions performance of DLC coatings is limited by their resistance to contact damage [110, 111] and typically they behave poorly at higher load despite being hard and elastic. Jiang and Arnell have shown that in pin-on-disk tests DLC films that exhibited very low rates of wear under low contact pressure sliding were susceptible to abrupt increases in wear rate as the pressure increases beyond a critical threshold at higher load [112].

Beake and co-workers have investigated how the nanomechanical and nano/micro-scratch behaviour of Si-doped DLC and WC/C coatings compared to a typical hard hydrogenated DLC [40]. WC/C (a-C:H:W) coatings have been shown to perform well in a wide range of tribological conditions [113–115]. With lower friction Si-doped DLC has been explored for a range of applications although its wear resistance can be inferior to undoped DLC [116]. All the coatings studied were 3–4 μm thick with multilayered (Cr/W-C:H/a-C:H, Cr/W-C:H/Si-a-C:H and CrN/a-C:H:W) architectures to improve their adhesion to the hardened M2 steel substrate. They are referred to by their top-layer composition below for simplicity. The a-C:H and Si-a-C:H were deposited by PECVD and a-C:H:W was a commercial coating, Balinit C Star. In the a-C:H and Si-a-C:H coating systems the

TABLE 4.4

Mechanical properties of DLC coatings

	H (GPa)	E (GPa)	H/E	H^3/E^2 (GPa)
a-C:H	23.4	234	0.10	0.23
Si-a-C:H	16.8	151	0.11	0.21
a-C:H:W	13	149	0.09	0.10

Coating-only values from nanoindentation.

adhesion layer is a thin Cr and then gradient layers are applied to adapt the elastic modulus of the soft substrate to the elastic modulus of the hard top coating, thus giving the coating both abrasive wear resistance and impact fatigue wear resistance (flexibility/toughness). In a-C:H:W the hard CrN sublayer provides load support and improved adhesion. Coating-only hardness and elastic modulus (Table 4.4) were determined by the approach described in 4.3.1. Nano- and micro-scratch tests were performed to 500 mN and 5 N, respectively, using a spheroconical diamond probe with an end radius of 5 µm for the nano-scratch tests and 25 µm for the micro-scratch tests. All tests were carried out as 3-scan procedures involving a pre-scan surface profile, ramped load scratch and post-scan surface profile. The critical loads in the nano-scratch tests with $R = 5$ µm and micro-scratch tests with $R = 25$ µm are shown in Table 4.5. The Table shows that probe radius plays an important role in the yield behaviour. With the sharper probe there is a very clear correlation between the coating properties and critical load for yield, but this is absent for the larger radius probe.

The Surface Stress Analyzer (from SIO) was used to provide a more detailed assessment of the stress distributions [117]. Input parameters were (i) mechanical properties of the coating (taken as monolayered) and substrate, i.e. H, E and H/Y ($H/Y = 1.2$ for coatings, $H/Y = 2.5$ for the steel substrate), (ii) coating thickness, (iii) Poisson ratios and (iii) probe radius, applied load and measured friction coefficient in the nano-scratch test. Figure 4.5(a) shows the 2D projections (through the centre-line of scratch) of simulated Von Mises stress distributions at yield in the nano-scratch test with the overstressed areas (where the magnitude of the von Mises stress is greater than the yield stress at that point) shown by hashed region. The scratch direction is left to right. With the $R = 5$ µm scratch probe the initial yield occurs within coating for Si:a-C:H and a-C:H:W but in the substrate for the a-C:H. The stresses in the a-C:H coating are extremely high.

Simulations of sliding contact at 100 mN provide an explanation of the differences in wear resistance reported in 4500 cycle nano-fretting (reciprocating sliding) tests on these coatings with a probe of same nominal geometry [118]. In that study the best wear resistance was found for a-C:H and the worst for a-C:H:W. The simulations show contact is elastic for a-C:H, a small isolated zone of plasticity for Si:a-C:H with more extensive yielded region for a-C:H:W. In practice contact is not

TABLE 4.5

Nano- and micro-scratch critical loads

Coating	$R = 5$ µm nano-scratch			$R = 25$ µm micro-scratch		
	L_y (mN)	L_{c1} (mN)	L_{c2} (mN)	L_y (mN)	L_{c1} (mN)	L_{c2} (mN)
a-C:H	206 ± 5	422 ± 4	> 500	356 ± 9	2179 ± 120	2612 ± 127
Si:a-C:H	110 ± 10	445 ± 12	> 500	383 ± 52	1827 ± 111	2830 ± 367
a-C:H:W	68 ± 4	>500	> 500	375 ± 49	2256 ± 116	3695 ± 132

Peak load 500 mN in nano-scratch; 5 N in micro-scratch.

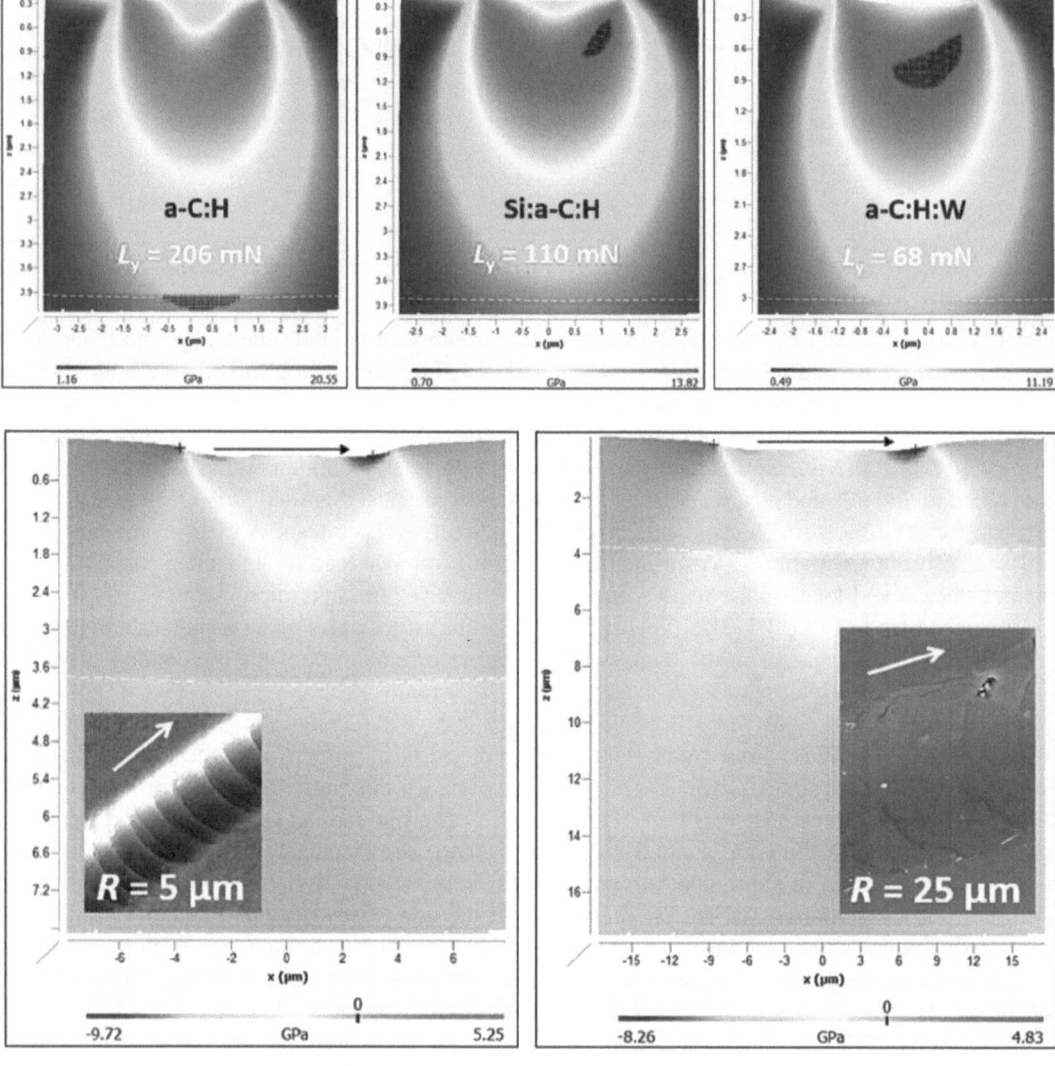

FIGURE 4.5 (a) 2D projections (through the centre-line of scratch) of simulated Von Mises stress distributions at yield in the nano-scratch test. The overstressed areas (where the magnitude of the von Mises stress is greater than the yield stress at that point) shown by hashed region. The scratch direction is left to right. (b) Simulated normal stress in scratch direction at Lc1 on Si:a-C:H shown for $R = 5$ and 25 μm probe geometries. The insets show SEM images of the crack patterns.

TABLE 4.6

Influence of probe geometry on stresses on Si:a-C:H at L_{c1}

Probe radius (μm)	L_{c1} (mN)	Max Von Mises stress in coating (GPa)	Max Von Mises stress in substrate (GPa)	Max normal stress at surface (GPa)
5	445	7.7	5.0	5.3
25	1830	6.7	5.1	4.8

completely wear-free even on a-C:H due to higher contact pressure at asperities as the probe and sample are not perfectly smooth and flat, respectively.

In the nano-scratch tests with the 5 μm probe, there were correlations between crack resistance and the mechanical properties of the coatings. On a-C:H and Si-a-C:H semi-circular cracks at the rear of the contact extended across the entire scratch track from 422 and 445 mN, respectively, but cracking was not observed on the softer and higher plasticity index (lower H/E) a-C:H:W. Plastic flow can be considered as a major source of stress relaxation in the coating system, so more extensive plastic flow in the softer coating system reduces its susceptibility to cracking. A more quantitative treatment is possible from simulating stress distributions, particularly the normal stress in the scratch direction by analytical modelling. The results show that the maximum stress is off-axis, but only slightly lower at the centre line, so for convenience 2D profiles are shown. As the load increased, the location of yield moved into the substrate (over-stressed from coating–substrate interface) for all the coatings. For the a-C:H and Si:a-C:H coating systems, there is arc cracking at rear of contact zone due to high tensile stress – coatings cannot relieve this by plastic flow. The normal stress in scratch direction at L_{c1} on Si:a-C:H is shown for both probe geometries in Figure 4.5(b). The comparative data are summarised in Table 4.6.

Modelling was particularly informative in micro-scratch tests with the 25 μm probe as the higher maximum loads in the micro-scratch test produced L_{c2} failure for all the coatings. With the $R = 25$ μm scratch probe yield began in the substrate for all the coatings, consistent with the almost identical probe depth at that load. Hertzian calculations show the maximum von Mises stress is in the coating at yield, but its magnitude is under the coating yield stress. At 1800 mN, i.e. before cracking occurs in any of the coatings, the stress distributions were very similar and there was more extensive substrate yield. Differences in stress distributions are clearer at L_{c1}. Peak tensile stresses at the rear of the sliding contact at L_{c1} were ~5–7 GPa. For Si-a-C:H the tensile stresses were very similar for the 5 and 25 μm probes (Figure 4.5(b)). a-C:H:W was capable of sustaining higher tensile stresses before cracking. More extensive plastic flow in the substrate in the a-C:H:W system is beneficial. At L_{c2} substrate yield over a greater area on a-C:H:W is consistent with less cracking around the scratch track on this coating than the others. In the micro-scratch tests, interfacial shear stresses at L_{c2} failure were around 2 GPa for all the coatings despite the large differences in critical load. The simulations suggest the failure mechanism on all these DLC coatings was a combination of high tensile stress with plastic flow in the substrate adjacent to the coating–substrate interface.

4.5.3 NITRIDES ON CEMENTED CARBIDES FOR METAL CUTTING

Coated tools in metal cutting operate in tribologically extreme conditions with tangential loads, and in interrupted cutting intermittent high strain rate mechanical contacts also occur. The components in machining applications operate in the region of their elastic limit, or above it, as stresses in cutting can be 2 GPa or more, with Bouzakis and co-workers estimating peak stresses >4 GPa [119]. It is desirable to mimic these high stresses in the laboratory tests. Micro-scratch tests provide severe tests for coatings that can simulate high contact pressure sliding/abrasive contacts. Analytical modelling of scratch tests of hard PVD coatings on nitrides used in high-performance cutting applications is described in more detail below. Nano- and micro-impact tests (Section 4.6) can simulate the intermittent highly loaded contact in milling and interrupted turning operations.

4.5.3.1 Thicker (>10 μm) PVD Coatings

Schwarzer and co-workers studied a range of monolayer and bilayer PVD coatings on WC-Co. TiAlN, AlTiN and CrN coatings were deposited with and without an oxide top layer ($Al_{1.34}Cr_{0.66}O_3$) to a total thickness of 10–11 μm [6]. Scratch tests were performed with 20, 50 and 200 μm diamond probes. Results were correlated to turning tests on C45 steel. Although not directly mentioned in that publication, it is probable that $H/Y = 1$ was used in the analysis as the specific yield strength was given as 22 GPa for the WC-Co and 15–30 GPa for the coating systems. While low values of

H/Y are possible for high *Y/E* materials [120], using *H/Y* = 1 will considerably overestimate the yield strength of the WC-Co substrate (where *Y/E* is rather low) which is considered to be ~10 GPa. Notwithstanding this, the authors were able to (i) show that the location of initial yield was dependent on probe radius, (ii) propose a mechanism and (iii) suggest a route for coating improvement by decreasing the normal stress at the coating surface. The mechanism involved weakening by plastic deformation. They noted that plastic flow occurs wherever the critical von Mises stress is exceeded in the scratch test (initially in the substrate). With increasing load this plastic zone grows until it reaches the interface between the coating and the substrate. This weakens the integrity of the system. If additionally high tensile stresses at the surface coincide with this weakening then mode-I fractures could propagate to this interface resulting in global coating failure by shearing off large areas at the observed critical load [64].

4.5.3.2 3 μm AlCrN, TiAlN and AlTiN Monolayers

Critical load data from micro-scratch tests to 10 N of 3 μm monolayer AlCrN (with Al:Cr = 0.7), TiAlN and AlTiN (with Al:Ti = 0.5 and 0.67, respectively) PVD coatings on cemented carbide with R = 25 μm spheroconical diamond and a 3-scan (topography-scratch-topography) procedure are summarised in Table 4.7 [121, 122]. The AlCrN and AlTiN showed improved resistance to cracking (high L_{c1}). As has been reported elsewhere for AlTiN and AlCrN [123], unloading failures behind the probe were observed which are due to the high maximum tensile stress at the rear of the sliding probe. The maximum shear stresses acting on the coating and substrate sides of the interface at L_{c1} and L_{c2} were estimated as 3.5–3.9 GPa. The Surface Stress Analyzer (SIO) was used to provide the full 3D simulated von Mises, normal stress and shear stress distributions. The maximum stresses on the scratch centre-axis at 1 N and L_{c2} are summarised in Table 4.8. As the load increases the peak stresses in coating reduce but the substrate is overloaded over a much greater area, as illustrated for AlCrN in Figure 4.6. In the ramped load micro-scratch tests significant substrate deformation was required prior to L_{c1} failure of the coating systems. Hertzian analysis predicts that the maximum

TABLE 4.7

Critical loads and position of maximum Von Mises stress below the surface

Coating	L_{c1} (N)	Depth of maximum Von Mises stress at L_{c1} (μm)	L_{c2} (N)	Depth of maximum Von Mises stress at L_{c2} (μm)
AlCrN	2.8 ± 0.2	4.5	5.7 ± 0.2	6.5
AlTiN	3.5 ± 0.8	5.7	4.5 ± 0.1	7.5
TiAlN	2.1 ± 0.4	4.5	4.2 ± 0.5	6.5

R = 25 μm

TABLE 4.8

Maximum stresses on the scratch centre-axis at 1 N and L_{c2} at 25 °C

Coating	Applied load or critical load	Maximum normal stress at surface (GPa)	Maximum von Mises stress in coating (GPa)	Maximum von Mises stress in substrate (GPa)
AlTiN	1 N	11.8	13.1	10.0
	L_{c2}	9.2	11.1	9.8
AlCrN	1 N	10.9	12.3	9.7
	L_{c2}	9.0	10.6	9.8
TiAlN	1 N	10.9	12.3	9.7
	L_{c2}	10.0	11.5	9.8

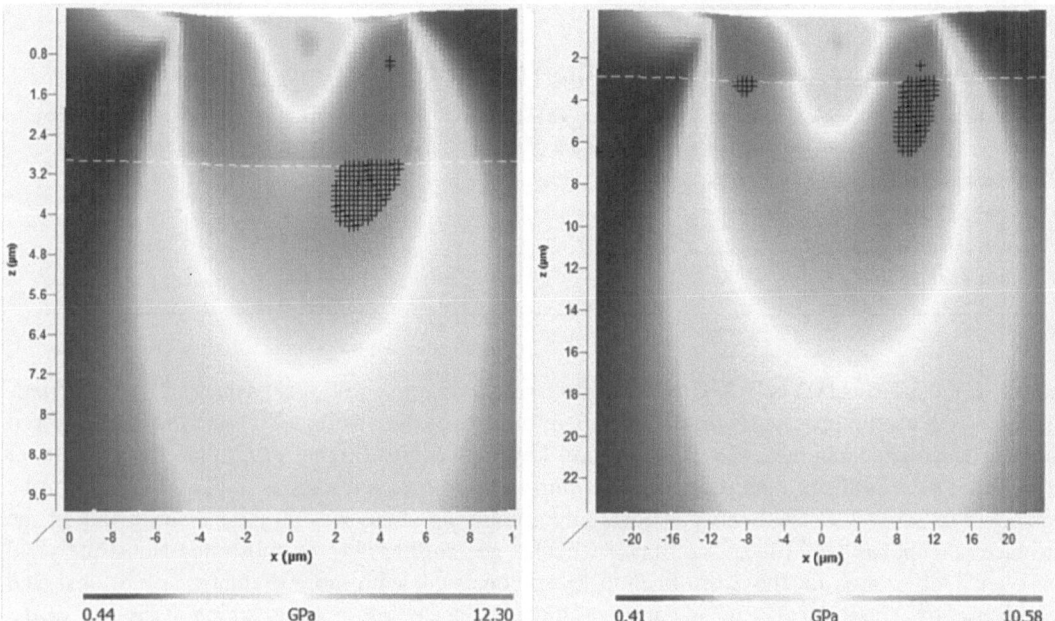

FIGURE 4.6 Simulated von Mises stress distribution for AlCrN at 1 N and L_{c2}. Sliding direction is left to right. Overstressed areas are indicated by the hashed regions.

von Mises stress is located in the substrate at L_{c1} for all three coatings and moves deeper as the load increases. This is consistent with the experimentally observed residual depth at the point of failure exceeding the coating thickness. Repetitive constant load micro-scratch tests were performed at 1 N where the behaviour is more coating-dominated and less influenced by the deformation of the substrate. At this load there were high stresses near the coating–substrate interface. Repetitive scratch tests at this load can be considered as a sub-critical load micro-scale wear test which is potentially more sensitive to adhesion differences than the ramped load scratch test. The on-load depth during the first scratch was almost identical for the three coatings but with repeat scratches differences between them emerged. AlTiN showed the best wear resistance. There was a gradual increase in scratch depth on TiAlN and more abrupt failure on AlCrN. AlTiN has the highest L_{c1} of the three coatings at 3.5 N so micro-fracture events are less likely at the load in the repetitive test (1 N) than on the other coatings which have L_{c1} values closer to this. Grain refinement is promoted in high Al-fraction coatings. The grain size was much smaller in the nanocrystalline AlTiN and AlCrN (~10–20 nm) than in TiAlN (~200 nm) [124–126]. The different deformation mechanisms observed in the repetitive micro-scratch test are consistent with microstructural differences with the Al-rich nanocrystalline coatings initially more wear resistant than the coarser grained TiAlN. The almost identical on-load depth during the first scratch for the different coatings reflected the dominant role of the substrate stiffness on the initial deformation when using a 25 μm probe. Hertzian calculations show that the maximum von Mises stress at 1 N was located in the vicinity of the coating–substrate interface. The repetitive micro-scratch test at this load should be more sensitive to the presence of any areas of weaker adhesion. Analytical modelling (Surface Stress Analyzer from SIO) showed that the maximum von Mises stress on the substrate side of the interface at 1 N exceeded the substrate yield stress although there was no coating plasticity at this load.

4.5.3.3 2–3 μm TiAlCrSiYN-Based Nano-Multilayers

Chowdhury and co-workers studied the influence of coating architecture (monolayer, nano-multilayer, nano-multilayer with interlayer) on tool life for TiAlCrSiYN-based coatings [127]. The coatings studied included (i) 3 μm TiAlCrN monolayer, (ii) 2 μm TiCrAlSiYN/TiAlCrN nano-multilayer

TABLE 4.9

Simulated stresses at L_{c2} failure of TiAlCrSiYN-based coatings

Coating	Max. normal stress at surface (GPa)	Max. VM stress in coating (GPa)	Max. VM stress in interlayer (GPa)	Max. VM stress in substrate (GPa)
2 μm multilayer	9.0	10.8	-	9.8
3 μm monolayer	9.9	11.3	-	9.7
2 μm bi-multilayer	9.2	11.0	11.0	10.2
3 μm bi-multilayer	9.7	11.1	10.5	9.8

and (iii) 2 and 3 TiCrAlSiYN/TiAlCrN nano-multilayers with an additional ~100 nm TiAlCrN bonding layer with elastic properties closely matched to those of the coating. All the nano-multilayered samples had higher hardness (28–31 GPa) and lower plasticity (higher H/E) than the monolayer ($H = 25$ GPa). Simulated maximum stresses during micro-scratch tests with a $R = 25$ μm probe are shown in Table 4.9. The analysis showed the maximum stress was on the substrate side of the interface (except for the 2 μm bi-multilayer where it was further into the substrate). Substrate yield was over a larger area for the 2 μm bi-multilayer, consistent with more dramatic failure observed by microscopy. The 9–10 GPa tensile stresses at the surface result in arc-cracking at the rear of the moving probe. L_{c2} failure requires high tensile stresses at the surface in combination with appreciable interface weakening due to coating bending and substrate yield [64]. Simulation of the shear stress distribution along the coating–substrate interface showed that shear stresses were ~4 GPa at L_{c2}. The calculated stresses were slightly greater for thinner coatings, consistent with thinner coatings being able to accommodate substrate deformation more effectively.

It is useful to compare and contrast the deformation in the micro-scratch tests on the coating systems described above, and to summarise what additional information the stress analysis provides. When deposited on Si wafers, unloading failures occur which result in delamination. On WC-Co unloading failures also can occur, but they are associated with chipping not delamination. On hardened steel these unloading failures were not seen. Provided the coating is sufficiently hard the mechanism of coating failure is consistent with a combination of high tensile stress with plastic flow in the substrate adjacent to the coating–substrate interface, although a different mechanism was found for softer coatings. In this case the interface is weakened from coating side, as since the tensile stresses are lower there is little/no delamination outside of the scratch track. The 4 GPa shear stresses for the hard nitrides on WC-Co were around twice as large as those in the DLC-steel coating systems. Nevertheless, although the critical loads for the DLC coatings and the nitrides varied strongly within each set of coatings (e.g. the critical load of the W-doped DLC was ~50% higher than on Si-doped DLC, and the AlTiN and AlCrN critical loads were ~25 % higher than TiAlN), the calculated shear stresses varied little within each sample set. This result provides further evidence that coating behaviour in the progressive load scratch test is dominated by mechanical factors than interfacial adhesion once the latter is sufficiently good (i.e. while it can be a useful test to show up very poorly produced coatings, otherwise it is not particularly sensitive to adhesion).

4.6 IMPACT RESISTANCE OF PVD NITRIDES ON CEMENTED CARBIDE

High-performance metal cutting operations, such as end and face milling or interrupted turning, involve intermittent high strain rate contact. Intensive adhesive and oxidation wear can occur and some plasticity to dissipate energy generated during friction is important with the fatigue properties of the surface engineered layers also critical for the tool life. Typical failure mechanisms are thermal fatigue during face milling and low cycling fatigue during end milling [128]. To help improve coating durability for these applications, impact tests have been used [129]. Similar arguments to those described above for scratch tests also apply to the importance of contact size in impact tests. Fatigue mechanisms can vary with the ratio of coating thickness to the indenter radius [26]. However, macro-scale tests (typically with WC-Co or hardened steel indenters of 1–3 mm end radius) of thin PVD coatings are performed over a limited range of t/R values which are very low (~0.001). Under these conditions the peak stresses are well into the substrate and the coating contribution to the overall fatigue behaviour may be reduced and the results more strongly influenced by the substrate hardness and toughness as has also been reported in erosion testing under severe conditions [130]. In comparison, fully instrumented nano- and micro-scale impact tests with diamond probes have distinct benefits. Nano- and micro-impact testing utilises the depth sensing capability of a nanoindentation system (NanoTest) to perform impact indentation testing at strain rates several orders of magnitude higher than in quasi-static indentation [131–133]. In the test the probe is retracted a set distance above the sample surface and then rapidly accelerated to produce a high strain rate impact event. Once the probe has come to rest it is retracted and reaccelerated to produce a repetitive test where the impacts occur at the same position on the sample surface. Bouzakis and co-workers have shown the importance of strain rate on the fatigue failure of coatings [134]. The high strain rate contact in nano- and micro-impact tests can provide much closer simulation of the performance of coating systems under highly loaded intermittent contact and the evolution of wear under these conditions. When testing hard PVD coatings a sharp cube corner diamond indenter whose geometry induces high contact strain is commonly used as the test probe in the nano-impact test, to fracture coatings within a short test time.

Benefits of assessing coating fatigue resistance by nano- or micro-scale tests include (i) short duration of the experiments compared to conventional, high-cycle macro-scale tests enabling rapid screening to evaluate the performance of novel coating compositions, (ii) multiple rapid tests possible on single samples, (iii) flexibility to alter loading level and severity of impact loading, (iv) cycle-by-cycle monitoring of the impact-induced deformation providing accurate recording of cycles to failure and information on the fatigue failure mechanism and (v) opportunity to study the stochastic nature of the fracture process.

In the nano-impact test the transition to more severe test conditions with increasingly crack-dominated behaviour is achieved by increasing the impact force. Coatings with higher H^3/E^2 often perform better than coatings with lower H^3/E^2 at low force but less well as the impact force is increased. At high force higher H^3/E^2 coatings may display greater cracking compared to coatings with lower H^3/E^2 [93]. Studies have reported strong correlation between fracture resistance in the nano-impact test and wear of coated tools in high-speed machining [135–138]. Ceramic hard coatings are also applied to compressor blades in aero-engines to reduce sand erosion which for helicopters and transports taking off and landing in desert environments can be life-limiting [139]. Chen and co-workers have reported that the influence of thermal ageing on the solid particle erosion testing of columnar EB-PVD TBCs for aero-engines correlated with rapid nano-impact tests, which was considered to be due to the similar contact footprints in both types of test [140].

(Ti,Al)N-based coatings that are Al-rich (>50 at.%) have shown enhanced fracture resistance and have been consistently reported to be more durable in nano-impact testing [137, 138]. Even a small increase in Al:Ti ratio can result in greater resistance to cracking and lower final impact wear depths than Ti0.50Al0.50N. The nano-impact data is consistent with cutting data, with studies reporting that (Ti,Al)N coatings with %Al >50 at.% showed longer tool life than TiAlN [141, 142]. In machining strong aerospace alloys such as Ti6Al4V and Waspaloy TiAlCrN perform poorly in comparison with AlTiN. Repetitive nano-impact tests suggest a possible reason, since although both coatings initially behave similarly on impact, with repetitive impact the TiAlCrN is much less fracture resistant [135].

In high-speed cutting the optimum combination of coating hardness and plasticity to minimise wear varies with the severity and nature of the cutting conditions and the material being machined. Plasticity index measurements from nanoindentation tests and nano-impact tests both can provide a useful indication of likely fracture resistance. In machining hardened steels, a correlation between room temperature plasticity index measurements and tool life was found for monolayer nitrides [38, 136]. In end milling AlCrN, AlTiN and TiAlCrN all have longer tool life than the harder, lower plasticity index TiAlN. However, the requirement for fracture resistance (rather than high hardness) is generally lower in continuous turning. In continuous turning, TiAlN showed longer life than TiAlCrN, and annealing AlTiN at 700–900 °C (which reduces plasticity index) outperforms non-annealed AlTiN [136]. Several other factors also influence fracture resistance and cutting performance, including the coating nano/microstructure and its compressive stress. The topic of intrinsic stress in thin films has been comprehensively reviewed by Windischmann [143] with the current state of the art in stress measurement recently reviewed by Abadias and co-workers [144]. A pronounced columnar structure (as in TiAlN) is not generally beneficial since the column boundaries can act as lines of weakness for the through-thickness cracks that can result in extensive chipping. To optimise coating durability, a structural advantage is beneficial, such as multilayering, grain refinement/suppression of weak columnar boundaries or introduction of a sublayer.

Multilayer nitride coatings have shown enhanced performance in tribological tests and machining applications [145–154]. Multilayering in Al-rich nitrides has proved effective in producing coatings of relatively low plasticity and yet better fracture resistance than monolayer compositions. Bouzakis and co-workers reported that varying through-thickness multilayered microstructure of 8 μm $Al_{0.54}Ti_{0.46}N$ coatings by increasing the number of layers significantly enhanced resistance to repetitive nano-impact [138, 155]. Figure 4.7 shows the (inverse) correlation between impact depth and cutting life from their data. Impact resistance was also improved in multilayered TiAlCrSiYN/ TiAlCrN with higher H^3/E^2 in comparison to monolayered TiAlCrSiYN coatings with lower H^3/E^2 [156–158]. Higher H^3/E^2 can result in more abrupt failures in repetitive tests. This has been investigated in monolayer TiAlCrSiYN and in TiAlCrSiYN/TiAlCrN with 55–60% Al [157]. Grain size decreases in TiAlCrSiYN multilayers at higher Al %. Monolayers and multilayers with 60% Al were more susceptible to fracture, although it occurred more gradually. Multilayering has also proved effective on other substrates. Chen and co-workers studied the response of multilayer TiAlSiN and monolayer TiN coatings on hardened tool steel in nano-impact tests [153]. A higher load was required for chipping in the multilayered TiAlSiN which was considered to be due to a combination of microstructural (less columnar with multilayer structure to aid crack deflection) and mechanical (higher H^3/E^2) advantages over the monolayered columnar TiN. The nano-impact test has been used to investigate the influence of compressive stresses developed during micro-blasting on the brittleness of $Al_{0.6}Ti_{0.4}N$ on WC-Co [159]. High impact resistance was strongly correlated

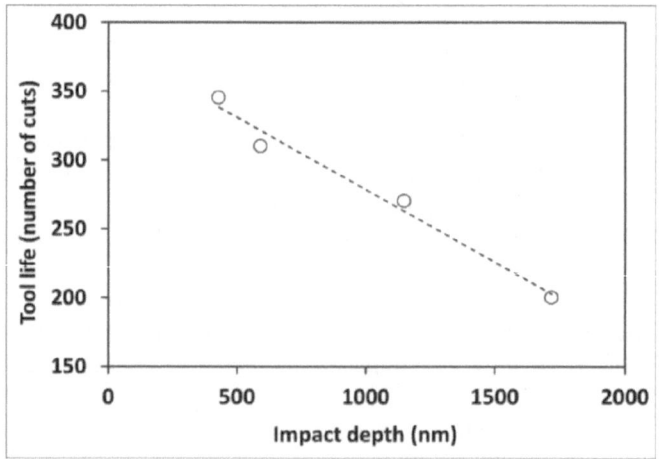

FIGURE 4.7 Relationship between impact depth and cutting life for 8 μm $Al_{0.54}Ti_{0.46}N$ coatings with differing microstructures.

with longer cutting tool life when milling hardened steel (AISI 4140), and trends in tool life with micro-blasting pressure were also well reproduced by the nano-impact test. The relative ranking of cutting performance after micro-blasting with ZrO_2 and Al_2O_3 and a switch in relative performance between 0.2–0.4 MPa were reproduced in the impact test data [101, 159].

Changing the applied load and/or probe geometry in the impact test alters the severity of the test and moves the positions of peak impact-induced stresses relative to the coating–substrate interface. Recently a micro-impact test has been developed enabling tests at higher load (0.5–5 N) and greater accelerating distances than in nano-impact [133]. The energy supplied per impact is the product of the impact load and accelerating distance. The maximum energy supplied per impact with the micro-impact technique is around 2 orders of magnitude greater than the maximum possible in the nano-impact technique enabling indenters of less sharp geometry to be used and cause rapid coating failure.

Nano- and micro-impact tests have been used to study the comparative resistance to impact fatigue of Al-rich PVD nitride coatings ($Al_{0.67}Ti_{0.33}N$, $Ti_{0.1}Al_{0.7}Cr_{0.2}N$ and $Ti_{0.25}Al_{0.65}Cr_{0.1}N$) on cemented carbide. $Ti_{0.25}Al_{0.65}Cr_{0.1}N$ with high H^3/E^2 performed best although it was not the hardest coating studied [133]. There were some differences in the impact fatigue mechanisms observed in nano- and micro-scale impact tests due to the different stress distributions involved. In the nano-impact test with the sharp cube corner probe the failure is more abrupt, with the depth increasing more rapidly over a smaller number of cycles. The abrupt depth change is due to cohesive chipping outside of the impact crater, which is the predominant failure mode for nitride coatings on WC-Co in the nano-impact test, with substrate exposure typically not observed. This absence of delamination is because the stresses are concentrated within the coating. In contrast to nano-impact, the micro-impact test probes the composite behaviour of the entire coating and substrate. The indentation depth at failure is much larger than in a nano-impact test due to greater elastic deformation of the substrate and coating bending to accommodate this. Changing the ratio between coating thickness t and indenter radius R can influence the dominant failure mechanisms in deformation of a coated system by a spherical indenter. The t/R ratio varies between the micro- and nano-impact tests (typically t/R ~1–10 in nano-impact and ~0.1–0.15 in micro-impact), which affects the dominant deformation mechanism. In contrast to the chipping without delamination observed in nano-impact, in the micro-impact chipping is accompanied by substrate exposure and degradation of the hard WC crystals at the periphery of the impact crater. Potential benefits of the micro-impact test lie in its ability to (1) use probes of less sharp geometry, which are more durable than sharper probes, and cause rapid coating failure due to the higher impact energy (2) use different t/R to investigate the role of the substrate on the coating degradation mechanism.

4.7 HIGH TEMPERATURE TESTING

In many applications of hard coatings they, operate at high temperature [136, 141, 160]. Although coatings can reduce the thermal load on the body of the tool nevertheless temperatures can be high in high-speed metal cutting. In developing advanced wear-resistant coatings for this application a critical requirement is to study their behaviour at elevated temperature since the cutting process generates frictional heat which can raise the temperature in the cutting zone to 700–900 °C or more. For this application mechanical properties determined at room temperature may be less relevant than those measured in high-temperature tests. Coatings with the highest hardness at room temperature may soften more rapidly with temperature. For example, Jinhal and co-workers studied the temperature dependence of the hardness of 8 µm PVD TiAlN, TiCN and TiN coatings [161]. Throughout the cutting speed/temperature range the TiN coating had the lowest hardness and poorest machining performance in continuous turning of medium carbon steel, ductile cast iron and Inconel 718. TiCN showed higher hardness than TiAlN from room temperature until 600 °C. Between 600 and 800 °C its hardness decreased more rapidly than TiAlN so that at ≥800 °C TiAlN had the higher hardness. This high hot hardness, along with its lower thermal conductivity, resulted in TiAlN having longer tool life than TiCN in continuous turning of inconel and carbon steel. As further evidence for the importance of temperature the relative improvement of TiAlN over TiCN was more significant at higher cutting speeds where the temperature is highest [161]. Similar behaviour has been observed for CVD coatings. In a study of tool life vs. cutting speed in turning 1045 steel with CVD Al_2O_3, TiN and TiC coated tools Inspektor and Salvador reported a crossover in behaviour as the cutting speed (i.e. and simultaneously the temperature in the cutting zone) increases. At low speed TiC and TiN have longer life but at >300 m/min the alumina coating performed best due to its higher hot hardness [141].

High temperature mechanical/tribological test methods can be divided into the same two main categories as at room temperature: (i) tests for characterisation of mechanical properties and (ii) tests for simulating contact conditions. The general approach taken is the same as at room temperature. High temperature nano-tribological tests provide severe tests for coatings that can simulate high contact pressure sliding/abrasive contacts at elevated temperature. Analytical modelling shows that the behaviour in the high-temperature test can explained by temperature-dependent changes to the stress distribution in the highly loaded sliding contact which influence the critical load.

4.7.1 ELEVATED TEMPERATURE NANOMECHANICS

In addition to providing measurements of hardness and elastic modulus, high temperature nano-mechanical tests can be designed to reveal the temperature dependence of the strain rate sensitivity and apparent activation volume for the process governing plastic flow which can provide evidence for changing dominant mechanisms of plasticity with temperature since their magnitudes can vary strongly for different rate limiting processes [162–164]. Accurate elevated temperature nanomechanical measurements require more careful instrumental and experimental design than at room temperature, considering (i) instrumental stability and thermal drift, (ii) test environment and its influence on the stability of the indenter and sample (i.e. tip and shape wear due to chemical reactions, mechanical wear of the indenter) and (iii) modifications to the experimental load history and analysis procedures to minimise the effect of greater time-dependency. Effective strategies for mitigating these effects to achieving reliable, validated data to 1000 °C are discussed in in the work of Beake and Harris [165].

For hard PVD and CVD coatings the most critical challenge in acquiring reliable data is often not related to instrument stability but to the durability of the hard material chosen as the indenter. Diamond is used almost exclusively at room temperature but at high temperature it is subject to oxidative degradation resulting in dramatic area function changes (i.e. tip blunting) [166, 167]. Other candidate hard materials for indenters, such as cubic boron nitride, have been used, particularly above 500 °C. However, they are more susceptible to mechanical wear at high temperature. This is due to their lower hardness at elevated temperature which can be more important when testing hard materials so that it is difficult to maintain sufficiently higher hardness on the indenter than the sample being tested. For this reason, it is recommended to check for possible indenter area function changes and, if necessary, to correct data for the correct indenter geometry.

High temperature hardness of hard coatings has been measured by Vickers or Knoop micro-hardness hot testing, though this is limited to using high loads and thicker coatings to avoid substrate influence. Quinto and co-workers investigated the variation in microhardness with temperature of a range of 8–15 μm thick PVD and CVD coatings on cemented carbide at 500 mN applied load [168]. Differences in high temperature mechanical properties between PVD and CVD coatings were correlated with microstructure. The CVD coatings possessed larger, defect free, thermally equilibrated grains while the PVD coatings were finer grain size with non-equilibrated microstructure which resulted in a more rapid thermal recovery due to its stored energy (i.e. relaxation of high compressive residual stresses), together with grain boundary sliding.

Most nanoindentation studies with diamond indenters have been limited to a few hundred degrees. Using a cBN indenter and argon purging atmosphere extends the temperature range. Figure 4.8 shows the temperature dependence of H, H/E and H^3/E^2 vs. T for studies with a maximum temperature of 600–750 °C where a cBN indenter and Ar purging have been used. For all the coatings studied there was a clear decrease in mechanical properties with increasing temperature, which is particularly pronounced from 400 to 500 °C. The decrease in H^3/E^2 is even more marked. For these coatings it is notable that there are large differences in how their hardness, modulus and H/E vary with increasing temperature, with the TiAlSiN coatings in particular exhibiting improved thermal stability.

Since its development in the mid-eighties [169] the (Ti,Al)N coating system has been widely studied due to its industrial importance [170–173]. The incorporation of Al into the cubic (fcc) TiN matrix results in a metastable solid solution [174]. Excellent tool life of c-$Ti_{1-x}Al_xN$ coatings has been explained by isostructural decomposition via spinodal decomposition into coherent c-Al and

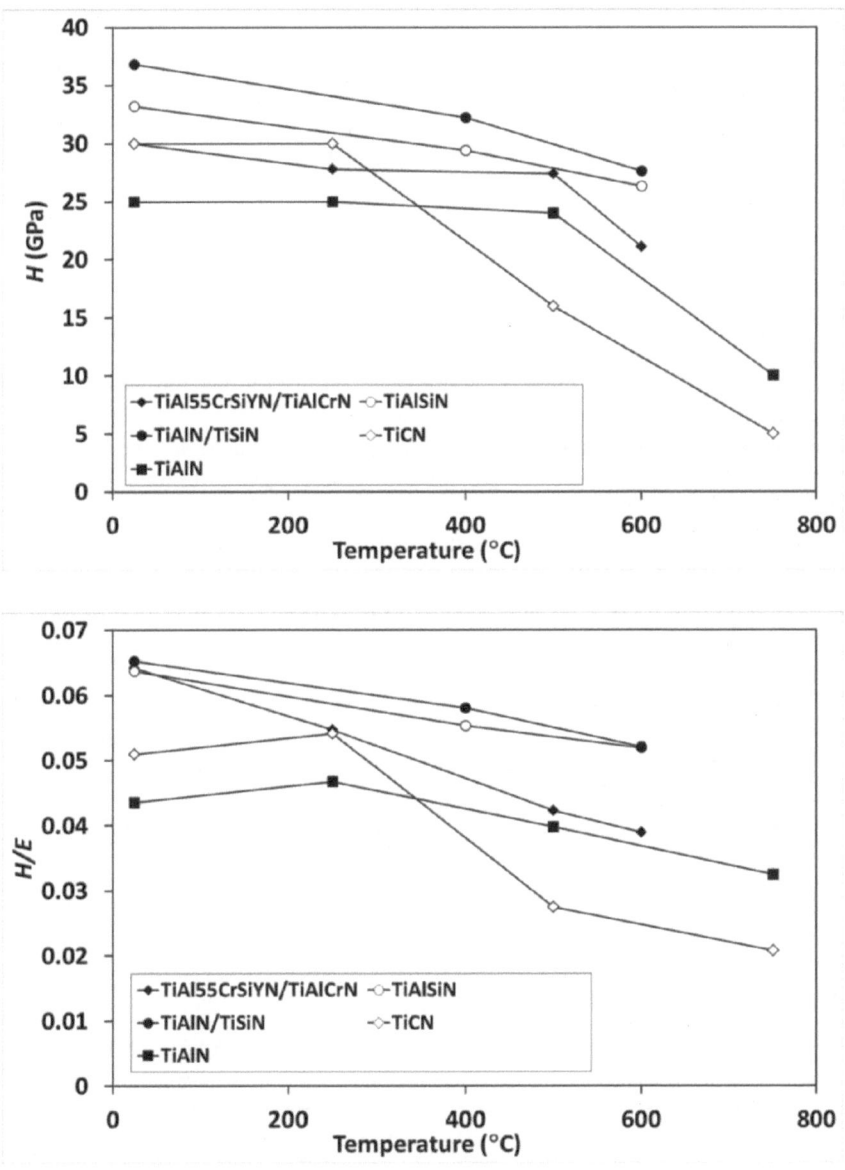

FIGURE 4.8 Temperature dependence of (a) hardness, (b) H/E.

(Continued)

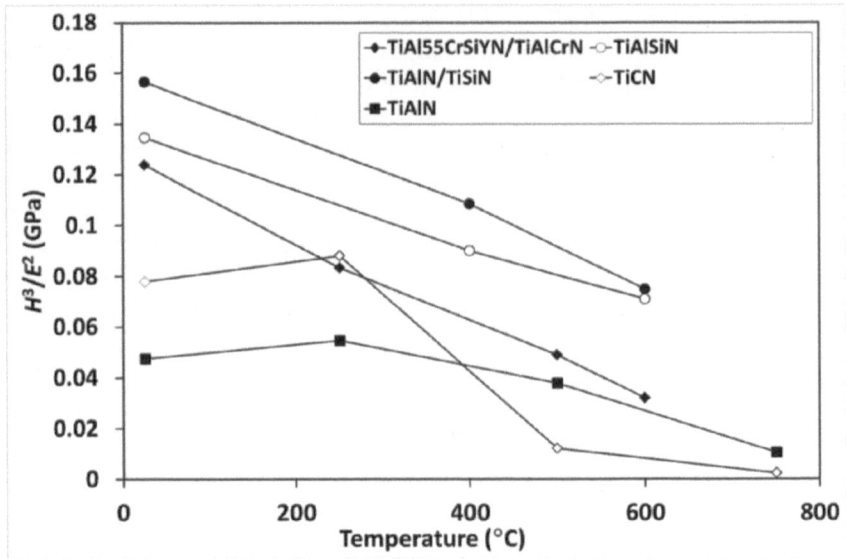

FIGURE 4.8 (CONTINUED) (c) H^3/E^2 for selected PVD and CVD hard coatings.

c-Ti enriched domains at ≥700 °C resulting in increased hardness [173, 175, 176]. The combination of elastic anisotropy and lattice mismatch between the coherent c-AlN and c-TiN rich domains results in age-hardening of ~15–20%. Age-hardening is more significant for $x = 0.66$ [172]. Figure 4.9 shows the influence of composition and annealing on the high temperature hardness of c-Ti$_{1-x}$Al$_x$N. A diamond indenter was used for all these tests with the exception of tests on Al$_{0.6}$Ti$_{0.4}$N to 750 °C that were performed with a cBN indenter under high vacuum. The reduction in hardness is due to relaxation of high compressive residual stresses, grain boundary sliding and thermally activated dislocations [176]. Giuliani and co-workers determined that lattice resistance controlled dislocation glide was the rate-controlling mechanism of TiAlN coatings over the temperature range they studied (25–300 °C) [175]. The hardness of TiN and Ti$_{0.66}$Al$_{0.34}$N coatings decreased rapidly with increasing temperature but increasing Al-fraction resulted in improved stability. The Ti$_{1-x}$Al$_x$N

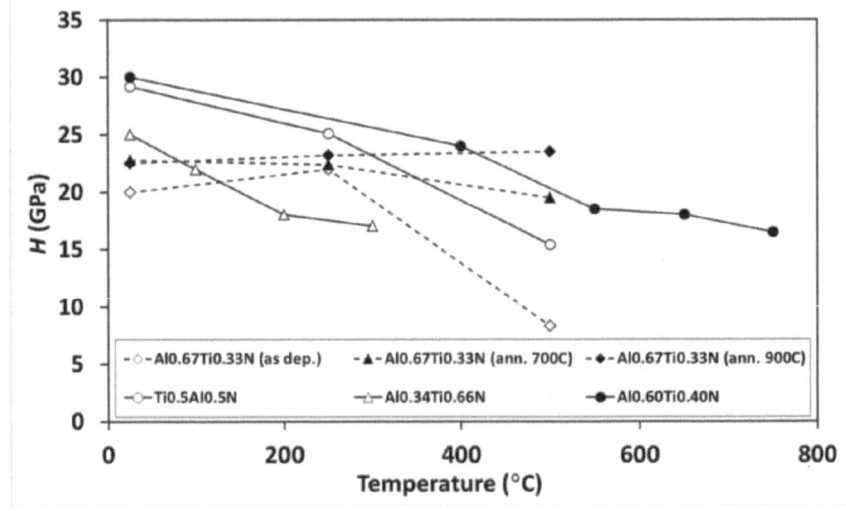

FIGURE 4.9 Influence of composition and annealing on the high temperature hardness of c-Ti$_{1-x}$Al$_x$N coatings.

matrix was more stable than TiN (and high Ti- TiAlN) at elevated temperatures by providing a smaller decrease in lattice resistance to the dislocation flow with increasing temperature. At $x = 67\%$ the composition is close to that where the coatings have dual-phase character with h-AlN and the thermal stability decreases again. However, annealing at 700–900 °C increases the thermal stability in $x = 67\%$. The differences in the hardness measured at temperature in age-hardened AlTiN are more significant than those measured at room temperature. The relatively high hardness of the $Al_{0.6}Ti_{0.4}N$ coating measured in vacuum at 750 °C may be indicative of some hardening at temperature due to spinodal decomposition.

4.7.2 Elevated Temperature Micro-Scratch and Impact Tests

Tribological tests at high temperature show that the test temperature plays a key role in determining wear rate and the dominant deformation mechanism [177–185]. In cutting tests high stresses and temperatures exist in the contact zone. In continuous high-speed cutting applications (e.g. turning) the temperatures can be as high as 700–1000 °C due to frictional heating [172, 173]. In interrupted cutting of hard-to-cut materials, such as Ti6Al4V, they can be significantly lower, and depending on cutting speed may be well under 400 °C [186–188]. Coated components in machining applications operate in the region of their elastic limit, or above it, as stresses in cutting can be >2 GPa so these high stresses should be mimicked in the laboratory tests. Instrumentation developments and key results from high-temperature micro-/nano-scratch and tribology tests have been reviewed by Chavoshi and Xu [189]. Elevated temperature nano- and micro-scale tribological tests can simulate highly loaded sliding/abrasive contacts (scratch) or interrupted contacts in milling (impact). Maximum test temperatures have been usually limited to 500–600 °C due to indenter durability issues at higher temperatures. Although testing at 500 °C cannot replicate all of the coating adaptive mechanisms operative at much higher temperatures (e.g. age-hardening by spinodal decomposition, oxidation and tribo-film formation), it has proved effective in revealing features of the coating deformation at high temperature that could not be predicted from tests at room temperature.

4.7.2.1 High Temperature Micro-Scratch Tests on AlCrN, TiAlN and AlTiN Monolayers at 500 °C

The coatings described in Section 4.5.3 were also micro-scratch tested to 500 °C with the same diamond indenter [121]. L_{c2} increased slightly in comparison to measurements at room temperature for AlCrN (from 5.7 N at 25 °C to 6.0 N at 500 °C) and more significantly on AlTiN (from 4.5 to 5.8 N), but decreased for TiAlN (from 4.2 to 3.7 N). Table 4.10 shows that the contact stresses are lower than those calculated from room temperature (Table 4.7). The maximum shear stresses at the interface at L_{c2} also reduced from 3.9 GPa at 25 °C to 2.2–3.4 GPa at 500 °C. The temperature dependence of the micro-scratch test critical load is primarily a result of the different stress distribution developing a high temperature rather than an improvement in adhesion strength with increasing

---

TABLE 4.10
Maximum stresses on the scratch centre-axis at 1 N and L_{c2} at 500 °C

Coating	Applied load or critical load	Maximum normal stress at surface (GPa)	Maximum von Mises stress in coating (GPa)	Maximum von Mises stress in substrate (GPa)
AlTiN	1 N	5.2	6.0	5.0
	L_{c2}	4.8	5.9	5.5
AlCrN	1 N	8.9	10.5	8.3
	L_{c2}	7.7	9.0	8.4
TiAlN	1 N	9.9	11.2	8.6
	L_{c2}	7.9	9.2	8.4

TABLE 4.11

Repetitive micro-scratch tests at 500 °C

Coating	Scratch pass #1 h_l (µm)	Scratch pass #10 h_l (µm)	Increase in h_l in 10 cycles (µm)	h_r (µm)
AlCrN	1.2	3.5	2.3	3.2
AlTiN	1.2	1.5	0.3	0.5
TiAlN	1.1	3.1	2.0	2.6

h_l = depth under load, h_r = residual depth.

temperature. The maximum shear stress at the L_{c2} failure is lower at 500 °C (compare Tables 4.7 and Table 4.10) implying a reduction in the coating–substrate bonding strength at this temperature.

The temperature-dependent changes in L_{c2} were also seen in repetitive micro-scratch tests at 1 N. The studied TiAlN, with lower L_{c2} at 500 °C than at 25 °C, performed less well in 1 N tests at 500 °C than at 25 °C. In contrast, AlCrN and AlTiN show improved performance at 500 °C, with more cycles required before failure and lower residual depth at the end of the test (Table 4.11). The lower residual depth is likely to be due to reduced fracturing of the WC-Co substrate which also softens slightly at 500 °C. The improved behaviour of the AlTiN coating may be related to its significantly higher L_{c1} at 500 °C. At 1 N there are high stresses in the vicinity of the coating–substrate interface, with maximum von Mises stress estimated by Hertzian mechanics at a depth of 3 µm below the surface (i.e. very close to the interface, assuming the coatings deform elastically). Analytical modelling of the main stresses acting at the interface has shown how the clear differences in coating high temperature mechanical properties influence the initial interface weakening (i.e. by initial substrate or coating yielding or both) and thus determine the deformation failure mechanism at elevated temperature. The differences in the behaviour in the 1 N tests at 500 °C between AlTiN and the other coatings (Table 4.11) can be explained by differences in stress distributions at the interface. For all three coatings at 25 °C, and for AlCrN at 500 °C, the maximum von Mises stress on the substrate side of the interface at 1 N exceeds the substrate yield stress although there is no coating plasticity at this load. For TiAlN at 500 °C the interface is weakened by plastic flow of coating and substrate. On AlTiN at 500 °C there was plastic flow in the coating but von Mises stress in the substrate was much lower than its yield stress, although this is lower at temperature. The higher L_{c1} on AlTiN at 500 °C is a consequence of lower interfacial stresses, plastic flow in the coating and reduced substrate deformation. Von Mises stress distributions in the 1 N wear tests at 500 °C reveal marked differences in how the different coating systems are overloaded. At this temperature and load (i) the TiAlN coating system is overloaded in both the coating and the substrate, (ii) the AlCrN coating system is overloaded only in the substrate and (iii) the AlTiN coating system is overloaded only in the coating. These simulated stress distributions are able to explain the experimental results, i.e. that a change in yield location and deformation mechanism can occur at high temperature due to the reduction in high temperature mechanical properties of the coatings.

The results of the elevated temperature ramped and repetitive micro-scratch tests are consistent with cutting tests where high Al-fraction coatings outperform TiAlN in cutting tests [121, 190–193]. Fox-Rabinovich and co-workers have reported longer tool life for AlTiN compared to TiAlN in face milling of 1040 steel, end milling of 4340 steel and Ti6Al4V, with other groups reporting moderate improvement for AlTiN over TiAlN in face milling of low carbon steel [142] and turning of medium carbon steel [64]. Inspektor and Salvador have reported a gradual improvement in tool life with increasing Al:Ti ratio in face milling of 4140 steel [141]. Fox-Rabinovich reported better performance of AlCrN than TiAlN in end milling of 1040 steel, interrupted turning 42CrMo4V steel and deep hole drilling of hardened structural steel. AlCrN showed longer tool life than AlTiN in hobbing case hardened 5115 steel and in turning Inconel 718 but shorter tool life in end milling 316 austenitic stainless steel.

The higher Al-fraction coatings display multifunctional and adaptive behaviour in high-speed metal cutting. Reasons for their improved tool life include [121]:

(i) age-hardening by spinodal decomposition more effectively than more Ti- or Cr-rich compositions
(ii) forming more protective alumina-based tribo-films than the less protective rutile and chromia tribo-films
(iii) lower thermal conductivity at elevated temperature than TiAlN which (together with the tribo-film) protects the tool from thermal softening, i.e. acting as a thermal shield
(iv) low brittleness at elevated temperature

AlTiN has generally performed well when cutting conditions require relatively high coating plasticity to minimise intensive adhesive-fatigue interaction with workpiece materials (e.g Ti6Al4V, Ni-based superalloys or austenitic stainless steel). Its lower brittleness helps provide more favourable, low wear conditions for effective tribo-film formation protecting the cutting tool surface against chipping. The high temperature micro-scratch data and the analytical modelling described in this section fully support this. Although it has relatively poor high temperature stability at >250 °C, it retains sufficient load support below this temperature, for example as in interrupted cutting of Ti alloys. However, in applications requiring high hot hardness at the cutting temperature, such as continuous turning or end milling of hardened steel, AlTiN is outperformed by other Al-rich coatings that have higher hot hardness (e.g. AlCrN, annealed AlTiN, TiAlCrN, TiAlCrSiYN and TiAlCrSYN/TiAlCrN multilayers).

4.7.2.2 High Temperature Impact Testing

The extent of correlation between properties such as plasticity index and nano-impact fracture resistance when measured at room temperature and the tool life implies either (1) that any trend in mechanical properties with temperature stays broadly similar and/or (2) plasticity (toughness) is significantly more important than the hot hardness/load support in the application – i.e. cracking is life-limiting. Depending on cutting speed, temperatures in interrupted cutting of hard-to-cut materials, such as Ti6Al4V, can be well under 400 °C [186]. Although the hardness of TiAlN and AlTiN reduce considerably by 500 °C, at 250 °C the decrease is not too great [136].

Elevated Temperature Nano-Impact of TiAlN and AlTiN

Elevated temperature nano-impact tests with a sharp cube corner indenter on TiAlN and AlTiN deposited on H10A cemented carbide revealed a reduced probability of coating fracture during the nano-impact test at 500 °C than at 25 °C for both coatings (Table 4.12) [137]. This result was consistent with enhanced plasticity at temperature shown by elevated temperature nanoindentation.

TABLE 4.12

Variation in plasticity index and fracture probability with temperature for TiAlN and AlTiN

Coating	Temperature (°C)	Plasticity index	Fracture probability in nano-impact*
TiAlN	25	0.48	0.8
	500	0.70	0.5
AlTiN	25	0.52	0.4
	500	0.73	<0.2

* Fracture probability, $P(f) = n(N+1)$ at end of test for nth ranked fracture event in total sample size N.

Reduced fracture of AlTiN in elevated temperature nano-impact tests is also consistent with this coating having a higher cracking threshold (L_{c1} critical load) in the micro-scratch test at elevated temperature than at 25 °C [121]. As TiAlN and AlTiN both soften considerably, the impact stresses in the high-temperature tests are lower resulting in reduced fracture. There was qualitative agreement between the nano-impact results and the results of cutting tool tests with these coatings. The AlTiN coating exhibited better resistance to fracture in the elevated temperature nano-impact test and improved cutting performance. When compared to TiAlN it showed: (i) 90% higher tool life in face milling AISI 1040 steel, (ii) 20% higher tool life end milling AISI 4340 steel and (iii) 70% higher tool life end milling Ti6Al4V alloy.

High Temperature Micro-Impact Testing of Monolayer and Multilayer TiAlSiN Coatings

TiAlSiN coatings have better mechanical properties, thermal stability and tool life than TiAlN [194–196]. The addition of Si induces grain refinement producing a dense microstructure resistant to oxidation and contact fatigue [197]. The improved properties are believed to result from the formation of a nanocomposite TiAlSiN structure with cubic TiAlN embedded in an amorphous Si3N4 matrix. Recently nanoindentation, micro-scratch and micro-impact tests have been performed to 600 °C on 2 μm PVD monolayer TiAlSiN and nano-multilayer TiAlN/TiSiN coatings on P30 cemented carbide (10 wt.% Co) [198]. The elevated temperature nanoindentation tests were performed with a cBN Berkovich indenter in a purged Ar atmosphere. Temperature-dependent changes in the elastic properties of the cBN indenter were accounted for in calculating the coating elastic modulus. High temperature nanoindentation data (Figure 4.8) show the TiAlSiN coatings have excellent high temperature mechanical properties in comparison to other ternary and quarternary nitride coatings without Si incorporation. The probes used for micro-impact and micro-scratch tests were spheroconical diamonds with end radii of 17 μm and 25 μm, respectively. In the scratch and impact tests the larger applied loads and greater indenter radius used mean that the high temperature properties of the substrate also become important. Milman and co-workers noted that WC-Co grades with high Co fraction (such as the P30 substrate) have lower room temperature hardness and more rapid decrease in hardness with temperature than grades with lower %Co (6 wt.% in H10A cemented carbide) [199]. In micro-scratch test there was reduced elastic recovery at elevated temperature. The critical load was relatively insensitive to temperature reflecting increased coating bending due to lower load support from substrate softness at high temperature being offset by reduced brittleness. The behaviour of the TiAlSiN coatings in the micro-impact test was dependent on the impact load, coating microstructure, coating and substrate mechanical properties and their temperature dependence. At room temperature both coatings show a brittle response with a transition to more rapid wear. At higher temperatures there was a change in the dominant fatigue mechanism from fracture-dominated to more plasticity-dominated behaviour. This transition to a milder wear mode occurred by 600 °C for the monolayer and by 500 °C for the nano-multilayer. The improvement in behaviour at high temperature is a result of reduced coating and substrate brittleness. SEM images reveal a transition between fracture of WC grains resulting in sub-micron sized WC debris at lower temperature and a more ductile deformation mechanism at higher temperature. Goéz and co-workers noted that for applications requiring greater reliability cemented carbide grades should be chosen for their damage tolerance (i.e. plastic deformation prevailing over brittle fracture as the major damage mode) [200]. Raising the temperature in the micro-impact test appears to have a similar effect as moving to a tougher cemented carbide grade, changing the main deformation mechanism by reducing brittle fracture. Studies by Best and co-workers on CrN coatings have provided further support for temperature-dependent changes in deformation in small-scale tests with flat punch indenters [201, 202]. In high frequency impact testing of a 5 μm CrN coating with a 5 μm flat punch, Best and co-workers reported enhanced plastic deformation at 500 °C [202]. In micro-pillar tests at 500 C they reported that the fracture toughness of a 10 μm CrN was similar to that determined at room temperature, where intergranular fracture was the mechanism, although there was evidence for some transgranular fracture at elevated temperature [201].

4.7.3 Correlation to Application Performance: Design Rules for PVD Coatings on WC-Co

The studied PVD nitride coatings exhibited large differences in how much their hardness, modulus, H/E and H^3/E^2 vary with increasing temperature (e.g. see Figures 4.8 and 4.9).This has a significant influence on their behaviour in high temperature mechanical contact applications (in various cutting tests and small-scale tests – scratch and impact – used to simulate them) which have differing relative requirements in terms of hot hardness (and yield stress) and plasticity (and fracture resistance). The requirements of continuous and interrupted high-speed machining tests are discussed below, illustrated by the effect of annealing on the life of AlTiN coated tools. Fox-Rabinovich and co-workers studied the influence of annealing for 2 hr in vacuum at 700–900 °C on the tool life of AlTiN coated WC-Co in continuous and interrupted contact cutting tests on steels [117, 203]. The AlTiN coating was deposited at 600 °C. Comparative tool life vs. the as-deposited AlTiN coating shown in Table 4.13 illustrates the differing requirements for coating properties in the different cutting situations. The data in the Table show that annealing at 700 °C improved tool life in these applications while annealing at 900 °C was beneficial for continuous turning but detrimental for interrupted turning.

Cutting temperatures in continuous high-speed turning operations are very high and can reach >1000 °C under certain conditions, so consequently high temperature hardness is important for cutting tool life. Coatings with high hot hardness (e.g. AlTiN after annealing at 900 °C, multilayer TiAlCrSiYN/TiAlCrN) have shown longer tool life in high-speed turning of hardened steel and ME 16 superalloy. The requirement for plasticity appears to be less important in the continuous high-speed turning of structural steel. Interrupted cutting conditions alter the relative importance of high temperature hardness (or high temperature yield stress) and plasticity (or high temperature fracture resistance). Here the cutting temperatures are typically lower and toughness and plasticity can be at least as important. In end milling of hardened steels for example, high hot hardness should be combined with improved plasticity for longer tool life.

High Al-fraction monolayer PVD nitride coatings with higher plasticity than TiAlN, such as TiAlCrN, AlTiN and AlCrN outperform it under interrupted cutting conditions. For monolayer nitride coatings optimum values of hardness and plasticity at 450–500 °C to achieve longer tool life in interrupted cutting conditions such as end milling were found to be H ~20 GPa, H/E_r ~0.06 and Plasticity index ~0.6 [38, 126]. The idea of an optimum hardness together with an optimum H/E may be due to an optimum H^3/E^2 since the latter can be considered as H x $(H/E)^2$. The presence of an optimum value at intermediate hardness and plasticity is due to the requirement for both high hot hardness/yield stress and toughness/fracture strength in this application. Coatings with values far outside of these typically have significantly lower tool life under these conditions. After annealing at 900 °C AlTiN does not display an optimum balance of these properties. Annealing at 700 °C results in the AlTiN coating with an optimum combination of properties for both interrupted cutting tests (Table 4.13). In the high-speed machining of hardened steels, high temperature properties of the coating are critical as they influence the environment where the tribo-films form. Coating

---

TABLE 4.13

Influence of annealing vs. tool life of as-deposited AlTiN

	Annealing at 700 °C	Annealing at 800 °C	Annealing at 900 °C
High-speed turning of 1040 steel	+ 85 %		+102%
Interrupted turning of 4340 steel	+31%		-18%
End milling of H13 steel	+64 %	+31%	

Tool life quoted relative to that for the as-deposited AlTiN coating

mechanical properties are important in providing a stable environment for the system to display adaptive behaviour. TiAlCrN, a coating with metal-covalent character, high hot hardness and low plasticity outperforms AlTiN with its more metallic bonding, low hot hardness and higher plasticity under these conditions. Extending the above ideas, in a recent review Fox-Rabinovich, Gershman and Veldhuis suggested that the wear-resistant PVD coatings for high-speed cutting could be considered as *functionally graded nanomaterials* that display adaptive behaviour with features similar to that shown in natural systems [204]. The nanomaterials systems consist of two major layers: the underlying PVD coating and the surface on which an outer nanoscale layer of dynamically re-generating tribo-films is produced as a result of self-organization during friction. The tribo-film nanolayer can efficiently protect the frictional surface, thereby enabling control over the performance of the entire coating layer. The coating layer functions as a catalytic medium for tribo-film generation and replenishment. To promote this function, its architecture, structure and properties have to be carefully optimized.

The mechanical properties alone do not control tool life and it is important to consider the micro/nanostructure of the coating, with nano-multilayer coatings showing relatively low plasticity but yet improved fracture resistance in impact tests and longer tool life in cutting tests. This is due to a combination of factors including (i) higher H^3/E^2 providing enhanced load carrying capability and resistance to crack initiation, (ii) dense microstructure, eliminating/minimising weak columnar boundaries and (iii) nano-multilayer structure providing resistance to through-thickness cracking by crack deflection between layers, with this crack deflection along interfaces providing an additional route to relieving strain accumulation in heavy loaded cutting, (iv) enhanced high temperature mechanical properties. In contrast, the microstructure in columnar monolayer coatings means they cannot as readily combine high H^3/E^2 with a mechanism that avoids large-scale fracture under increasingly severe conditions (i.e. fast crack propagation) and the cutting speed range is more limited.

Simulated stress distributions show reduced stresses are present in high temperature contact than at room temperature, particularly when the coating or substrate properties are highly temperature dependent. At high temperature plasticity becomes more important than fracture, influencing the tool life through the relative importance of these in the cutting test.

4.8 CONCLUSIONS/OUTLOOK

Nano- and micro-scale mechanical test methods can be divided into: (i) tests for characterisation of mechanical properties and (ii) tests for simulating contact conditions.

Recent developments in nanomechanical testing for engineering applications have looked beyond room temperature measurements of coating hardness and focussed on test method development (both experimental capability and improved data modelling) to provide more application-relevant data. This trend offers the possibility to dramatically increase the pace of materials development for demanding applications involving high temperatures and/or contact stresses and/or complex mechanical loading conditions. In particular, the development of high vacuum nanomechanical test systems [9, 165] capable of performing a range of nano- and micro-mechanical tests to 1000 °C and beyond will provide direct linkages between material nano-structure and performance in these applications and the development of nano/microstructure–properties–performance relationships.

Supporting nanomechanical and nano/microtribological testing with simulated stress distributions has been shown to improve our understanding of coating system performance in different tribological contact and to design coating systems for enhanced performance. The importance of contact size on the deformation has been shown with the aid of simulated stress distributions.

High temperature nanomechanical testing has been successfully applied to the study of PVD coatings to help explain why certain coating compositions work well in some applications but not others. Testing at the correct temperature is particularly important where coating and/or substrate properties are strongly temperature dependent as this can result in a change in deformation mechanism from room temperature. For example, plastic deformation can be promoted over fracture at higher temperature.

When used as part of performance prediction or coating screening research the nanomechanical data should ideally be supported by other small-scale laboratory tests simulating high stresses and temperatures in different applications. Changing the contact radius in these tests alters the test sensitivity to coating, interface or substrate properties. In metal cutting continuous turning can be simulated by repetitive high temperature micro-scratch testing and end milling simulated by high temperature nano- and micro-impact with good correlation observed between the coating ranking in the small-scale accelerated tests and the coating performance in the cutting tests.

REFERENCES

[1] J.F. Archard, Contact and rubbing of flat surfaces, *J. Appl. Phys.* 24 (1953) 981.

[2] A. Leyland and A. Matthews, Design criteria for wear-resistant nanostructured glassy-metal coatings, *Surf. Coat. Technol.* 177–178 (2004) 317–324.

[3] A. Leyland and A. Matthews, On the significance of the H/E ratio in wear control: a nanocomposite coating approach to optimised tribological behaviour, *Wear* 246 (2000) 1–11.

[4] C. Rebholz, H. Ziegele, A. Leyland and A. Matthews, Structure, mechanical and tribological properties of nitrogen-containing chromium coatings prepared by reactive magnetron sputtering, *Surf. Coat. Technol.* 115 (1999) 222–229.

[5] J.A. Greenwood and J.P. Williamson, Contact of nominally flat surfaces, *Proc. Royal Soc. London A* 295 (1966) 300–319.

[6] X. Chen, Y. Du, Y.-W. Chung, Commentary on using H/E and H3/E2 as proxies for fracture toughness of hard coatings, *Thin Solid Films* 688 (2019) 137265.

[7] Y.-T. Cheng and C.-M. Cheng, Relationships between hardness, elastic modulus, and the work of indentation, *Appl. Phys. Lett.* 73 (1998) 614–616.

[8] Y.-T. Cheng and C.-M. Cheng, Scaling, dimensional analysis, and indentation measurements, *Mater. Sci. Eng. R* 44 (2004) 91–149.

[9] A.J. Harris, B.D. Beake, D.E.J. Armstrong, M.I. Davies, Development of high temperature nanoindentation methodology and its application in the nanoindentation of polycrystalline tungsten in vacuum to 950 °C, *Exp. Mech.* 57 (2017) 1115–1126.

[10] T.Y. Tsui, G.M. Pharr, W.C. Oliver, C.S. Bhatia, R.L. White, S. Anders, A. Anders, I.G. Brown, Nanoindentation and nanoscratching of hard carbon coatings for magnetic disks, *Mater. Res. Soc. Symp. Proc.* 383 (1995) 447.

[11] K.L. Johnson, *Contact Mechanics*, Cambridge University Press, London, UK, 1985, ISBN: 0-521-34796-3, p. 464.

[12] Y.T. Pei, D. Galvan and J. Th. M. De Hosson, Nanostructure and properties of TiC/a-C:H composite coatings, *Acta Materialia* 53 (2005) 4505–4521.

[13] S. Hassani, M. Bielawski, W. Beres, L. Martinu, M. Balazinski and J.E. Klemberg-Sapieha, Predictive tools for the design of erosion resistant coatings, *Surf. Coat. Technol.* 203 (2008) 204–210.

[14] E. Bousser, M. Benkahoul, L. Martinu and J.E. Klemberg-Sapieha, Effect of microstructure on the erosion resistance of Cr-Si-N coatings, *Surf. Coat. Technol.* 203 (2008) 776–780.

15. D. Jianxin, W. Fengfang, L. Yunsong, X. Youqiang, L. Shipeng, Erosion wear of CrN, TiN, CrAlN and TiAlN PVD nitride coatings, *Int. J. Refract. Met. Hard Mater.* 35 (2012) 10–16.

[16] Y.X. Xu, H. Riedl, D. Holec, L. Chen, Y. Du and P.H. Mayrhofer, Thermal stability and oxidation resistance of sputtered Ti-Al-Cr-N hard coatings, *Surf. Coat. Technol.* 324 (2017) 48–56.

[17] M. Bartosik, C. Rumeau, R. Hahn, Z.L. Zhang and P.H. Mayrhofer, Fracture toughness and structural evolution in the TiAlN system upon annealing, *Sci. Rep.* 7 (2017) 16476.

[18] J. Musil, Physical and Mechanical Properties of Hard Nanocomposite Films Prepared by Reactive Magnetron Sputtering. in *Nanostructured coatings*, eds. A. Cavaleiro and J.Th.M. De Hosson, Springer, New York 2006, pp 407–463.

[19] J. Musil and M. Jirout, Toughness of hard nanostructured ceramic thin films, *Surf. Coat. Technol.* 201 (2007) 5148–5152.

[20] J. Musil, Advanced hard coatings with enhanced toughness and resistance to cracking, Chapter 7, pp 378–463 in *Thin Films and Coatings: Toughening and Toughness Characterisation*, ed. S Zhang, CRC Press, July 2015, p 383.

[21] J. Musil, Hard nanocomposite coatings: thermal stability, oxidation resistance and toughness, *Surf. Coat. Technol.* 207 (2012) 50–65.

[22] M. Tiegel, R. Hosseinabadi, S. Kuhn, A. Hermann, C. Russel, Young's modulus, Vickers hardness and indentation fracture toughness of alumino silicate glasses, *Ceram. Int.* 41 (2015) 7267–7275.

[23] K. Shi, C. Wang, C. Gross, Y.W. Chung, Reversing the inverse hardness-toughness trend using W/VC multilayer coatings, *Surf. Coat. Technol.* 284 (2015) 80–84.

[24] K.-D. Bouzakis, A. Siganos, T. Leyendecker and G. Erkens, Thin hard coatings fracture propagation during the impact test, *Thin Solid Films* 460 (2004) 181–189.

[25] S.J. Bull, Using work of indentation to predict erosion behaviour in bulk materials and coatings, *J. Phys. D: Appl. Phys.* 39 (2006) 1626–1634.

[26] J. Michler, E. Blank, Analysis of coating fracture and substrate plasticity induced by spherical indentors: diamond and diamond-like carbon layers on steel substrates, *Thin Solid Films* 381 (2001) 119–134.

[27] Q. Yang, D.Y. Seo, L.R. Zhao, X.T. Zeng, Erosion resistance performance of magnetron sputtering deposited TiAlN coatings, *Surf. Coat. Technol.* 188–189 (2004) 168.

[28] T. Liskiewicz, K. Kubiak, T. Comyn, Nano-indentation mapping of fretting-induced surface layers, *Tribol. Int.* 108 (2017) 186–193

[29] Y. Liu, T. Liskiewicz, B.D. Beake, Dynamic changes of mechanical properties induced by friction in the Archard wear model, *Wear*, 428–429 (2019) 366–375.

[30] D.B. Marshall, R.F. Cook, N.P. Padture, M.L. Oyen, A. Parajes, J.E. Bradby, I.E. Reimanis, R. Tandon, T.F. Page, G.M. Pharr and B.R. Lawn, The compelling case for indentation as a functional exploratory and characterization tool, *J. Am. Ceram. Soc.* 98 (2015) 2671–2680.

[31] M. Sebastiani, K.E. Johanns, E.G. Herbert, G.M. Pharr, Measurement of fracture toughness by nanoindentation methods: recent advances and future challenges, *Current Opinion in Solid State and Materials Science* 19 (2015) 324–333.

[32] ISO 14577, Metallic Materials — Instrumented Indentation Test for Hardness and Materials Parameters, Part 4 — Test Method for metallic and non-metallic coatings, 2007.

[33] M.G.J. Veprek-Heijman, S. Veprek, The deformation of the substrate during indentation into superhard coatings: Bückle's rule revised, *Surf. Coat. Technol.* 284 (2015) 206–214.

[34] N.M. Jennett, G. Aldrich-Smith and A.S. Maxwell, Validated measurement of Young's modulus, Poisson ratio and thickness for thin coatings by combining instrumented nanoindentation an acoustical measurements, *J Mater Res* 19 (2004) 143–147.

[35] G.M. Pharr, A. Bolshakov, Understanding nanoindentation unloading curves, *J. Mater. Res.* 17 (2002) 2660–2671.

[36] N. Schwarzer, Analysing nanoindentation unloading curves using Pharr's concept of the effective indenter shape, *Thin Solid Films* 494 (2006) 168–172.

[37] Y. Choi, H.-S. Lee and D. Kwon, Analysis of sharp-tip indentation load-depth curve for contact area determination taking into account pile-up and sink-in effects, *J. Mater. Res.* 19 (2004) 3307–3315.

[38] B.D. Beake, G.S. Fox-Rabinovich, S.C. Veldhuis, S.R. Goodes, Coating optimisation for high-speed machining with advanced nanomechanical test methods, *Surf. Coat. Technol.*, 203 (2009) 1919–1925.

[39] B.D. Beake, G.S. Fox-Rabinovich, Y. Losset, K. Yamamoto et al, Why can TiAlCrSiYN-based adaptive coatings deliver exceptional performance under extreme frictional conditions? *Faraday Discussions* 156 (2012) 267–278.

[40] B.D. Beake, T.W. Liskiewicz, V.M. Vishnyakov and M.I. Davies, Development of DLC coating architectures for demanding functional surface applications through nano- and micro-mechanical testing, *Surf. Coat. Technol.* 284 (2015) 334–343.

[41] J. Chen, S.J. Bull, Relation between the ratio of elastic work to the total work of indentation and the ratio of hardness to Young's modulus for a perfect conical tip, *J. Mater. Res.* 24 (2009) 590–598.

[42] D. Tabor, *The Hardness of Metals*, Clarendon Press, Oxford, UK, 1951, ISBN: 0-521-29183-6.

[43] H. Huang, B.R. Lawn, R.F. Cook, D.B. Marshall, Critique of materials-based models of ductile machining in brittle solids, *J. Am. Ceram. Soc.* (2020) DOI: 10.1111/jace.17344.

[44] P.A. Steinmann, Y. Tardy and H. E. Hintermann, Adhesion testing by the scratch test method: influence of intrinsic and extrinsic parameters on the critical load, *Thin Solid Films* 154 (1987) 333–349.

[45] ASTM C1624-05 (2015), Standard Test Method for Adhesion Strength and Mechanical Failure Modes of Ceramic Coatings by Quantitative Single Point Scratch Testing.

[46] M.G. Gee, Low load multiple scratch tests of ceramics and hard metals, *Wear* 250 (2001) 264–281.

[47] J.C.P. Zuñega, M.G. Gee, R.J.K. Wood and J. Walker, Scratch testing of WC/Co hardmetals, *Tribol. Int.* 54 (2012) 77–86.

[48] N.X. Randall, G. Favaro, C.H. Frankel, The effect of intrinsic parameters on the critical load as measured with the scratch test method, *Surf. Coat. Technol.* 137 (2001) 146–151

[49] N.X. Randall, The current state-of-the-art in scratch testing of coated systems, *Surf. Coat. Technol.* 380 (2019) 125092.

[50] J. Von Stebut, R. Rezakanlou, K. Anoun, H. Michel, M. Gantois, Major damage mechanisms during scratch and wear testing of hard coatings on hard substrates, *Thin Solid Films* 181 (1989) 555–564.

[51] R. Rezakhanlou and J. von Stebut, Damage mechanisms of hard coatings on hard substrates: a critical analysis of failure in scratch and wear testing, in D. Dowson, C. M. Taylor and M. Goder (eds.), *Proc. 16th Leeds-Lyon Symposium, INSA Lyon, 1989, Tribology Series 17*, Elsevier, Amsterdam, 1990.

[52] R. Rezakhanlou, A. Billard, M. Foos, C. Frantz and J. von Stebut, Influence of the intrinsic coating properties on the contact mechanical strength of perfectly adhering carbon-doped AISI 310 PVD films, *Surf. Coat. Technol.* 43–44 (1990) 907-919.

[53] H. Ollendorf and D. Schneider, A comparative study of adhesion test methods for hard coatings, *Surf. Coat. Technol.* 113 (1999) 86–102.

[54] S.J. Bull and D.S. Rickerby, New developments in the modelling of the hardness and scratch adhesion of thin films, *Surf. Coat. Technol.* 142 (1990) 149–164.

[55] P.J. Burnett and D.S. Rickerby, The relationship between hardness and scratch adhesion, *Thin Solid Films* 154 (1987) 403–416.

[56] S.J. Bull, Failure modes in scratch adhesion testing, *Surf. Coat. Technol.* 50 (1991) 25–32.

[57] W. Heinke, A. Leyland, A. Matthews, G. Berg, C. Friedrich, E. Broszeit, Evaluation of PVD nitride coatings, using impact, scratch and Rockwell-C adhesion tests, *Thin Solid Films* 270 (1995) 431–438.

[58] M. Larsson, M. Olsson, P. Hedenqvist and S. Hogmark, Mechanisms of coating failure as demonstrated by scratch and indentation testing of TiN coated HSS – On the influence of coating thickness, substrate hardness and surface topography, *Surf. Eng.* 16 (2000) 436–444.

[59] H. Ichimura and Y. Ishii, Effects of indenter radius on the critical load in scratch testing. *Surf. Coat. Technol.* 165 (2003) 1–7.

[60] C.T. Wang, A. Escudeiro, T. Polcar, A. Cavaleiro and R.J.K. Wood, Indentation and scratch testing of DLC-Zr coatings on ultrafine-grained titanium processed by high-pressure torsion, *Wear* 2013 (306) 304–310.

[61] K. Holmberg, A. Laukkanen, H. Ronkainen, K. Wallin, S. Varjus, J. Koskinen, Tribological contact analysis of a rigid ball sliding on a hard coated surface, Part I: Modelling stresses and strains, *Surf. Coat. Technol.* 200 (2006) 3793–3809.

[62] K. Holmberg, A. Laukkanen, H. Ronkainen, K. Wallin, S. Varjus, J. Koskinen, Tribological contact analysis of a rigid ball sliding on a hard coated surface, Part II: Material deformations, influence of coating thickness and Young's modulus, *Surf. Coat. Technol.* 200 (2006) 3810–3823.

[63] Laukkanen, K. Holmberg, J. Koskinen, H. Ronkainen, K. Wallin, S. Varjus, Tribological contact analysis of a rigid ball sliding on a hard coated surface, Part III: Fracture toughness calculation and influence of residual stresses, *Surf. Coat. Technol.* 200 (2006) 3824–3844.

[64] N. Schwarzer, Q.-H. Duong, N. Bierwisch, G. Favaro, M. Fuchs, P. Kempe, B. Widrig and J. Ramm, Optimisation of the scratch for specific coating designs, *Surf. Coat. Technol.* 206 (2011) 1327–1335.

[65] NANOINDENT-PLUS project Standardising the nano-scratch test, FP7/2007-2013, grant agreement NMP-2012-CSA-6-319208.

[66] Nanotechnologies: Nano- and micro-scale scratch testing, draft CEN standard in preparation through CEN 352.

[67] B.D. Beake, A.A. Ogwu and T. Wagner, Influence of experimental factors and film thickness on the measured critical load in the nanoscratch test, *Mater. Sci. Eng. A* 423 (2006) 70–74.

[68] B.D. Beake, M.I. Davies, T.W. Liskiewicz, V.M. Vishnyakov and S.R. Goodes, Nano-scratch, nanoindentation and fretting tests of 5-80 nm ta-C films on Si(100), *Wear*, 301 (2013) 575.

[69] B.D. Beake, A.J. Harris and T.W. Liskiewicz, Review of recent progress in nanoscratch testing, *Tribology* 7 (2013) 87.

[70] T.W. Wu, R.A. Burn, M.M. Chen and P.S. Alexopoulos, in *Thin Films: Stresses and Mechanical Properties*, (eds. J.C. Bravman, W.D. Nix, D.M. Barnett, D.A. Smith) Mat. Res. Soc. Symp. Proc., 1989, 130, 117–122.

[71] T.W. Wu, Microscratch and load relaxation tests for ultra-thin films, *J. Mater. Res.* 6 (1991) 407.

[72] B.D. Beake, V.M. Vishnyakov, R. Valizadeh, J.S. Colligon, Influence of mechanical properties on the nanoscratch behaviour of hard nanocomposite TiN/Si_3N_4 coatings on Si, *J Phys D: Appl Phys* 39 (2006) 1392–1397.

[73] B.D. Beake, S.R. Goodes and B. Shi, Nanomechanical and nanotribological testing of ultra-thin carbon-based and MoST films for increased MEMS durability, *J Phys D: Appl Phys* 42 (2009) 065301.

[74] G.M. Hamilton and L. E. Goodman, *J. Appl. Mech.*, 33 (1966) 371.

[75] B.D. Beake and T.W. Liskiewicz, Nanomechanical characterization of carbon films pp 19–68, *Applied Nanoindentation in Advanced Materials*, Wiley, Eds. Dr A Tiwari and S. Natarajan, 2017.

[76] Film Doctor, SIOMEC, http://www.siomec.de/FilmDoctor

[77] T. Chudoba, N. Schwarzer and F. Richter, Steps towards a mechanical modelling of layered systems. *Surf. Coat. Technol.* 154 (2002) 140.

[78] N. Schwarzer, Modelling of the contact mechanics of thin films using analytical linear elastic approaches, Habilitation thesis, *TU Chemnitz.* 74 (2004).

[79] K. Holmberg, H. Ronkainen, A. Laukkanen and K. Wallin, Friction and wear of coated surfaces – scales, modelling and tribomechanisms, *Surf. Coat. Technol.* 202 (2007) 1034–1049.

[80] K. Holmberg, H. Ronkainen, A. Laukkanen and K. Wallin, A. Erdemir and O. Eryilmaz, Tribological analysis of TiN and DLC coated contacts by 3D FEM modelling and stress simulation, *Wear* 264 (2008) 877-884.

[81] K. Holmberg, A. Laukkanen, E. Turunen, T. Laitinen, Wear resistance optimisation of composite coatings by computational microstructural modelling, *Surf. Coat. Technol.* 2014.

[82] K. Holmberg, A. Laukkanen, A. Ghabchi, M. Rombouts, E. Turunen, R. Waudby, T. Suhonen, K. Valtonen, E. Sarlin, Computational modelling based wear resistance analysis of thick composite coatings, *Tribol. Int.* 72 (2014) 13-30.

[83] S.J. Bull and D.S. Rickerby, Multi-pass scratch testing as a model for abrasive wear, *Thin Solid Films*, 181 (1989) 545–553.

[84] N.N. Gosvami, J.A. Bares, F. Mangolini, A.R. Konicek, D.G. Yablon, R.W. Carpick, Mechanisms of antiwear tribofilm growth revealed in situ by single asperity sliding contact, *Science*, 348 (2015) 102–106.

[85] B. Bhushan, K.J. Kwak, Effect of temperature on nanowear of platinum-coated probes sliding against coated silicon wafers for probe-based recording technology, *Acta Materialia* 56 (2008) 380–386.

[86] B. Shi, J.L. Sullivan and B.D. Beake, An investigation into which factors control the nanotribological behaviour of thin sputtered carbon films, *J Phys D:Appl Phys* 41 (2008) 045303.

[87] F.A. Smidt, Use of ion beam assisted deposition to modify the microstructure and properties of thin films, *Int. Mater. Rev.* 35 (1990) 61–128.

[88] W. Ensinger, A. Schröer, G.X. Wolf, A comparison of IBAD films for wear and corrosion protection with other PVD coatings, *Nuclear Instruments and Methods in Physics Research* B80 81 (1993) 445–454.

[89] C.M. Cotell, J.K. Hirvonen, Effect of ion energy on the mechanical properties of ion beam assisted deposition (IBAD) wear resistant coatings, *Surf. Coat. Technol.* 81 (1996) 118–125.

[90] D.M. Mattox, Handbook of Physical Vapor Deposition (PVD) Processing (2nd Ed) 2010, Ch. 9, *Ion Plating and Ion Beam-Assisted Deposition*, pp 301–331, *Elsevier*, ISBN: 9780815520375.

[91] J. Colligon, V. Vishnyakov, Thin Films: Sputtering, PVD Methods, and Applications, Ch. 61, pp 1-55, Surface and Interface Science: Volume 9: Applications of Surface Science I, Volume 9, Ed: Klaus Wandelt, First published: 17/1/2020, ISBN: 9783527413829, DOI:10.1002/9783527822492. *Wiley-VCH Verlag GmbH & Co.*

[92] J.S. Colligon, V.M. Vishnyakov, R. Valizadeh, S.E. Donnelly and S. Kumashiro, Study of nanocrystalline TiN/Si_3N_4 thin films deposited using a dual ion beam method. *Thin Solid Films* 485 (2005) 148–154.

[93] B.D. Beake, V.M. Vishnyakov and J.S. Colligon, Nano-impact testing of TiFeN and TiFeMoN films for dynamic toughness evaluation, *J Phys D: Appl Phys* 44 (2011) 085301.

[94] B.D. Beake, V.M. Vishnyakov, A.J. Harris, Relationship between mechanical properties of thin nitride-based films and their behaviour in nano-scratch tests, *Tribol. Int.* 44 (2011) 468.

[95] B.D. Beake, V.M. Vishnyakov and A.J. Harris, Nano-scratch testing of $(Ti,Fe)N_x$ thin films on silicon, *Surf. Coat. Technol.* 309 (2017) 671–679.

[96] S. Zhang, X.L. Bui, Y. Fu and H. Du, Development of carbon-based coating on extremely high toughness with good hardness, *Int. J. Nanosci.* 3 (2004) 571–578.

[97] S. Zhang, X.L. Bui, Y. Fu, D.L. Butler and H. Du, Bias-graded deposition of diamond-like carbon for tribological applications, *Diam. Relat. Mater.* 13 (2004) 867–871.

[98] Y.X. Wang and S. Zhang, Toward hard yet tough ceramic coatings, *Surf. Coat. Technol.* 258 (2014) 1–16.

[99] S. Zhang, D. Sun, Y. Fu and H. Du, Effect of sputtering target power on microstructure and mechanical properties of nanocomposite $nc-TiN/a-SiN_x$ thin films, *Thin Solid Films*, 447–8 (2004) 462–467.

[100] Y.X. Wang and S. Zhang, Present status of hard-yet-tough ceramic coatings, Chapter 1, pp 1–45 in *Thin Films and Coatings: Toughening and Toughness Characterisation*, ed. S Zhang, CRC Press, July 2015, p 15.

[101] X. Zhang, B.D. Beake and S. Zhang, Toughness Evaluation of Thin Hard Coatings and Films, Chapter 2, pp 48–121 in Thin Films and Coatings: Toughening and Toughness Characterisation, ed. S Zhang, CRC Press, July 2015, p 53.

[102] Y.X. Wang, S. Zhang, J.-W. Lee, W.S. Lew and B. Li, Toughening effect of Ni on $nc-CrAlN/a-SiN_x$ hard nanocomposite, *Appl. Surf. Sci.* 265 (2013) 418–423.

[103] A.A. Voevodin and J.S. Zabinski, Load-adaptive crystalline-amorphous nanocomposites, *J. Mater. Sci.* 33 (1998) 319–327.

[104] K. Holmberg, P. Andersson, A. Erdemir, Global energy consumption due to friction in passenger cars, *Tribol. Int.* 147 (2012) 221–234.

[105] K. Holmberg, A. Erdemir, Influence of tribology on global energy consumption, costs and emissions, *Friction* 5 (2017) 263–284.

[106] M. Woydt, R. Luther, T. Gradt, A. Rienäcker, T. Hosenfeldt, F.-J. Wetzel and C. Wincierz, Tribology in Germany; Interdisciplinary technology for the reduction of CO_2-emissions and the conservation of resources, pp 1–34, 2019, Gesellschaft für Tribologie e.V. (German Society for Tribology e.V. www.gft-ev.de) Jülich, *Germany*.

[107] A. Erdemir, C. Donnet, Tribology of diamond-like carbon films: recent progress and future prospects, *J. Phys. D: Appl. Phys.* 39 (2006) R311–R327.

[108] S.D.A. Lawes, S.V. Hainsworth, M.E. Fitzpatrick, Impact wear testing of diamond-like carbon films for engine valve-tappet surfaces, *Wear* 268 (2010) 1303–1308.

[109] R. Gåhlin, M. Larsson, P. Hedenqvist, Me-C:H coatings in motor vehicles, *Wear* 249 (2001) 302–309.

[110] D. Bernoulli, A. Rico, A. Wyss, K. Thorwarth, J.P. Best, R. Hauert, R. Spolenak, Improved contact damage resistance of hydrogenated diamond-like carbon (DLC) with a ductile α-Ta interlayer, *Diam. Relat. Mater.* 58 (2015) 78–83.

[111] D. Bernoulli, A. Wyss, R. Raghavan, K. Thorwarth, R. Hauert, R. Spolenak, Contact damage of hard and brittle thin films on ductile metallic substrates: an analysis of diamond-like carbon on titanium substrates, *J. Mater. Sci.* 50 (2015) 2779–2787.

[112] J. Jiang and R.D. Arnell, On the running-in behaviour of diamond-like carbon coatings under the ball-on-disk contact geometry, *Wear* 217 (1998) 190–199.

[113] O. Wänstrand, M. Larsson, P. Hedenqvist, Mechanical and tribological evaluation of PVD WC/C coatings, *Surface and Coatings Technology* 111 (1999) 247–254.

[114] G. Ramírez, E. Jiménez-Piqué, A. Mestra, M. Vilaseca, D. Casellas, L. Llanes, A comparative study of the contact fatigue behaviour and associated damage micromechanisms of TiN- and WC:H- coated cold-work tool steel, *Tribol. Int.* 88 (2015) 263–270.

[115] P. Mutafov, J. Lanigan, A. Neville, A. Cavaleiro, T. Polcar, DLC-W coatings tested in combustion engine — Frictional and wear analysis, *Surf. Coat. Technol.* 260 (2014) 284–289.

[116] J.L. Lanigan, C. Wang, A. Morina, A. Neville, Repressing oxidative wear within Si doped DLCs, *Tribol. Int.* 93 (2016) 651–659.

[117] B.D. Beake, T.W. Liskiewicz, V.M. Vishnyakov, Stress and friction modelling for improved nano-scratch testing of hard DLC coatings, presentation at ICMCTF, San Diego, USA, 24–28 April 2017.

[118] T.W. Liskiewicz, B.D. Beake, N. Schwarzer and M.I. Davies, Short note on improved integration of mechanical testing in predictive wear models, *Surf. Coat. Technol.* 237 (2013) 212.

[119] K.-D. Bouzakis, N. Michailidis, G. Skordaris, E. Bouzakis, D. Biermann, R. M'Saoubi, Cutting with coated tools: coating technologies, characterization methods and performance optimisation, *CRIP Ann. Manuf. Technol.* 61 (2012) 703–723.

[120] A. Clausner, F. Richter, Determination of yield stress from nano-indentation experiments, *Eur. J. Mech. A/Solids* 51 (2015) 11–20.

[121] B.D. Beake, J.L. Endrino, C. Kimpton, G.S. Fox-Rabinovich, S.C. Veldhuis, Elevated temperature repetitive micro-scratch testing of AlCrN, TiAlN and AlTiN PVD coatings, *Int. J. Refract. Metal. Hard Mater.* 69 (2017) 215–226.

[122] G.S. Fox-Rabinovich, B.D. Beake, S.C. Veldhuis, J.L. Endrino, R. Parkinson, L.S. Shuster, M.S. Migranov, Impact of mechanical properties measured at room and elevated temperatures on wear resistance of cutting tools with TiAlN and AlCrN coatings, *Surf. Coat. Technol.* 200 (2006) 5738–5742.

[123] B.D. Beake, L. Ning, C. Gey, S.C. Veldhuis, A. Komarov, A. Weaver, M. Khanna, G.S. Fox-Rabinovich, Wear performance of different PVD coatings during wet end milling of H13 tool steel, *Surf. Coat. Technol.* 279 (2015) 118–125.

[124] J.M. Anderson, J. Vetter, J. Müller, J. Sjölén, Structural effects of energy input growth of Ti1-xAlxN ($0.55 \leq x \leq 0.66$) coatings by cathodic arc evaporation, *Surf. Coat. Technol.* 240 (2014) 211–220.

[125] E. Le Bourhis, P. Goudeau, M.H. Staia, E. Carrasquero, E.S. Puchi-Cabrera, Mechanical properties of hard AlCrN-based coated substrates, *Surf. Coat. Technol.* 203 (2009) 2961–2968.

[126] G.S. Fox-Rabinovich, J.L. Endrino, B.D. Beake, M.H. Aguirre, S.C. Veldhuis, D.T. Quinto, C.E. Bauer, A.I. Kovalev, A. Gray, Effect of annealing below 900°C on structure, properties and tool life of an AlTiN coating under various cutting conditions, *Surf. Coat. Technol.* 202 (2008) 2985.

[127] S. Chowdhury, B.D. Beake, K. Yamamoto, B. Bose, M. Aguirre, G.S. Fox-Rabinovich, S.C. Veldhuis, Improvement of Wear Performance of Nano-Multilayer PVD Coatings under Dry Hard End Milling Conditions Based on Their Architectural Development, *Coatings* 8 (2018) 59 (15pp).

[128] P. Wright, E.M. Trent, (2000) *Metal Cutting, 4th Edition* (Butterworth-Heinmann, Boston, USA).

[129] G. Skordaris, A. Bouzakis and K.-D. Bouzakis, Impact Test Applications Supported by FEA Models in Surface Engineering for Coating Characterization, *Mater. Proc.* 2 (2020) 13.

[130] M. Bromark, P. Hedenqvist, S. Hogmark, The influence of substrate material on the erosion resistance of TiN coated steels, *Wear* 186–187 (1995) 189–194.

[131] B.D. Beake, S.R. Goodes and J.F. Smith, Micro-impact testing: a new technique for evaluating fracture toughness, adhesion, erosive wear resistance and dynamic hardness, *Surf. Eng.* 17 (2001) 187.

[132] B.D. Beake and J.F. Smith, Nano-impact testing – an effective tool for assessing the resistance of advanced wear-resistant coatings to fatigue failure and delamination, *Surf Coat Technol* 188–189C (2004) 594.

[133] B.D. Beake, L. Isern, J.L. Endrino, G.S. Fox-Rabinovich, Micro-impact testing of AlTiN and TiAlCrN coatings, *Wear* 418–419 (2019) 102–110.

[134] K.-D. Bouzakis, G. Maliaris, S. Makrimallakis, Strain rate effect on the fatigue failure of thin PVD coatings: an investigation by novel impact tester with adjustable repetitive force, *Int. J. Fatigue* 44 (2012) 89–97.

[135] G.S. Fox-Rabinovich, S.C. Veldhuis, K. Yamamoto, M.H. Aguirre, A. Kovalev, D.L. Wainstein, B.D. Beake, J.L. Endrino, D.L. Wainstein, A.Y. Rashkovskiy, Design and performance of AlTiN and TiAlCrN PVD coatings for machining of hard to cut materials, *Surf. Coat. Technol.* 204 (2009) 489–496.

[136] B.D. Beake, G.S. Fox-Rabinovich, Progress in high temperature nanomechanical testing of coatings for optimising their performance in high speed machining, *Surf. Coat. Technol.* 255 (2014) 1021115.

[137] B.D. Beake, J.F. Smith, A. Gray, G.S. Fox-Rabinovich, S.C. Veldhuis, J.L. Endrino, Investigating the correlation between nano-impact fracture resistance and hardness/modulus ratio from nanoindentation at 25-500°C and the fracture resistance and lifetime of cutting tools with $Ti_{1-x}Al_xN$ (x=0.5 and 0.67) PVD coatings in milling operations, *Surf. Coat. Technol.* 201 (2007) 4585.

[138] G. Skordaris et al, Brittleness and fatigue effect of mono- and multi-layer PVD films on the cutting performance of coated cemented carbide inserts, *CIRP Annals - Manuf Tech* 63 (2014) 93.

[139] G. He, D. Sun, J. Chen, X. Han, Z. Zhang, Z. Fang et al, Key problems affecting the anti-erosion coating performance of aero-engine compressor: a review, *Coatings* 9 (2019) 821.

[140] J. Chen, B.D. Beake, R.G. Wellman, J.R. Nicholls, H. Dong, An investigation into the correlation between nano-impact resistance and erosion performance of EB-PVD thermal barrier coatings on thermal ageing, *Surf. Coat. Technol.* 206 (2012) 4992–4998.

[141] A. Inspektor and P.A. Salvador, Architecture of PVD coatings for metalcutting applications: a review, *Surf. Coat. Technol.* 257 (2014) 138–153.

[142] A. Hörling, L. Hultman, M. Odén, J. Sjölén, L. Karlsson, Mechanical properties and machining performance of Ti1-xAlxN-coated cutting tools, *Surf. Coat. Technol.* 191 (2005) 384–392.

[143] H. Windischmann, Intrinsic stress in sputter-deposited thin films, *Critical Reviews in Solid State and Materials Sciences*, 17 (1992) 547–596; doi:10.1080/10408439208244586.

[144] G. Abadias, E. Chason, J. Keckes, M. Sebastiani, G.B. Thompson, E. Barthel, G.L. Doll, C.E. Murray, C.H. Stoessel, L. Martinu, Stress in thin films and coatings: Current status, challenges, and prospects, *J. Vac. Sci. Technol. A* 36 (2018) 020801 (49pp); doi:10.1116/1.5011790.

[145] Q. Zhang, Y. Xu, T. Zhang, Z. Wu, Q. Wang, Tribological properties, oxidation resistance and turning performance of AlTiN/AlCrSiN multilayer coatings by arc ion plating, *Surf. Coat. Technol.* 356 (2018) 1–10.

[146] X. Zha, F. Jiang, X. Xu, Investigating the high frequency fatigue failure mechanisms of mono and multilayer PVD coatings by the cyclic impact tests, *Surf. Coat. Technol.* 344 (2018) 689–701.

[147] J.J. Roa, E. Jiménez-Pique, R. Martínez, G. Ramírez, J.J. Tarragó, R. Rodríguez, L. Llanes, Contact damage and fracture micromechanisms of multi-layered TiN/CrN coatings at micro- and nano-length scales, *Thin Solid Films* 571 (2014) 308–315.

[148] Y.-Y. Chang, C.-J. Wu, Mechanical properties and impact resistance of multi-layered TiAlN/ZrN coatings, *Surf. Coat. Technol.* 231 (2013) 62–66.

[149] Y.X. Ou, J. Lin, H.L. Che, W.D. Sproul, J.J. Moore, M.K. Lei, Mechanical and tribological properties of CrN/TiN multilayer coatings deposited by pulsed dc magnetron sputtering, *Surf. Coat. Technol.* 276 (2015) 152–159.

[150] C. Mendibide, P. Steyer, J. Fontaine, P. Goudeau, Improvement of the tribological behaviour of PVD nanostratified TiN/CrN coatings – an explanation, *Surf. Coat. Technol.* 201 (2006) 4119–4124.

[151] C. Mendibide, J. Fontaine, P. Steyer, C. Esnoul, Dry sliding wear model of nanometer scale multi-layered TiN/CrN PVD hard coatings, *Tribol. Lett.* 17 (2004) 779–789.

[152] A.A. Vereschaka, S.N. Grigoriev, Study of cracking mechanisms in multi-layered composite nano-structured coatings, *Wear* 378–379 (2017) 43–57.

[153] J. Chen, R. Ji, R.H.U. Khan, X. Li, B.D. Beake and H. Dong, Effects of mechanical properties and layer structure on the cyclic loading of TiN-based coatings Surf. *Coat. Technol.* 206 (2011) 522–529.

[154] A. Vereschaka, V. Tabakov, S. Grigoriev, N. Sitnikov, F. Milovich, N. Andreev, C. Sotova, N. Kutina, Investigation of the influence of the thickness of nanolayers in wear-resistant layers of Ti-TiN-(Ti,Cr,Al)N coating on destruction in the cutting and wear of carbide cutting tools, *Surf. Coat. Technol.* 385 (2020) 125402.

[155] G. Skodaris, K.-D. Bouzakis, P. Charalampous, A dynamic FEM simulation of the nano-impact test on mono- or multi-layered PVD coatings considering their graded strength properties determined by experimental-analytical procedures, *Surf. Coat. Technol.* 265 (2015) 53–61.

[156] G.S. Fox-Rabinovich, B.D. Beake, K. Yamamoto, M.H. Aguirre, S.C. Veldhuis, G. Dosbaeva, A. Elfizy, A. Biksa, and L.S. Shuster, A.Y. Rashkovskiy, Structure, properties and wear performance of nano-multilayered TiAlCrSiYN/TiAlCrN coatings during machining of Ni-based aerospace superalloys, *Surf Coat Technol* 204 (2010) 3698–3706.

[157] G.S. Fox-Rabinovich, K. Yamamoto, B.D. Beake, A.I. Kovalev, M.H. Aguirre, S.C. Veldhuis, G.K. Dosbaeva, D.L. Wainstein, A. Biksa and A.Y. Rashkovskiy, Emergent behavior of nano-multilayered coatings during dry high speed machining of hardened tool steels, *Surf. Coat. Technol.* 204 (2010) 3425–3435.

[158] G.S. Fox-Rabinovich, J.L. Endrino, M.H. Agguire, B.D. Beake, S.C. Veldhuis, A.I. Kovalev, I.S. Gershman, K. Yamamoto, Y. Losset, D.L. Wainstein and A.Y. Rashkovskiy, Mechanism of adaptability for the nano-structured TiAlCrSiYN-based hard physical vapor deposition coatings under extreme frictional conditions, *J. Appl. Phys.* 111 (2012) 064306.

[159] K.-D. Bouzakis, F. Flocke, G. Skordaris, E. Bouzakis, S. Geradis, G. Katirtzoglou and S. Makrimallakis, Influence of dry micro-blasting grain quality on wear behaviour of TiAlN coated tools, *Wear* 271 (2011) 783–791.

[160] M. Tkadletz, N. Schalk, R. Daniel, J. Keckes, C. Czettl, C. Mitterer, Advanced characterisation methods for wear resistant hard coatings: a review on recent progress, *Surf. Coat. Technol.* 285 (2016) 31–46.

[161] P.C. Jinhal, A.T. Santhanam, U. Schleinkofer, A.F. Shuster, Performance of PVD TiN, TiCN and TiAlN coated cemented carbide tools in machining, *Int. J. Refract. Met. Hard Mater.* 17 (1999) 163.

[162] J.M. Wheeler, V. Maier, K. Durst, M. Göken, J. Michler, Activation parameters for deformation of ultra-fine-grained aluminium as determined by indentation strain rate jumps at elevated temperature, *Mater. Sci. Eng. A* 585 (2013) 108–113.

[163] S. Koch, M.D. Abad, S. Renhart, H. Antrekowitsch, P. Hosemann, A high temperature nanoindentation study of Al-Cu wrought alloy, *Mater. Sci. Eng. A* 6 (2015) 218–224.

[164] M.-M. Primorac, M.D. Abad, P. Hosemann, M. Kreuzeder, V. Maier, D. Kiener, Elevated temperature mechanical properties of novel ultra-fine grained Cu-Nb composites, *Mater. Sci. Eng. A* 625 (2015) 296–302.

[165] B.D. Beake, A.J. Harris, Nanomechanics to 1000 °C for high temperature mechanical properties of bulk materials and hard coatings, *Vacuum* 159 (2019) 17–28.

[166] J.M. Wheeler, R.A. Oliver, T.W. Clyne, AFM observation of diamond indenters after oxidation at elevated temperatures, *Diam. Relat. Mater.* 19 (2010) 1348–1353.

[167] J.M. Wheeler and J. Michler, Indenter materials for high temperature nanoindentation, *Rev. Sci. Instr.* 84 (2013) 101301.

[168] D.T. Quinto, G.J. Wolfe, P.C. Jinhal, High temperature microhardness of hard coatings produced by physical vapour deposition, *Thin Solid Films* 153, (1987) 19.

[169] W.D. Munz, Titanium aluminium nitride films: a new alternative to TiN coatings, *J. Vac. Sci. Technol. A* 4 (1986) 2717.

[170] S. Paldey and S.C. Deevi, Single later and multilayer wear resistant coatings of (Ti,Al)N: a review, *Mater. Sci. Eng. A* 342 (2003) 58.

[171] W. Kalss, A. Reiter, V. Derflinger, C. Gey, J.L. Endrino, Modern coatings in high performance cutting applications, *Int. J. Refract. Met. Hard Mater.* 24 (2006) 399–404.

[172] N. Norrby, M. P. Johansson, R. M'Saoubi, M. Odén, Pressure and temperature effects on the decomposition of arc evaporated Ti0.6Al0.4N coatings in continuous turning, *Surf. Coat. Technol.* 209 (2012) 203–207.

[173] M. P. Johansson Jõesaar, N. Norrby, J. Ullbrand, R. M'Saoubi and M. Odén, Anisotropy effects on the microstructure and properties in decomposed evaporated $Ti_{1-x}Al_xN$ coatings during metal cutting, *Surf. Coat. Technol.* 235 (2013) 181–185.

[174] P.H. Mayrhofer, A. Hörling, L. Karlsson, J. Sjölén, T. Larsson, C. Mitterer, and L. Hultman Self-organized nanostructures in the Ti-Al-N system, *Appl. Phys. Lett.* 83 (2003) 2049–2051.

[175] F. Giuliani, C. Ciurea, V. Bhakhri, L.J. Werchota, L.J. Vandeperre, P.H. Mayrhofer, Deformation behaviour of TiN and Ti-Al-N coatings at 295 to 573 K, *Thin Solid Films* 688 (2019) 137363.

[176] Hultman, L. and Mitterer, C. (2006) Nanostructured coatings Cavaleiro, A. and De Hosson, J.Th.M. eds. Chapter 11, *"Thermal stability of advanced nanostructured wear-resistant coatings"* (Springer, New York) pp 464–510.

[177] K. Kutschej, P.H. Mayrhofer, M. Kathrein, P. Polcik, R. Tessadri, C. Mitterer, Structure, mechanical and tribological properties of sputtered coatings with $Ti_{1-x}Al_xN$ (0.5 ≤ x ≤ 0.75) *Surf. Coat. Technol.* 200 (2005) 2358–2365.

[178] T. Polcar, R. Novak, P. Siroky, The tribological characteristics of TiCN coating at elevated temperatures, *Wear* 260 (2006) 40.

[179] A. Pauschitz, M. Roy, F. Franek, Mechanisms of sliding wear of metals and alloys at elevated temperatures, *Tribol. Int.* 41 (2008) 584–602.

[180] J. Veverkova, S.V. Hainsworth, Effect of temperature and counterface on the tribological properties of W-DLC on a steel substrate, *Wear* 264 (2008) 518–523.

[181] O. Jantschner, C. Walter, C. Muratore, A.A. Voevodin, C. Mitterer, V-alloyed ZrO2 coatings with temperature homogenization function for high-temperature sliding contacts, *Surf. Coat. Technol.* 228 (2013) 76–83.

[182] A.A. Voevodin, C. Muratore, S.M. Aouadi, Hard coatings with high temperature adaptive lubrication and contact thermal management: review, *Surface and Coating Technol.* 257 (2014) 247–265.

[183] M.H. Staia, M. D'Alessandria, D.T. Quinto, F. Roudet, M. Marsal Astort, High-temperature tribological characterisation of commercial TiAlN coatings, *J. Phys.: Condens. Matter* 18 (2006) S1727–S1736.

[184] J. Pujante, M. Vilaseca, D. Casellas and M.D. Riera, High temperature scratch testing of hard PVD coatings deposited on surface treated tool steel, *Surf. Coat. Technol.* 257 (2014) 352–357.

[185] D.N. Allsopp and I.M. Hutchings, Micro-scale abrasion and scratch response of PVD coatings at elevated temperatures, *Wear* 251 (2001) 1308–1314.

[186] N. Michailidis, Variations in the cutting performance of PVD-coated tools in milling Ti6Al4V, explained through temperature-dependent coating properties, *Surf. Coat. Technol.* 304 (2016) 325–329.

[187] P. Zgorniak, A. Grdulska, Investigation of temperature distribution during milling process of Az91hp magnesium alloys, *Mechanics and Mechanical Engineering*, 16 (2012) 33–40.

[188] U. Karaguzel, M. Bakkal, E. Budak, Modeling and measurement of cutting temperatures in milling, *Procedia CIRP* 46 (2016) 173–176.

[189] S.Z. Chavoshi and S. Xu, A review on micro- and nanoscratching/tribology at high temperatures: instrumentation and experimentation, *JMEPEG* 27 (2018) 3844–3858.

[190] J.L. Endrino, G.S. Fox-Rabinovich, C. Gey, Hard AlTiN, AlCrN PVD coatings for machining of austenitic stainless steel, *Surf. Coat. Technol.* 200 (2006) 6840–6845.

[191] G.K. Dosbaeva, S.C. Veldhuis, K. Yamamoto, D.S. Wilkinson, B.D. Beake, N. Jenkins, A. Elfizy, G.S. Fox-Rabinovich, Oxide scales formation in nano-crystalline TiAlCrSiYN PVD coatings at elevated temperature, *Int. J. Refract. Met. Hard Mater.* 28 (2010) 133–141.

[192] G.S. Fox-Rabinovich, K. Yamamoto, B.D. Beake, I.S. Gershman, A.I. Kovalev, S.C. Veldhuis, M.H. Aguirre, G. Dosbaeva, J.L. Endrino, Hierarchical adaptive nanostructured PVD coatings for extreme tribological applications: the quest for nonequilibrium states and emergent behaviour, *Sci. Technol. Adv. Mater.* 13 (2012) 043001 (26pp).

[193] G.S. Fox-Rabinovich, S.C. Veldhuis, G.K. Dosbaeva, K. Yamamoto, A.I. Kovalev, D.L. Wainstein, I.S. Gershman, L.S. Shuster, B.D. Beake, Nanocrystalline coating design for extreme applications based on the concept of complex adaptive behaviour, *J. Appl. Phys.* 103 (2008) 083510.

[194] G. Li, J. Sun, Y. Xu, J. Gu, L. Wang, K. Huang et al, Microstructure, mechanical properties, and cutting performance of TiAlSiN multilayer coatings prepared by HiPIMS, *Surf. Coat. Technol.* 353 (2018) 274–281.

[195] W. Tillmann, M. Dildrop, Influence of Si content on mechanical properties and tribological performance of TiAlSiN coatings at elevated temperatures, *Surf. Coat. Technol.* 321 (2017) 448–454.

[196] D. Yu, C. Wang, X. Cheng, F. Zhang, Microstructure and properties of TiAlSiN coatings prepared by hybrid PVD technology, *Thin Solid Films* 517 (2009) 4950–4955.

[197] G.G. Fuentes, E. Almandoz, R.J. Rodríquez, H. Dong, Y. Qin, S. Mato et al, Vapour deposition technologies for the fabrication of hot-forming tools, *Manufacturing Rev.* 1 (2014) 20.

[198] B.D. Beake, A. Bird, L. Isern, J.L. Endrino, F. Jiang, Elevated temperature micro-impact testing of TiAlSiN coatings produced by physical vapour deposition, *Thin Solid Films* 688 (2019) 137358 (9pp).

[199] Yu.V. Milman, S. Luyckx, I.T. Northrop, Influence of temperature, grain size and cobalt content on the hardness of WC-Co alloys, *Int. J. Refract. Met. Hard Mater.* 17 (1999) 39–44.

[200] A. Góez, D. Coureaux, A. Ingebrand, B. Reig, E. Tarrés, A. Mateo, E. Jiménez-Piqué, L. Llanes, Contact damage and residual strength in hardmetals, *Int. J. Refract. Metal. Hard Mater.* 30 (2012) 121–127.

[201] J.P. Best, J. Zechner, J.M. Wheeler, R. Schoeppner, M. Morstein, J. Michler, Small-scale fracture toughness of ceramic thin films: the effects of specimen geometry, ion beam notching and high temperature on chromium nitride toughness evaluation, *Philos. Mag.* 2016, doi:10.1080/14786435.2016.1223891.

[202] J.P. Best, G. Guillonneau, S. Grop, A.A. Taylor, D. Frey, Q. Longchamp, T. Schär, M. Morstein, J.-M. Breguet, J. Michler, High temperature impact testing of a thin hard coating using a novel high-frequency in situ micromechanical device, *Surf. Coat. Technol.* 333 (2018) 178–186.

[203] G.S. Fox-Rabinovich, J.L. Endrino, B.D. Beake, A.I. Kovalev, S.C. Veldhuis, L. Ning, F. Fotaine and A. Gray, Impact of annealing on the microstructure, properties, and cutting performance of AlTiN coating, *Surf. Coat. Technol.* 201 (2006) 3524.

[204] G. S. Fox-Rabinovich, I. S. Gershman and S. Veldhuis, Thin-Film PVD Coating Metamaterials Exhibiting Similarities to Natural Processes under Extreme Tribological Conditions, *Nanomaterials* 10 (2020) 1720.

5 Corrosion and Tribo-Corrosion of Hard Coating Prepared by Advanced Magnetron Sputtering

Deen Sun
Southwest University, China

CONTENTS

5.1 INTRODUCTION

Physical Vapor Deposition (PVD) technology is a modern surface treatment technology carried out in a vacuum environment. Compared with other methods, it has no waste water, exhaust gas, and noise emissions. It is an environmentally friendly coating technology. PVD technology includes arc ion plating, evaporation plating, and magnetron sputtering methods. There are large metal droplets during the arc ion plating process, resulting in a number of large particles on the surface and inside of the coating [1]. Large particles are often the weak region in the occurrence of corrosion. Arc ion plating coatings can improve the corrosion resistance of alloys. However, the effect is limited. Due to its low metal ionization rate and lack of ion bombardment effect during the coating process, evaporation plating coating shows loose texture, poor film/base bonding strength, and poor corrosion resistance. Magnetron sputtering (MS) technology can prepare a uniform and smooth coating, and the coating composition and microstructure are controllable and adjustable. Based on the previous research, it is found that the magnetron sputtering technology has the advantages of low deposition temperature in the preparation process and high process controllability. The prepared coating has the advantages of smooth surface, dense structure, and high coating/substrate bonding strength [2–5].

5.1.1 MAGNETRON SPUTTERING-RELATED PROCESS

One of the drawbacks of the conventional magnetron sputtering is the "shadowing effect" in the deposition process [6] because of the low energy delivered to the growing film. There are some areas on the growing surface of film, which will attract and deposit sputtering ions preferentially than the adjacent areas, forming "mountains". This will cause the adjacent area to trap fewer ions from the deposition flux and be shielded, forming a structure similar to a "valley". The shadowing effect will cause the coating structure to be loose, resulting in defects such as coarse columnar crystals and through holes penetrating the coating, which will directly affect the corrosion and wear resistance of the prepared coating [7]. High Power Impulse Magnetron Sputtering (HiPIMS) technology is a new technology developed in recent years [8]. It was first proposed by Kouznetsov et al. in 1999 that HiPIMS technology can greatly increase the ionization rate of metal targets. The coating prepared using this technology shows a higher density [9–10]. Therefore, this subject has launched a preliminary study on high-power pulsed magnetron sputtering technology.

In addition, the research results show that the ion source can greatly increase the gas ionization rate, and the ion source assisted magnetron sputtering deposition technology can improve the density of the film, reduce the columnar crystal structure, and effectively improve the corrosion resistance of the coating [11–12]. M. Fenker et al. [11] have systematically studied the effects of ion source assisted deposition on the surface morphology, hardness, bonding strength, and corrosion resistance of the as-prepared coating. And the results showed that there is a certain linear relationship between ion source bombardment and the improvement of coating corrosion resistance. Ion source assisted deposition can greatly improve the corrosion resistance of NbN coating. The introduction of ion source bombardment during the coating growth process can densify the coating, interrupt the growth of columnar grains, provide more crystal nucleation spot, promote equiaxed crystal nucleation, and inhibit grain growth [13]. During the ion source assisted magnetron sputtering coating process, the kinetic energy and thermal energy of the deposited particles on the film surface are the key factors that determine the growth structure and morphology of the film. The structure area model of the coating growth shows that by controlling the coating temperature, gas pressure, substrate bias, power density, and other process parameters, the coating microstructure can be precisely controlled [14–16]. Metal surface protective coatings usually grow in the form of columnar crystals. The penetrating pores formed between the columnar crystals are an open structure. That kind of microstructure defects, such as grain boundaries and dislocations, will become diffusion path for corrosive medium. The diffusion channel will accelerate the diffusion of the corrosive medium through the coating to the metal substrate during the corrosion process [17]. In order to improve the corrosion resistance and wear resistance of the metals, the surface protective coating is usually improved by adjusting the process parameters of the ion source assisted magnetron sputtering coating process, increasing the density of the coating, breaking the diffusion channel of the corrosive medium, and reducing the diffusion path.

5.1.2 MULTILAYER STRUCTURE

In addition to improving the corrosion resistance of materials by introducing auxiliary means to obtain dense films and interrupting the growth of columnar crystals, many materials researchers have also discussed from the perspective of material microstructure design. At present, many scholars have done a lot of research work on the electrochemical oxidation mechanism of Ti/TiN "metal/ceramic" single-layer structure coating on Mg substrate. Generally, in the Ti/TiN single-layer structure, Ti is used as a bonding layer and is in close contact with the Mg alloy substrate. The corrosion cell composed of Ti/Mg has a large reaction driving force, and the Ti bonding layer is easily destroyed during the anodic polarization process. Its corrosion resistance is not obvious [18]. Although the TiN ceramic coating has good corrosion resistance and mechanical properties, the TiN film prepared with conventional PVD process has a columnar structure, and the corrosive medium can penetrate the TiN film through the columnar crystal gap. Therefore, the corrosion resistance of the traditional PVD Ti/TiN single-layer structure is not ideal.

In order to improve the corrosion resistance of the PVD coating, the "metal/ceramic" multilayer structure coating has attracted great attention. $(Ti/TiN)_n$ is a common "metal/ceramic" multilayer structure coating [19–23] which is usually obtained by alternately depositing Ti metal sub-layer and TiN ceramic sub-layer. Studies have shown that in the corrosive medium, the single set of (Ti/TiN) sub-layers in the multilayer structure is easy to fail due to the existence of the Ti metal sub-films, which resulting in $(Ti/TiN)_n$ multilayer structure coating fails overall.

In order to further improve the corrosion resistance of PVD coatings, "ceramic/ceramic" multilayer coatings have gradually attracted the attention of researchers. Uslu et al. [24] successfully prepared a "ceramic/ceramic" multilayer structure coating on the surface of a magnesium alloy by constructing a $(TiN/VN)_n$ "ceramic/ceramic" multilayer structure system. The results showed that compared with the "metal/ceramic" multilayer structure, the "ceramic/ceramic" multilayer structure coating possesses better corrosion resistance. Liang et al. [25] interrupted the growth of columnar crystals in the TiAlN coating through cyclic bombardment of the ion source and successfully prepared a multi-interface (multi-interface) Al-TiAlN "ceramic/ceramic" multilayer structure coating on the AZ91D magnesium alloy. When the modulation period is 100nm, the corrosion current density can be as low as 9.37×10^{-8} A•cm^{-2}, and the coating hardness can reach 31.3GPa.

In addition, "Ceramic/ceramic" multilayer structure coating can obtain higher hardness and exhibit excellent wear resistance. Wang et al. [26] used magnetron sputtering technology to prepare the "ceramic/ceramic" multilayer structure pc-CrAlN/(nc-CrAlN/a-SiNx) coating and studied the effect of modulation cycles on the hardness and toughness. The results show that when the modulation period is 20nm, the coating hardness is 33GPa, showing good wear resistance. Zhang et al. [27] employed magnetron sputtering method to prepare "ceramic/ceramic" multilayer structure CrAlN/ZrN coating. Fixing the thickness ratio of CrAlN sub-layer to ZrN sub-layer at 2.6, and keeping the coating thickness and modulation ratio unchanged, the results show that change in the modulation period can change the hardness and toughness of the "ceramic/ceramic" multilayer structure coating effectively, and thus can accurately and effectively control the wear resistance of the coating. He et al. [28] used ion source assisted magnetron sputtering technology to prepare CrSiN/SiN nano-multilayer coatings with different modulation cycles on the surface of TC4 titanium alloy. When the modulation cycle is 90nm and 360nm, the coatings show improved corrosion resistance, and corrosion current densities are 1.31×10^{-8} A•cm^{-2} and 1.20×10^{-8} A•cm^{-2}, respectively. When the modulation period is 45nm, the coating shows the highest hardness and elastic modulus, 22.5 ± 0.6 GPa and 226.4 ± 6.3 GPa, respectively. And the wear rate achieved can be as low as 9.67×10^{-7} mm^3•(Nm)$^{-1}$. The "ceramic/ceramic" multilayer structure coating significantly improves both the corrosion resistance and wear resistance of the substrate.

In this chapter, deposition methods which can obtain coatings with dense structure thereafter good corrosion resistance are reviewed and discussed. Besides the deposition technique, coating microstructure design plays an important role in developing coating with improved corrosion resistance, which will be discussed in detail. Finally, tribo-corrosion which differs from corrosion will be introduced.

5.2 DEPOSITION METHOD

HiPIMS is a promising PVD technique, which produces high density plasma and ionized metal particles during sputtering [29]. The high pulse peak power can reach 2 to 3 orders of magnitude beyond conventional magnetron sputtering and the metal ionization rate achieved can be greater than 50%. It has significant technical advantages in reducing the internal stress and improving the denseness and uniformity of the coating [30–32]. HiPIMS technique is capable of offering high target peak current, high peak power density, and the degree of ionization that are much higher than the conventional direct current (DC), radio frequency (RF), and middle frequency (MF) magnetron sputtering systems. However, the most serious problem with HiPIMS is the low deposition rate. With the same average power density, the deposition rate is only 25% to 35% that of DC magnetron sputtering (DCMS). Both experimental and numerical simulation studies have shown that the low deposition rate in HiPIMS results from the reversed attraction of ionized species and the gas exhaustion during deposition [33].

It becomes very essential to overcome this drawback. Keep that in mind, different HiPIMS-related processes were conducted in present study to improve the deposition rate. The hybrid HiPIMS-related process includes HiPIMS combined with RF power system (HiPIMS+RF), HiPIMS combined with DC magnetron sputtering plus RF power system (HiPIMS+DCMS+RF). Another approach is by superimposing MF powers into HiPIMS pulsing. This superimposition concept has also been shown to improve the deposition rate in pure Cr [34], and reactive deposition of TiO$_2$ [35, 36]. Superimposition concept in HiPIMS is very promising but its further application may be associated with many factors which require more research on the optimization of deposition parameters, such as the role of MF pulse duration, the influence of average MF power on the deposition process [10].**

Deferent from HiPIMS which produces high ionized metal particles during sputtering, ion beam source can increase the ionization rate of gas. The electrons in the anode ion source form Hall current under the action of magnetic field, which increase the probability of collision between electrons and neutral gas molecules or atoms, and then increase the ionization rate of gas. The ionized particles can bombard the growing surface thus increase the density of the coatings [37].

5.2.1 SUPERIMPOSED HIPIMS WITH MF MS

To study the effect of MF duration, eight TiN films were reactively deposited on p-type Si (100) wafer and hardened AISI420 disc substrates by sputtering a rectangular Ti target (5 in. × 12 in.) under a gas mixture of Ar and N_2 in a chamber with the size of Φ 660 mm × 600 mm (LJ-UHV Technology Co., Ltd., Taiwan). Three deposition modes including single DC, single HiPIMS, and superimposed HiPIMS and MF were used in this work to fabricate eight films (c.f. Table 5.1). For the superimposed HiPIMS and MF mode, the Ti target was powered by a HiPIMS power system (SIPP2000USB Dual, MELEC GmbH, Germany) consisting of two DC power sources, two entirely separated 4000 µF capacitor banks, HiPIMS and MF pulse generating devices.

For the single HiPIMS mode, the duty cycle and repetition frequency were maintained at 2% and 200 Hz, respectively. The pulsing configuration in the superimposed HiPIMS and MF mode is illustrated in Figure 5.1.

In the superimposed HiPIMS and MF mode, each superimposed cycle consists of one HiPIMS pulse and 5 to 80 MF pulses. The duty cycle and repetition frequency of HiPIMS and MF were set at 2%, 200 Hz and 60%, 20,000 Hz, respectively. The pulse t_{on} and t_{off} of HiPIMS were fixed at 100 and 4900 µs (duty cycle of 2%), respectively, while t_{on} and t_{off} of MF were set to be 30 and 20 µs (duty cycle of 60%), respectively. To reveal the role of MF pulse duration, the ratio of total duration time, Σt_{on}, of MF to one duration time, t_{on}, of HiPIMS in one superimposed cycle was varied to be 150/100, 300/100, 600/100, 1200/100, 1800/100, and 2400/100, while the average powers of HiPIMS and MF were both fixed at 1000W (defined as 1.5×1000W, 3×1000W, 6×1000W,

TABLE 5.1

Deposition time of TiN films with different MF duration

Deposition mode	DC	HiPIMS	1.5×1000 W	3×1000 W	6×1000 W	12×1000 W	18×1000 W	24×1000 W
Deposition time (min)	80	240	50	50	50	50	40	40

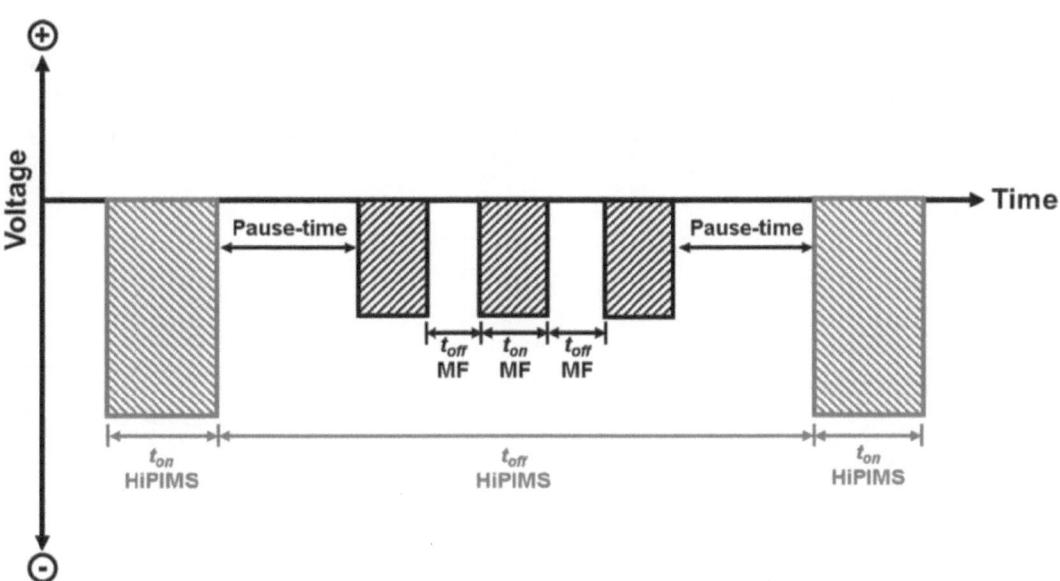

FIGURE 5.1 Schematic drawing of pulsing configuration of superimposed HiPIMS and MF.

TABLE 5.2

Parameter setup with different MF duration

Deposition mode	HiPIMS	1.5×1000 W	3×1000 W	6×1000 W	12×1000 W	18×1000 W	24×1000 W
Pause-time (μs)	N/A	2325	2200	1950	1450	950	450
MF pulses	N/A	5	10	20	40	60	80
$\dfrac{\Sigma t_{on}\,MF\,(\mu s)}{t_{on}\,MF\,(\mu s)}$	$\dfrac{0}{100}$	$\dfrac{150}{100}$	$\dfrac{300}{100}$	$\dfrac{600}{100}$	$\dfrac{1200}{100}$	$\dfrac{1800}{100}$	$\dfrac{2400}{100}$

12×1000W, 18×1000W, and 24×1000W hereafter (c.f. Table 5.2)). For instance, 6× 1000W (six times) means that the total duration (Σt_{on}) of MF pulses was set to be six times longer than the ton of HiPIMS pulsing in one cycle of superimposed HiPIMS-MF. In order to have the Σt_{on} of MF to be six times longer than the ton of HiPIMS, the Σt_{on} of MF was set at 600 μs. To achieve this 600 μs, the time for introducing MF pulses before and after one HiPIMS pulse (defined as the pause-time) was adjusted to 1950 μs. In other words, there were 5, 10, and 20 cycles of MF pulsing in 1.5×1000W, 3×1000W, and 6×1000W deposition modes, respectively, by adjusting the pause-time intervals. In this case, both HiPIMS and MF powers were run in the unipolar negative mode.

5.2.1.1 MF Duration

5.2.1.1.1 Effect of MF Duration on Peak Power Density

The peak power densities of HiPIMS and MF during different deposition modes are plotted in Figure 5.2. The peak target power density was calculated by dividing the peak power by the ion eroded target area of 200 cm². Single HiPIMS mode has the greatest peak power density of 726 W/cm². The peak power density of HiPIMS in superimposed modes decreases from 686 W/cm² (1.5×1000W) to 460 W/cm² (24×1000W) with increasing MF pulse duration. Compared with single HiPIMS mode, the peak power density of superimposed HiPIMS decreases 5.5% (686 W/cm²), 24.9% (545 W/cm²), 28.9% (516 W/cm²), 32.9% (487 W/cm²), 34.8% (473 W/cm²), and 36.6% (460 W/cm²). The pulse voltage changes from 698V for single HiPIMS to 799 V (1.5×1000W), and then decreases to 678 V (24× 1000 W) for superimposed HiPIMS. At the same time, the current changes from 208 A (HiPIMS) to 143 A (3×1000W) and then remains constant for superimposed HiPIMS (c.f. Figure 5.2(a)). Nevertheless, increasing MF pulse duration lowers the peak power density of MF pulsing. The increase in MF pulse duration significantly decreases MF peak power density from 306 24 W/cm² (1.5×1000W) to 24 W/cm² (24×1000W) (c.f. Figure 5.2(b)).

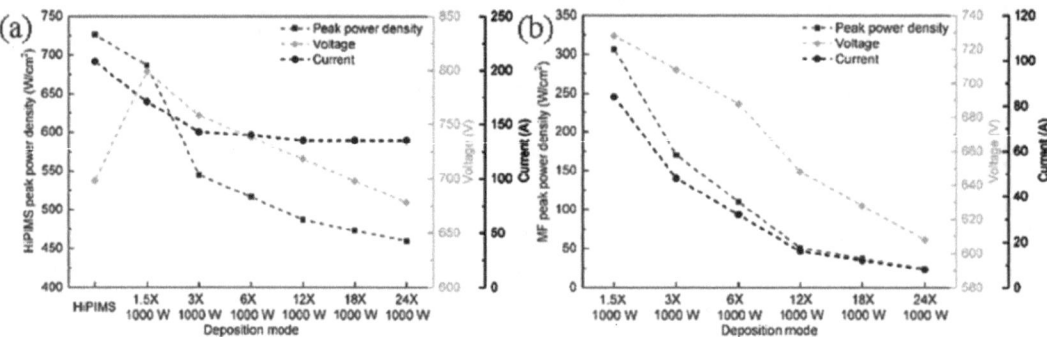

FIGURE 5.2 The diagram of target voltage, current and peak power density in different MF duration, (a) HiPIMS and (b) MF.

5.2.1.1.2 Effect of MF Duration on Ionization

Figure 5.3 shows the typical OES spectra of neutral Ti and Ti^+ ions observed in the plasma of single HiPIMS and other superimposed modes. OES spectra of the Ti^+ line at 376 nm and Ti line at 453 nm are selected for further study and their intensities are listed in Table 5.3. In single HiPIMS mode, the intensity of Ti^+ ion and neutral Ti are 18,708 and 12,790 count/s, respectively. It can be seen that only single HiPIMS mode has the Ti^+ ion intensity higher than that of neutral Ti, which is attributed to a significant amount of the sputtered metal species being ionized by HiPIMS power [9]. For 3×1000W superimposed modes, the intensities of Ti^+ ion (376 nm) and Ti (453 nm) are highest among samples tested, with increase the MF duration, the intensity for both Ti+ ion (376 nm) and Ti (453 nm) increase little and then decrease drastically. However, all are higher than that of DC mode. For 1.5×1000W, 3×1000W and 6×1000W superimposed modes, the intensities of Ti^+ ion are higher than these in the single HiPIMS mode, but their neutral Ti intensities are also much higher than their Ti^+ ion intensities.

Furthermore, comparing the intensity ratio of $Ti^+/(Ti^++Ti)$ yields a qualitative measure for degree of ionization. This ratio can also be used as a parameter to reveal the discharge effectiveness [38, 39]. The intensity ratios are listed in Table 5.3. As compared with the data in Table 5.3 and Figure 5.2,

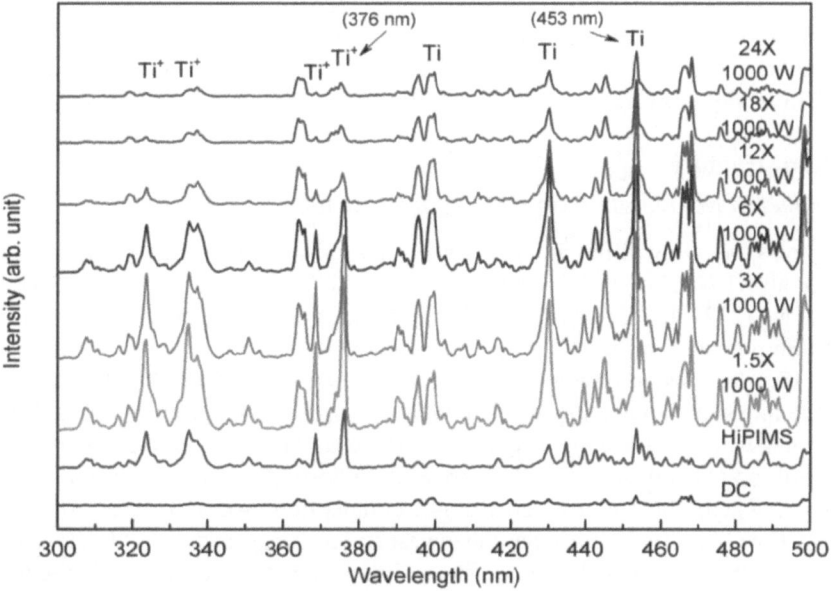

FIGURE 5.3 OES spectra of Ti^+ and Ti observed in different MF duration.

TABLE 5.3
Plasma intensity of Ti^+ and Ti with different MF duration

Wavelength (nm)	Intensity (count/s)							
	DC	HiPIMS	1.5×1000 W	3×1000 W	6×1000 W	12×1000 W	18×1000 W	24×1000 W
Ti^+ (376.2)	190	18708	39890	40372	22311	7384	2403	1639
Ti (453.4)	3052	12790	63070	64770	60970	37690	20240	14496
Ratio of $Ti^+/(Ti^++Ti)$	0.06	0.59	0.39	0.38	0.27	0.16	0.11	0.10

the single HiPIMS mode has the greatest intensity ratio of 0.59 and the highest HiPIMS peak power density of 726 W/cm^2. For the superimposed modes with a fixed 1000W of average MF power, the Ti$^+$/(Ti$^+$+Ti) ratio decreases from 0.39 in 1.5×1000 W mode to 0.10 in 24×1000 W mode which is about 17% of the intensity ratio in single HiPIMS mode. Since a lower ratio represents a lower amount of ionized Ti atoms and vice versa, it is therefore suggested that the superimposing MF pulses into HiPIMS pulsing lowers the ionization degree in the plasma.

5.2.1.2 Chemical Composition

Figure 5.4 shows the field emission electron probe microanalysis (FE-EPMA) composition test results of TiN films under different MF durations. The O content of TiN films prepared under different modes is less than 0.40 at.%, which indicates that TiN films prepared under high background vacuum present higher purity. The N/Ti ratio of TiN film prepared in DC mode is 0.98, indicating that the content of N and Ti are basically the same. The TiN film prepared in HiPIMS mode has a N content of 53.43 at.%, a Ti content of 46.48 at.%, and a N/Ti ratio of 1.15. Because HiPIMS mode has a longer pause time (4900 μs) and a shorter pulse time (100 μs), the effective sputtering time is short, and the sputtered particles have sufficient time and sufficient nitrogen to react with, the N content is slightly higher than that of Ti. In the superimposed HiPIMS mode, as the duration of MF increases, the Ti content gradually increases, from 47.45 at.% (1.5×1000W) to 57.69 at.% (12×1000W) and then stabilized at around 58.12 at.% for 18× 1000 W and 58.13 at.% for 24× 1000 W. This indicates that when the MF duration increases to a certain extent, the Ti content no longer increases with the increase of the MF duration. The N/Ti ratio of TiN films prepared by the superimposed HiPIMS mode is reduced from 1.09 (1.5×1000W) to 0.72 (18, 24×1000W).

5.2.1.3 Microstructure

5.2.1.3.1 Effect of MF Duration on TiN Crystal Structure

Figure 5.5 shows the X-ray diffraction patterns of TiN films prepared under different MF durations. TiN films are all B1 NaCl crystal structures, with five diffraction characteristic peaks (111), (200), (220), (311), and (222) (JCPDS: 87-0633). The intensities of the two crystal planes (111) and (200) of the TiN film prepared in the DC mode are relatively close, while the TiN film prepared in the single HiPIMS mode has an obvious preferred orientation, with the (111) diffraction peak dominated, and the intensity is much higher than other peaks. In the superimposed HiPIMS mode, with the increase of MF duration, the intensity of the (200) peak of the TiN film gradually weakened, and the intensity of the (111) peak gradually increased, indicating that the preferred orientation of the TiN film has changed. The preferred growth is a competition process between surface energy and strain energy. The (111) crystal plane has the lowest strain energy, and the (200) crystal plane has the lowest surface energy [40]. HiPIMS mode has a high peak power density (726 W/cm^2) and ionization rate (0.59), which results in high-energy particles, in addition, the pause time is long enough. Thus, the film growth process has enough time for energy migration and strain release. Because of the low particle energy, DC has little effect on surface energy and strain energy, and has no obvious preferred orientation. In the superimposed HiPIMS mode, with the increase of the MF duration, the (200) peak intensity decreases and the (111) peak intensity increases, indicating that the TiN film gradually shifts from the lowest surface energy to the lowest strain energy. The change of MF duration significantly affects the superimposed type.

The grain sizes calculated based on different crystal planes are listed in Table 5.4. In order to minimize the calculation error, it is calculated by substituting the integral width of each diffraction peak into the Scherer formula. The grain size calculated based on the (111) diffraction pattern of the TiN film prepared under different modes varies between 6.2~7.8 nm, 1.8~8.3 nm for (200) crystal and 6.2~7.8 nm for (220) one. Among them, 1.5×1000 W has the strongest diffraction peak on the (200) crystal plane and the largest grain size of 8.3 nm. During film growth, high-energy particles bombarding the surface will refine the grain size [41]. In addition, in order to obtain film with same

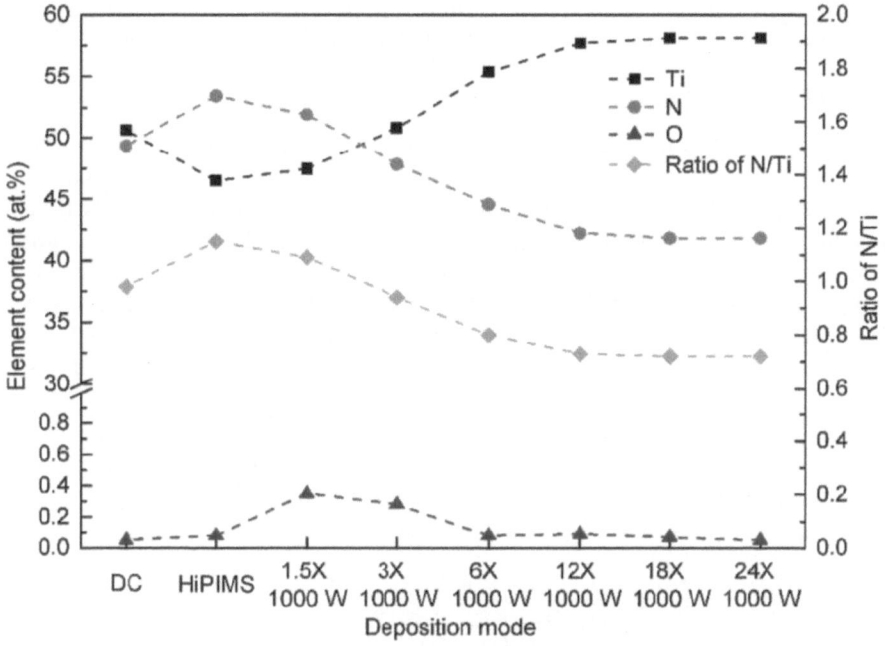

FIGURE 5.4 Element content of TiN films deposited with different MF duration.

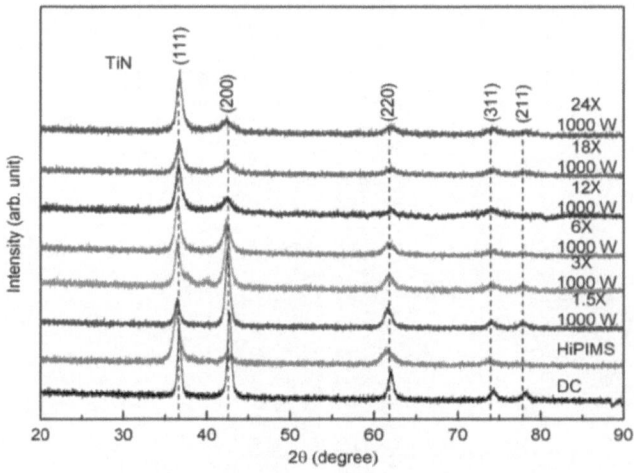

FIGURE 5.5 XRD patterns of TiN films deposited in different MF duration.

TABLE 5.4
Grain size of TiN films deposited with different MF duration

Deposition mode		DC	HiPIMS	1.5×1000 W	3×1000 W	6×1000 W	12×1000 W	18×1000 W	24×1000 W
Grain size (nm)	(111) grain	7.0	6.7	6.5	7.0	7.5	6.8	6.2	7.8
	(200) grain	3.7	1.8	8.3	1.4	2.4	3.2	3.0	2.9
	(220) grain	3.2	2.2	3.9	1.6	2.1	2.8	2.4	3.1

thicknesses, the deposition time of TiN film is different for different modes. This also leads to an insignificant change in the grain size of TiN films prepared under different modes.

5.2.1.3.2 Effect of MF Duration on TiN Surface Morphology

Figure 5.6 presents the SEM images of surface morphologies of TiN films grown by different deposition modes. The TiN film prepared in the DC mode has a characteristic triangular morphology (c.f. Figure 5.6(a)), which is consistent with the surface morphology of many TiN films prepared by conventional magnetron sputtering. Single HiPIMS grown sample shows the smoothest surface morphology with the smallest particle size as compared with the superimposed grown samples. This is because HiPIMS has a high peak power density during the film deposition process, and the sputtered particles have high energy. The high-energy particles bombard the growing surface of the

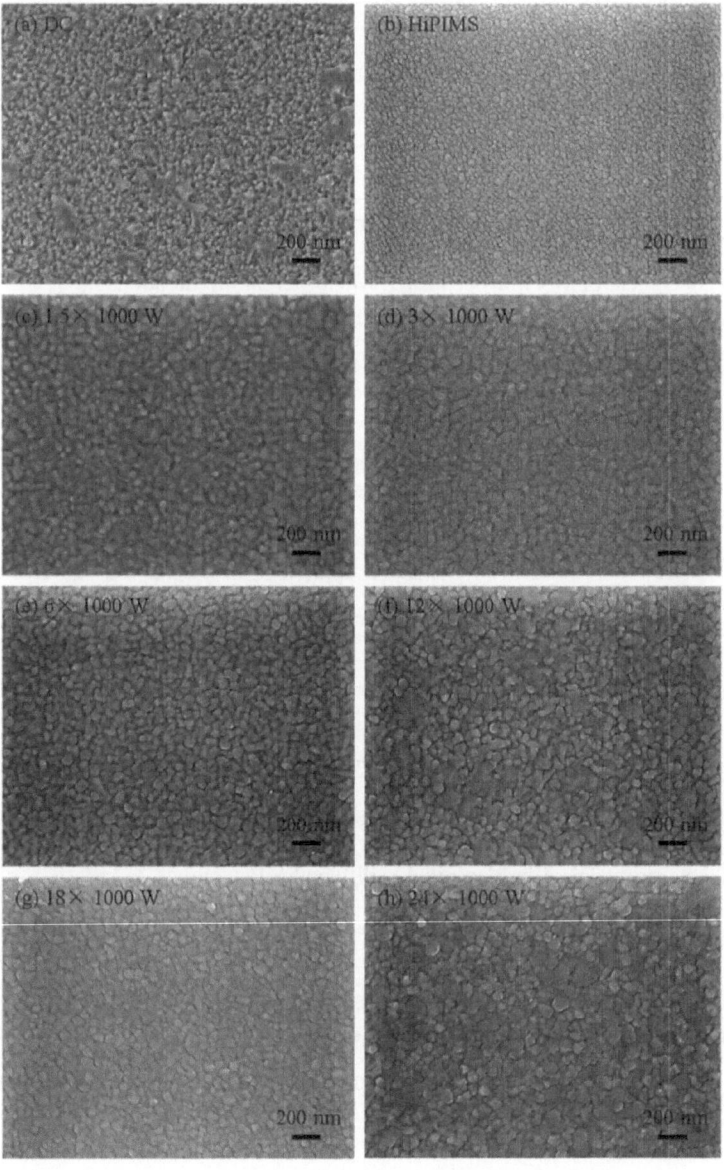

FIGURE 5.6 Surface morphologies of TiN films deposited in different MF duration.

film during deposition and form fine particles and smooth surface. The comparison of the surface morphology of TiN films prepared by DC and HiPIMS modes shows that the bombardment of high-energy particles will change the surface morphology of TiN films. The higher the particle energy, the finer and smoother the surface particles. The TiN films prepared by the superimposed HiPIMS mode with different MF durations have similar particle morphology to the HiPIMS mode, but the particles gradually increase as the MF duration increases. This is because as the MF duration increases, the power density of both HiPIMS and MF decreases, the energy of the sputtered particles decreases, the bombardment effect on the film is weakened, and the surface particles gradually grow.

5.2.1.4 Effect of MF Duration on TiN Surface Roughness

Figure 5.7 shows the surface morphology and roughness measured with AFM. For the DC magnetron sputtering sample, surface roughness is 4.1nm, single HiPIMS one is much smoother, about 1.5nm. While the roughness of superimposed samples is within the range of 1.5~4.1 nm. Compared with DC magnetron sputtering, the single HiPIMS modes provide TiN films with relatively smooth surface.

5.2.1.4.1 Effect of MF Duration on TiN Cross-Sectional Morphology

Figure 5.8 displays the SEM images of cross-sectional microstructure of deposited TiN films. Apparently, the single HiPIMS and superimposed modes grown sample show a fine and dense microstructure owing to the use of HiPIMS pulsing in all deposition mode. The TiN film prepared in the DC mode has a disorderly growth columnar morphology, indicating that the growth is

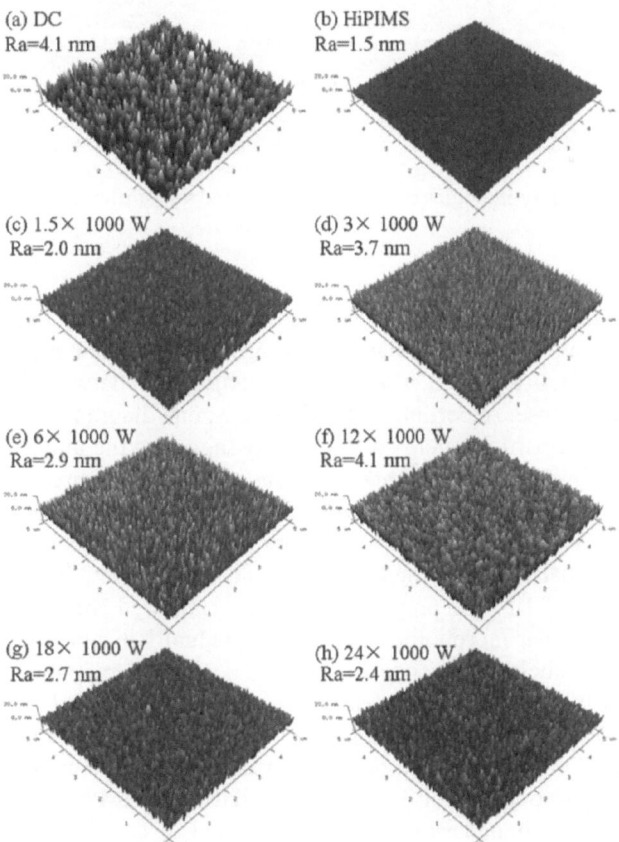

FIGURE 5.7 AFM images of TiN films deposited in different MF duration.

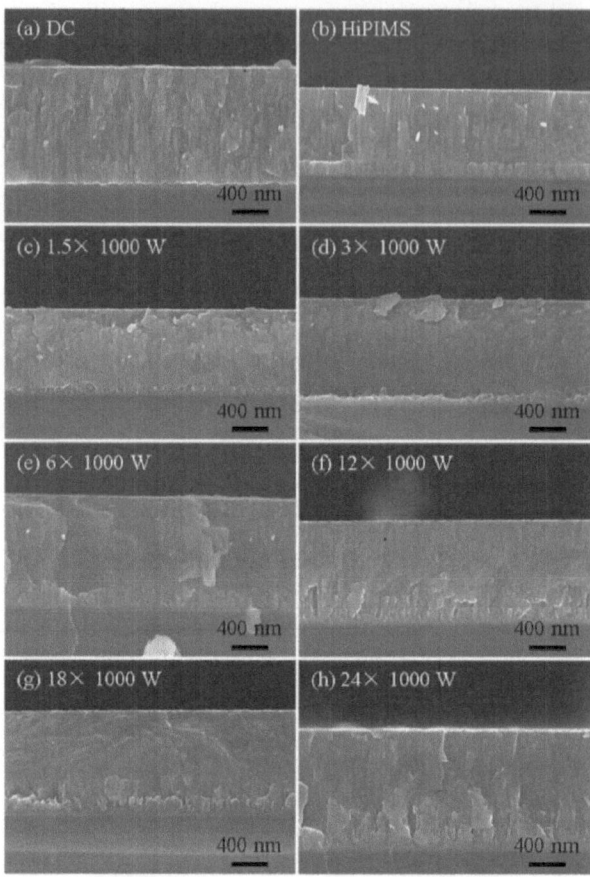

FIGURE 5.8 Cross-sectional morphologies of TiN films deposited with different MF duration.

disordered, and there are gaps between the columnar tissues, which results in the loose and porous film. According to the structure zone diagram, it belongs to Zone 1[16]. The TiN film prepared in HiPIMS mode presents an obvious columnar morphology, the columnar structure grows neatly and vertically, and the directionality is good. There is no obvious gap in between the columnar interface, and the film layer is very dense. According to the structure zone diagram, it belongs to Zone 2. The deposition temperature and particle energy are relatively high, and the uniform columnar organization runs through the entire film. The HiPIMS mode has better growth directionality than that of the DC mode. This is because the high-energy particles produced by the high peak power density of the HiPIMS pulse have good directionality and can grow well along the incident direction; while the DC energy is low and the particle movement direction is dispersed, which result in disorder growth. The cross-sectional image of TiN film prepared by the superimposed HiPIMS mode showed a dense morphology, and the columnar morphology of the film gradually disappeared with the increase of MF duration. The structure belongs to Zone T.

For further detailed microstructure analysis, TEM images were used to examine the DCMS, single HiPIMS, and 1.5×1000W grown samples as shown in Figure 5.9. In the bright field image, the direction of crystal growth of TiN film prepared in DC mode is inconsistent, and the film is not dense; the columnar grain boundary of TiN film prepared in HiPIMS mode is clear and the film is dense; the columnar grain boundary of TiN film prepared in 1.5×1000 W mode is clearer than that of DC, and the film grows along the deposition direction, and the columnar crystals have good directionality. Single HiPIMS mode has high peak power density and ionization rate, 1.5× 1000W

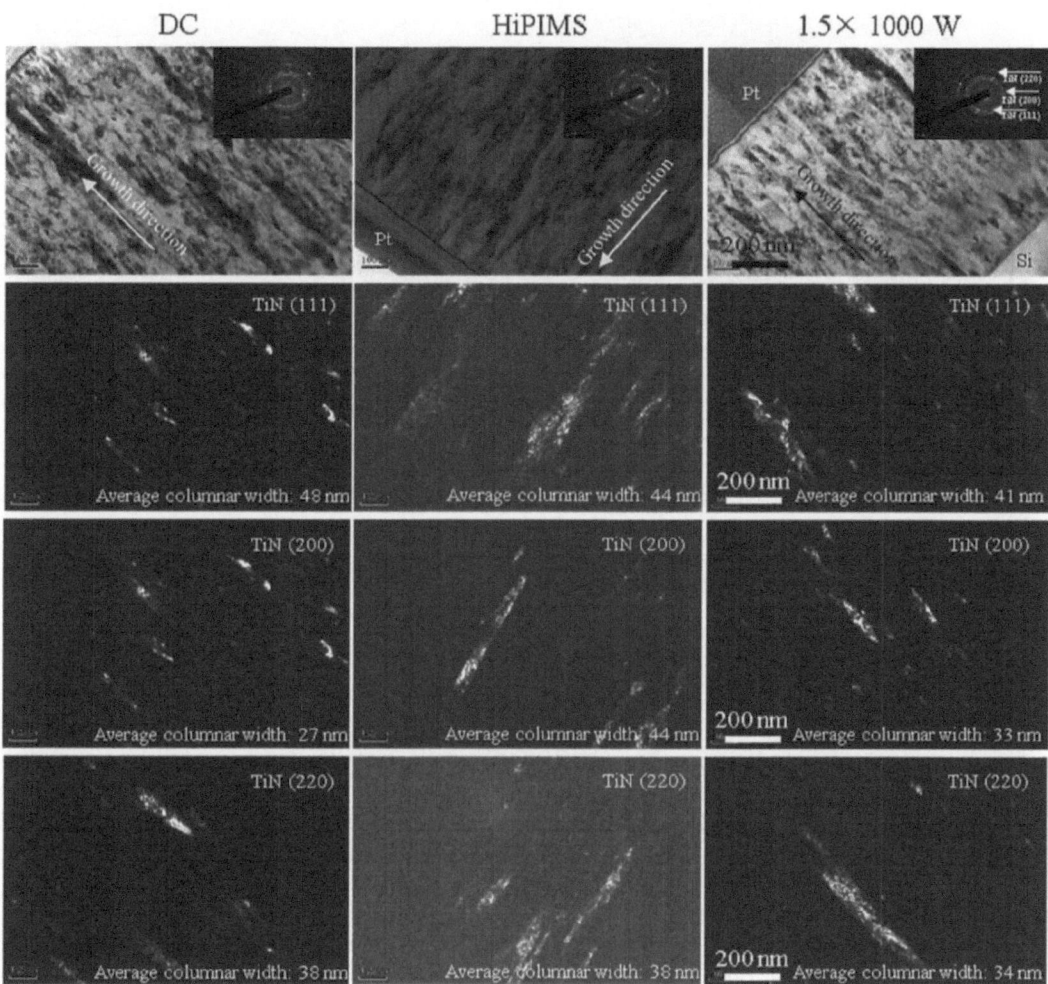

FIGURE 5.9 TEM images of TiN films deposited with different deposition mode.

mode HiPIMS and MF pulse both have higher peak power density, but the ionization rate is lower than that of HiPIMS mode, while DC mode power density and ionization rate are lower than both HiPIMS and superimposed HiPIMS mode. The HiPIMS technique yields a high ion bombardment effect leading to a fine and dense microstructure, which is actually a well-known mechanism. For instance, the beneficial effect of high ion bombardment on the microstructure densification effect of TiN films has been studied by Machunze et al. [42]. The presence of TiN phase is confirmed by the selected area electron diffraction (SAED) patterns of each sample (c.f. Figure 5.9). The measured average columnar widths for (200) diffraction in DCMS, single HiPIMS, and 1.5×1000W grown samples are 27±6 nm (DCMS), 44±16 nm (HiPIMS), and 33±7 nm (1.5×1000W), respectively. A slightly wider columnar width in single HiPIMS grown sample is possibly due to much longer deposition time, 240 min. Surprisingly, the high densification effect due to high ion bombardment also took place during deposition of Ti interlayer in all samples, which further shows the beneficial effect of superimposed HiPIMS and MF modes.

Figure 5.10 is the TEM image of Ti bonding layer of TiN film prepared in DC, HiPIMS and 1.5×1000W three modes. In the Ti bonding layer, the DC mode columnar crystals are wider, and the average columnar crystal width is 47±10 nm. White gaps can be observed in the TiN film.

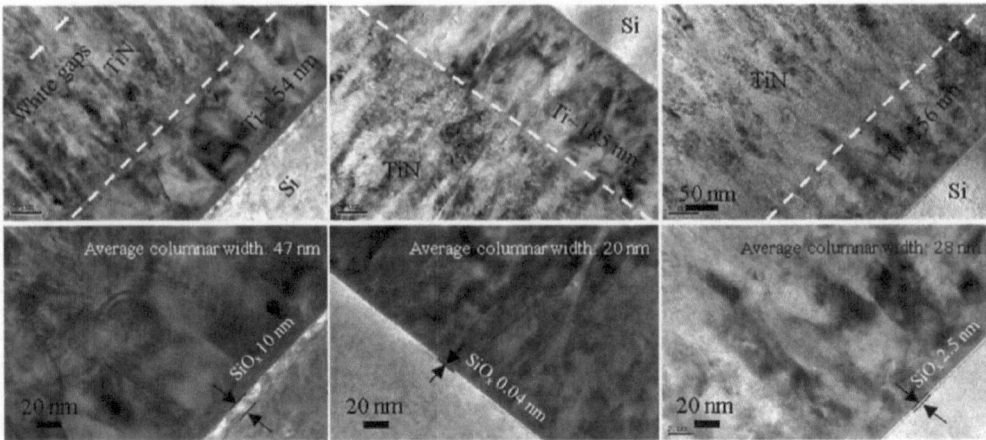

FIGURE 5.10 TEM images of Ti interlayer of selected TiN samples.

There are similar results in other research literature [42], which shows TiN film is not dense enough [43]. Under the bombardment of high-energy particles, the Ti bonding layer of HiPIMS and 1.5×1000 W mode is dense, the average columnar crystal width of HiPIMS mode is 20±5 nm, and the average columnar crystal width of 1.5×1000W mode is 28±7 nm, indicating high-energy particle bombardment can refine grains. In addition, there is a layer of SiOx between the Si substrate and the Ti bonding layer. In the DC mode, the thickness is at most 10 nm, the HiPIMS mode has the smallest thickness, approaching 0 nm, and the 1.5×1000W mode has a thickness of 2.5 nm. This shows that the sputter etching between Ti bonding layer deposition does not completely remove the oxide layer on the Si wafer, but the oxide layer thickness of HiPIMS and superimposed HiPIMS systems is significantly reduced, which shows that HiPIMS and superimposed HiPIMS can be used as an effective method to remove surface oxide on Si wafers further expands the application range of HiPIMS and superimposed HiPIMS systems.

5.2.1.4.2 Effect of MF Duration on TiN Density

In order to quantitatively analyze the changes in density of the TiN film prepared under different modes, the density of the TiN film was measured by the XRR method, and the results listed in Table 5.5. The density of bulk TiN is 5.43 g/cm³ [44], and the density of TiN film prepared in DC mode is 4.79 g/cm³, which is 88.2% of that of bulk one. The density of TiN film prepared by HiPIMS mode is 5.27 g/cm³, which is 97.0% that of bulk one, which is higher than these from other superimposed HiPIMS and DC modes except for 1.5×1000W mode. The density of the TiN film prepared in the 1.5×1000W mode is 5.35 g/cm³, which is 98.5% of the bulk TiN, which is very close to the density of the bulk material and higher than that from the HiPIMS mode. This may be due to the high MF pulse power density of 306 W/cm² for the 1.5×1000W sample, which has a similar effect to HiPIMS high-energy particle bombardment, and further improves the density of the film. As the MF duration increases, the density of the TiN film gradually decreases, and the longer the MF duration, the more slowly the density decreases, and it remains stable in the 18×1000W mode and no longer

TABLE 5.5

Density of TiN films deposited with different MF duration

Deposition mode	DC	HiPIMS	1.5×1000 W	3×1000 W	6×1000 W	12×1000 W	18×1000 W	24×1000 W
Density (g/cm³)	4.79	5.27	5.35	5.22	5.20	5.10	5.02	5.02

TABLE 5.6

Thickness and deposition rate of TiN films deposited with different MF duration

Deposition mode	DC	HiPIMS	1.5×1000 W	3×1000 W	6×1000 W	12×1000 W	18×1000 W	24×1000 W
Thickness (nm)	1395	1022	1050	1310	1375	1182	1200	1377
Deposition time (min)	80	240	50	50	50	40	40	40
Deposition rate (nm/·min)	17.4	4.3	21.0	26.2	27.5	29.6	30.0	34.4
Deposition rate (nm/kW·min)	17.4	4.3	10.5	13.1	13.8	14.8	15.0	17.2

decreases. This is because the MF duration increases, the power density of HiPIMS and MF pulses gradually decreases, and the particle energy gradually decreases.

5.2.1.5 Deposition Rate

The differences in pulsing configurations alter the film growth and thus the deposition rate. The film thickness and deposition rate of different samples are measured based on Figure 5.8 and the data are tabulated in Table 5.6. The deposition rate can be normalized to accommodate the total applied power. The power normalized deposition rates are thus calculated in the unit of nm/kW·min to accommodate the total power output of 2000W for 1.5×1000W through 24×1000W samples using HiPIMS (1000 W) and MF (1000 W). Meanwhile, the total power of single HiPIMS mode is restricted to 1000 W. DCMS reach the highest deposition rate of 17.4 nm/min, while, single HiPIMS mode exhibits the lowest deposition rate of 4.3 nm/kW·min. And the deposition rate can be successfully improved up to 17.2 nm/kW·min for 24× 1000 W sample in superimposed mode. Figure 5.2 illustrates the relationship between deposition rate and peak power density of HiPIMS and MF in different deposition modes. Although the peak power density of either HiPIMS or MF keeps decreasing, increasing MF pulse duration significantly prolongs the working duration, reduces the pause-time between HiPIMS and MF. Thereafter, an enhanced deposition rate can be obtained.

5.2.1.6 Mechanical Properties

5.2.1.6.1 Effect of MF Duration on Hardness and Modulus

Figure 5.11 presents the hardness, elastic modulus, H/E and H^3/E^2 of TiN films prepared in different modes. In Figure 5.11(a), the hardness of TiN films prepared in different modes are 18.3±0.4 GPa

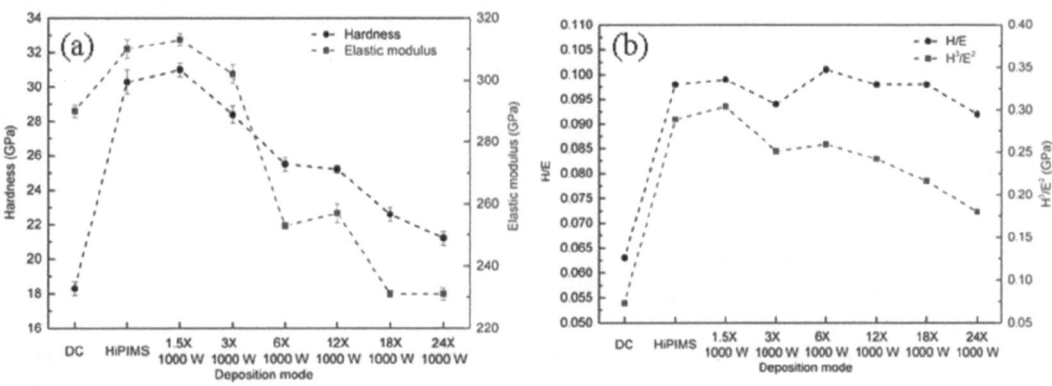

FIGURE 5.11 Mechanical properties of TiN films deposited with different deposition mode, (a) H and E, (b) ratios of H/E and H^3/E^2.

(DC), 30.3±0.7 GPa (HiPIMS), 31.0±0.4 GPa (1.5×1000W), 28.4±0.5 GPa (3×1000W), 25.5±0.4 GPa (6×1000W), 25.2±0.2 GPa (12×1000W), 22.6±0.4 GPa (18×1000W), and 21.2±0.4 GPa (24×1000W). The TiN film prepared in the DC mode shows the lowest hardness and the 1.5×1000 W mode has the highest hardness, but there is no significant difference from the HiPIMS mode (considering the standard deviation). In the superimposed HiPIMS mode, as the MF duration increases, the hardness of the TiN film decreases, but all of them are higher than that of the DC mode. The microstructure and density of the film are the main reasons that affect the hardness of the TiN film [45]. The microstructure of the TiN film prepared in the DC mode has white gaps (c.f. Figure 5.9), which results in the looseness of the structure. And low density will result in the lowest hardness of TiN film. The HiPIMS mode and the superimposed HiPIMS mode show the high density, and the TiN film presents high hardness, and the hardness change trend is consistent with that of the density.

The elastic modulus of TiN films prepared in different modes are 290±2 GPa (DC), 310±3 GPa (HiPIMS), 313±2 GPa (1.5× 1000W), 302±3 GPa (3× 1000W), 253± 1 GPa (6×1000W), 257±3 GPa (12×1000W), 231±1 GPa (18×1000W), and 231±2 GPa (24×1000W). The change trend of elastic modulus is similar to that of hardness, but the elastic modulus of TiN film prepared in DC mode is higher than that after the MF duration is increased to 6×1000 W. On the one hand, it is from the composition changes, because the deposition rate of the superimposed HiPIMS mode increases significantly with the increase of MF duration (c.f. Table 5.6). Nitrogen vacancies are formed in the crystal lattice and the chemical composition ratio of the TiN film is changed (c.f. Figure 5.4). The ratio of N/Ti is 0.98 for TiN prepared in DC mode, while for the superimposed HiPIMS mode, it gradually decreases from 1.09 to 0.72. Therefore, the elastic modulus of TiN in the DC mode is higher than that of the TiN film prepared by the partially superimposed HiPIMS mode. On the other hand, it is from density variation. The density of TiN films prepared by the superimposed HiPIMS mode gradually decreases, and the elastic modulus also decreases. This shows that when the components are close, the density has a significant influence on the elastic modulus; when the density is close, the components have a significant influence on the elastic modulus.

Figure 5.11(b) shows the trend of H/E and H^3/E^2 for TiN films prepared with different modes. H/E and H^3/E^2 show the similar changing trends. The H/E ratios of different modes are 0.063, 0.098, 0.099, 0.094, 0.101, 0.098, 0.098, and 0.092 for DC, single HiPIMS and 1.5~24×1000W superimposed HiPIMS mode. The H^3/E^2 ratios are 0.073, 0.289, 0.304, 0.251, 0.259, 0.242, 0.216, and 0.180, for DC, single HiPIMS, 1.5~24×1000W superimposed HiPIMS mode. H/E represents the toughness of the material. The larger the ratio, the better the toughness of the material, the less likely it is to produce cracks [46], and it is also conducive to improving the tribological properties [47]. H^3/E^2 represents the material's ability to resist plastic deformation, and the larger the ratio, the stronger the material's ability to resist plastic deformation [48]. In summary, the change of MF duration affects HiPIMS and MF pulses, which in turn causes changes in the composition, structure and density of TiN films, and ultimately affects and determines the mechanical properties of TiN films.

5.2.1.6.2 Effect of MF Duration on Residual Stress

The residual stress of TiN films prepared by different modes is listed in Table 5.7. The negative value of the stress indicates that the stress belongs to the type of compressive stress, which helps

TABLE 5.7

Residual stress of TiN films deposited with different MF durations

Deposition mode	DC	HiPIMS	1.5×1000 W	3×1000 W	6×1000 W	12×1000 W	18×1000 W	24×1000 W
Residual stress (GPa)	−0.92	−1.76	−3.73	−2.67	−2.24	−1.46	−1.39	−0.95

to improve the load capacity of the film. The residual stress of TiN film prepared in DC mode is the lowest one −0.92 GPa, while in 1.5×1000W mode is the highest one −3.73 GPa which is much higher than that in single HiPIMS mode. This may be due to the fact that both HiPIMS and MF pulses in the 1.5×1000W mode possess high peak power density and high particle energy. Although the single HiPIMS mode displays the highest peak power density and high particle energy, the long pause time during deposition process allows the growing surface to have enough time to dissipate the energy. In addition, the deposition time for TiN in single HiPIMS mode is 240 min, and the deposition temperature is 200 °C, which can further eliminate the stress accumulated during deposition, therefore the internal stress for the single HiPIMS is not the highest one. With the increase of the MF duration, the residual stress of TiN films prepared by the superimposed HiPIMS mode gradually decreased from −3.73 GPa (1.5×1000W) to −0.95 GPa (24×1000W), and a reduction of 74.5% is achievable. The HiPIMS and MF pulse power densities were reduced by 36.6% and 92.2%, respectively, which promoted the reduction of the residual stress of the TiN film. This shows that the MF duration has a greater influence on the residual stress origin for the TiN film by the superimposed HiPIMS power system.

5.2.1.6.3 Effect of MF Duration on Adhesion

In addition to the mechanical properties, having acceptable adhesion of TiN films to the steel substrates is also important. The surface morphologies of the indentation craters of TiN films deposited on AISI420 substrates after the HRC-DB adhesion test are illustrated in Figure 5.12. In HRC-DB test, the adhesion quality of the films is classified into six grades, HF1–HF6. The adhesion in HF1-HF4 is acceptable according to the VDI 3198-1992 standard. Single HiPIMS grown sample present HF1 grade while superimposed grown samples with different MF pulse durations (except the 24×1000W) show HF2 grade and the sample from DC mode have HF3 grade. The absence of spallation or chipping is found for the samples with HF1 and only circular cracks and very tiny chipping are observed around the indentation craters of the samples with HF2. Nevertheless, the HRC-DB test suggests that all samples have acceptable adhesion strength quality ranging from HF1 to HF3. The spalling and cracks in the HRC indentation crater morphology are related to the microstructure, stress, and density of the TiN film. A small number of cracks can be observed in the partial enlarged image of the HRC indentation morphology of the TiN film prepared in HiPIMS mode. This is because the TiN film prepared in HiPIMS mode has good comprehensive performance, few structural defects, high density, and moderate residual stress, making it very good coating/substrate bonding. On the other hand, peeling off and cracks can be observed in the local magnification of the HRC indentation of the TiN film prepared in the DC mode. This is due to the defects in the microstructure, low density and low comprehensive properties of films obtained in the DC mode. There are spallation and cracks in the HRC indentation craters morphology of TiN film prepared in 1.5×1000W mode. This is because the residual stress of TiN film prepared in HiPIMS mode is too large, which offsets some of the advantages of defect free structure and high density. The overall performance is degraded, and the coating/substrate adhesion is reduced. Based on the above analysis, the coating/substrate adhesion of TiN film is comprehensively affected by the film structure, density, residual stress and other properties. Only the best comprehensive performance can achieve good coating/substrate adhesion.

5.2.2 Hybrid HiPIMS with DC/RF MS

CrSiN coatings were prepared by different power supply-assisted magnetron sputtering. There are three targets in the magnetron sputtering system. The schematic diagram of deposition system is shown in Figure 5.13. Two Cr targets (purity of 99.999 wt.%) are controlled by a DC power supply and a HiPIMS power supply, respectively. The Si_3N_4 target (purity of 99.999 wt.%) is controlled by the RF power supply. Mirror polished 304 stainless steel (20 mm×20 mm×4 mm) and silicon (100) were used as substrates. The substrates were ultrasonically cleaned by ethanol and deionized water

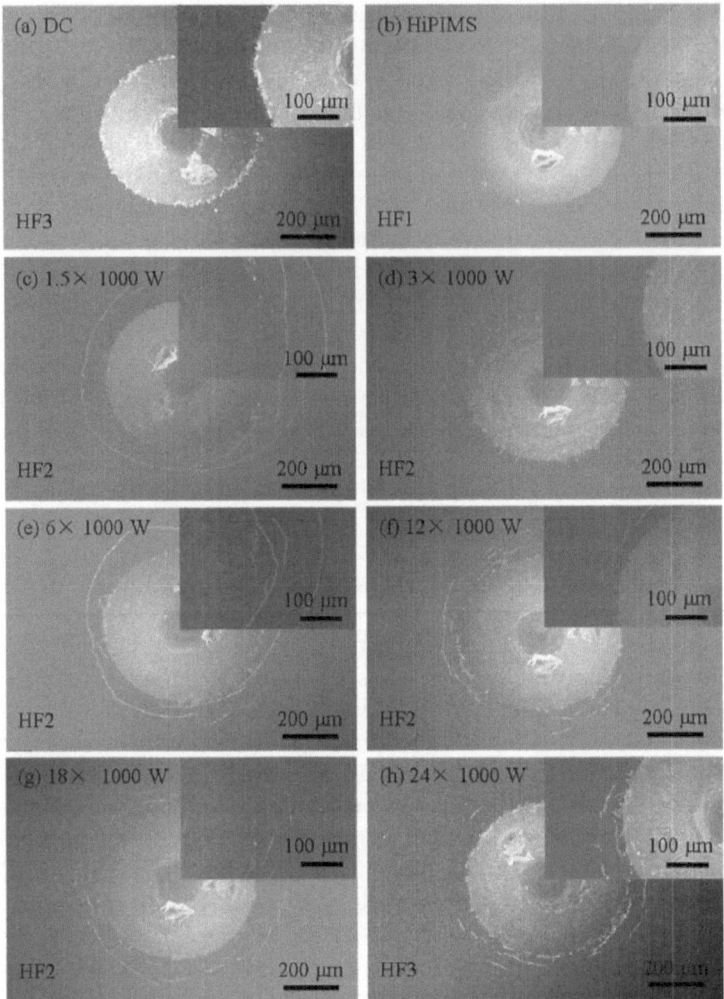

FIGURE 5.12 HRC adhesion test images of TiN films deposited with different deposition mode.

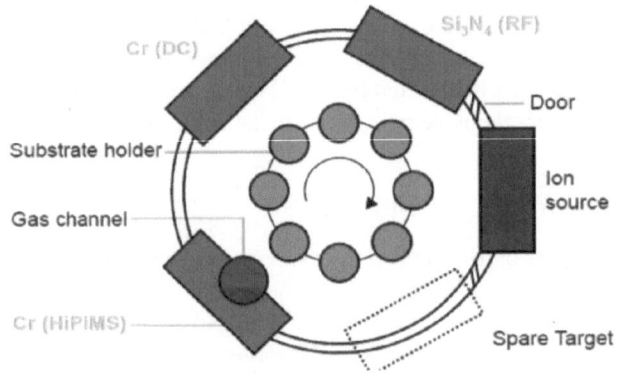

FIGURE 5.13 Schematic diagram of the coating deposition system (top view) [49].

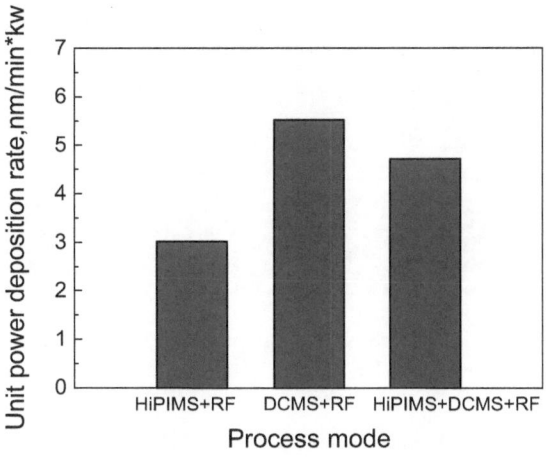

FIGURE 5.14 Deposition rate of CrSiN coatings prepared with different process mode.

in sequence for 20 min at 20 °C, then dried at lab ambient. One type of CrSiN was deposited using a Cr target in HiPIMS mode and a Si_3N_4 target in RF mode (HiPIMS+RF). The second type of CrSiN was deposited using a Cr target in DC mode and a Si_3N_4 target in RF mode (DCMS+RF). The last one type was produced by using such a magnetron configuration that two Cr targets in HiPIMS mode and DC mode, respectively, and Si_3N_4 target in RF mode (HiPIMS+DCMS+RF).

5.2.2.1 Deposition Rate

Considering the power inconsistency of the three processes, and in order to rule out the effect of different deposition power on the deposition rate, unit power deposition rate is defined as coating thickness divided by deposition time and the average power. The deposition rate of CrSiN coatings under three different processes is shown in Figure 5.14. The unit power deposition rates for process of HiPIMS+RF, DCMS+RF, and HiPIMS+DCMS+RF are 3.0 nm/(min·kW), 5.5 nm/(min·kW), and 4.7 nm/(min·kW), respectively. As reported in [31], HiPIMS process has the lowest deposition rate. It is verified by our results. The HiPIMS+RF process shows the lowest unit power deposition rate, while DCMS+RF has the highest one. The CrSiN coating deposited under the DCMS+RF process has the highest unit power deposition rate, which is about 1.8 times that of the HiPIMS+RF process. HiPIMS+DCMS+RF hybrid deposition process possesses a greater improvement in unit power deposition rate than that of HiPIMS+RF.

5.2.2.2 Microstructure

The surface morphology and cross-section morphology of CrSiN coatings prepared with different processes are shown in Figure 5.15 and Figure 5.16, respectively. The cross-sectional images show that the CrSiN coatings from different deposition mode present a columnar structure, and the thicknesses of the bonding layer Cr and the function layer CrSiN are about 0.1μm and 1μm, respectively. The morphology of CrSiN coating prepared by HiPIMS+RF process shows cauliflower-like appearance with fine and flat dome, reveals a dense structure (c.f. Figure 5.15(a) & Figure 5.16(a)), which is due to the fact that during the deposition process, the peak power density of HiPIMS and the energy of sputtered particles are very high, and the growing surface was bombarded by the high-energy particle. The enhanced surface diffusion decreases the defects. Nucleation is promoted while grain growth compressed, leading to epitaxial growth. In the DCMS+RF mode, the CrSiN coating shows triangular in shape with high surface porosity, the pore between each accumulated particle is large and the surface is not dense enough (c.f. Figure 5.15(b)&Figure 5.16(b)). According to the Thornton's Structure Zone Model (SZM) [16], the coating structure prepared by

FIGURE 5.15 Surface morphologies of CrSiN coatings prepared with different process mode (a) HiPIMS +RF, (b) DCMS+RF, (C) HiPIMS +DCMS+RF [49].

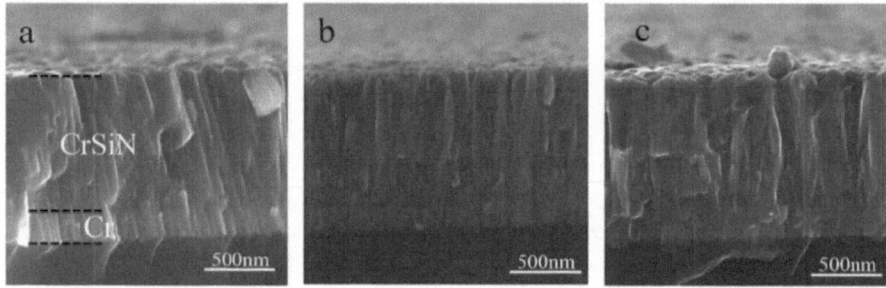

FIGURE 5.16 Cross section SEM image of CrSiN coatings prepared with different process mode (a) HiPIMS +RF, (b) DCMS+RF, (C) HiPIMS +DCMS+RF [49].

DCMS+RF conforms to Zone 1 due to the poor adatom mobility. Therefore, there is a small gap between the columnar structures. The size of dome on the surface of CrSiN coating prepared in HiPIMS +DCMS+RF mode is larger than that in the HiPIMS +RF mode (c.f. Figure 5.15(c)&Figure 5.16(c)). That is reasonable considering the low particle energy from DCMS where the ion bombardment effect is weakened which hinder the nucleation and promote grain growth.

Figure 5.17 shows the GIXRD patterns of CrSiN coatings prepared with different process mode. All the CrSiN coatings mainly exhibit single-phase NaCl-type cubic structure and strong (111) orientation. It can be seen from the Figure 5.17, the CrSiN coatings prepared with the participation of HiPIMS have strong (111) preferred orientation. The (111) diffraction peak is dominant, and the intensity is much higher than other peaks. The preferred orientation follows the lowest

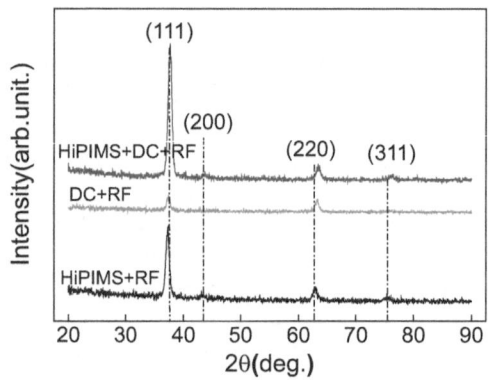

FIGURE 5.17 GIXRD pattern of CrSiN coatings prepared with different process mode [49].

energy principle and is determined by the competing between surface energy and strain energy. The (200) plane is the plane of the lowest surface energy in CrN crystal (which has the NaCl-type structure). Thus, at the very beginning of the deposition, it should be a surface energy-controlled process because the strain energy has not kick in yet. Since the total strain energy is proportional to the layer thickness and depends on the mean elastic moduli acting in the (hkl) plane parallel to the interface, competition from strain energy kicks in as the film grows. For CrN, the strain energy is minimized when (111) plane is parallel to the interface [40, 50]. For the HiPIMS-related process, the peak power density and ionization rate of HiPIMS are high, so the particles have high energy and have a long pause time. There is enough time for energy migration during coating growth to keep the strain energy at the lowest state. Therefore, CrSiN coatings by HiPIMS-related process show (111) preferred orientation.

Each diffraction pattern contains only the information from the CrSiN coating and does not include the diffraction peaks from the substrate and the metallic Cr interlayer. There are no peaks from Si contained compounds in the CrSiN coating, such as $CrSi_2$, Si_3N_4, etc. From literature [51, 52], usually the Si exists as amorphous SiNx.

5.2.2.3 Mechanical Properties

5.2.2.3.1 Hardness and Modulus

Figure 5.18(a) shows the hardness and elastic modulus of CrSiN coatings with different process mode. The coating with HiPIMS+RF process has the highest elastic modulus and hardness values of 241.7 GPa and 11.3 GPa, respectively. Both values have declined for coatings with the other two processes involving DCMS. The elastic modulus and hardness values of the CrSiN coating grown by DCMS+RF are only 160.9 GPa and 4.8 GPa, respectively. It has been pointed out that the growth structure and surface state of coatings play an important role on its mechanical and electrochemical properties [45]. CrSiN coatings prepared by hybrid HiPIMS show a higher hardness because the surface mobility will be improved with the increase of ion energy. However, the coating prepared by DCMS is loose and not compact, which results in the lowest hardness. Therefore, the change trend of hardness is consisted with that of density.

The ratio of hardness to elastic modulus H/E (resistance to elastic strain) and H^3/E^{*2} (resistance to plastic deformation, where $E^* = E / (1 - v)^2$, E is the modulus of elasticity and v is the Poisson's ratio) has been proposed as a key parameter for controlling wear resistance [53]. The H/E and H^3/E^{*2} of the CrSiN coatings prepared with different process mode are shown in Figure 5.18(b), which shows an evolution trend similar to that of the hardness and elastic modulus. The H/E and the H^3/E^{*2} of the CrSiN coating grown by the HiPIMS+RF process are the highest.

5.2.2.3.2 Tribological Properties

The friction coefficient and wear rate of CrSiN coatings prepared with different process are shown in Figure 5.19. It can be seen from Figure 5.19(a) that in the initial stage of wear, the friction coefficients of all the CrSiN coatings have an upward trend. The gradual increase of friction from a low value to a relatively steady value obeys the general rule of the running-in friction of most transition metal nitride coatings in unlubricated sliding wear. That is mainly due to the solid–solid friction pair. The origin of friction in the running-in period comprises the resistant forces arising from the generation of wear particles and the subsequent complex interactions between the wear particles within the sliding contact zone, namely their sliding and rolling motions. And it can be known from Figure 5.19(a) that the friction coefficient of all three coatings with different processes is low. That is because during pin on disk wear test, SiO_x or $Si(OH)_4$ for lubrication was synthesized by the reaction of Si and oxygen and steam. The results suggest that, friction coefficient is mainly determined by intrinsic properties of CrSiN coatings and is not affected by HiPIMS process.

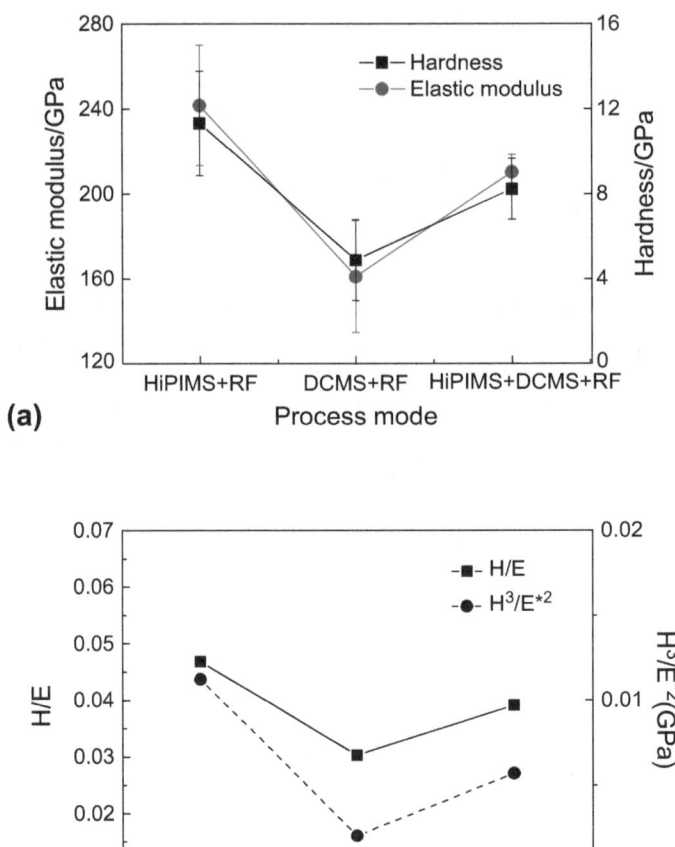

FIGURE 5.18 (a) Hardness and Young's modulus, and (b) H/E and H^3/E^{*2} ratios of CrSiN coatings prepared with different process mode [49].

The wear rate of CrSiN coatings prepared with different process mode is shown in Figure 5.19(b). The wear rate of CrSiN coatings grown with HiPIMS+RF process is one order of magnitude lower than that of the other two coatings, which is 5.0×10^{-6} mm³·N⁻¹·m⁻¹, showing a better wear resistance. That is obviously related to the high hardness of the coating and the H/E and H^3/E^{*2} ratio.

Figure 5.20 shows the cross-sectional profile of the wear track of the CrSiN coatings prepared with different process mode. The depths of the wear tracks created on the coatings less than the coating thickness. Thus, the friction and wear properties measured represent properties of the coatings only, without contribution from the substrate. The wear depth and width of CrSiN coating grown under HiPIMS+RF process are significantly less than that grown under DCMS+RF mode.

5.2.2.4 Corrosion Resistance

The polarization curves of CrSiN coatings prepared with different process mode in 3.5 wt.% NaCl aqueous solution are shown in Figure 5.21. Corrosion performance of CrSiN coatings under different processes is deviated. The corrosion current density of CrSiN coating grown under HiPIMS+RF process is the lowest, which is 3.2×10^{-7} A·cm⁻². The corrosion resistance of CrSiN coatings grown under DCMS+RF process is poor, and the corrosion current density is 2.4×10^{-6} A·cm⁻². And the corrosion potential of CrSiN coatings grown by HiPIMS+RF process is significantly higher than that

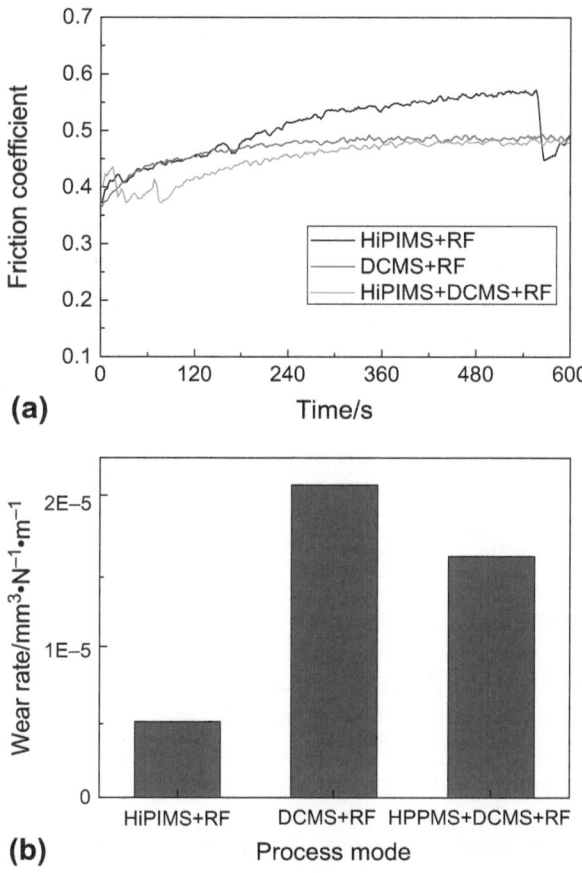

FIGURE 5.19 (a) Friction coefficient and (b) Wear rate of CrSiN coatings prepared with different process mode [49].

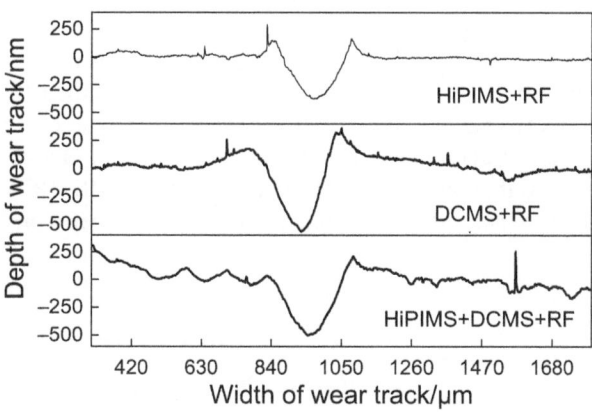

FIGURE 5.20 Cross section profiles of wear tracks of CrSiN coatings prepared with different process mode [49].

of the other two processes. CrSiN coatings deposited using the HiPIMS power source demonstrate better corrosion resistance. This result corresponds to the microscopic morphology (Figure 5.15 and Figure 5.16). In order to improve the corrosion performance, a finer and denser interface morphology is required [54]. As shown in Figure 5.15(b) & Figure 5.16(b), the CrSiN coating prepared under the condition of DCMS+RF shows a thick columnar structure with a large grain gap, and the

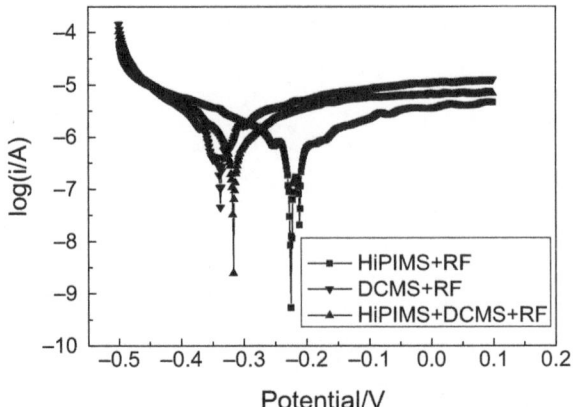

FIGURE 5.21 Polarization curves of CrSiN coatings prepared with different mode in 3.5 wt.% NaCl aqueous solution [49].

NaCl solution can penetrate into the substrate through the channel between the columns to corrode the substrate. This is undoubtedly harmful to the corrosion resistance of the coating. It can be seen in Figure 5.16(a) & (c) that the columnar structure of the CrSiN coating with HiPIMS power supply is interrupted, especially the CrSiN coating prepared by HiPIMS+RF, which is more compact and denser. The substrate is better protected from exposure to the solution, thereby increasing its corrosion resistance.

The typical cross-sectional structure of a PVD coating is a columnar crystal, as shown in Figure 5.16. The grain boundary where the columnar crystal grows is composed of a plurality of open pores, and a large number of independent closed pores exist in the crystal. These defects show a negative effect on corrosion performance because the corrosive media can directly contact the substrate itself through these defects, corroding the substrate. Therefore, the porosity and protection efficiency of the film are important criteria for characterizing the corrosion resistance of the coating. Table 5.8 lists the values of porosity and protection efficiency for all coatings and substrate. It can be found that the protection efficiency of the CrSiN film prepared by the DCMS+RF process is only 42.99%, but the film prepared by the HiPIMS+RF process is as high as 92.42%, which greatly increases the protection efficiency. At the same time, it can be known that the film prepared by the HiPIMS+RF process also has the lowest porosity. The lower value of the porosity factor suggests that the better barrier effect against the corrosive medium. In fact, it can be confirmed from Figure 5.15 and Figure 5.16 that the CrSiN film prepared by HiPIMS+RF shows a denser structure without continuous holes which effectively prevent contact between corrosive solution and substrate, thereby playing a protective role.

TABLE 5.8
Potentiodynamic polarization data of the specimens in 3.5 wt.% NaCl solution

Specimens	E_{corr}(V versus SCE)	i_{corr}(μA cm^{-2})	b_a(V^{-1})	Rp(Ω)	P(%)	P_i(%)
Substrate	−0.473	4.236	4.72	261.2	—	—
DC + RF	−0.346	2.415	3.61	456.9	53.7	42.99
HiPIMS + DC + RF	−0.319	1.932	3.97	578.7	41.9	54.39
HiPIMS + RF	−0.221	0.321	4.40	3376.8	6.8	92.42

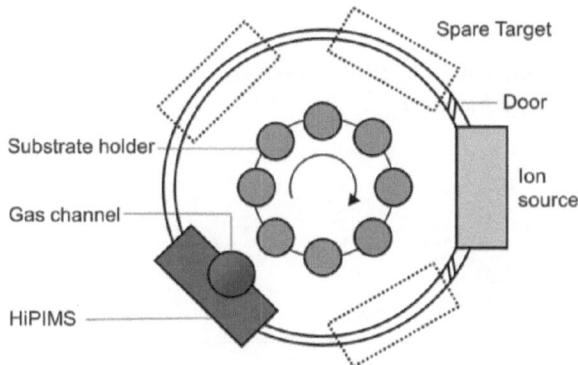

FIGURE 5.22 Schematic diagram of the coating deposition setup (top view) [37].

5.2.3 Ion Beam Assistant MS

Titanium nitride (TiN$_x$) films were deposited on 304 stainless steel and silicon (100) wafers by ion source assisted magnetron sputtering system as shown in Figure 5.22. Four targets are installed on the chamber and an anode ion source is equipped. The gas channel is located below the target, and the argon gas mixed with nitrogen introduce into the cavity. Ti (99.99% purity, 400 mm × 100 mm) target is connected with HiPIMS power supply. The effect of substrate bias voltage on the preparation and properties of TiN$_x$ films was investigated.

5.2.3.1 Deposition Rate And Microstructure

The deposition rate of TiN$_x$ films under different bias voltages is shown in Figure 5.23. It can be seen that the deposition rate of films decreases sharply from 7.69 nm/min to 4.55 nm/min with substrate bias voltage increasing from 0 V to −400 V. The impurities (oxygen and carbon) are removed from growing surface, since the increase of ions energy, meanwhile some deposits with poor adhesion to the growing surface will also be sputtered off, then film densification can be achieved [55]. All those factors contribute to the decrease of deposition rate and lead to a denser structure of TiN$_x$ film. In addition, the stoichiometry x was measured by EDS and the result shows that the stoichiometry x = N/Ti ranges from 0.89 to 1.05 (c.f. Table 5.9). During the film deposition process, there are two main

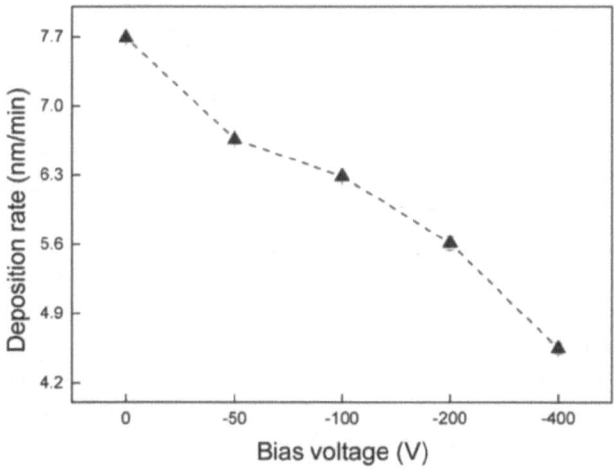

FIGURE 5.23 The deposition rates of TiN$_x$ films obtained at different substrate bias voltages [37].

TABLE 5.9

Thickness and toughness data of TiN$_x$ films prepared with different substrate bias voltages; here x = N/Ti is the film stoichiometry

Samples	Thickness (nm)	x	C (μm)	H (GPa)	E* (GPa)	K$_{IC}$ (MPa·m$^{1/2}$)	H/E*
0V	928.9	0.91	24.75	6.1	174.8	1.37	0.0346
−50V	828.1	0.97	26.52	6.2	197.6	1.29	0.0325
−100V	788.5	0.95	26.15	6.6	202.5	1.31	0.0327
−200V	724.3	1.05	18.96	12.9	260.8	1.72	0.0496
−400V	614.6	0.89	13.05	12.5	225.4	2.82	0.0555

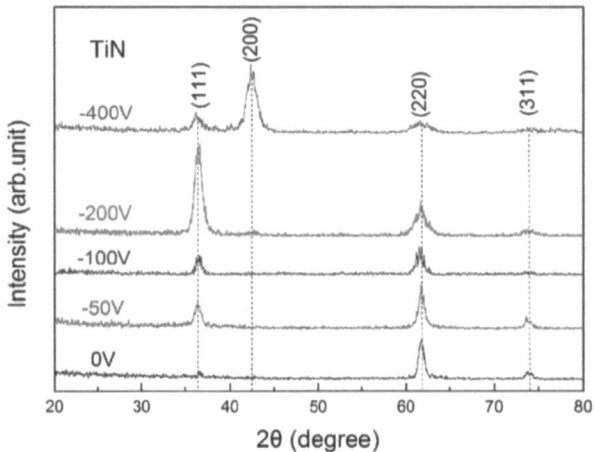

FIGURE 5.24 GIXRD diffraction patterns of TiN$_x$ films obtained at different substrate bias voltages [37].

mechanisms. On the one hand, the Ti-N bond is destroyed by ion bombardment [55]. On the other hand, more gas ions bombard the growing film, because of the assistant deposition of ion source and the effect of substrate bias voltage. Therefore, the change of stoichiometry x is the result of competition between the two mechanisms.

Figure 5.24 shows the grazing incidence X-ray diffraction patterns of TiN$_x$ films as a function of the substrate bias voltage. It is known that there is main four orientations and the preferred orientation of TiN$_x$ films varies from (111) orientation to (200) obviously as the energy delivered into the growing film increasing. So far, most scholars explain this experimental result through the surface energy and strain energy of the film [56, 57]. TiN$_x$ crystal has a NaCl type structure, the lowest surface energy plane is (200) and the lowest strain energy plane is (111). We always expect that the film grows in the direction with the lowest sum of surface energy and strain energy. The peak intensity of (111) plane becomes higher with the raise of bias voltage form 0 V to −200 V, caused by the increased bombardment of ions. However, with the further increase of substrate bias voltage, strain energy will no longer dominant the growth of the film. Instead, surface energy can play an important role. Growing with (200) plane possess the lowest surface energy.

Figure 5.25 shows the surface morphology of TiN$_x$ films with different substrate bias voltage. Without bias voltage, the surface morphology appears as a typical triangular cone. With the substrate bias changing from −50 V, −100 V to −200 V, the surface morphology appears as a laminated structure. Granular structure appeared, as the bias voltage further increases to −400 V. From the surface morphology evolution, it can be deduced that the density of the as-prepared film increases

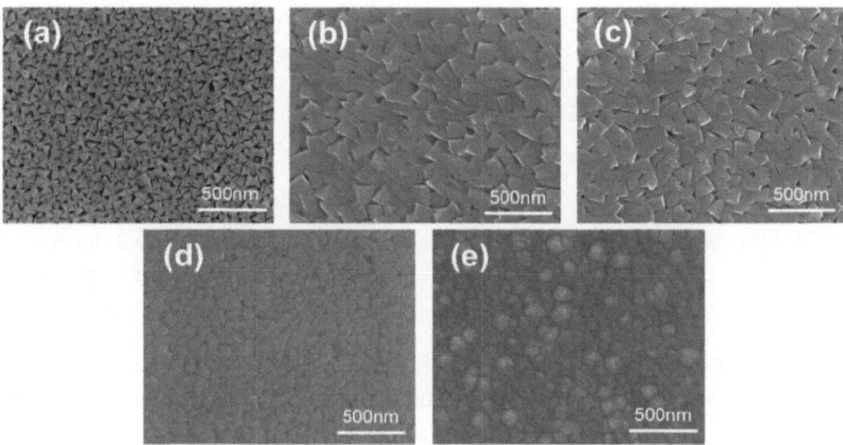

FIGURE 5.25 Surface morphology of TiN_x films with the substrate bias voltage of: (a) 0 V, (b) −50 V, (c) −100 V, (d) −200 V, (e) −400 V [37].

and the porosity decreases. The energy of ions can be calculated by $E \propto V_b \cdot P^{-\frac{1}{2}}$, where E, V_b, and P are the ion energy, substrate bias voltage and process pressure, respectively [58]. It proves that the ions have high energy under the high substrate bias voltage. In addition, J. Musil [59, 60] proposes that the energy delivered to the growing film relates to the film deposition rate. The energy of ions is inversely proportional to the deposition rate. Therefore, it can deduce that the energy of ions increases with the increase of substrate bias voltage. The re-sputtering effect, enhanced with the increase of ions energy, evidently reduces the shadowing effect and makes the film denser.

Figure 5.26 shows the cross-sectional microstructure of the TiN_x films. The TiN_x films exhibit columnar-type structure obviously, with neat vertical growth of columnar tissue and good directional

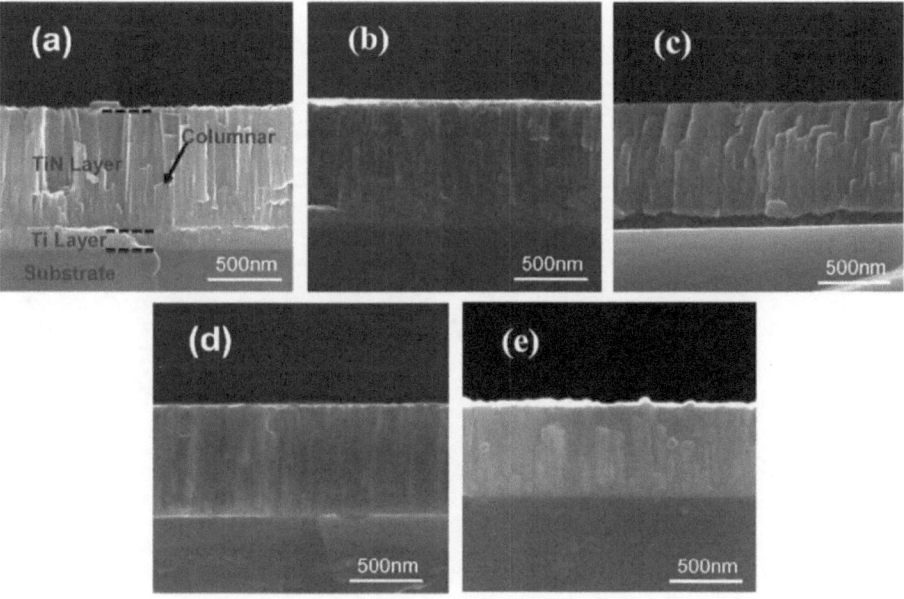

FIGURE 5.26 The cross-sectional microstructure of TiN_x films with the substrate bias voltage of: (a) 0 V, (b) −50 V, (c) −100 V, (d) −200 V, and (e) −400 V [37].

property. It can be explained by the high-energy particles, generated by the high peak power density of HiPIMS pulse. High-energy particles have good directionality and can grow well along the incident direction. According to structural zone model (SZM), introduced by Movchan and Demchishin [14], the homologous temperature $T_h=T/T_m$ (where T is substrate temperature and T_m is the film melting point, all in Kelvin) can be used to illustrate the microstructure of as-prepared films. In zone one, at $T_h < 0.3$, the mobility of atom is low leading to continuous grain nucleation, which results in a columnar structure with voids in the grain boundaries. In this study, T_h is lower than 0.3, so the structure of TiN_x films belongs to the structure of zone one. It is well known that the growth structure and surface state have a profound effect on the mechanical and electrochemical properties of these films [61].

5.2.3.2 Mechanical Properties

Figure 5.27(a) shows the hardness and Young's modulus of TiN_x films with various substrate bias voltages. **The value of hardness (H) and Young's modulus (E*) were determined by nanoindentation. The Young's modulus (E*) can be calculated through $E^* = E/(1 - v^2)$, where v is the Poisson's ratio [62]. The results indicate that the hardness of the TiN_x film increases before the substrate bias arrives −200 V and decreases slightly at −400 V. The film prepared under −200 V bias voltage reaches the maximum hardness of 12.9 GPa and a Young's modulus of 260.8 GPa, respectively. As substrate bias voltage increases, the bombardment of ions enhances, which can eliminate impurities, densify the film structure, and increase residual compressive stress [63]. Thus, the hardness of TiN_x film increases. In addition, the thickness of the film also affects the hardness of the film. Within a certain range, with the decrease of film thickness, the grain size is refined due to the insufficient time for grain growth [64]. The change of hardness and grain size satisfies the Hall–Petch relation. Figure 5.27(b) shows the load-displacement curves. With the increase of experimental load, the corresponding indentation depth increases. When the indentation depth reaches the defaults, the load also reaches the maximum value, which is consistent with the result shown in Figure 5.27(a). In addition, it should be emphasized that the peak intensity of the (111) plane reach the maximum at -200 V, which can also be used to explain the film hardness test results [65, 66].

Under the similar load indentation condition (1.96 N), the surface crack pattern of the films is shown in Figure 5.28. The crack length can be measured by micrograph and K_{IC} can be calculated through $K_{IC} = \delta \left(\dfrac{E}{H} \right)^{1/2} \left(\dfrac{P}{c^{3/2}} \right)$. The values obtained in Table 5.9 and the curve of the K_{IC} variation

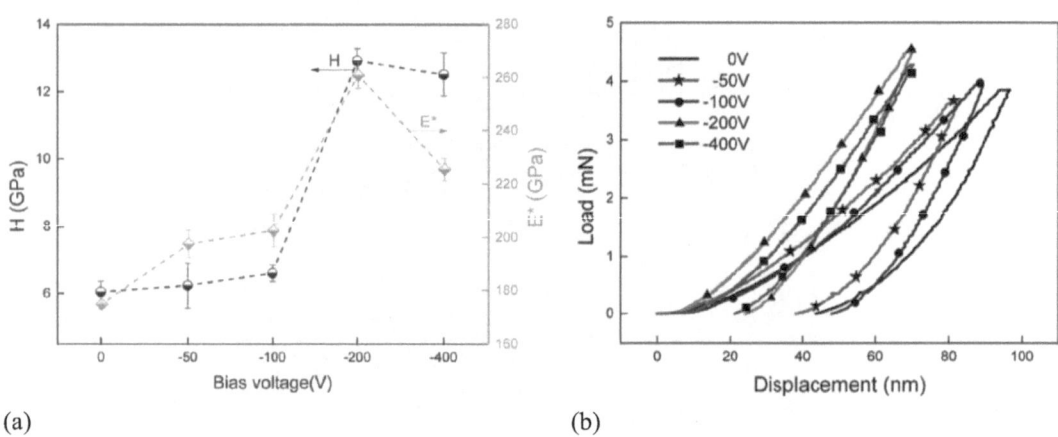

(a)   (b)

FIGURE 5.27 Experimental results of hardness test of TiN_x films obtained at different substrate bias voltages (a) hardness and elastic modulus of TiN_x films and (b) the load-displacement curves of TiN_x films [37].

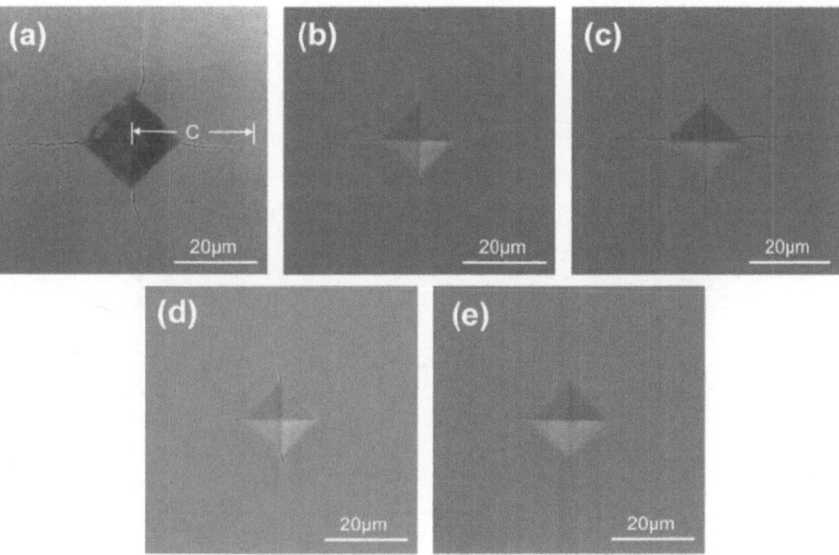

FIGURE 5.28 Morphology of TiN$_x$ films with different bias voltage after loading by diamond indenter at the same load L = 1.96 N are compared: (a) 0 V, (b) −50 V, (c) −100 V, (d) −200 V, and (e) −400 V [37].

TABLE 5.10
Potentiodynamic polarization data of TiN$_x$ films prepared with different substrate bias voltages

Samples	E_{corr} (V)	I_{corr} (A/cm²)	b_a (1/V)	b_c (1/V)
0V	−0.47	1.70×10^{-6}	3.16	12.15
−50V	−0.55	1.06×10^{-6}	4.62	6.01
−100V	−0.51	9.00×10^{-7}	4.03	5.55
−200V	−0.43	8.88×10^{-7}	4.95	5.44
−400V	−0.25	7.92×10^{-8}	3.83	7.78

with substrate bias voltages is shown in Figure 5.29. Therefore, it's known that when the bias voltage varies from 0 V to −100 V, the bias voltage has little effect on the toughness of the film, and the value of K_{IC} has no obvious variation tendency. When the bias voltage is −200 V and −400 V, the value of K_{IC} is significantly increased. The maximum value of K_{IC} is 2.82 MPa•m$^{1/2}$. Combining with the results presented in Figure 5.27(a), it can be concluded that the substrate bias voltage has little effect on the hardness and toughness of TiN$_x$ film, when the substrate bias voltage is lower than −100 V. The value of H/ E* also can be used to characterize the toughness of TiN$_x$ film [47]. The greater the ratio is, the better the material toughness [67]. It can be explained by the mechanism of film crack propagation. Cracking is caused by tensile stress; before forming a crack, the residual compressive stress should be overcome firstly. Therefore, by increasing the substrate bias voltage and introducing appropriate residual compressive stress into the film, the film needs to use more tensile stress before breaking, so as to improve the toughness of the film [68, 69].

5.2.3.3 Corrosion Resistance
The polarization curves for TiN$_x$ coating with different substrate bias voltages in 3.5 wt.% NaCl solution are shown in Figure 5.30. The corrosion potential E_{corr} (V), corrosion current density I_{corr} −400 V, the TiN$_x$ film shows the least corrosion current density and a nobler corrosion potential. It

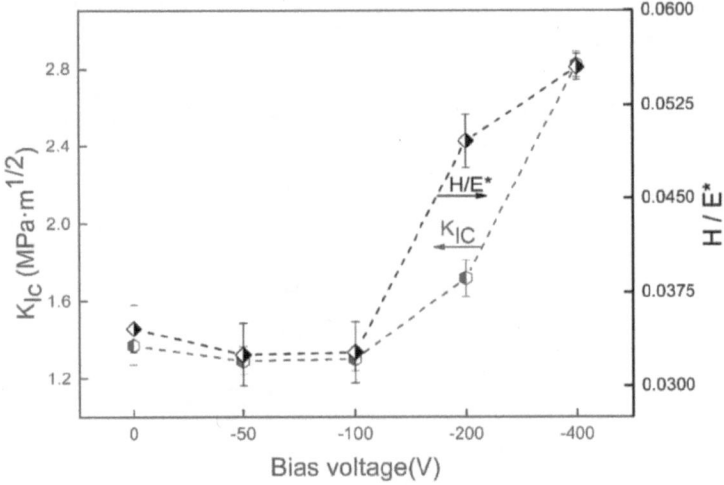

FIGURE 5.29 Variation curve of K_{IC} and H/E* with substrate bias voltages [37].

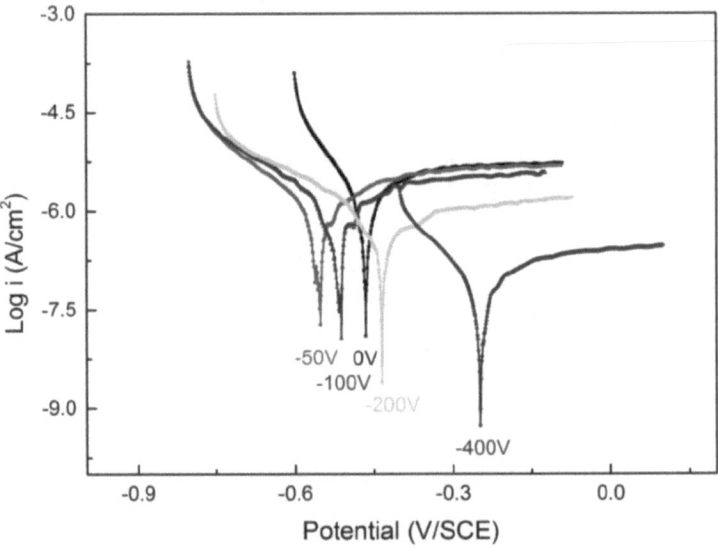

FIGURE 5.30 Potentiodynamic polarization curves for TiN$_x$ films with different substrate bias voltages in a 3.5 wt.% NaCl solution [37].

indicates that it has the best corrosion resistance. It is mainly caused by film densification. Increasing the substrate bias voltage, the shadowing effect of the film during deposition is reduced. The higher the density of the film is, the more difficult for the etching solution to contact with the substrate is, and TiN$_x$ film plays a protective role.

The Nyquist plots recorded for TiN$_x$ coating with different substrate bias voltages in 3.5 wt.% NaCl solution are shown in Figure 5.31. It is clear from the plots that TiN$_x$ films have a significant change in the impedance response of the corrosion solution. Simultaneously, it can be seen that the impedance loops do not yield perfect semicircles as expected from the EIS theory. Due to the "dispersing effect", roughness, film density and surface attachments have been related to the state of the film surface[70, 71]. We usually use Q to represent the ideal capacitor. It will be used to build

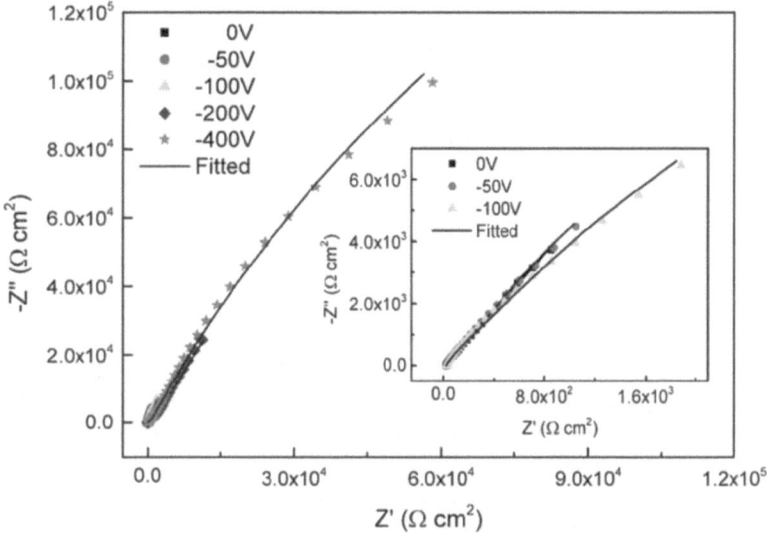

FIGURE 5.31 Nyquist plots recorded for TiN$_x$ films with different substrate bias voltages in a 3.5 wt.% NaCl solution [37].

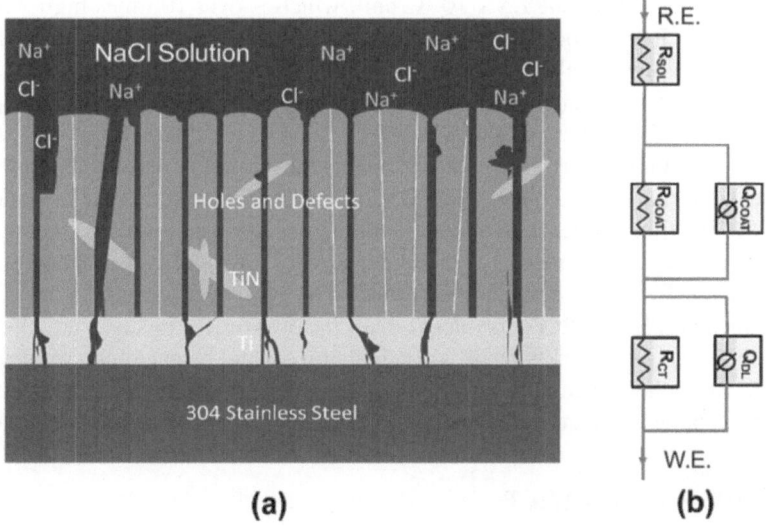

FIGURE 5.32 Corrosion mechanism diagram for TiN$_x$ films: (a) schematic presentation of corrosion protection mechanism provided by TiN$_x$ layer and Ti sub-layer deposited on 304 stainless steel in a 3.5 wt.% NaCl solution and (b) the corresponding equivalent circuits [37].

etch equivalent circuit diagrams subsequently. Figure 5.31 shows that the TiN$_x$ film prepared with a substrate bias voltage of −400 V has the biggest semicircle, it has the highest overall resistance.

In order to quantitatively compare the corrosion resistance of TiN$_x$ films and further analyze the corrosion mechanism, equivalent circuit simulation was conducted and the related results are shown in Figure 5.32. Figure 5.32(a) shows a corrosion mechanism diagram and Figure 5.32(b) is an equivalent circuit diagram based on the corrosion mechanism and impedance spectrogram. The fitted results by this model have the lowest chi-squared value compared with other models. The circuit elements are represented as solution resistance, R$_{sol}$; film resistance, R$_{coat}$; capacitance of TiN$_x$ film;

TABLE 5.11

EIS data obtained by equivalent circuit simulation of different TiN_x films studied in the present work

Samples	Rsol (Ω cm²)	Rcoat (Ω cm²)	Ccoat ($\Omega^{-1}s^{n1}$cm²)	n1	Rct (Ω cm²)	Cdl ($\Omega^{-1}s^{n1}$cm²)	n2	Chsq/E⁻⁴
0V	20.59	67650	0.002734	0.9105	106.9	0.0211	0.8618	6.58
-50V	19.71	79310	0.002243	0.9129	127.7	0.01566	0.8734	5.23
-100V	22.03	622430	0.001586	0.9228	366.1	0.00717	0.8455	4.9
-200V	21.76	689700	0.0002359	0.7912	542.6	0.000212	0.6304	16.1
-400V	22.23	751700	0.00005909	0.7746	623.5	0.0001068	0.6453	5.79

$C_{coat} = Q_{coat} \times (\omega'')^{1-n_1}$; the charge transfer resistance, R_{ct}; capacitance of double layer, $C_{dl} = Q_{dl} \times (\omega'')^{1-n_2}$. In general, Q is the constant phase element (CPE) [72], which is used to describe the physical quantity, when the parameter of capacitor C deviates, because in most cases there exist dispersing effect between the electrode and the solution, n is the exponent and "ω" is the angular frequency. CPE can be a resistor (n = 0), capacitor (n = 1), a Warburg's impedance (n = 0.5) or an inductor (n = −1) [73]. The value of each component in the equivalent circuit diagram (c.f. Figure 5.32(b)) was analyzed by using the computer program ZSimpwin. Fitted results are given in Table 5.11.

As shown in Table 5.11, with the raise of substrate bias voltage, the value of R_{coat} increases. The corrosion resistance of the TiN_x film is increased. The sample with a substrate bias voltage of 400 V exhibits highest impedances about 7.5×10^5 Ω cm², which is over 10 times higher than the sample prepared without bias. It is attributed to the film densification. Microcracks and pinholes presented in the film and make it prone to galvanic corrosion. While the value of Q_{CPE} decreases with increasing substrate bias voltage which can be attributed to the change of the electrical double layer thickness [74], that is to say, the difference of Q_{CPE} is mainly related to the thickness of the TiN_x film. Figure 5.32(a) is used to explain how the film structure affects the corrosion resistance of materials. First of all, in the entire corrosion system, there exists a solution resistance, which plays a major role at high frequencies [75]. The film surface is in contact with the corrosion solution and the interface between them constitutes a capacitor, namely Q_{coat} in Figure 5.32(b). Secondly, the contact between the substrate surface and the TiN_x film forms another double-layer Q_{ct}, which is mainly used to block corrosion at low frequencies. Hence this equivalent circuit has two times constants and the two capacitors are connected in series. When the material is applied with current in an etching solution, as the film is not dense, there are pinholes and defects. The corrosive Cl⁻ in the etching solution slowly penetrates into the film through these defects and eventually comes into contact with the substrate and corrodes.

5.3 CORROSION RESISTANCE

5.3.1 VACANCIES EFFECT

N vacancies as one of primary point defects are easily formed during nitride films growth at lower temperatures, which influence the stoichiometry of the as-prepared films and thereafter their properties [76]. In general, it's difficult to form perfect nitride crystal with full stoichiometric ratio (1:1) to keep structure equilibrium. Furthermore, formation energies for interstitial defects and metal vacancy are larger than N vacancy formation energy [77]. In these situations, it is relatively easier to prepare thin films with nitrogen vacancies at lower temperatures. It was reported that N vacancy had effects on the corrosion resistance [78]. It's found that the energy band, electronic structure and chemical bonds have changed with the variation of nitrogen content in the compound of variable composition, thereby affecting the properties. In this study, ZrN_x films were prepared by adjusting the N/Zr stoichiometric ratio to introduce different N vacancy concentration to explore the relationship between N vacancy concentration and corrosion resistance.

TABLE 5.12

Deposition parameters of the three sets of ZrN$_x$ films

Samples	Base pressure (Pa)	Deposition temperature (°C)	Ar flow rate (sccm)	N$_2$ flow rate (sccm)	Substrate bias (V)	Working pressure (Pa)	Thickness (um)
Z1	1×10^{-3}	100	54	14	-20	~ 0.5	0.42
Z2	1×10^{-3}	100	54	11	-20	~ 0.5	0.54
Z3	1×10^{-3}	100	54	8	-20	~ 0.5	0.65
Z4	1×10^{-3}	100	54	7	-20	~ 0.5	0.64
M1	1×10^{-3}	250	54	18	-20	~ 0.5	0.68
M2	1×10^{-3}	250	54	15	-20	~ 0.5	0.67
M3	1×10^{-3}	250	54	14	-20	~ 0.5	0.66
M4	1×10^{-3}	250	54	11	-20	~ 0.5	0.67
N1	1×10^{-3}	250	42	18	-20	~ 0.4	0.71
N2	1×10^{-3}	250	42	13	-20	~ 0.4	0.72
N3	1×10^{-3}	250	42	11	-20	~ 0.4	0.70

Three sets of ZrN$_x$ films with different V_N were obtained as listed in Table 5.12. The first set (sampling as Z1, Z2, Z3, Z4) without binding layers was prepared with different N$_2$ flow rate and at deposition temperature 100 °C. The Ar flow rate was kept at 54 sccm constantly in a mixed Ar (99.999%) and N$_2$ (99.999%) atmosphere at a pressure of ~0.5 Pa. To further confirm the relationship between the N vacancy concentration and corrosion resistance, the second set (M1, M2, M3, M4) with Zr binding layers were prepared with Ar flow rate of 54 sccm at 250 °C. Zirconium binding layer was deposited to enhance the adhesion strength between the films and the substrate. After getting the sample M2, about 1 sccm of nitrogen flow was turned down to obtain the sample M3. The third set (N1, N2, N3) was obtained at Ar flow rate of 42 sccm to verify whether the ability of corrosion resistance can be improved. For all samples, the Zr target power was kept at 1.8 kW unchanged to obtain a similar coating thickness.

5.3.1.1 Chemical Composition

Stoichiometric ratio of ZrN$_x$ films were listed in Table 5.13. The measured N/Zr ratio of first set was 0.98, 0.90, 0.76 and 0.70, respectively. The N/Zr ratio of second set was 0.89, 0.77, 0.74, and 0.69, respectively. The N/Zr ratio of third set was 0.97, 0.76, and 0.70, respectively. The V_N for M2 and M3 was 0.23 and 0.26, respectively. The two samples were studied for corrosion efficiency near V_N=0.25. Due to the change of Ar flow rate, even the sample N1 was obtained at same N$_2$ flow rate to M1, and the N/Zr ratio was not close to that.

Although the all samples are N deficient, it is conceivable that small amount of Frenkel pairs can thus be formed under RF growth conditions. The effective power density during ZrN$_x$ thin film formation was 11.6 W cm^{-2}. For instance, it can be speculated that N interstitial atom can be trapped

TABLE 5.13

N/Zr ratio and V_N of three sets of samples

Group No.	The first set				The second set				The third set		
Sample No.	Z1	Z2	Z3	Z4	M1	M2	M3	M4	N1	N2	N3
N/Zr ratio	0.98	0.90	0.76	0.70	0.89	0.77	0.74	0.69	0.97	0.76	0.70
V_N	0.02	0.08	0.24	0.30	0.11	0.23	0.26	0.31	0.03	0.24	0.30

below one atomic layer in the growing films. Such reasoning is consistent with the observed N deficiency in $Cr_{0.78}Al_{0.22}N_{0.93}$ film where the target's power density during deposition was about 6.3 W cm^{-2} [79]. For most of samples, N/Zr ratios were substoichiometric and far less than 1. Therefore, the N vacancy was only considered within the corrosion experiments and DFT based approach in this study.

5.3.1.2 Mechanical Properties

The hardness of sample N1, N2 and N3 is about 28.5 ± 2.0 GPa, 37.9 ± 1.4 GPa, and 35.3 ± 1.8 GPa, respectively. The hardness values are much higher than that of the first and second set (c.f. Table 5.14). That was because the sufficient energy was delivered to the growing film by bombarding ions when a lower sputtering gas pressure was used. The residual stress of three sets of ZrN_x films are listed in Figure 5.33. It is obvious that the residual stress of third set is larger than that of the other two sets. In such a situation, the films with relatively large residual stress were formed which directly influences the hardness and resistance to cracking, and the harder films would enhance their corrosion resistance efficiency.

TABLE 5.14

Corrosion potential (E_{corr}), corrosion current density (I_{corr}), EIS date and hardness of the three groups of samples

Samples	E_{corr} (V vs SCE)	I_{corr}(A/cm²)	R_S (Ω cm²)	CPE-1 (Ω⁻¹ sⁿ cm−2)	R_P (Ω cm²)	CPE-2 (Ω⁻¹ sⁿ cm−2)	R_{ct} (Ω cm²)	Hardness (GPa)
304 SS	−0.376	5.6×10^{-7}	22.56	2.06×10^{-5}	2.53×10^{4}	1.01×10^{-5}	2.59×10^{5}	/
Z1	−0.381	2.04×10^{-8}	28.38	3.58×10^{-6}	6.11×10^{4}	7.04×10^{-6}	6.22×10^{6}	17.9 ± 1.3
Z2	−0.357	3.74×10^{-8}	26.40	1.15×10^{-6}	5.78×10^{4}	7.11×10^{-6}	6.02×10^{6}	19.7 ± 1.9
Z3	−0.359	8.70×10^{-8}	23.15	3.15×10^{-6}	3.90×10^{4}	7.04×10^{-6}	4.27×10^{6}	23.9 ± 1.2
Z4	−0.358	9.48×10^{-9}	24.56	6.17×10^{-6}	8.66×10^{4}	7.25×10^{-6}	6.28×10^{6}	19.2 ± 2.2
M1	−0.328	2.7×10^{-8}	24.83	1.99×10^{-6}	3.72×10^{4}	7.72×10^{-6}	3.68×10^{6}	22.6 ± 2.3
M2	−0.320	4.8×10^{-8}	22.58	1.39×10^{-5}	1.71×10^{4}	6.88×10^{-6}	3.44×10^{6}	26.3 ± 2.1
M3	−0.237	8.7×10^{-8}	24.87	1.73×10^{-5}	4.32×10^{4}	3.55×10^{-6}	6.91×10^{6}	27.9 ± 1.2
M4	−0.225	1.4×10^{-8}	24.60	6.17×10^{-6}	8.66×10^{4}	7.25×10^{-6}	6.82×10^{6}	26.8 ± 1.6
N1	−0.234	3.4×10^{-8}	24.43	2.5×10^{-5}	3.48×10^{4}	4.52×10^{-6}	1.97×10^{6}	28.5 ± 2.0
N2	−0.248	2.3×10^{-8}	23.87	1.03×10^{-5}	1.72×10^{4}	4.35×10^{-6}	1.84×10^{6}	35.3 ± 1.4
N3	−0.212	9.6×10^{-9}	23.93	1.72×10^{-5}	4.42×10^{4}	3.10×10^{-6}	4.08×10^{6}	34.8 ± 1.8

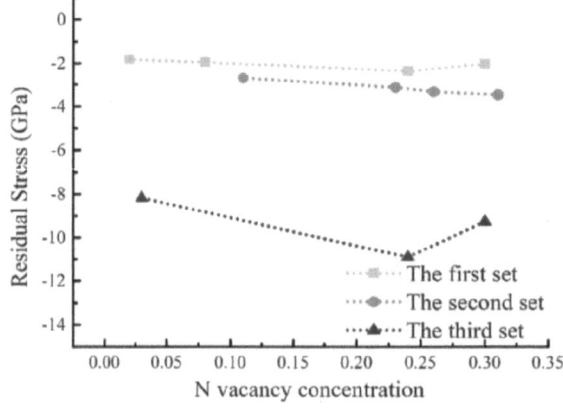

FIGURE 5.33 Residual stress of ZrN_x films as a function of N vacancy concentration [80].

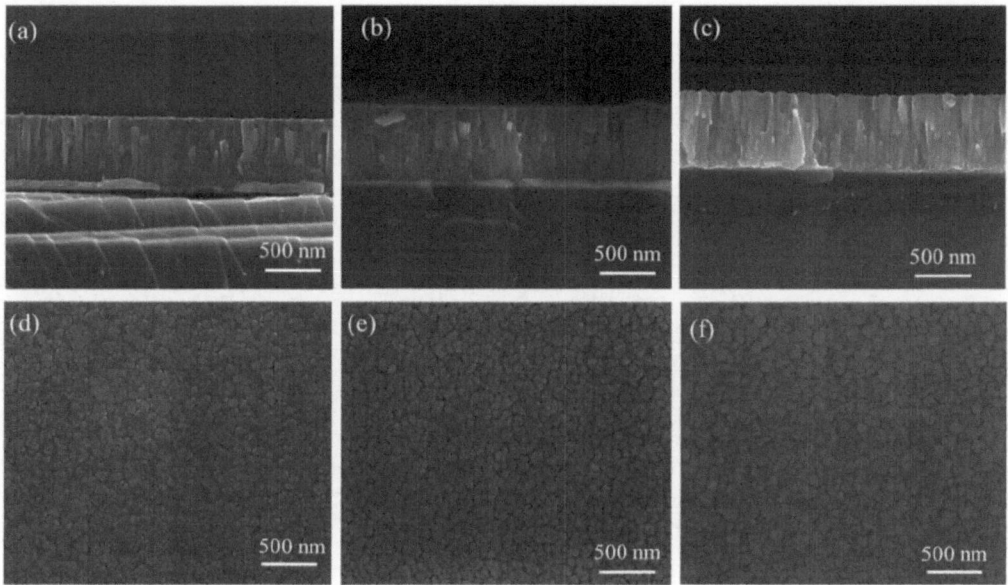

FIGURE 5.34 Cross-section and surface morphology SEM images of third set of samples, (a) and (d) sample N1, (b) and (e) sample N2, (c) and (f) sample N3 [80].

Figure 5.34 shows the cross-section and surface morphology of the samples from N1 to N3. The microstructure had greatly changed. Though all samples present columnar structure, it was clearly seen that every sample had slight difference. The sample N2 at $V_N=0.24$ produced significant morphological changes, where the particles were more aggregated than N1. With the increase of N vacancy concentration, the pronounced cauliflower-like patterned surface transformed to a densely granular structure. That morphology transformation would cause variation of corrosion properties [81]. All above of the reasons, the corrosion resistance may little be affected by N vacancy concentration, especially for the sample N2 at $V_N=0.24$.

5.3.1.3 Corrosion Resistance

5.3.1.3.1 Polarization Measurements

The corrosion behavior for the uncoated and the coated 304 SS was evaluated by potentiodynamic polarization tests. Table 5.14 shows the corrosion current density (I_{corr}) and corrosion potential (E_{corr}) obtained by the polarization curves and the Tafel extrapolation method. Figure 5.35 displays the variation of I_{corr} for ZrN_x films with different V_N deposited on stainless steel substrates compared with uncoated polished substrates. Figure 5.35(a) shows the polarization curve of bare polished 304 SS substrate. Figure 5.35(b) display the polarization curves of the first set of ZrN_x films with no binding layer. It was evident from the curves that sample Z4 had the lowest value of I_{corr} at $V_N=0.31$. The values of I_{corr} followed the order of Z3 >Z2 >Z1 >Z4, where Z3 ($V_N=0.24$) presented the worst corrosion resistance and Z4 ($V_N=0.31$) showed the opposite result. The corrosion efficiency was also affected by thickness of films [82]. Nevertheless, some researchers proposed that ZrN_x film would be less defect sensitive and the coating thickness didn't control the corrosion behavior [83]. The influence of thickness on the corrosion resistance has not been fully discussed so far. The second set of ZrN_x films with binding layer and similar thickness values between samples were measured for analysis. The polarization curves of the second set were presented in Figure 5.35(c). The values of I_{corr} for the set followed the order of M3>M2>M1>M4 and presented the same variation as the first set. The sample M4 at $V_N=0.31$ had the minimum corrosion rate, and the sample M3 at $V_N=0.26$ had the maximum corrosion rate. The third set of ZrN_x film showed different results from the other two

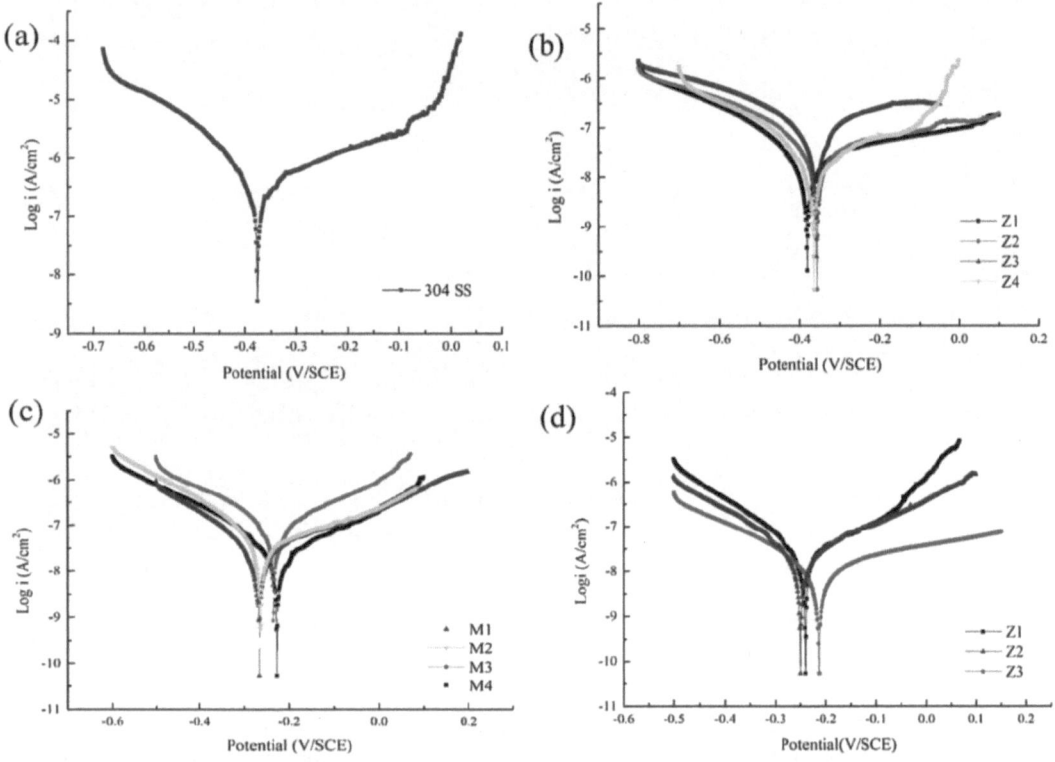

FIGURE 5.35 Potentiodynamic polarization curves for (a) uncoated 304 SS, (b) the first set of samples, (c) the second set of samples, (d) the third set of samples [80].

groups in Figure 5.35(c) and Table 5.14. The corrosion current density values of ZrN_x films with different N vacancy concentration are 3.4×10^{-8}, 2.3×10^{-8} and 9.6×10^{-9}A/cm^2, respectively. Compared to the other two sets, the corrosion resistance had improved and the gap of corrosion current density values also decreased.

5.3.1.3.2 Electrochemical Impedance Measurements

Electrochemical impedance spectroscopy (EIS) was performed to further elucidate the pathways of corrosion reaction and quantify the presence of defects in films. The impedance plots were used to evaluate the corrosion behavior and then compare the ZrN_x film resistance. Figure 5.36 shows the

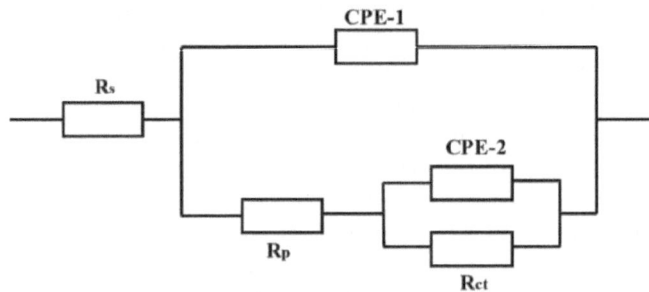

FIGURE 5.36 Electrical equivalent circuit diagram used for modeling the metal/solution interface [80].

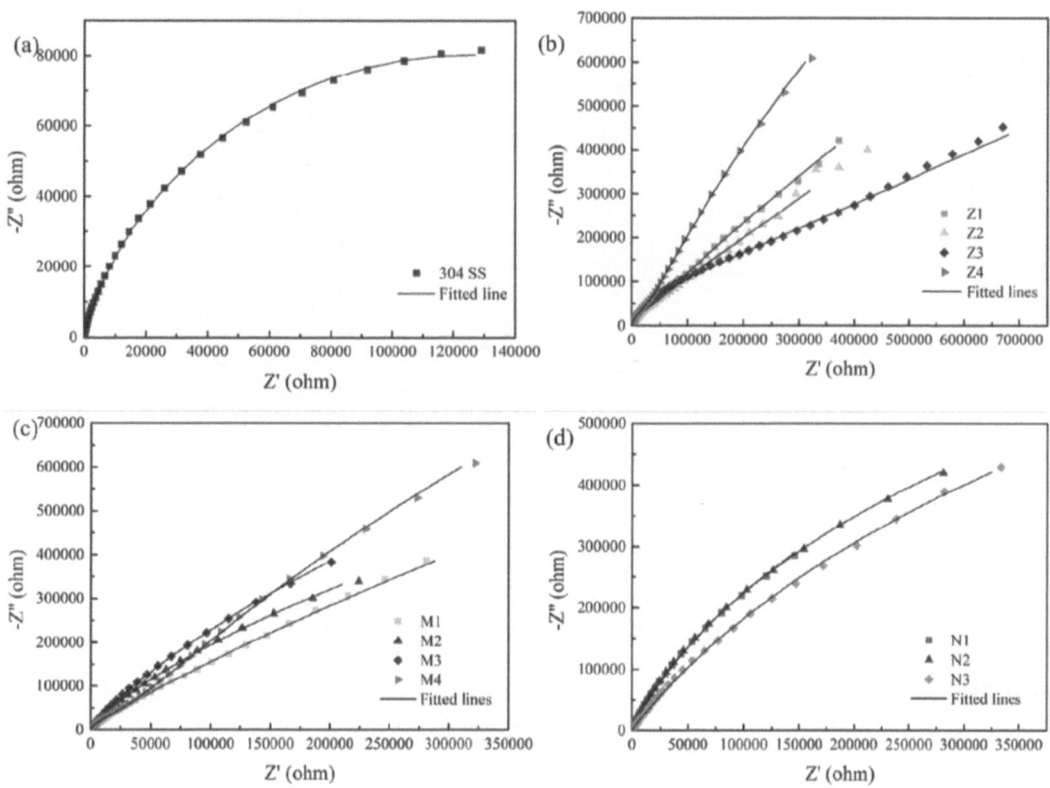

FIGURE 5.37 Nyquist plots for (a) uncoated 304 SS, (b) the first set of samples, (c) the second set of samples, (d) the third set of samples [80].

equivalent circuit model used to fit the experimental impedance data, which scheme corresponded to the behavior of corroding uncoated 304 SS substrate and coated surface layers of two sub-electrochemical interfaces: electrolyte/coating and electrolyte/metal interface, where R_s is the solution resistance, CPE-1 and CPE-2 are the constant phase element (CPE) representing film and coating capacitance, R_p is the pore resistance and R_{ct} is the charge transfer resistance. The values of R_s, R_p, and R_{ct} are analyzed using ZSim Demo software and the mean values of R_s, R_p, R_{ct} for the three sets are reported in Table 5.14. The Nyquist plots and fitting lines of the bare 304 SS and the working electrodes covered with ZrN$_x$ films are shown in Figure 5.37. The resistance efficiency is usually determined by the values of R_p. The first set of ZrN$_x$ films with different N vacancy concentration related to charge transfer showed variation trend that were consistent with corrosion current density. The sample Z3 at V_N=0.24 showed the minimum polarization resistance, and Z4 at V_N=0.31 showed the maximum polarization resistance. The values of R_p did not increase with the increase of V_N and followed the order of Z4 >Z1 >Z2 >Z3.

The second set of films also showed the same tendency of results as the first set. The values of R_p followed the order of M4 >M3 >M1 >M2. However, The Nyquist plots of M2 and M3 measured at V_N =0.23 and at V_N =0.26 showed distinct differences, thereby revealing that the ZrN$_x$ film corrosion resistance was different depending on the N vacancy concentration. The values of R_p for samples N1, N2, and N3 with different V_N presented no significant difference. The sample N3 (V_N=0.30) still had the best corrosion resistance. Compared to the other two sets, the overall corrosion resistance had been improved. In order to further explore the reasons, the three samples were analyzed by structure characterization.

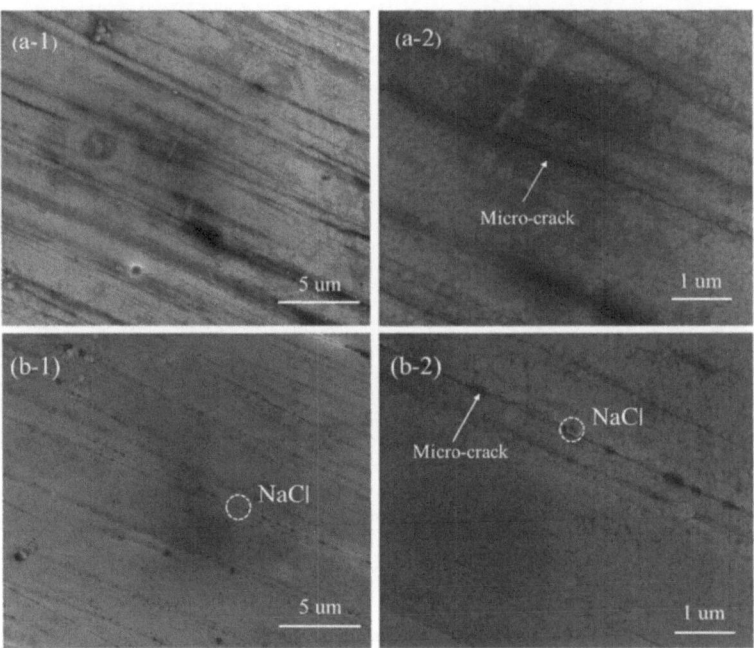

FIGURE 5.38 The eroded surface of samples M3 and M4 after polarization test: (a-1) and (a-2) the sample M coated 304 SS, (b-1) and (b-2) the sample M4 coated 304 SS [80].

5.3.1.3.3 Corrosion Surface Morphologies

Figure 5.38 shows the surface morphologies of samples M3 and M4 after corrosion test in 3.5 wt.% NaCl solution. In the original state, the films were in contact with 3.5 wt.% NaCl solution, and the solution would erode mainly along the microcracks and other macro-defects which were formed along the polishing lines. For all films, the erosion channels were in a similar situation, that is to say the path of corrosion medium to substrate mainly depended on microcracks and other defects. The Figure 5.38(a-1) showed more terrible surface morphology than that of Figure 5.38(b-1). The corrosion pits and new cracks did not appear in Figure 5.38. It can be seen from the polarization curves that no pitting corrosion occurred as well. The Figure 5.38(b-1) displayed a comparatively better corroded surface, and the damage to the surface was obviously reduced. This illustrates that sample M4 has better corrosion resistance and protective efficiency as discussed in previous discussion than M3.

5.3.2 MULTILAYER STRUCTURE EFFECT

In order to understand the relationship between microstructure evolution and corrosion resistance of coatings with different interface period. Four Al-TiAlN nanocomposite films with different interface period were deposited with a PVD technique named as ion source hybrid magnetron sputtering, the interface was introduced by cycle plasma etching technique. Figure 5.39 shows the deposition process of multi-interfaces Al-TiAlN nanocomposite films. After a certain thickness films deposition, interface was introduced by the plasma etching, and then the films grown on the interface for a thickness, and plasma etching was applied again. At the repetition of plasma etching and thereafter film deposition, the thick films with multi-interfaces were prepared. The interval thickness between two plasma etchings was named as interface period. The thickness of interface period is varied by changing the deposition time. The multi-interfaces were introduced by plasma etching for 2 min every time after the growing films with different thickness. The energy of ion etching for each interface

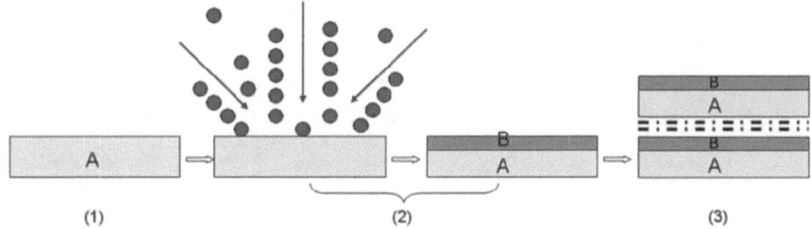

FIGURE 5.39 Process of interface introducing by plasma etching the growing surface after each sublayer deposition.

TABLE 5.15

Processing parameters of multi-interfaces Al-TiAlN nanocomposite films

Process	Power (kW)	Current (A)		Gas flow rate (mL/min)		Bias (V)	Deposition rate (nm/min)
		DC	Ion source	Ar	N$_2$		
Plasma etching	0	0	2	56	0	−200	NA
Al	0	4	1	56	0	0	26
TiAl	1.8	0	1	56	0	0	33
TiAlN	1.8	0	1	56	14	0	14

---

is controlled by the current of ion source and substrate bias. The interface period was varied to be 100 nm, 200 nm and 400 nm, and a single interface layer without plasma etching was also deposited as a contrast. The samples were respectively defined as their interface period and single interface hereafter. The processes and detailed parameters are listed in Table 5.15.

The evolution of microstructure with different interface period was studied. And the effect of microstructure evolution on hardness and corrosion resistance of AZ91D magnesium alloy was investigated.

5.3.2.1 Chemical Composition

A typical XPS survey spectrum is shown in Figure 5.40(a), photoelectron peaks from Ti, Al, N, O, Ar, and C were detected apparently. The presence of trace Ar 2p is due to the Ar$^+$ ion clean, and the detected spectra of O 1s and C 1s at the surface are a consequence of molecules being adsorbed from the lab ambient [84]. The Ti 2p spectra (c.f. Figure 5.40(b)) are resolved into one shoulder and three spin doublets. The additional shoulder is observed at a binding energy of 454.9 eV, which is attributed to the presence of TiN [85]. The binding energy at the range of 454 to 460 eV is corresponding to Ti 2p$_{3/2}$ state and more than 460 eV in Figure 5.40(b) is corresponding to Ti 2p$_{1/2}$ state. In this range, Ti 2p$_{3/2}$ and Ti 2p$_{1/2}$, locked at 454.9 and 455.8 eV, are assigned to Ti–N (TiN) bonds and the medium peak of 457.0 eV (Ti 2p$_{3/2}$) and 462.6 eV (Ti 2p$_{1/2}$) are attributed to Ti-N-O (TiN$_x$O$_y$) bonds; and the high peak of 458.3 eV (Ti 2p$_{3/2}$) and 464.1 eV (Ti 2p$_{1/2}$) are related to Ti-O (TiO$_2$ and Ti$_2$O$_3$) bonds [86]. Al 2p spectra (c.f. Figure 5.40(c)) are fitted with two components, the value of 74.2 eV is close to standard Al 2p 74.5 eV in AlN compound, and unstoichiometric AlN$_x$ compound may lead the shift of binding energy [87]. Another peak of 75.6 eV is associate with Al-O (α- Al$_2$O$_3$) bonds [88]. N 1s spectra (c.f. Figure 5.40(d)) are also fitted with two parts, the dominant peak of N 1s, centered in 396.76 eV, is corresponded to N-Al (AlN) bonds and/or N-Ti (TiN) bonds, and it is hard to distinguish them due to the closely correlated binding energy of 396.7 eV (AlN) and 396.8 eV (TiN) [85]. The shoulder peak of 398.3 eV is related to N-Ti-O (TiN$_x$O$_y$) bonds [89].

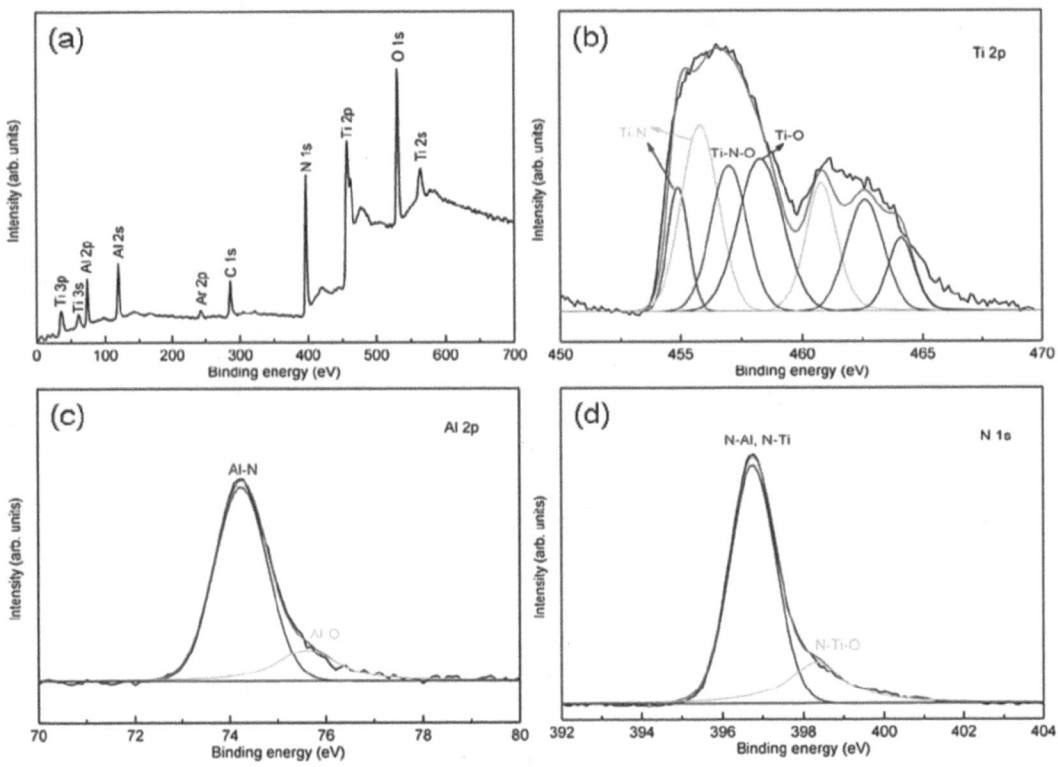

FIGURE 5.40 XPS spectra of Al-TiAlN nanocomposite films with 100 nm interface period, (a) survey spectrum, (b) Ti 2p, (c) Al 2p, and (d) N 1s [25].

5.3.2.2 Microstructure Evolution

The measured elemental composition of TiAlN layer is Ti 21 at.%, Al 27 at.% and N 52 at.%. The content of Ti and Al of TiAlN layer is a little different with the stoichiometric ratio of Ti_{50} $at._{\%}Al_{50}$ $at._{\%}$ target, but the TiAlN layer is near stoichiometric in nitrogen. Figure 5.41 presents the microstructure evolution of multi-interfaces Al-TiAlN nanocomposite films with different interface period deposited by ion source hybrid magnetron sputtering. Four multi-interfaces Al-TiAlN nanocomposite films with different interface period are all constituted by 3 μm Al layer, 0.15 μm TiAl layer and 2 μm TiAlN layer. The interface is introduced by cycle plasma etching as shown in Figure 5.39. The multi-interfaces of Al-TiAlN nanocomposite films with 100 nm interface period can be observed

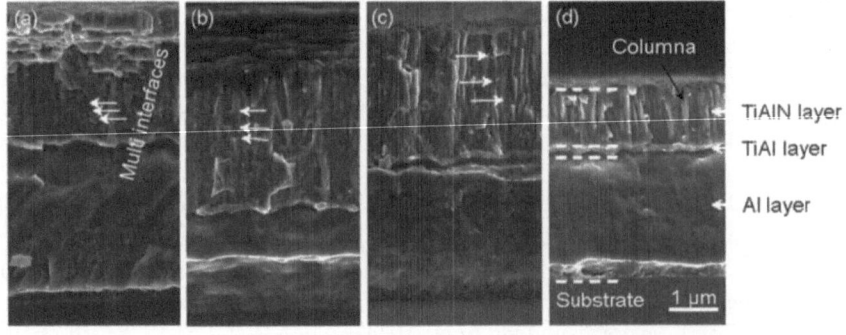

FIGURE 5.41 Cross-sectional microstructure of multi-interfaces Al-TiAlN nanocomposite films with different period, (a) 100 nm, (b) 200 nm, (c) 400 nm and (d) single interface [25].

in Figure 5.41(a), and there is no obvious columnar microstructure. In Figure 5.41(b), the interface period increased to 200 nm, and the columnar microstructure can be observed clearer than that of the 100 nm one. As the interface period further increased to 400 nm (c.f. Figure 5.41(c)), the column structure is much clearer than the interface. For single interface film (c.f. Figure 5.41(d)), there is no plasma etching after film deposition, and the film show a strong column structure. That means with decrease of interface period, the column structure weakened. Based on the microstructure evolution, it can be concluded that though the columnar microstructure cannot be eliminated by cycle plasma etching process, the microstructure changes from column to multi-interfaces with the decreasing of period. As compared with the typical columnar structure in the single interface sample (c.f. Figure 5.41(d)), the sample with 100 nm interface period exhibits multi-interfaces microstructure with a fine and dense microstructure (c.f. Figure 5.41(a)). That indicates that the films would possess the improved mechanical properties and corrosion resistance when compared with other samples.

The X-ray diffraction patterns in Figure 5.42 reveal the crystallinity of the as-deposited films. All multi-interfaces Al-TiAlN samples are found to have a crystal TiN single phase B1 NaCl structure with typical (111), (200), (220), (311), and (222) diffraction peaks according to JCPDS No. 87-0633. No other crystal phases are detected. Compared to the standard spectra, the peaks of TiN are shifted to the high diffraction angle, indicating that Al causes the lattice distortion [90]. The grain size of multi-interfaces Al-TiAlN nanocomposite films, calculated using the Scherrer's formula, are also listed in Table 5.16. Though there is lack of obvious trend, the TiN grains size are around 5–6 nm and fall in the nano-scale.

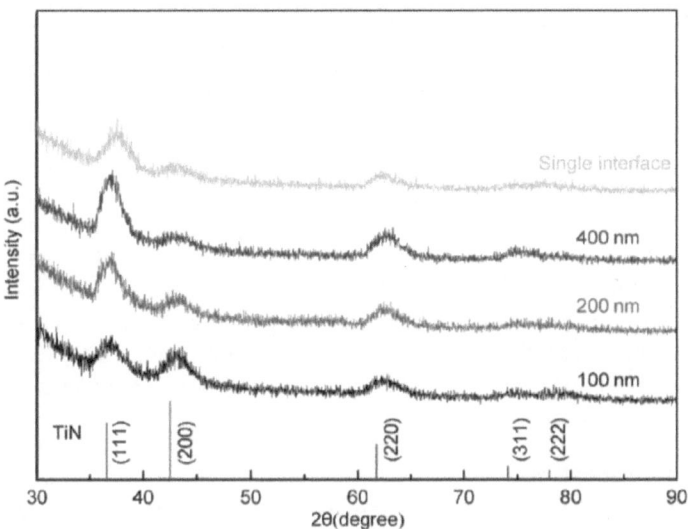

FIGURE 5.42 GIXRD patterns of multi-interfaces Al-TiAlN nanocomposite films with different interface period [25].

TABLE 5.16

Grain size of multi-interfaces Al-TiAlN nanocomposite films with different period

Sample	100 nm	200 nm	400 nm	Single interface
Grain size (nm)	4.8±0.2	5.9±0.1	5.8±0.2	5.5±0.4

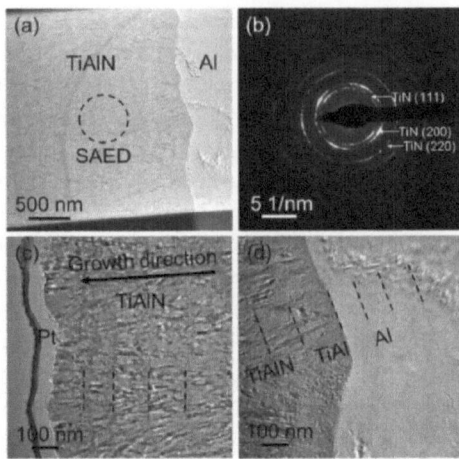

FIGURE 5.43 Cross-sectional TEM images of Al-TiAlN nanocomposite films with 100 nm interface period, (a) bright-field images, (b) selected area electron diffraction (SAED) patterns, (c) top layer and (d) transition zone of TiAlN-TiAl-Al layers [25].

Figure 5.43 shows the cross-sectional TEM bright-field images, selected area electron diffraction (SAED) patterns and enlarged scale images of 100 nm interface period sample. A selected area is circled out in Figure 5.43(a), and the corresponding diffraction patterns are present in Figure 5.43(b). The feature pattern in Figure 5.43(b) is matched well with crystal TiN, and that is double confirmed with GIXRD result (c.f. Figure 5.42). The TiAlN zone and TiAlN-TiAl-Al transition zone with high magnification are shown in Figure 5.43(c) and Figure 5.43(d), respectively. The column growth of TiAlN nanocomposite films is interrupted by multi-interfaces, and the grain size is small. The measured period of interface is about 116 ± 20 nm, which differ little from that of the ideal one. Both TiAlN and Al sublayer exhibit a fine and dense microstructure, which is good for improvement of mechanical properties and corrosion resistance.

Results from XPS show that there is Ti-N bond and Al-N bond, which indicate the existing of TiN and AlN. Results from the GIXRD and TEM show that there is crystal TiN phase, and the grains size calculated with Scherrer's formula are within 5~6 nm, and there is no other crystal phase. That means the AlN should be in amorphous phase. Therefore, based on the results of GIXRD, TEM and XPS, it can be deduced that the structure of the films should be nanocrystal TiN (nc-TiN) surrounded by amorphous AlN phase (a-AlN), in other words, there is nanocomposite microstructure (nc-TiN/a-AlN).

5.3.2.3 Mechanical Properties

Figure 5.44 displays the hardness, load-displacement of the as-prepared multi-interfaces Al-TiAlN nanocomposite films with different period. Along with the increase of displacement, the hardness of samples ascends in the first stage, and then descends (c.f. Figure 5.44(a)). Some breakpoints appear on the load-displacement curves, and the displacement of first breakpoint is improved by the decreasing of period (c.f. Figure 5.44(a)). That indicates that the multi-interfaces films exhibit better fracture toughness due to their numerous interfaces [91]. In order to avoid the substrate effect, the average values of 100 to 150 nm are chosen to characterize the hardness of films. The averaged hardness of films with 100, 200, 400 nm, and single interface are 31.3 ± 0.8, 29.6 ± 0.4, 19.4 ± 1.2, 23.5 ± 0.8 GPa, respectively. The 100 nm sample shows the highest hardness of 31.3 GPa due to the numerous interfaces effect brought by plasma etching during the deposition process. Meanwhile, the turning point of 400 nm may be caused by a large period of interfaces, the microstructure of 400 nm is columnar microstructure as is shown in Figure 5.41, and the crack is easy to be formed in the interlayer and column boundaries under the indentation load and degraded the hardness. The

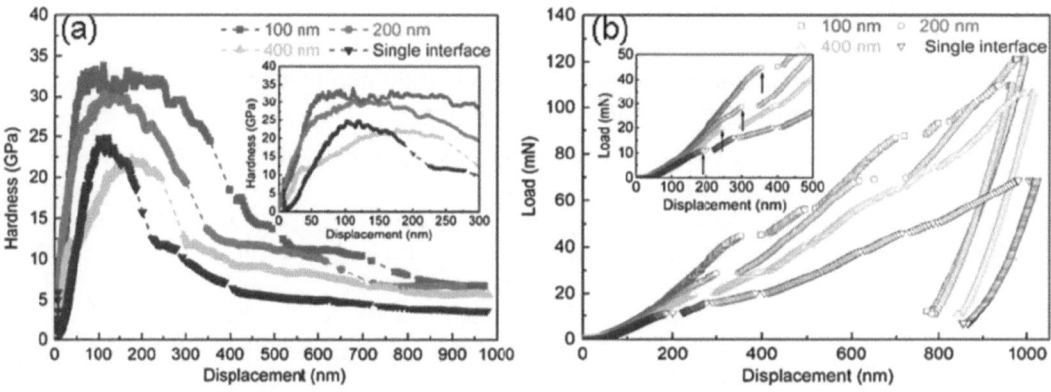

FIGURE 5.44 Nanoindentation test of multi-interfaces Al-TiAlN nanocomposite films with different period, (a) hardness and (b) load-displacement curves [25].

phenomenon of breakpoints and hardness changing can be explained by interfaces deflection effect which is similar to multilayer [92].

5.3.2.4 Corrosion Resistance

5.3.2.4.1 Polarization Test

The corrosion behavior of multi-interfaces Al-TiAlN nanocomposite films in 3.5 wt.% NaCl solution is displayed in Figure 5.45. The estimated values of corrosion potential (E_{corr}) and corrosion current density (I_{corr}) by Tafel extrapolation method are listed in Table 5.17 [93]. The cathodic branch of AZ91D is activation controlled; hydrogen evolution reaction is going on substrate. And in the anodic branch, a platform illustrates the formation of passive films. But in breakdown potential at −1.30 V/SCE, passive films dissolution leads the linear increase of corrosion current density and indicates the destruction of passive protection. For single interface and 400 nm samples, the polarization curves appear activation-controlled behavior without obvious passive feature. Partial pitting corrosion occurs in this region. Their anodic branches are similar to the transpassive region of substrate, which indicates they have near anodic dissolution tendency. As to 200 and 100 nm samples, their cathodic branches are slower and the anodic branches show an incipient passive region. With

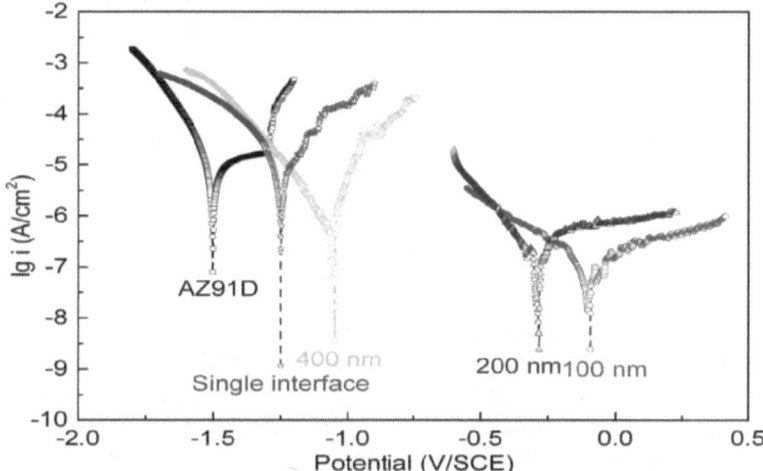

FIGURE 5.45 Polarization curves of multi-interfaces Al-TiAlN nanocomposite films in different period [25].

TABLE 5.17

Electrochemical parameters of AZ91D substrate and coated with multi-interfaces Al-TiAlN nanocomposite films

Sample	E_{corr} (V/SCE)	I_{corr} (A/cm²)	R_p (kΩ/cm²)	P (%)	Pi (%)
100 nm	−0.10	9.37×10^{-8}	270.92	0.80	99.29
200 nm	−0.28	1.46×10^{-7}	182.44	1.20	98.89
400 nm	−1.05	1.06×10^{-6}	13.59	16.10	91.97
Single interface	−1.25	8.70×10^{-6}	4.01	54.71	34.09
AZ91D	−1.50	1.32×10^{-5}	2.19	N/A	N/A

the gradual advance of the anode Tafel slope, the coated substrates are typical of passive surface. Hence, the multi-interfaces Al-TiAlN nanocomposite films, especially lower period, have a definitive impact on the corrosion behavior of AZ91D substrates.

The values of E_{corr} and I_{corr} in Table 5.17 are remarkably shifted to anodic potentials and lower side, correspondingly, with decreasing the period of multi-interfaces Al-TiAlN nanocomposite films. Furthermore, the values of polarization resistance vary with I_{corr} tendency. E_{corr} is improved from -1.50 V/SCE of AZ91D substrate to -0.10 V/SCE of 100 nm, and I_{corr} of 100 nm is reduced to 9.37×10^{-8} A/cm². Higher E_{corr} indicates better corrosion resistance of samples and lower I_{corr} attributes the less corrosion rate. The potential of multi-interfaces Al-TiAlN nanocomposite films shifts toward positive side and a decrease in corrosion current density is observed for the single interface to ~100 nm period Al-TiAlN nanocomposite films coated AZ91D samples suggesting that the films with multi-interfaces microstructure have better corrosion resistance. The improved corrosion protection effect is strongly associated with the microstructure evolution of multi-interfaces Al-TiAlN nanocomposite films as shown in Figure 5.41 [94].

5.3.2.4.2 Corrosion Protection Characterization

Magnesium alloy is very sensitive to corrosion medium in corrosion tests due to its lower standard electrode potential in contrast to other metals. The presence of defects such as pinhole as well as the columnar microstructure boundaries of PVD films would induce the galvanic corrosion as a result of the potential difference, and promote the corrosion procedure and reduce the level of protection [95]. Therefore, the porosity and protective efficiency factors are in relation to the barrier effect of PVD films. The values of film porosity (P) and protective efficiency (Pi) are also listed in Table 5.17. They are determined by Equation (5.1) and (5.2), individually [95, 96]:

$$P = \left(\frac{R_p^s}{R_p} \right) \times 10^{-\frac{|\Delta E|}{b_a}} \times 100\% \qquad (5.1)$$

$$P_i = \left(1 - \frac{I_{corr}}{I_{corr}^s} \right) \times 100\% \qquad (5.2)$$

Where, R_p^s is the polarization resistance of AZ91D substrate, R_p is the polarization resistance of four multi-interfaces Al-TiAlN nanocomposite films coated AZ91D, ΔE is the difference between corrosion potential of coated and uncoated samples, b_a is the anodic Tafel slope of AZ91D, I_{corr} is the corrosion potential of four coated samples and I_{corr}^s is the corrosion potential of AZ91D. It is worth noting that the values of porosity reduce significantly by the reduction of period of interfaces, and the protective efficiency presents significant growing tendency. This phenomenon reveals that the good barrier effect is closely related to the microstructure of films.

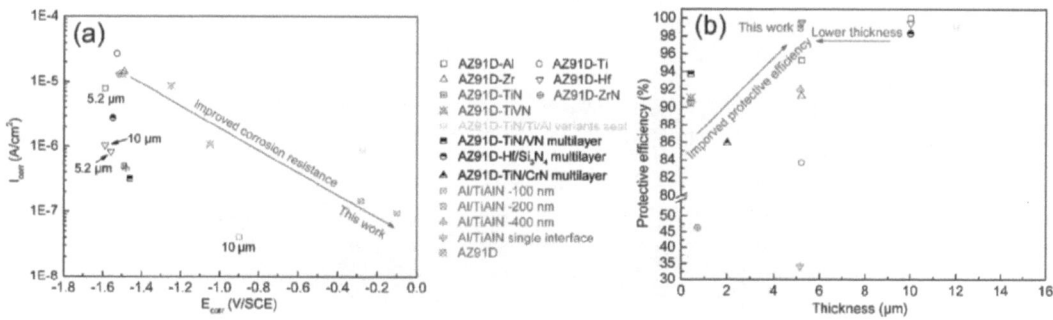

FIGURE 5.46 Comparative electrochemical corrosion results of AZ91D based different films in 3.5 wt.% NaCl solution, (a) corrosion resistance, (b) corrosion protective efficiency [25].

A comparative study of present work and other works [24, 97–102] is shown in Figure 5.46. Corrosion resistance of pure metal films coated AZ91D in 3.5 wt.% NaCl solution is influenced by metal electrode potential and thickness (c.f. Figure 5.46(a)). Improved corrosion resistance is achieved by the reduction of period of multi-interfaces Al-TiAlN nanocomposite films. The best one of 100 nm sample has lower I_{corr} than AZ91D based TiN/Ti/Al variants seal treatment films, and higher E_{corr} than 10 µm Al films. Otherwise, Al-TiAlN nanocomposite films are also more effective than various metal nitride and multilayer in corrosion resistance. Multi-interfaces Al-TiAlN nanocomposite films can overcome the drawback of lower hardness of pure metal films while improving corrosion resistance, and improve the corrosion resistance of hard films. Calculated protective efficiency versus different thick films is presented in Figure 5.46(b). Pure Al films have a lower electrode potential than Ti, Zr, and Hf films, and a thick film would reduce the defects of films, which is contributed to the improving of protective efficiency. Al-TiAlN composite films with 100 nm period attain improved protective efficiency than other films.

5.3.2.4.3 Corrosion Morphology

To obtain further information about the protective efficiency of multi-interfaces Al-TiAlN nanocomposite films, selected corrosion morphologies of 100 and 400 nm samples are shown in Figure 5.47. In low magnification of SEM images, the surface of sample with 100 nm interface period is smooth and clean (c.f. Figure 5.47(a)), but some pitting corrosion and filiform corrosion are produced in 400 nm sample (c.f. Figure 5.47(b)). There is no obvious corrosion phenomenon of 100 nm sample in enlarged scale (c.f. Figure 5.47(a)), this illustrates that 100nm sample has good corrosion resistance and protective efficiency as discussed in previous discussion. In Figure 5.47(d), corrosion products and microcrack of 400 nm sample corrosion surface with enlarged scaled are clear. The microcrack is formed at the edge of corrosion products boundary and extend along the filiform corrosion. In addition, the corrosion morphologies of samples are consistent with E_{corr} and I_{corr} (c.f. Table 5.17).

An EDS mapping technique is conducted to characterize the composition of corrosion products (c.f. Figure 5.48). Pitting corrosion destroys the completeness of 400 nm period Al-TiAlN nanocomposite films (c.f. Figure 5.48(a)), and a linear microcrack is also observed in filiform corrosion surface (similar to Figure 5.47(d)) which is formed by the increasing volume dilatation of corrosion products underneath films. The elements distribution of Ti, Al and N elements (c.f. *Figure* 5.48(b-d)) are mainly concentrated in the complete protected area, and the filiform corrosion region is also included because of the films are not peeling off from the substrate. Therefore, it is hard to distinguish the filiform corrosion from EDS mapping. But for Mg and O elements (c.f. Figure 5.48(e) and (f)), the composition as corrosion products is mainly distributed over the destruction zone, and a less obvious Cl element also appears in this region (c.f. Figure 5.48(g)). The corrosion procedure and products of AZ91D in 3.5 wt.% NaCl solution are well known [103], the overall corrosion

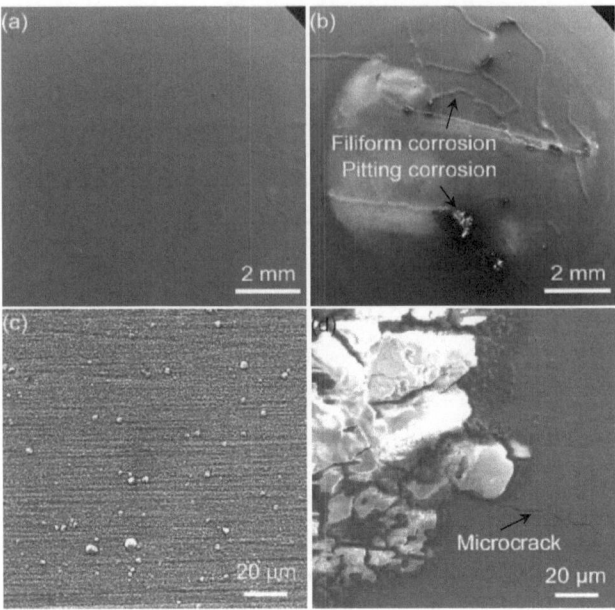

FIGURE 5.47 Macroscopic corrosion morphologies of selected multi-interfaces Al-TiAlN nanocomposite films with interface period, macroscopic corrosion morphologies of (a) 100 nm and (b) 400 nm, detail information with enlarged scale of (c) 100 nm and (d) 400 nm [25].

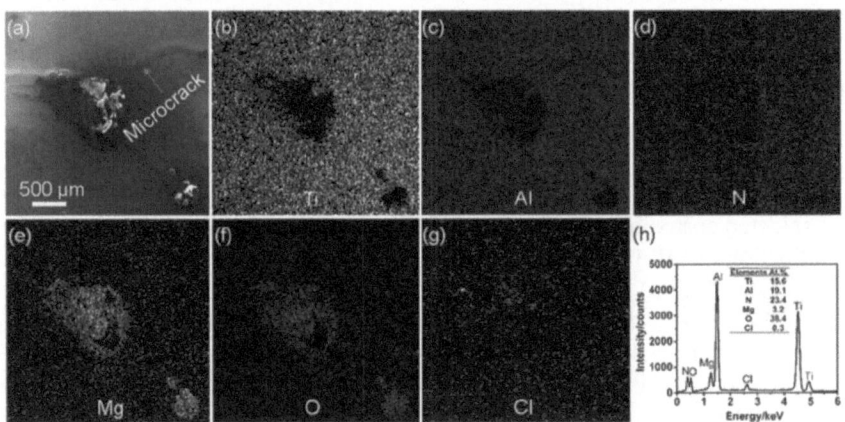

FIGURE 5.48 Corrosion products analysis of multi-interfaces Al-TiAlN nanocomposite films with 400 nm interface period, (a) corrosion morphology, elements distribution of (b) Ti, (c) Al, (d) N, (e) Mg, (f) O and (g) Cl, and (h) EDS mapping spectrum results [25].

reaction follows Equation (5.3), the Mg dissolution and H_2 evolution reaction in different polarity are summarized in Equation (5.4) and (5.5), and the insoluble $Mg(OH)_2$ product formed follow Equation (5.6) [104]. The presence of Cl element (c.f. Figure 5.48(g) & (h)) indicates that the $MgCl_2$ product is formed in high chloride ion concentration (Equation (5.7)) [105], parts of the corrosion products transform insoluble $Mg(OH)_2$ into soluble $MgCl_2$ [106]. From elements content in Figure 5.48 (h), O is reached 38.4 at.% and much higher than Mg (3.2 at.%), this illustrates that in addition to $Mg(OH)_2$, there are other oxides in this region. For the AZ91D substrate, the porous $Mg(OH)_2$ and soluble $MgCl_2$ corrosion products dissolve and make the fresh substrate to expose to corrosion medium. Then the reaction of Equations (5.4–5.7) would cycle until the substrate is exhausted [107], and the protection of films is prevented.

$$Mg + 2H_2O \rightarrow Mg(OH)_2 \downarrow + H_2 \uparrow \left(\text{Overall reaction}\right) \qquad (5.3)$$

$$Mg \rightarrow Mg^{2+} + 2e^- \left(\text{Anodic reaction}\right) \qquad (5.4)$$

$$2H_2O + 2e^- \rightarrow H_2 \uparrow + 2OH^- \left(\text{Cathodic reaction}\right) \qquad (5.5)$$

$$Mg^{2+} + 2OH^- \rightarrow Mg(OH)_2 \downarrow \left(\text{Insoluble product formation}\right) \qquad (5.6)$$

$$Mg(OH)_2 + 2Cl^- \rightarrow MgCl_2 \downarrow + 2OH^- \left(\text{Soluble product formation}\right) \qquad (5.7)$$

5.3.2.4.4 Corrosion Mechanism

Based on the above cross-sectional microstructure observations and the results of corrosion behavior, a schematic of the corrosion mechanism and the failure of multi-interfaces Al-TiAlN nanocomposite films is shown in Figure 5.49. In the original state, the multi-interfaces microstructure films (c.f. Figure 5.49(a)) and columnar microstructure films (c.f. Figure 5.49(b)) are in contact with 3.5 wt.% NaCl solution, the Cl⁻ would erode along the weak interfaces and boundaries. In corrosion state, the multi-interfaces microstructure with a denser and defect-free films can restrain the penetration of Cl⁻ (c.f. Figure 5.49(c)). The microstructure of column is interrupted by introducing multi-interfaces (c.f. Figure 5.41 and Figure 5.43); interfaces and boundaries are enhanced significantly and improve the corrosion resistance and protective efficiency (c.f. Figure 5.46). Cl⁻ erosion channel is deflected by the lots of irregular interfaces and boundaries, that is to say the path of corrosion medium to substrate is blocked. This kind of phenomenon is similar to deflection effect of multilayer

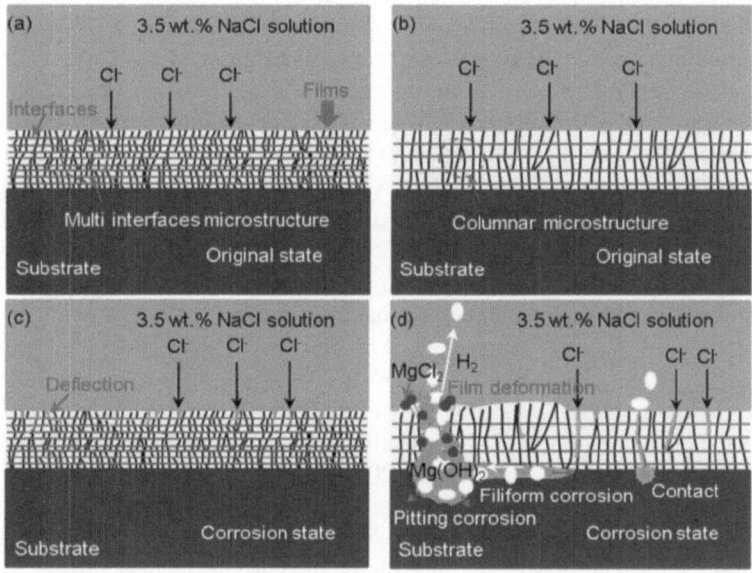

FIGURE 5.49 Corrosion mechanism of multi-interfaces Al-TiAlN nanocomposite films with different microstructure, the original states of (a) multi-interfaces microstructure and (b) columnar microstructure, the corrosion state of (c) and (d) are contrasted to (a) and (b) [25].

toughening mechanism in mechanical properties. Al-TiAlN nanocomposite films with columnar microstructure are terrible in corrosion prevention (c.f. Figure 5.49(d)). The strong boundaries are eroded easily; the corrosion medium infiltrates the films and contacts with substrate. Pitting corrosion and filiform corrosion are reacted in the interface between films and substrate, and promote the process of each other. During corrosion process, Mg dissolves along with the direction of red arrows and H_2 is released through corrosion holes and filiform cracks. As the corrosion time increases, the increased corrosion products of volume result in the film deformation, and cooperated with fast extended filiform corrosion, the films would be destroyed quickly. Therefore, the improved corrosion resistance of microstructure evolution is remarkable.

5.4 TRIBO-CORROSION RESISTANCE

Protecting load-bearing surfaces with PVD hard coating has been pursued for improved tribological performance in diverse applications. However, in corrosive environment the coatings are subjected simultaneously to both wear and corrosion phenomena, which be termed **tribo-corrosion**. The conjoint action of wear and corrosion is often more intense than pure corrosion and pure wear effects alone [107, 108]. It is necessary to fabricate coating with excellent resistance against both wear and corrosion and study the tribological behavior under corrosive environment.

In comparison with the purely mechanical wear or the purely chemical corrosion, the dominant factors of the tribo-corrosion resistance are generally more complicated for the PVD hard coatings. (i) The PVD hard coatings always contain intrinsic micro-defects (columnar structures, pores, discontinuities) as well as extrinsic macro-defects (craters, cone structures), which are unable to completely prevent the galvanic coupling with metal substrates [109, 110]. (ii) Moreover, the hard coatings are subjected to fracture or delamination from the substrates under mechanical contacts, whereby the extensive cracks can act as new diffusional paths for corrosive medium and cause severe surface and subsurface damage [111, 112]. The tribo-corrosion resistance of the coatings is improved to a great extent when the propagation of the cracks is suppressed effectively by employing the structure of interlayers or multilayers [113–115]. (iii) Additionally, the passive layers on the worn surfaces can resist the penetration of corrosive medium and decrease the coefficients of friction during the sliding, which has been evidenced especially in some metals (such as nickel, TA6V4, and stainless steel) [116, 117]. Overall, the tribo-corrosion resistance of PVD hard coatings greatly depends on their compactness, cracking resistance, and self-passivation of the worn surfaces.

However, enhancing the tribo-corrosion resistance of PVD hard coatings is still dominated by trial-and-error empiricism. More comprehensive studies are required to distinguish the tribo-corrosion mechanisms for the coatings with different microstructures. In this work, we focused on the effect of microstructural factors on the tribo-corrosion behavior of the CrSiN coatings in a 3.5 wt.% NaCl solution. Three sets of CrSiN were prepared by magnetron sputtering with different substrate bias, details can be found in Table 5.18. The tribo-corrosion behavior was compared among three typical structures of the magnetron sputtered CrSiN coatings by the combination of tribo-electrochemical measurements and microscopic examinations on the worn regions.

5.4.1 COMPOSITION

Table 5.19 lists the elemental composition, the thickness, stress, and the density of the CrSiN coatings prepared at various bias voltages (-10 V, -20 V, and -50 V). The three coatings have a similar elemental composition, namely, ~42 at.% Cr, ~1 at.% Si, and ~57 at.% N. The microstructural parameters, however, depends critically on the processing conditions. Under the same deposition time, the thickness of the coatings decreases with the increase of the bias voltage, i.e., 3.2 μm ± 0.1 μm, 2.8 μm ± 0.1 μm, and 2.6 μm ± 0.1 μm. Moreover, the coatings became more smoother and more compact when the negative bias voltage increased from -10 V to -50 V. The density values are 4.91 g/cm^3, 5.37 g/cm^3, and 5.41 g/cm^3 for the three coatings, respectively. The densification of the

TABLE 5.18

Deposition parameters for CrSiN coating

Parameters	S1	S2	S3
Cr target power (100 kHz, 80% duty)		280 W	
Si target power		300 W (RF)+100 W (DC)	
Work pressure		1.0 Pa	
Ar partial pressure		0.7 Pa	
Ar/N$_2$ flow rate		32/24 sccm	
Deposition temperature		573 K	
Bias voltage	−10 V (DC)	−20 V (DC)	−50 V (DC)
Substrate holder rotation		16 rpm	

TABLE 5.19

Elemental composition, density and stress for CrSiN coating

Sample ID	Bias (V)	Composition (at.%) Cr	Si	N	Density (g/cm³)	Stress (GPa)	Thickness (μm)
S1	−10	42.5	1.0	56.5	4.91	--	3.2
S2	−20	41.7	1.1	57.2	5.37	0.2	2.9
S3	−50	42.0	1.0	57	5.41	−1.9	2.6

coatings could be related to the improved surface mobility of atoms in the case of the high substrate bias, ensuring the full diffusion on the film surface and reducing the defects.

5.4.2 MICROSTRUCTURE AND MECHANICAL PROPERTIES

Figure 5.50 shows the cross-sectional SEM images of the three kinds of coatings. When the bias voltage is −10 V, the cross-sectional morphology appears as penetrating columnar crystals; when the bias voltage is increased to −20 V, the cross-section of the coating still consists of columnar crystals, while the gap in-between columnar crystals is very blurred, and the cross-section appears to be dense V-shaped crystals. When the bias voltage is −50 V, the cross-section of the coating can no longer see the trace of columnar structure, which is obvious featureless.

Figure 5.51 shows the cross-sectional TEM images of the three kinds of coatings. The "−10 V" coating is composed of columns that extend throughout the coating with well-defined boundaries (c.f. Figure 5.51(a)). Many inter- and intra-columnar pores (as denoted by arrowheads) are visible in the higher-magnification image of the selected area (as denoted by a square). The columns, by

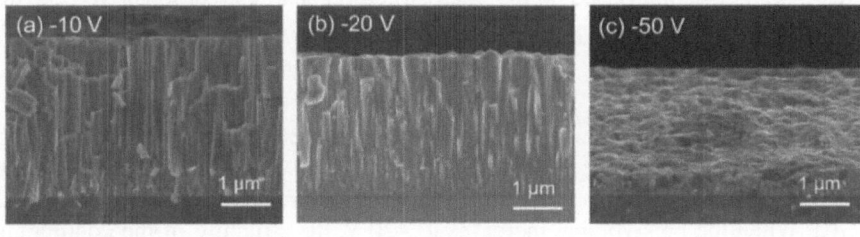

FIGURE 5.50 Cross-sectional SEM images for three CrSiN coatings.

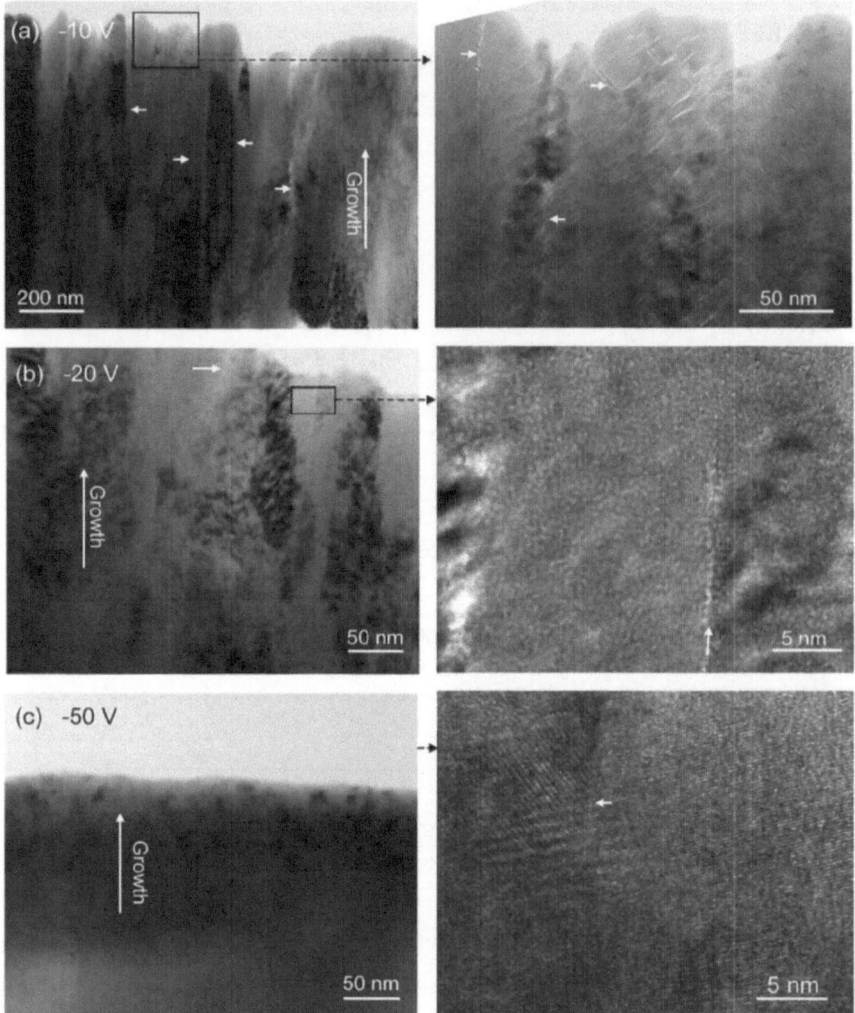

FIGURE 5.51 Cross-sectional TEM images for three coatings with various microstructures. (a) The "−10 V" coating: porous and columnar, (b) the "−20 V" coating: columnar but dense, (c) and (d) the "−50 V" coating: almost column-free [118].

contrast, are densely packed in the "−20 V" coating (c.f. Figure 5.51(b)). When the bias voltage increased to -50 V, the growth structure evolved into a denser one without visible large or long columns in its TEM image (c.f. Figure 5.51(a)). In a high-resolution TEM image (HRTEM) shown in Figure 5.51(d), the adjacent columns arrange tightly, where the boundary (as denoted by an arrowhead) is hardly distinguishable. Based on the microstructural characterization above, the three coatings were described as "columnar and porous", "columnar but dense", and "almost column-free", respectively.

5.4.2.1.1 Hardness and Young's Modulus

The hardness and modulus of the coatings are shown in Figure 5.52. When the bias voltage is −10 V, the structure of the coating is "columnar and porous structure", and the hardness of the coating is 5.8±0.8 GPa. When the bias voltage increases to −20 V, the structure of the coating changes to "columnar but dense structure", and the hardness of the coating increases to 15.5±2.1 GPa. Further

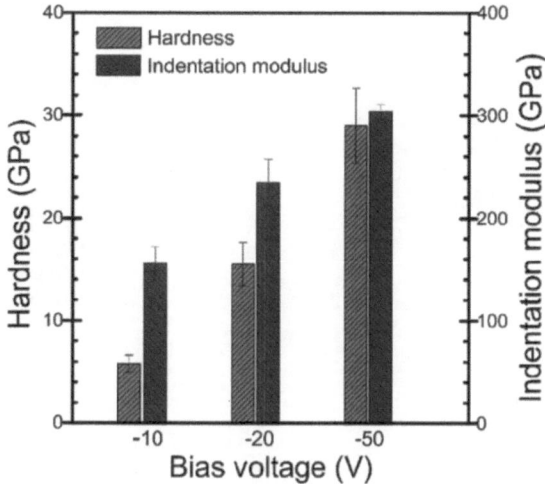

FIGURE 5.52 Hardness and elastic modulus for the three coatings.

increasing the bias to −50 V, the structure of the coating changes to "dense no columnar structure", and the hardness of the coating arrives to 29.0±3.7 GPa. As the bias voltage increases from −10 V to −50 V, the hardness and modulus of the coating increase. It can be seen from the SEM and TEM cross-sectional images that as the bias voltage increases, the degree of densification of the coating increases, which may be a main reason for the increase in the hardness of the coating. Relevant studies have shown that increasing the density is an effective means to increase the hardness of the coating [119, 120]. Additional, when the bias voltage is −50 V, the coating has a larger compressive stress (−1.9 GPa), which also contributes to the higher hardness [121].

5.4.2.1.2 Toughness

Plasticity (δ_H) defined as $\delta_H = \dfrac{\varepsilon_P}{\varepsilon} = \dfrac{Plastic\ displacement}{Plastic\ displacement + Elastic\ displacement}$ was conducted to quantitatively indicate the coatings toughness [122, 123]. Figure 5.53 shows the plasticity calculated based on the nanoindentation data for three kinds of coatings. It can be seen that the plasticity of the "columnar but dense structure" coating is the highest, about 0.58, indicating that it possesses the best toughness. While the "compact structure" coating and the "columnar structure" coating show the poor plasticity index and lower toughness.

The difference in film toughness can also be confirmed by comparing the crack behavior caused during indentation. Holleck and Schulz et al. [124] compared the cracking length of the film under the same load to determine the toughness of the coatings. Kustas et al. [125] measured the peeling diameter of the damaged area around the indentation to characterize toughness. Musil and Jirout's research indicated [53] that the substrate also plays a key role in the measurement of film toughness using the indentation method. In order to be close to the actual working conditions, Vickers indentation method was used to compare the toughness of the three coatings on the stainless steel substrate.

Figure 5.54 shows the SEM images of Vickers indentation under load of 0.49 N and 4.9N. It can be seen that, for the "columnar and porous" sample (S1), under the condition of low load (0.49 N), lots of ring-shaped cracks and radical cracks appeared in the indentation area. In the case of high load (4.9N), the cracking situation intensifies, the length of the radiation crack further increases, showing poor toughness. For the "columnar but dense" sample (S2), in the case of low load, no radioactive cracks occurred in the indentation area, and the internal cracks were stepped. In the case of high load, the conical edge of the indentation was curved and cracked and no radiation cracks

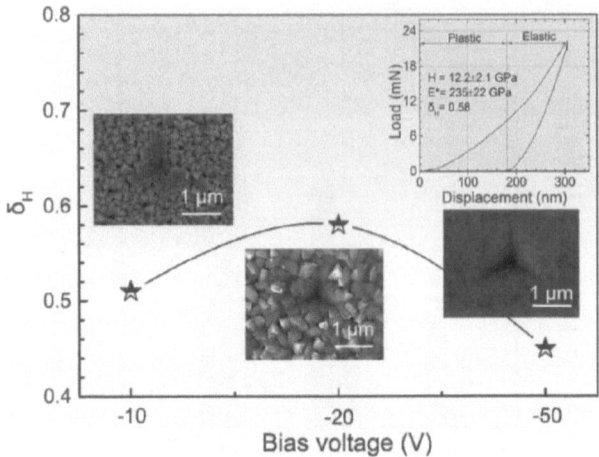

FIGURE 5.53 Plasticity and corresponding nanoindentation impression for the three coatings.

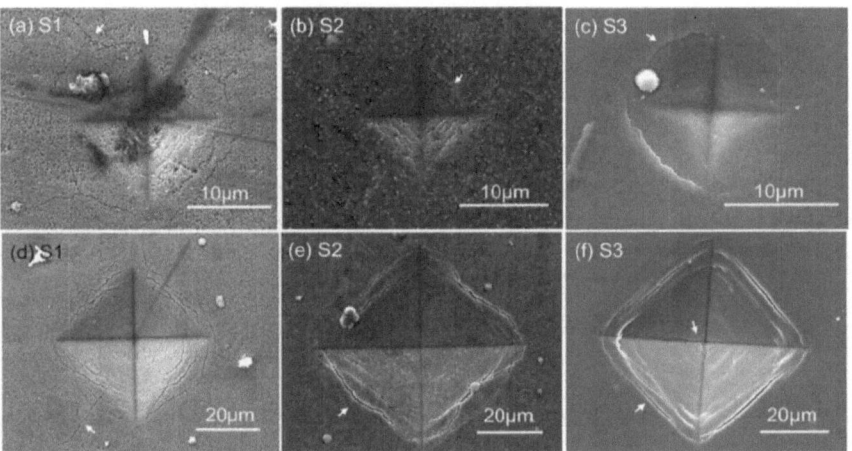

FIGURE 5.54 SEM images of Vickers indents with load 0.49 N (a, b and c) and 4.9 N (d, e and f), respectively. (a) and (d) Sample S1 (−10 V); (b) and (e) Sample S2 (−20 V); (c) and (f) Sample S3 (−50V).

were found, indicating a better toughness. While for the "dense non-columnar" sample (S3), under low load, the indentation cone edge cracks are very obvious, similar to brittle fracture characteristics. In addition, under high load, the cone edge cracks straightly, and there are some cracks around cone center (as shown by the arrow), the brittle fracture feature is more obvious. Therefore, judging by indentation, it is believed that the "columnar but dense" coating (S2) has the best toughness on the stainless steel substrate, which is consistent with the results of the plasticity (c.f. Figure 5.53).

5.4.3 TRIBO-CORROSION RESISTANCE

The polarization curves of the three coatings are shown in Figure 5.55. The "columnar and porous" coating (S1) has a corrosion current density of 7.18×10^{-7} A/cm^2 and a corrosion potential of −0.93 V. The "columnar but dense" coating (S2) has a corrosion current density of 2.06×10^{-7} A/cm^2, the corrosion potential is −0.36 V, and the "dense non-columnar" coating (S3) has a corrosion current density of 1.14×10^{-8} A/cm^2 and the corrosion potential is −0.22 V. The "dense and column-free" coating (S3) exhibits the best corrosion resistance. Its corrosion current density is about 1/63 of that

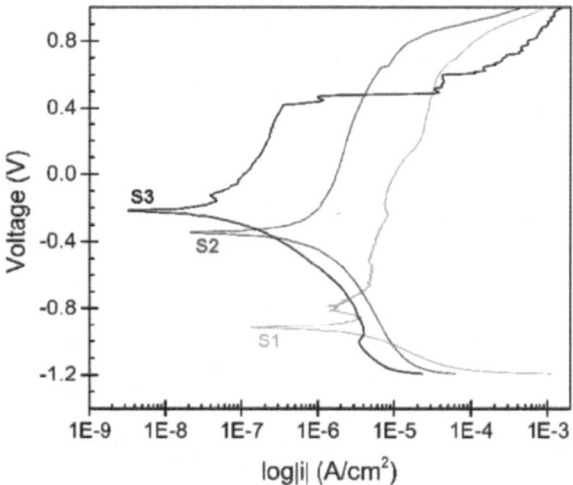

FIGURE 5.55 Potentiodynamic polarization curves for the three coatings in the corrosion condition.

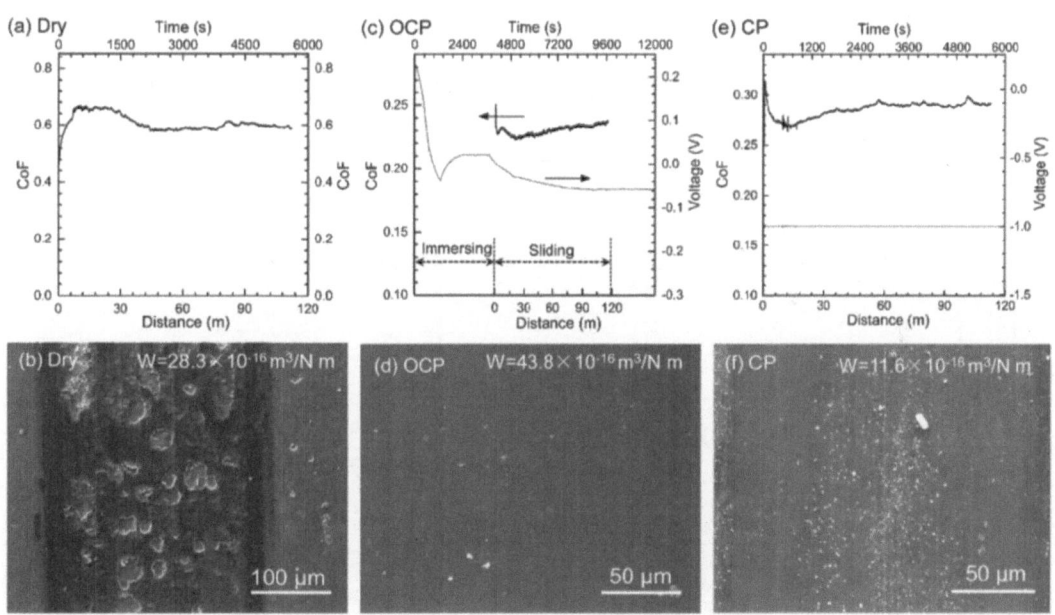

FIGURE 5.56 Evolution of the tribo-corrosion parameters as the function of sliding distance (immersion time) for the porous and columnar coating under the three conditions. (a), (c), and (e) for the CoF and the OCP; (b), (d), and (f) for morphology of the corresponding wear tracks after the sliding [118].

of S1 coating and about 1/18 of S2 coating. The corrosion potential is also the highest level among the three kinds of coatings. Therefore, with increase of density, a better corrosion resistance can be obtained. The S3 coating is the densest and the columnar gaps are almost invisible, and the corrosion resistance is the best.

Though the S3 sample shows the best corrosion resistance in the corrosive medium, the tribo-corrosion properties may not be the same story. Therefore, tribo-corrosion properties for the three kinds of coatings which show three typical microstructures were studied under both wear and corrosion condition.

5.4.3.1 Tribo-Corrosion Behavior of the "Columnar and Porous" Coating

Figure 5.56 presents the result of the tribo-corrosion tests for the "columnar and porous" coating under the three conditions (dry, OCP and CP). Under the dry condition (c.f. Figure 5.56(a)), the coefficient of friction (CoF) reaches a relatively stable value of ~0.6 after the running-in period. A relatively high specific wear rate (approximately 28.3×10^{-16} m³/N m) was obtained; and severe delamination and spallation were seen from the morphology of the wear tracks in Figure 5.56(b). The size of the broken spots is relatively large, with a diameter in the range of 10 μm ~ 30 μm. For some regions, the delamination and the spallation took place at the interface between the coating and the substrate, resulting in the local exposure of the substrate to the air.

Figure 5.56(c) shows the CoF value as a function of the sliding distance under the OCP, together with the evolution of the OCP in the prior immersion and the following sliding process. The average value of the CoF is ~0.23, much lower than that under the dry condition (~0.6 in Figure 5.56(a)). The specific wear rate is estimated to be approximately $43.8 \pm 1.8 \times 10^{-16}$ m³/N m, and the worn surface is relatively smooth in Figure 5.56(d)). Moreover, prior to the tribo-corrosion test, the OCP decreases from 220 mV to 20 mV ($\delta_{OCP} = -200$ mV) when the immersion lasts 3600 s. The negative shift of the OCP could mean the penetration of the ion solution into the coating, which is prone to occur in some highly porous coatings. During the sliding of 108 m, the OCP progressively decreases

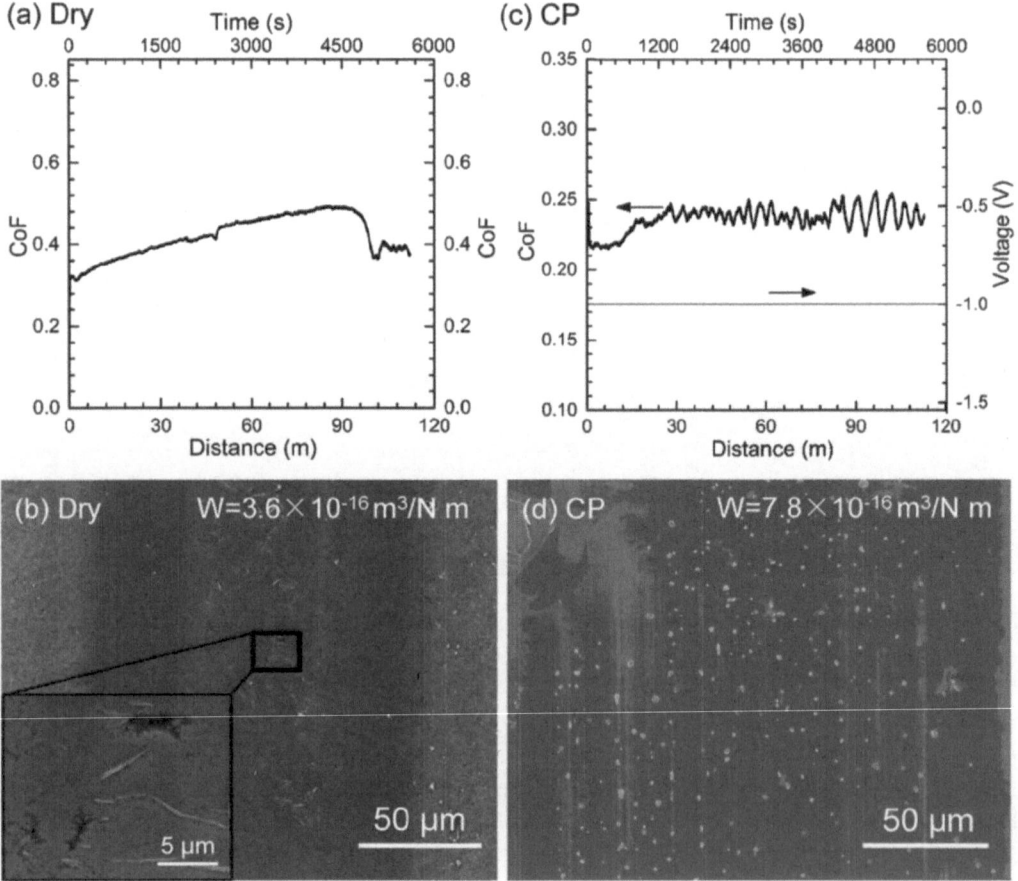

FIGURE 5.57 Evolution of the tribo-corrosion parameters as the function of sliding distance (immersion time) for the columnar but dense coating under the dry and the CP conditions. (a) and (c) for the CoF and the OCP; (b) and (d) for morphology of the corresponding wear tracks after the sliding [118].

from 20 mV to –60 mV (δ_{OCP} = –80 mV), indicating a reduction of the passive area on the worn surface. After the sliding ceases, the OCP keeps stable at the value of –60 mV, unable to recover to the initial one. As for the CP ((c.f. Figure 5.56(e)-(f)), the CoF is ~0.28. The specific wear rate (11.6 ± 1.9 ×10^{-16} m^3/N) is lower in comparison with the dry and the OCP, since only the mechanical wear was involved in the CP. The wear track appears uniform without asperities, except that some salt crystals distribute on the middle of the wear track.

5.4.3.2 Tribo-Corrosion Behavior of the "Columnar But Dense" Coating

Figure 5.57 displays the result of the tribo-corrosion tests for the "columnar but dense" coating under the dry and the CP conditions, respectively. In the case of the dry sliding (c.f. Figure 5.57(a)), the CoF value varies in the range from 0.50 to 0.35, and its specific wear rate (~3.6 × 10^{-16} m^3/N m) is about one order of magnitude lower than that of the "columnar and porous" coating. The corresponding wear track (c.f. Figure 5.57(b)) appears smooth, without noticeable flaking or spallation. More in-depth examination reveals, however, some micro-cavities and roll-like debris on the worn surface, as shown in the inset (c.f. Figure 5.57(b)). The low wear rate, together with the roll-like debris and the smooth worn surface, points to a mild wear by the layer-by-layer removal mechanism [126]. Under the CP condition (c.f. Figure 5.57(c)), the average CoF and the specific wear rate are ~0.24 and ~7.8 × 10^{-16} m^3/N m, respectively. In Figure 5.57(d), the wear track is relatively rough, on which longitudinal micro-grooves are arranging along the sliding direction.

5.4.3.3 Tribo-Corrosion Behavior of the "Almost Column-Free" Coating

Figure 5.58 shows the result of the friction-wear tests of the "almost column-free" coating under the three conditions. During the dry sliding (c.f. Figure 5.58(a) & (b)), the coating displays an average CoF of ~0.49 and a specific wear rate of 2.4 × 10^{-16} m^3/N m. Some features, such as parallel grooves, local pits, and roll-like debris, were found on the wear track. In the case of the OCP (c.f. Figure 5.58(c) & (d)), the average CoF and the specific wear rate are ~0.19 and 9.7 × 10^{-16} m^3/N m, respectively. In the initial immersion, the OCP keeps a stable value of ~0 mV. When the sliding

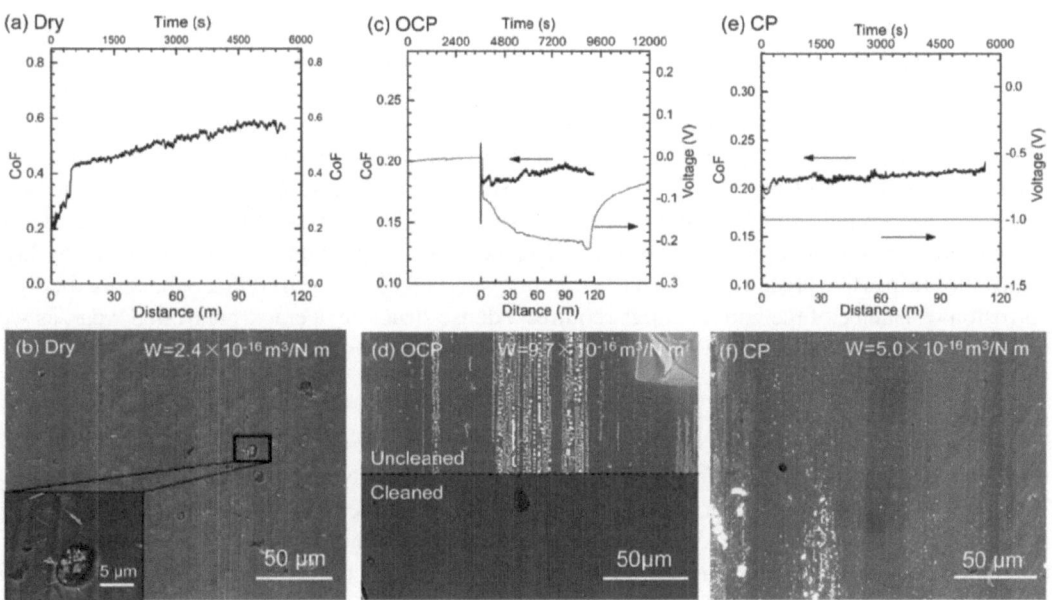

FIGURE 5.58 Evolution of the tribo-corrosion parameters as the function of sliding distance (immersion time) for the almost column-free coating under the three conditions. (a), (c), and (e) for the CoF and the OCP; (b), (d), and (f) for morphology of the corresponding wear tracks after the sliding [118].

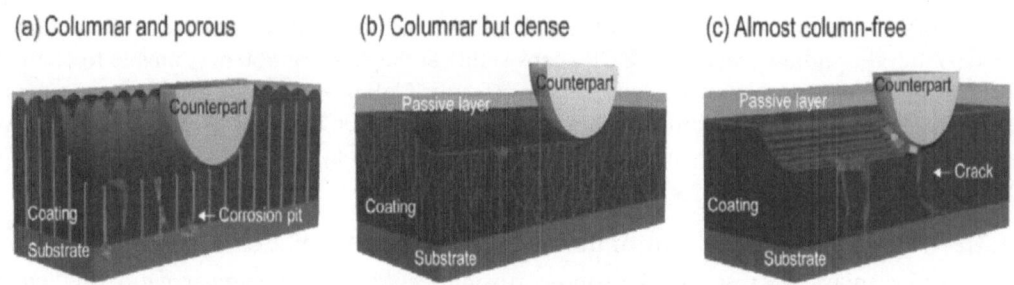

FIGURE 5.59 Schematics of tribo-corrosion process of three coatings. (a) columnar and porous coating was predominated with corrosion due to its poor intrinsic corrosion resistance; (b) columnar but dense coating exhibited best tribo-corrosion resistance, where proceed mainly by removal of passive layer in layer-by-layer mechanism; (c) almost column-free coating failed in cracks formed under mechanical contact [118].

proceeds, however, the OCP progressively drops from 0 to −200 mV in 4500 s ($\delta_{OCP} \approx -200$ mV). Subsequently, the OCP rises from −200 mV to −60 mV ($\delta_{OCP} \approx 140$ mV) after the sliding test. Figure 5.58(d) presents the images of the original track as well as the cleaned track. The original track was covered with a large number of salt crystals, while extensive grooves and cracks were actually observed after cleaning the surface. In comparison, under the CP condition (c.f. Figure 5.58(e) & (f)), the average CoF and the specific wear rate are ~0.21 and 5.0×10^{-16} m³/N m, respectively. Correspondingly, the worn surface is rough with shallow grooves and micro-pits.

5.4.3.4 Tribo-Corrosion Mechanism

Figure 5.59 schematically shows the tribo-corrosion process of the three kinds of coating with typical microstructure. The "almost column-free" coating exhibited relatively lower wear-corrosion resistance than the "columnar but dense" coating, although the former had the highest density, the highest hardness, and the best intrinsic corrosion resistance. The cracking resistance of the hard coatings plays an important role in the protective ability in the wear-corrosion process, especially for those metal substrates with poor corrosion resistance. Rossi et al. reported that the (Ti,Cr)N coatings, which have the best intrinsic corrosion resistance, turn out to be the less resistant coatings in the wear-corrosion process, probably because the coatings are more brittle and poorly adhered to the substrate [127]. Ahmed et al. revealed that after the indenting tests, the TiSiN nanocomposite coating exhibits lower corrosion resistance than the columnar CrN coating, mainly due to the open cracks formed in the indented TiSiN coating [116, 128]. The fracture-dominated wear tends to occur in the "almost column-free" coating, due to its brittleness. By contrast, the "columnar but dense" coating presents the better tribo-corrosion resistance which is predominated by the layer-by-layer removal of its passive layer, rather than the fracture wear. It can be inferred that enhancing the wear-corrosion resistance of the hard coatings requires a dense structure, a crack-controlled wear, as well as a dynamic equilibrium of depassivation–repassivation.

5.5 CONCLUDING REMARKS

Hard coatings are widely used in modern manufacturing industry as an abrasion-resistant coating, which helps to improve the wear resistance and prolong its service life. However, in corrosive environment the coatings are subjected simultaneously to both wear and corrosion phenomena, which be termed **tribo-corrosion**. The conjoint action of wear and corrosion is often more intense than pure corrosion and pure wear effects alone. It is necessary to fabricate coating with excellent resistance against both wear and corrosion and study the tribological behavior under corrosive environment. In order to obtain good corrosion resistance, the coating should be dense, compact and in lack of medium diffusion channel.

HiPIMS technology is a new coating technology developed in recent year, which has the characteristics of high metal ionization rate, and thereafter a dense coating with excellent mechanical properties and anti-corrosion can be obtained. However, HiPIMS has a low deposition rate and low production efficiency, which limits its large-scale application in the industrial field. Therefore, it is necessary to improve the deposition rate of HiPIMS. There are two kinds of methods to improve the deposition rate. One is from the design of power source. The research shows that superimpose other power source (DC/MF/RF) with HiPIMS can significantly increase the deposition rate of coatings and maintain good performance at the same time. Among them, the characteristics of the DC and RF power sources are fixed, while the MF can flexibly be adjusted which is convenient for studying the superimpose HiPIMS power system. The other is from deposition process sequence. A hybrid HiPIMS with DC/RF process can be conducted to improve the deposition rate, where the HiPIMS, DCMS, RFMS power source are connected to the targets individually. The targets are separated with each other. During the deposition, they can run either simultaneously or in sequent depending on the process recipe. Another technology is the use of ion source assisted magnetron sputtering technology, which can also obtain densely structured coatings. The ion source can increase the gas ionization rate during the coating process. Additionally, it can provide ions to bombard the growing surface to obtain a dense coating.

In addition to considering the preparation techniques, the corrosion resistance of the coating can also be improved from the design of the coating structure. The commonly used coating structure is a multilayer structure, which can effectively interrupt the growth of columnar crystals and reduce the diffusion path of corrosive medium. In order to improve the corrosion resistance further, a ceramic layer with excellent corrosion resistance can be the candidates as the sub-layer of the multilayer structure coatings. Thus, a "ceramic/ceramic" multilayer structure is preferred.

REFERENCES

[1] Andre Anders, A review comparing cathodic arcs and high power impulse magnetron sputtering (HiPIMS), *Surface and Coatings Technology*, 257(2014)308–325.

[2] Sam Zhang, Deen Sun, Yongqing Fu, Hejun Du, Effect of sputtering target power on microstructure and mechanical properties of nanocomposite nc-TiN/a-SiNx thin films, *Thin Solid Films*, 447–448(2004)462–467.

[3] Sam Zhang, Deen Sun, Yongqing Fu, Hejun Du Qing Zhang, Effect of sputtering target power density on topography and residual stress during growth of nanocomposite nc-TiN/a-SiNx thin films, *Diamond and Related Materials*, 13(2004)1777–1784.

[4] Sam Zhang, Deen Sun, Yongqing Fu, Hejun Du, Qing Zhang, Effect of sputtering target power on preferred orientation of the nc-TiN crystallites in nanocomposite nc-TiN/a-SiNx thin films, *Journal of Metastable and Nanocrystalline Materials*, 23(2005)175–178.

[5] Huili Wang, Sam Zhang, Yibin Li, Deen Sun, Bias effect on microstructure and mechanical properties of magnetron sputtered nanocrystalline titanium carbide thin films, *Thin Solid Films*, 516(2008)5419–5423.

[6] J.C. Oliveira, F. Ferreira, A. Anders, A. Cavaleiro, Reduced atomic shadowing in HiPIMS: role of the thermalized metal ions, *Applied Surface Science*, 2018,433(2018)934–944.

[7] Saravanan KG, Ananthakumar R, Subramaniam B, Maruthamuthu S, *Anodic corrosion behavior of nanostructured TiN, TiO2 single layer and TiN/TiO2 multilayer coated 316L SS, ICANMEET-2013, Chennai (2013.7.24-2013.7.26) International Conference on Advanced Nanomaterials & Emerging Engineering Technologies*, 331–334

[8] F.J. Jing, T.L. Yin, K. Yukimura, H. Sun, Y.X. Leng, N. Huang, Titanium film deposition by high-power impulse magnetron sputtering: Influence of pulse duration, *Vacuum*, 86(2012)2114–2119.

[9] V. Kouznetsov, K. Macák, J.M. Schneider, U. Helmersson, I. Petrov, A novel pulsed magnetron sputter technique utilizing very high target power densities, *Surface and Coatings Technology*, 122(1999)290–293.

[10] Wahyu Diyatmika, Fei-Ke Liang, Bih-Show Lou, Jong-Hong Lu, De-En Sun, Jyh-Wei Lee, Superimposed high power impulse and middle frequency magnetron sputtering: Role of pulse duration and average power of middle frequency, *Surface and Coatings Technology*, 352(2018)680–689.

[11] Fenker M, Balzer M, Jehn HA, Kappl H, Lee J-J, Lee K-H, Park H-S, Improvement of the corrosion resistance of hard wear resistant coatings by intermediate plasma etching or multilayered structure, *Surface and Coatings Technology*, 150(1)(2002) 101–106.

[12] Hu Fang, Dai Mingjiang, Lin Songsheng, Hou Huijun, Shi Qian, Zhao Li, Influences of cycles argon ion bombardment on structure and properties of Al films deposited by magnetron sputtering, *China Surface Engineering*, 28(1)(2015)49–55.

[13] Chun S-Y, Nanosize-controlled titanium nitride films in pulsed dc magnetron sputtering, *Journal of Nanoscience and Nanotechnology*, 13(3)(2013)2021–2024.

[14] B.A. Movchan, A.V. Demchishin, Obtaining depositions during vacuum condensation of metals and alloys, *Fizika Metallov i Metallovedenie*, 28 (1969) 653–660.

[15] John A. Thornton, The microstructure of sputter-deposited coatings, *Journal of Vaccum Science and Technology A*, 4(6)(1986)3059–3065.

[16] Andre Anders, A structure zone diagram including plasma-based deposition and ion etching, *Thin Solid Films*, 518(2010)4087–4090.

[17] D. A. Porter, Phase transformations in metals and alloys: Secondary phase transformations in metals and alloys, *Higher Education Press*, 2011, 53–56.

[18] Wu G, Liu Y, Liu C, Tang QH, Miao XS, Lu J, Novel multilayer structure design of metallic glass film deposited Mg alloy with superior mechanical properties and corrosion resistance, *Intermetallics*, 62(2015)22–26.

[19] Delblanc Bauer A, Herranen M, Ljungcrantz H, Carlsson J-O, Sundgren J-E, Corrosion behaviour of monocrystalline titanium nitride, *Surface and Coatings Technology*, 91(3)(1997)208–214.

[20] Massiani Y, Gravier P, Fedrizzi L, Marchetti F, Corrosion behaviour in acid solution of (Ti,Cr)Nx films deposited on glass, *Thin Solid Films*, 261(1–2)(1995)202–208.

[21] Milosev I, Strehblow H-H, Navinsek B. Comparison of TiN, ZrN and CrN hard nitride coatings: Electrochemical and thermal oxidation, *Thin Solid Films*, 303(1-2)(1997)246–254.

[22] Milosev I, Stehblow H-H, Navinsek B, Metikos-Hukovic M, Electrochemical and thermal oxidation of TiN coatings studied by XPS, *Surface and Interface Analysis*, 23(7–8)(1995):529–539.

[23] Schroer A, Ensinger W, Wolf GK, A comparison of the corrosion behaviour and hardness of steel samples (100Cr6) coated with titanium nitride and chromium nitride by different institutions using different deposition techniques, *Materials Science and Engineering A*, 140(C)(1991)625–630.

[24] Uslu ME, Onel AC, Ekinci G, Toydemir B, Durdu S, Usta M, Arslan LC, Investigation of (Ti,V)N and TiN/VN coatings on AZ91D Mg alloys, *Surface and Coatings Technology*, 284(2015) 252–257.

[25] Feike Liang, Yufan Shen, Chenrui Pei, Bin Qiu, Jinglei, Deen Sun, Microstructure evolution and corrosion resistance of multi interfaces Al-TiAlN nanocomposite films on AZ91D magnesium alloy, Surfaces and Coatings *Technology*, 357(2019)83–92.

[26] Yu Xi Wang, Sam Zhang, Jyh-Wei Lee, Wen Siang Lew, Deen Sun, Bo Li, Hard yet tough ceramic coating: not a dream any more -i. via nanostructured multilayering, *Nanoscience and Nanotechnology Letters*, 4(2012)375–377.

[27] Zhang Wen-yong, Sun De-en, Pei Chen-rui, Zhang Shi-hong, Huang Jia-mu, Effect of modulation period on toughness of CrAlN/ZrN nano-multilayer films, *Surface Technology*, 45(1)(2016)55–61.

[28] He Qian, Sun De-en, Zeng Xian-guang, Wear corrosion resistance of CrSiN/SiN nano-multilayer coatings deposited on TC4 titanium alloy in 3.5% NaCl solution, *China Surface Engineering*, 31(1)(2018)74–80.

[29] J.T. Gudmundsson, N. Brenning, D. Lundin, U. Helmersson, High power impulse magnetron sputtering discharge, *Journal of Vacuum Science and Technology A* 30 (2012) 030801.

[30] G. Bräuer, B. Szyszka, M. Vergöhl, R. Bandorf, Magnetron sputtering – Milestones of 30 years, *Vacuum*, 84 (2010) 1354–1359.

[31] U. Helmersson, M. Lattemann, J. Bohlmark, A.P. Ehiasarian, J.T. Gudmundsson, Ionized physical vapor deposition (IPVD): a review of technology and applications, *Thin Solid Films*, 513 (2006) 1–24.

[32] K. Sarakinos, J. Alami, S. Konstantinidis, High power pulsed magnetron sputtering: a review on scientific and engineering state of the art, *Surface and Coatings Technology*, 204 (2010) 1661–1684.

[33] J. Alami, P. Eklund, J. Emmerlich, O. Wilhelmsson, U. Jansson, H. Högberg, L. Hultman, U. Helmersson, High-power impulse magnetron sputtering of Ti-Si-C thin films from a Ti_3SiC_2 compound target, *Thin Solid Films*, 515 (2006) 1731–1736.

[34] M. Samuelsson, D. Lundin, K. Sarakinos, F. Bjorefors, B. Walivaarra, H. Ljungcrantz, U. Helmersson, Influence of ionization degree on film properties when using high power impulse magnetron sputtering, *Journal of Vacuum Science and Technology* A30 (2012) 031507.

[35] V. Sittinger, O. Lenck, M. Vergohl, B. Szyszka, G. Brauer, Applications of HIPIMS metal oxides, *Thin Solid Films*, 548(2013) 18–26.

[36] J. Olejnick, Z. Hubicka, S. Kment, M. Cada, P. Ksirova, P. Adamek, I. Gregora, Investigation of reactive HiPIMS+MF sputtering of TiO_2 crystalline thin films, *Surface and Coating Technology* 232 (2013) 376–383.

[37] Zhen He, Sam Zhang, Deen Sun, Effect of bias on structure mechanical properties and corrosion resistance of TiNx films prepared by ion source assisted magnetron sputtering, *Thin Solid Films*, 676 (2019) 60–67.

[38] Konstantinidis S, Dauchot J P, Ganciu M, et al. Influence of pulse duration on the plasma characteristics in high-power pulsed magnetron discharges. *Journal of Applied Physics*, 99(1)(2006)013307.

[39] Poolcharuansin P, Bowes M, Petty T J, et al. Ionized metal flux fraction measurements in HiPIMS discharges. *Journal of Physics D: Applied Physics*, 45(32)(2012) 322001.

[40] J. Alami, Z. Maric, H. Busch, F. Klein, U. Grabowy, M. Kopnarski, Enhanced ionization sputtering: A concept for superior industrial coatings. *Surface and Coatings Technology*, 255(2014) 43–51.

[41] Greene J E, Sundgren J E, Hultman L, et al. Development of preferred orientation in polycrystalline TiN layers grown by ultrahigh vacuum reactive magnetron sputtering. *Applied Physics Letters*, 67(20)(1995) 2928–2930.

[42] Machunze R, Ehiasarian A P, Tichelaar F D, et al. Stress and texture in HIPIMS TiN thin films. *Thin Solid Films*, 518(5)(2009) 1561–1565.

[43] Paulitsch J, Schenkel M, Zufraß T, et al. Structure and properties of high power impulse magnetron sputtering and DC magnetron sputtering CrN and TiN films deposited in an industrial scale unit. *Thin Solid Films*, 518(19)(2010) 5558–5564.

[44] Liang H, Xu J, Zhou D, et al. Thickness dependent microstructural and electrical properties of TiN thin films prepared by DC reactive magnetron sputtering. *Ceramics International*, 42(2)(2016)2642–2647.

[45] Tung-Sheng Yeh, Jenn-Ming Wu, Long-Jang Hu, The properties of TiN thin films deposited by pulsed direct current magnetron sputtering, *Thin Solid Films*, 516 (2008) 7294–7298.

[46] Musil J, Zítek M, Fajfrlík K, et al. Flexible antibacterial Zr-Cu-N thin films resistant to cracking. *Journal of Vacuum Science & Technology A: Vacuum, Surfaces, and Films*, 34(2)(2016) 021508.

[47] A. Leyland, A. Matthews, On the significance of the H/E ratio in wear control: a nanocomposite coating approach to optimised tribological behaviour, *Wear*, 246 (2000) 1–11.

[48] Musil J, Kunc F, Zeman H, et al. Relationships between hardness, Young's modulus and elastic recovery in hard nanocomposite coatings. *Surface and Coatings Technology*, 154(2)(2002)304–313.

[49] Hongjin Liu, Xinzhu Wang, Chenrui Pei, Deen Sun, Tribological properties and corrosion resistance of CrSiN coatings prepared via hybrid HiPIMS and DCMS, *Materials Research Express*, 6(2019)086432.

[50] U.C. Oh, Jung Ho Je, Effects of strain energy on the pretreated orientation of TiN thin films, *Journal of Applied Physics*, 74(1993) 1692–1696

[51] D. Mercs, N. Bonasso, S. Naamane, Jean-Michel Bordes, C. Coddet, Mechanical and tribological properties of Cr-N and Cr-Si-N coatings reactively sputter deposited, *Surface and Coatings Technology*, 200 (2005) 403–407.

[52] Jianliang Lin, Bo Wang, Yixiang Ou, William D. Sproul, Isaac Dahan, John J. Moore, Structure and properties of CrSiN nanocomposite coatings deposited by hybrid modulated pulsed power and pulsed dc magnetron sputtering, *Surface and Coatings Technology*, 216 (2013) 251–258.

[53] Musil J, Jirout M. Toughness of hard nanostructured ceramic thin films. *Surface and Coatings Technololgy*, 201(9–11)(2007)5148–5152.

[54] Yu-Chen Chan, Hsien-Wei Chen, Pen-Shen Chao, Jeng-Gong Duh, Jyh-Wei Lee, Microstructure control in TiAlN/SiNx multilayers with appropriate thickness ratios for improvement of hardness and anti-corrosion characteristics, *Vacuum*, 87(2013)195–199.

[55] M.K. Lee, H.S. Kang, W.W. Kim, J.S. Kim, W.J. Lee, Characteristics of TiN film deposited on satellite using reactive magnetron sputter ion plating, *Journal of Materials Research*, 12 (1997) 2393–2400.

[56] J.H. Je, D.Y. Noh, H.K. Kim, K.S. Liang, Preferred orientation of TiN films studied by a real time synchrotron x-ray scattering, *Journal of Applied Physics*, 81 (1997) 6126–6133.

[57] J. Pelleg, L.Z. Zevin, S. Lungo, N. Croitoru, Reactive-sputter-deposited TiN films on glass substrates, *Thin Solid Films*, 197 (1991) 117–128.

[58] Y. Catherine. Preparation techniques for diamond-like carbon, in: R.E. Clausing, L.L. Horton, J.C. Angus, P. Koidl (Eds.), *Diamond and Diamond-like Films and Coatings. NATO ASI Series (Series B: Physics)*, vol. 266, Springer, Boston, MA, 1991, pp. 193–227.

[59] M. Jaros, J. Musil, R. Cerstvy, S. Haviar, Effect of energy on macrostress in Ti(Al,V)N films prepared by magnetron sputtering, *Vacuum*, 158 (2018) 52–59.

[60] J. Musil, Hard nanocomposite coatings: Thermal stability, oxidation resistance and toughness, *Surface & Coatings Technology*, 207 (2012) 50–65.

[61] H. Elmkhah, F. Attarzadeh, A. Fattah-Alhosseini, K.H. Kim, Microstructural and electrochemical comparison between TiN coatings deposited through HIPIMS and DCMS techniques, *Journal of Alloys and Compounds*, 735 (2018) 422–429.

[62] W.D. Sproul, Multilyer multicomponent and mutiphase physical vapor deposition coatings for enhanced performance, *Journal of Vacuum Science & Technology A*, 12 (1994) 1595–1601.

[63] Y.X. Ou, J. Lin, S. Tong, H.L. Che, W.D. Sproul, M.K. Lei, Wear and corrosion resistance of CrN/TiN superlattice coatings deposited by a combined deep oscillation magnetron sputtering and pulsed dc magnetron sputtering, *Applied Surface Science*, 351 (2015) 332–343.

[64] R. Prakash, D. Kaur, Effect of film thickness on structural and mechanical properties of AlCrN nanocompoite thin films deposited by reactive DC magnetron sputtering, in: M.S. Shekhawat, S. Bhardwaj, B. Suthar (Eds.) *International Conference on Condensed Matter And Applied Physics*, Amer Inst Physics, Melville, 2016, pp. 1–4.

[65] W.-J. Chou, G.-P. Yu, J.-H. Huang, Mechanical properties of TiN thin film coatings on 304 stainless steel substrates, *Surface and Coatings Technology*, 149 (2002) 7–13.

[66] H. Ljungcrantz, M. Odén, L. Hultman, J.E. Greene, J.E. Sundgren, Nanoindentation studies of single-crystal (001)-, (011)-, and (111)-oriented TiN layers on MgO, *Journal of Applied Physics*, 80 (1996) 6725–6733.

[67] J. Musil, J. Sklenka, R. Cerstvy, Protection of brittle film against cracking, *Applied Surface Science*, 370 (2016) 306–311.

[68] M. Jirout, J. Musil, Effect of addition of Cu into ZrOx film on its properties, *Surface and Coatings Technology*, 200 (2006) 6792–6800.

[69] M. Jaroš, J. Musil, R. Čerstvý, S. Haviar, Effect of energy on structure, microstructure and mechanical properties of hard Ti(Al,V)Nx films prepared by magnetron sputtering, *Surface and Coatings Technology*, 332 (2017) 190–197.

[70] K.F. Khaled, M.M. Al-Qahtani, The inhibitive effect of some tetrazole derivatives towards Al corrosion in acid solution: Chemical, electrochemical and theoretical studies, *Materials Chemistry and Physics*, 113 (2009) 150–158.

[71] N. Yilmaz, A. Fitoz, ÿ. Ergun, K.C. Emregül, A combined electrochemical and theoretical study into the effect of 2-((thiazole-2-ylimino)methyl)phenol as a corrosion inhibitor for mild steel in a highly acidic environment, *Corrosion Science*, 111 (2016) 110–120.

[72] A.U. Chaudhry, B. Mansoor, T. Mungole, G. Ayoub, D.P. Field, Corrosion mechanism in PVD deposited nano-scale titanium nitride thin film with intercalated titanium for protecting the surface of silicon, *Electrochimica Acta*, 264 (2018) 69–82.

[73] A. Döner, G. Kardaş, N-Aminorhodanine as an effective corrosion inhibitor for mild steel in 0.5M H2SO4, *Corrosion Science*, 53 (2011) 4223–4232.

[74] M.A. Amin, K.F. Khaled, S.A. Fadl-Allah, Testing validity of the Tafel extrapolation method for monitoring corrosion of cold rolled steel in HCl solutions – Experimental and theoretical studies, *Corrosion Science*, 52 (2010) 140–151.

[75] H. Hoche, S. Groß, M. Oechsner, Development of new PVD coatings for magnesium alloys with improved corrosion properties, *Surface and Coatings Technology*, 259 (2014) 102–108.

[76] L. Tsetseris, N. Kalfagiannis, S. Logothetidis, S.T. Pantelides, Role of N defects on thermally induced atomic-scale structural changes in transition-metal nitrides, *Physical Review Letters* 99(125503) (2007) 1–4.

[77] Z. Gu, C. Hu, X. Fan, L. Xu, M. Wen, Q. Meng, L. Zhao, X. Zheng, W. Zheng, On the nature of point defect and its effect on electronic structure of rocksalt hafnium nitride films, *Acta Materialia*, 81 (2014) 315–325.

[78] V.A. Lavrenko, A.D. Panasyuk, Effects of nitrogen deficiency on the kinetics and mechanism of electro-lytic ZrNxOxidation, *Powder Metallurgy and Metal Ceramics*, 42(7) (2003) 419–423.

[79] D. Music, L. Banko, H. Ruess, M. Engels, A. Hecimovic, D. Grochla, D. Rogalla, T. Brögelmann, A. Ludwig, A. von Keudell, K. Bobzin, J.M. Schneider, Correlative plasma-surface model for metastable Cr-Al-N: Frenkel pair formation and influence of the stress state on the elastic properties, *Journal of Applied Physics*, 121(21) (2017) 215108.

[80] Chenrui Pei, Lijun Deng, Hongjin Liu, Zhen He, Chengjie Xiang, Sam Zhang, Deen Sun, Corrosion inhibition behaviors of ZrNx thin films with varied N vacancy concentration, *Vacuum*, 162(2019)28–38.

[81] G.I. Cubillos, E. Romero, J.E. Alfonso, Influence of corrosion on the morphology and structure of ZrO_xN_y-ZrN coatings deposited on stainless steel, *Materials Chemistry and Physics*, 176 (2016) 167–178.

[82] W. Chou, G. Yu, J. Huang, Corrosion behavior of TiN-coated 304 stainless steel, *Corrosion Science*, 43 (2001) 2023–2035.

[83] R. Brown, M.N. Alias, Effect of composition and thickness on corrosion behavior of TiN and ZrN thin films, *Surface and Coating Technology*, 62 (1993) 467–473.

[84] N. Drnovšek, N. Daneu, A. Rečnik, M. Mazaj, J. Kovač, S. Novak, Hydrothermal synthesis of a nano-crystalline anatase layer on Ti6A4V implants, *Surface and Coatings Technology*, 203 (2009) 1462–1468.

[85] F. Cao, P. Munroe, Z. Zhou, Z. Xie, Influence of substrate bias on microstructural evolution and mechani-cal properties of TiAlSiN thin films deposited by pulsed-DC magnetron sputtering, *Thin Solid Films*, 639 (2017) 137–144.

[86] M.H. Park, S.H. Kim, Temperature coefficient of resistivity of TiAlN films deposited by radio frequency magnetron sputtering, *Transactions of Nonferrous Metals Society of China*, 23 (2013) 433–438.

[87] J.T. Chena, J. Wang, F. Zhang, G.A. Zhang, X.Y. Fan, Z.G. Wu, P.X. Yan, Characterization and tempera-ture controlling property of TiAlN coatings deposited by reactive magnetron co-sputtering, *Journal of Alloys and Compounds*, 472 (2009) 91–96.

[88] I. Shimizu, Y. Setsuhara, S. Miyake, M. Kumagai, K. Ogata, M. Kohata, K. Yamaguchi, Preparation of aluminum oxide films by ion beam assisted deposition, *Surface and Coatings Technology*, 131 (2000) 187–191.

[89] N. Schalk, J.F. T. S. Fotso, D. Holec, G. Jakopic, A. Fian, V. L. Terziyska, R. Daniel, C. Mitterer, Influence of varying nitrogen partial pressures on microstructure, mechanical and optical properties of sputtered TiAlON coatings, *Acta Materialia*, 119 (2016) 26–34.

[90] D. Li, W. Xie, C. Zou, Effects of RF inductively coupled plasma ion source on the microstructure and mechanical properties of Ti–Al–N nanocrystalline films, *Applied Physics A: Materials Science & Processing*, 122 (2016) 307.

[91] R. Hahn, M. Bartosik, R. Soler, C. Kirchlechner, G. Dehm, P.H. Mayrhofer, Superlattice effect for enhanced fracture toughness of hard coatings, *Scripta Materialia*, 124 (2016) 67–70.

[92] M. Schlögl, C. Kirchlechner, J. Paulitsch, J. Keckes, P.H. Mayrhofer, Effects of structure and interfaces on fracture toughness of CrN/AlN multilayer coatings, *Scripta Materialia*, 68 (2013) 917–920.

[93] E. McCafferty, Validation of corrosion rates measured by the Tafel extrapolation method, *Corrosion Science*, 47 (2005) 3202–3215.

[94] M.F. Pillis, G.A. Geribola, G. Scheidt, E.G. de Araújo, M.C.L. de Oliveira, R.A. Antunes, Corrosion of thin, magnetron sputtered Nb2O5 films, *Corrosion Science*, 102 (2016) 317–325.

[95] J.C. Caicedo, G. Zambrano, W. Aperador, L. Escobar-Alarcon, E. Camps, Mechanical and electrochemi-cal characterization of vanadium nitride (VN) thin films, *Applied Surface Science*, 258 (2011) 312–320.

[96] S.H. Ahn, J.H. Lee, H.G. Kim, J.G. Kim, A study on the quantitative determination of through-coating porosity in PVD-grown coatings, *Applied Surface Science*, 233 (2004) 105–114.

[97] D. Zhang, B. Wei, Z. Wu, Z. Qi, Z. Wang, A comparative study on the corrosion behaviour of Al, Ti, Zr and Hf metallic coatings deposited on AZ91D magnesium alloys, *Surface and Coatings Technology*, 303 (2016) 94–102.

[98] M. Tacikowski, J. Grzonka, T. Płociński, R. Jakieła, M. Pisarek, T. Wierzchoń, Composite titanium nitride layers produced on the AZ91D magnesium alloy by a hybrid method including hydrothermal modification of the layer, *Applied Surface Science*, 346 (2015) 394–405.

[99] S.R. Kiahosseini, A. Afshar, M. Mojtahedzadeh Larijani, M. Yousefpour, Structural and corrosion char-acterization of hydroxyapatite/zirconium nitride-coated AZ91 magnesium alloy by ion beam sputtering, *Applied Surface Science* 401 (2017) 172–180.

[100] L.H. Tian, E.Q. Liu, A.L. Fan, L. Qin, D.X. Liu, B. Tang, J.D. Pan, Effect of TiN/CrN Multilayer Coating by Cathodic Arc Deposition on Wear and Corrosion Behaviours of AZ91D Magnesium Alloy, *Materials Science Forum* 610–613 (2009) 870–873.

[101] D. Zhang, Z. Qi, B. Wei, Z. Wu, Z. Wang, Anticorrosive yet conductive Hf/Si3N4 multilayer coatings on AZ91D magnesium alloy by magnetron sputtering, *Surface and Coatings Technology*, 309 (2017) 12–20.

[102] M. Mikula, D. Plašienka, D.G. Sangiovanni, M. Sahul, T. Roch, M. Truchlý, M. Gregor, L.U. Čaplovič, A. Plecenik, P. Kúš, Toughness enhancement in highly NbN-alloyed Ti-Al-N hard coatings, *Acta Materialia* 121 (2016) 59–67.

[103] I.B. Singh, M. Singh, S. Das, A comparative corrosion behavior of Mg, AZ31 and AZ91 alloys in 3.5% NaCl solution, *Journal of Magnesium and Alloys* 3 (2015) 142–148.

[104] G.L. Song., A. Atrens., Corrosion mechanisms of magnesium alloys, *Advanced Engineering Materials* 1 (1999) 11–33.

[105] M. Jamesh, S. Kumar, T.S.N. Sankara Narayanan, Corrosion behavior of commercially pure Mg and ZM21 Mg alloy in Ringer's solution – Long term evaluation by EIS, *Corrosion Science*, 53 (2011) 645–654.

[106] Z. Li, X. Gu, S. Lou, Y. Zheng, The development of binary Mg-Ca alloys for use as biodegradable materials within bone, *Biomaterials*, 29 (2008) 1329–1344.

[107] Y.N. Kok, R. Akid, P.E. Hovsepian, Tribocorrosion testing of stainless steel (SS) and PVD coated SS using a modified scanning reference electrode technique, *Wear*, 259 (2005) 1472–1481.

[108] L. Shan, Y. Wang, Y. Zhang, Q. Zhang, Q. Xue, Tribocorrosion behaviors of PVD CrN coated stainless steel in seawater, *Wear*, 362–363 (2016) 97–104.

[109] S.H. Ahn, J.H. Lee, J.G. Kim, J.G. Han, Localized corrosion mechanisms of the multilayered coatings related to growth defects, *Surface and Coatings Technology*, 177–178 (2004) 638–644.

[110] Y.H. Yoo, J.H. Hong, J.G. Kim, H.Y. Lee, J.G. Han, Effect of Si addition to CrN coatings on the corrosion resistance of CrN/stainless steel coating/substrate system in a deaerated 3.5 wt% NaCl solution, *Surface and Coatings Technology*, 201 (2007) 9518–9523.

[111] L. Shan, Y. Wang, J. Li, J. Chen, Effect of N2 flow rate on microstructure and mechanical properties of PVD CrNx coatings for tribological application in seawater, *Surface and Coatings Technology*, 242 (2014) 74–82.

[112] E. Gracia-Escosa, I. García, J.C. Sánchez-López, M.D. Abad, A. Mariscal, M.A. Arenas, J. de Damborenea, A. Conde, Tribocorrosion behavior of TiBxCy/a-C nanocomposite coating in strong oxidant disinfectant solutions, *Surface and Coatings Technology*, 263 (2015) 78–85.

[113] W.Q. Bai, L.L. Li, Y.J. Xie, D.G. Liu, X.L. Wang, G. Jin, J.P. Tu, Corrosion and tri-bocorrosion performance of M (M=Ta, Ti) doped amorphous carbon multilayers in Hank's solution, *Surface and Coatings Technology*, 305 (2016) 11–22.

[114] Y.P. Purandare, A.P. Ehiasarian, M.M. Stack, P.E. Hovsepian, CrN/NbN coatings deposited by HIPIMS: a preliminary study of erosion–corrosion performance, *Surface and Coatings Technology*, 204 (2010) 1158–1162.

[115] L. Shan, Y. Wang, J. Li, X. Jiang, J. Chen, Improving tribological performance of CrN coatings in seawater by structure design, *Tribology International* 82 (2015) 78–88.

[116] M.S. Ahmed, P. Munroe, Z.-T. Jiang, X. Zhao, W. Rickard, Z.-f. Zhou, L.K.Y. Li, Z. Xie, Corrosion behaviour of nanocomposite TiSiN coatings on steel substrates, *Corrosion Science* 53 (2011) 3678–3687.

[117] P. Henry, J. Takadoum, P. Berçot, Depassivation of some metals by sliding friction, *Corrosion Science*, 53 (2011) 320–328.

[118] Tao Shao, Fangfang Ge, Yue Dong, Ke Li, Peng Li, Deen Sun, Feng Huang, Microstructural effect on the tribo-corrosion behaviors of magnetron sputtered CrSiN coatings, *Wear*, 416–417(2018) 44–53.

[119] Zhu P, Ge F, Li S, et al. Microstructure, chemical states, and mechanical properties of magnetron co-sputtered $V_{1-x}Al_xN$ coatings, *Surface and Coatings Technology*, 232(2013) 311–318.

[120] Haines J, Jm Léger A, Bocquillon G. Synthesis and design of superhard materials. *Annual Review of Materials Research*, 2001, 31(1): 1–23.

[121] Paldey S, Deevi S C. Properties of single layer and gradient (Ti,Al)N coatings. *Materials Science and Engineering A*, 2003, 361(1–22): 1–8.

[122] Zhang S, Sun D, Fu Y, et al. Toughness measurement of thin films: a critical review. *Surface and Coatings Technology*, 198(1–3)(2005) 74–84.

[123] Milman Y V, Galanov B A, Chugunova S I. Plasticity characteristic obtained through hardness measurement. *Acta Metallurgica et Materialia*, 41(9)(1993)2523–2532.

[124] Holleck H, Schulz H. Preparation and behaviour of wear-resistant TiC/TiB2, TiN/TiB2 and TiC/TiN coatings with high amounts of phase boundaries. *Surface and Coatings Technology*, 36(3)(1988)707–714.

[125] Kustas F, Mishra B, Zhou J. Metal/carbide co-sputtered wear coatings. *Surface and Coatings Technology*, 120–121(C)(1999)489–494.

[126] Ge F, Zhu P, Meng F, et al. Achieving very low wear rates in binary transition-metal nitrides: The case of magnetron sputtered dense and highly oriented VN coatings. *Surface and Coatings Technology*, 248(2014) 81–90.

[127] Rossi S, Fedrizzi L, Leoni M, et al. (Ti, Cr)N and Ti/TiN PVD coatings on 304 stainless steel substrates: wear-corrosion behavior, *Thin Solid Films*, 350(1999) 161–167.

[128] Ahmed M S, Munroe P, Jiang Z-T, et al. Corrosion and damage resistant nitride coatings for steel. *Journal of the American Ceramic Society*, 95(9)(2012) 2997–3004.

6 High-Entropy Alloy-Based Coatings
Microstructures and Properties

Yujie Chen
Southwest University, China
The University of Adelaide, Australia

Paul Munroe
The University of New South Wales, Australia

Zonghan Xie
The University of Adelaide, Australia

Sam Zhang
Southwest University, China

CONTENTS

6.1 INTRODUCTION

Coatings provide an effective way of enhancing the durability, performance and reliability of materials by protecting them against mechanical and chemical damage. A protective surface coating, with remarkable chemical and mechanical properties, can maintain the functionality of engineering components even under severe conditions and, hence, significantly enhance component lifetime (Zhang et al. 2003). Protective coatings have become an active research field of considerable economic importance. Binary transition metal nitrides, such as TiN and CrN, are the most widely used protective coatings in industry. In order to improve their performance, appropriate alloying elements are often added into binary nitrides. For example, the addition of Al into TiN coating greatly increases its hardness and fracture toughness simultaneously (Bartosik et al. 2017). Moreover, $Ti_{1-x}Al_xN$ coatings exhibit better oxidation resistance than binary TiN coatings (Münz 1986). The addition of other elements, such as Cr, Zr, and Si can also enhance the performance of TiN coatings (Panjan et al. 1996). Alloying has long been used to confer desirable properties to materials. Incorporation of multicomponents is considered to have great potential for the development of novel hard protective coatings.

High-entropy alloys (HEAs), often referred to as multicomponent alloys, have emerged over the last decade as a new class of metallic materials with excellent physical and mechanical properties (Yeh et al. 2004, Cantor et al. 2004, Zhang et al. 2014b, Zhang et al. 2018a). The original design concept of these novel alloys was proposed to maximize the configurational entropy for the formation of single-phase solid solution by mixing multiple principal alloying elements in relatively high, often equiatomic, concentrations (Gludovatz et al. 2014, Lin et al. 2017). This concept stands in sharp contrast to conventional alloys, in which relatively small amounts of secondary elements are normally added to a primary element, leading to limited number of possible elemental combinations (Figure 6.1). Recently, it has been increasingly realized that maximum entropy, achieved by equiatomic ratios of elements, is not the most essential parameter when designing HEAs with superior properties, enabling the relaxation of the original strict restriction on the equiatomic ratios of HEAs (Otto et al. 2013, Ma et al. 2015, Yang and Zhang 2012). These findings significantly increase compositional space for this kind of novel alloy (Figure 6.1), offering promise to discover exciting new alloys. Many attractive properties have been achieved in HEAs, such as high hardness and strength, excellent corrosion resistance, oxidation resistance, and good high-temperature stability (Miracle and Senkov 2017).

HEAs has attracted much attention and also generated much research interest on related coatings based on this alloying concept. In 2004, FeCoNiCrCuAlMn and $FeCoNiCrCuAl_{0.5}$ high-entropy nitride coatings were deposited by Chen et al. using magnetron sputtering; the first study expanding the HEA concept from bulk alloys to coatings (Chen et al. 2004). Since then, investigations concerning the high-performance HEA-based coatings have aroused more and more interest due to their excellent properties such as high hardness, and good wear and corrosion resistance, making them suitable candidates for protective coatings (Li et al. 2018). It is clear that as a result of the increased

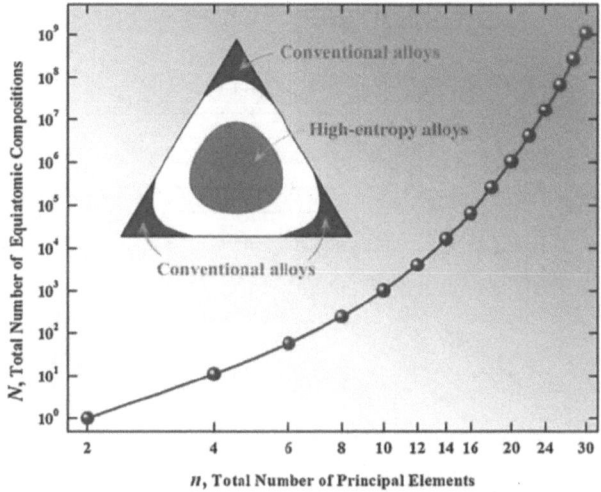

FIGURE 6.1 The relationship between the total number of equiatomic compositions (N) and the total number of principal elements (n). A ternary plot in the inset shows the difference between the design of conventional alloys and HEAs (Ye et al. 2016).

solid solubility in HEA-based coatings, the compositional space is vastly increased compared to conventional coatings. Meanwhile, these alloy coatings open up novel materials design concepts, since functional alloying elements can be added to a phase without compromising the desired crystal structure and properties, providing the hope of developing new coatings materials with increased functionality. This is essentially the reason for the growing interest in the HEA-based coatings. This chapter will review the research progress of surface coatings based on the HEA concept, including HEA coatings and high-entropy ceramic (HEC) coatings. The focus will be on the fabrication methods, microstructures, mechanical properties, including hardening and toughening mechanisms, and the corrosion resistance of HEA-based coatings. In addition, the current challenges and the future outlooks are discussed and addressed.

6.2 FABRICATION METHODS

HEA-based coatings can be easily fabricated using suitable conventional coating technologies, despite the fact that many components are involved during the coating process. At present, several methods have been applied to fabricate HEA-based coatings, including magnetron sputtering, vacuum arc deposition, laser cladding, plasma cladding, thermal spraying, cold spraying, plasma-transferred arc cladding, and other techniques based on powder metallurgy (Li et al. 2019). If a suitable reactive gas is used during deposition, high-entropy nitride, carbide, or oxide coatings could be deposited by magnetron sputtering or vacuum arc deposition. High-entropy nitride coatings can be deposited by allowing the atoms or ions from a single HEA target or split targets to react with N_2-containing Ar gas during deposition. Similarly, using CH_4-containing and O_2-containing gases during deposition will produce high-entropy carbide and oxide coatings, respectively. However, some other conventional coating technologies for thin film coating processing, including electroless plating and electroplating, have rarely been used to deposit HEA-based coating with multiple elements due to the differences in the reduction potentials among differing elements. The thickness of HEA-based coatings fabricated by vapor deposition technologies, including magnetron sputtering and vacuum arc deposition, are limited to within several micrometers, while laser cladding, thermal spraying, and their related methods can deposit much thicker coatings with a thickness from tens to hundreds of micrometers.

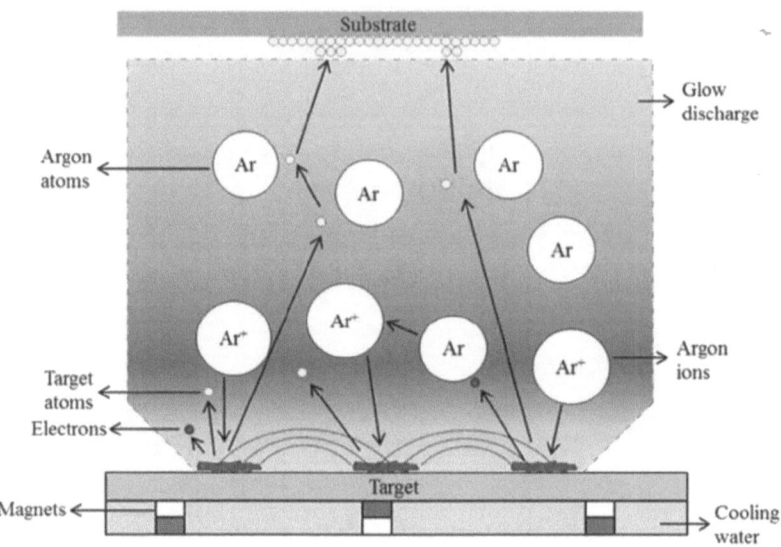

FIGURE 6.2 Schematic diagram illustrating the basic components of a magnetron sputtering system and the magnetron sputtering process (Maurya et al. 2014).

Among the various deposition methods, magnetron sputtering and laser cladding are the most commonly used methods for the preparation of HEA-based coatings. A schematic diagram highlighting the components of a magnetron sputtering system are shown in Figure 6.2. A gas, commonly Ar, is introduced into a vacuum chamber. A DC voltage is placed between the target and substrate to ionize Ar atoms. Positively charged energetic Ar ions are accelerated to bombard the target, the target atoms are ejected or "sputtered", and eventually being deposited on the surface of the substrate (Maurya et al. 2014). According to the published literature, HEA-based coatings fabricated by magnetron sputtering mainly include HEC coatings, including high-entropy nitride, carbide, and oxide coatings as well as HEA coatings.

Laser cladding has been widely used to fabricate HEA coatings. During the process, the material is fed into the substrate surface under the laser beam, and then melted by a high-power and high-speed laser. The strong metallurgical interfaces formed between the layer and matrix can improve the mechanical properties of laser clad coatings. However, cracks and pores formation is a common problem of the laser surface treatment (Chen et al. 2008, Ye et al. 2011). Furthermore, the effect of heat treatment on the previously clad layer induced by the overlapping of individual layers may generate adverse impacts on the performance of clad coatings.

6.3 MICROSTRUCTURE

The superior properties of HEA-based coatings are largely governed by their microstructures. Due to combination of the four core effects of HEAs and the rapid quenching that occurs during coating deposition, ether the face-centered cubic (FCC) or body-centered cubic (BCC) solid-solution phase, or amorphous phases, are commonly formed in HEA-based coatings. The phase structure and stability of HEA-based coatings can be different from their conventional coarse-grained bulk counterparts. The structure formed in HEA-based coatings is strongly dependent on their composition and the processing parameters used. In this section, the current study on the phase structure of HEA-based coatings is discussed.

6.3.1 General Concept of HEAs

For high-entropy materials, Yeh attributed their promising properties primarily to four so-called "core effects": the high-entropy effect, the lattice distortion effect, the sluggish diffusion effect and the "cocktail" effect (Yeh 2006). Generally, the compositional design for HEA-based coatings is consistent with that of bulk HEAs; that is, they also exhibit the four "core effects".

The high entropy effect is the signature concept of HEAs, which was suggested to be the factor promoting the formation of solid solutions in the original paper of Yeh et al. (Yeh 2006). The high-entropy effect provides an explanation for the thermodynamic principles for stabilizing a solid solution phase by the increased configurational entropy in near-equimolar alloys with five or more randomly distributed components (Miracle and Senkov 2017, Yeh et al. 2004). Therefore, HEA-based coatings tend to form random solid solutions, rather than intermetallic compounds. Moreover, relative to the HEA bulk alloy prepared by arc melting and casting, the cooling rates during the fabrication processes of HEA-based coatings are much faster, which thus increases the nucleation rate to reduce grain sizes and inhibit phase separation. This rapid quenching effect also exists in conventional coatings, but it could be further enhanced by sluggish diffusion and lattice distortion effects in HEA-based coatings.

The severe lattice distortion effect is due to the mixing of atoms with different atomic sizes in a single lattice structure. It provides extra strength and also contributes to the slow kinetics in HEAs (Yeh 2006, Zhang et al. 2014b). It has been shown that the consideration of atomic radii is important in designing multicomponent solid solution phases. The atomic-radius mismatch (δ) must be less than ~6.4% for a solid solution phase to form, as at higher δ-values the alloys tend to form either amorphous phases or intermetallic compounds (Sheng and Liu 2011). The sluggish diffusion effect also arises from a mix of different atomic sizes, which creates larger diffusion barriers and so retards diffusion. Thus, diffusion is proposed to be sluggish in HEAs compared to conventional alloys. This effect impedes the growth of second phase nuclei out of a single-phase solid solution, favoring the formation of simple solid solution phases. The cocktail effect is a phrase used to describe enhanced properties due to the synergetic effects of alloying, in which the end result is unpredictable and greater than the sum of its parts. As a result of the four core effects in HEA materials, as well as the rapid quenching effect in coatings, an amorphous phase or simple FCC or BCC solid solution phase is commonly formed in HEA-based coatings.

6.3.2 Different Phase Stabilities of HEA-Based Coatings and Their Bulk Counterparts

Although HEA-based coatings exhibit the same four core effects as bulk HEAs, the phase structure and stability of HEA-based coatings can be quite different from their conventional coarse-grained bulk counterparts, mainly due to the much smaller grain size and the effects of rapid quenching during deposition. The phase stability in nanostructured materials is quite different relative to their conventional coarse-grained counterparts (Owen and Jones 1954). Moreover, the cooling rates are much faster during the preparation processes for HEA-based coatings compared to that of HEA bulk alloys prepared by arc melting and casting, therefore, different cooling rates can lead to different phase structures between coatings and bulk alloys, even for the same alloy system (Yan et al. 2018). For example, it is generally believed that FCC is the predominant crystal structure for bulk CrCoNi medium entropy alloys, and the hexagonal close packed (HCP) phase is regarded as energetically unfavorable (Miao et al. 2017, Zhang et al. 2017a). However, for the CrCoNi coatings, shown in Figure 6.3, the HCP and FCC phases co-exist in a form of alternating nanolayers with a high density of twin boundaries and stacking faults (Chen et al. 2018, Chen et al. 2020b, Cao et al. 2018a). Another example is the $Al_{0.5}CrFeNiTi_{0.25}$ HEA coating, which exhibits an amorphous structure, while the bulk counterpart prepared by casting features a BCC phase structure (Zhang et al. 2018b). Therefore, the phase stability of HEA coatings are affected by various factors, including compositions, grain sizes and deposition conditions, making the phase formation in HEA coatings complicated.

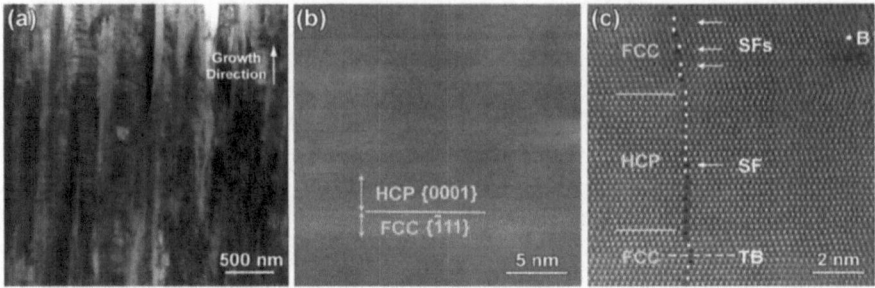

FIGURE 6.3 (a) A bright field transmission electron microscopy (TEM) image of an as-deposited CrCoNi alloy showing the columnar structure containing a high density of planar defects. (b–c) Scanning TEM (STEM) images of two separate grains showing the coexistence of HCP and FCC phases, as well as stacking faults (SFs) and twin boundaries (TBs) (Chen et al. 2020b).

The phase structure and stability for the HEA coatings are not only affected by their composition, but also the grain sizes and deposition conditions, making the phase transformation behaviors of HEA coatings more complex than those of their bulk counterparts.

6.3.3 Microstructure of HEC Coatings

HEC coatings, including high-entropy nitride, carbide, and oxide coatings, have garnered significant attention as protective coatings due to their superior properties. Most HEC coatings reported to date have been fabricated by reactive sputtering deposition or cathodic vacuum arc deposition.

6.3.3.1 High-Entropy Nitride Coatings

High-entropy nitride coatings are mainly composed of transition metals, but B, C, Si, and Al have also been included (Table 6.1). Most of the high-entropy nitride coatings are based on strong nitride-forming elements, such as Zr, Ti, Al, Ta, Nb, V, Cr, Hf, Mo, and Si, which tend to form binary nitrides (ZrN, TiN, VN, NbN, CrN, etc.) having a NaCl-type FCC structure. As shown in Figure 6.4, a high-entropy nitride is in fact a mixture of constituent binary nitrides, in which the metallic elements and N occupy the Na sites and Cl sites, respectively. Thus, as can be seen in Table 6.1, FCC phases with extensive solid solution nitrides can be formed for high-entropy nitride coatings containing strong nitride formers (Jansson and Lewin 2013). Although the FCC structure dominates the crystal structure of high-entropy nitride coatings, some other structures, such as amorphous and BCC phases, have also been observed. The first reported nitride coatings, (FeCoNiCrCuAlMn) N and (FeCoNiCrCuAl$_{0.5}$)N, exhibited an amorphous phase (Chen et al. 2004). Amorphous phases were also formed in FeCoNiCuVZrAl nitride coatings (Liu et al. 2013). The inclusion of a majority of weak-nitride forming elements, such as Fe, Cu, Mn, Ni, and Co, is proposed to be the main reason for the formation of amorphous nitrides (Yeh et al. 2004).

Generally, it is believed that high-entropy nitride coatings based on strong nitride forming elements tend to form a NaCl-type FCC phase, while coatings predominantly composed of weak nitride formers tend to form an amorphous phase; however, a few exceptions were observed. For example, NaCl-type nitride phases are formed in the (CoCrCuFeNi)N$_x$ system, in which all elements exhibit low nitride formability except Cr (Dedoncker et al. 2018). Amorphous phases were formed in coatings containing strong nitride forming elements-based nitride, such as NbTiAlSi and AlBCrSiTi nitride coatings (Sheng et al. 2016) (Tsai et al. 2012).

Deposition conditions, such as N$_2$ flow rate ratio, substrate bias voltage, and other deposition parameters play a significant effect on the microstructure of the high-entropy nitride coatings. N content, in particular, has a strong influence on the structure and mechanical behavior of high-entropy nitride coatings. However, the effects of N content on phase formation vary across different alloy

TABLE 6.1

Selected studies on high-entropy nitride coatings, sorted according to hardness values, showing compositions, fabrication methods, structures, hardnesses (*H*), elastic moduli (*E*), and other properties investigated in addition to mechanical response under nanoindentation

Composition	Method	Structure	H (GPa)	E (GPa)	Properties investigated	Reference
(TiVZrNbHf)N	Vacuum arc deposition	FCC	66	-	Thermal stability	(Firstov et al. 2014)
(TiHfZrVNb)N	Cathodic arc deposition	FCC	44.3	384	Oxidation and wear	(Pogrebnjak et al. 2014)
(AlCrNbSiTiV)N	Magnetron sputtering	FCC	41	360	-	(Huang and Yeh 2009a)
(AlCrMoTaTiZr)N	Magnetron sputtering	FCC	40.2	420	Wear	(Cheng et al. 2011a)
(AlCrNbSiTi)N	Magnetron sputtering	FCC	36.7	-	Oxidation	(Hsieh et al. 2013)
(AlCrTaTiZr)N	Magnetron sputtering	FCC	36	360	-	(Lai et al. 2006b)
(AlCrTaTiZr)N	Magnetron sputtering	FCC	35.2	-	Adhesion and wear	(Cheng et al. 2009)
(AlCrTaTiZr)N	Magnetron sputtering	FCC	35	350	-	(Lai et al. 2007)
(AlCrNbSiTiMo)N	Magnetron sputtering	FCC	34.5	228	Wear	(Lo et al. 2020)
(AlCrTiZrHf)N	Magnetron sputtering	FCC	33.1	347.3	Frication and wear	(Cui et al. 2020a)
(TiZrNbHfTa)N	Magnetron sputtering	FCC	32.9	-	Friction and wear	(Braic et al. 2012)
(AlCrTaTiZr)N	Magnetron sputtering	FCC	32	368	-	(Lai et al. 2006a)
(AlCrTaTiZrSi)N	Magnetron sputtering	FCC + Amorphous nanocomposite	30.2	258	-	(Lin et al. 2014)
(MoNbTaVW)N	Cathodic arc deposition and Magnetron sputtering	FCC	30	-	Electro-mechanical	(Xia et al. 2020)
(TiVCrZrHf)N	Magnetron sputtering	FCC	23.8	267.3	-	(Liang et al. 2011b)
(AlBCrSiTi)N	Magnetron sputtering	Amorphous	23	256.6	Thermal stability	(Tsai et al. 2012)
(AlCrSiTiZr)N	Magnetron sputtering	Amorphous	19.6	227.5	Corrosion	(Hsueh et al. 2012)
(FeMnNiCoCr)N	Magnetron sputtering	BCC	17.1	238.5	Wear and scratch	(Sha et al. 2020b)
(NbTiAlSiW)N	Magnetron sputtering	Amorphous	13.6	154	-	(Sheng et al. 2016)
(ZrTaNbTiW)N	Magnetron sputtering	FCC + BCC	13.5	178.9	-	(Feng et al. 2013)
(FeCoNiCuVZrAl)N	Magnetron sputtering	Amorphous	12	166	-	(Liu et al. 2013)
(AlCrMnMoNiZr)N	Magnetron sputtering	FCC	11.9	202	Wear	(Ren et al. 2013)
(FeCoNiCrCuAlMn)N	Magnetron sputtering	Amorphous	11.8	-	Resistivity	(Chen et al. 2004)
(FeCoNiCrCuAl$_{0.5}$)N	Magnetron sputtering	Amorphous	10.4	-	Resistivity	(Chen et al. 2004)

systems. In most of the high-entropy nitride coatings, transformation from an amorphous to a crystalline phase has been observed with increasing N content. For example, as shown in Figure 6.5a, the AlCrTaTiZr coatings exhibited an amorphous structure when deposited without N$_2$ flow (Lai et al. 2006a). With increasing N content, the phase structure of the (AlCrTaTiZr)N$_x$ coatings transformed from amorphous to a simple FCC solid solution structure (Lai et al. 2006a). Such a microstructural change is common in high-entropy nitride coatings (Chang et al. 2010, Liang et al. 2011b, Cheng et al. 2011a, Huang and Yeh 2009a, Ren et al. 2013, Zhang et al. 2018b, Cui et al. 2020a,

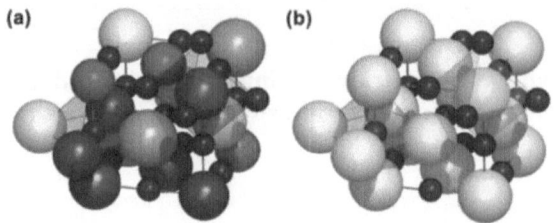

FIGURE 6.4 Schematic illustration of the atomic structure of the NaCl-type FCC crystal structure for (a) a high-entropy nitride containing five different metallic elements, in which the metallic elements and N occupy the Na sites and Cl sites, respectively, and (b) a binary nitride. Small atoms represent N; larger atoms represent metallic elements (Lewin 2020).

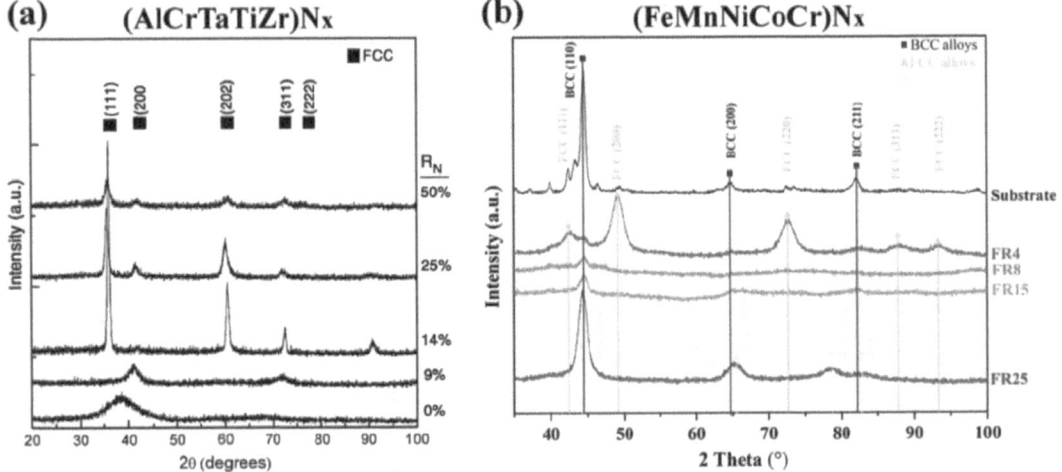

FIGURE 6.5 XRD patterns of multielement (a) AlCrTaTiZr nitride coatings (Lai et al. 2006a) and (b) FeMnNiCoCr nitride coatings deposited under different N$_2$ flow ratios (Sha et al. 2020b).

Pogrebnjak et al. 2016). This is because the formation of FCC structured binary nitrides is enhanced with increasing N$_2$ flow. In addition, the increased N$_2$ flow rate increases the mobility of the atoms, and the reduced sputtering frequency caused by reduced Ar ions slows down the deposition rate, ensuring the growth and diffusion of the grains to form a single solid solution phase rather than an amorphous phase.

With increasing N content, besides the above mentioned crystallization, a transformation from FCC to BCC structure (Figure 6.5b) was observed in (FeMnNiCoCr)N$_x$ coatings (Sha et al. 2020b). This transformation is mainly because the incorporation of N significantly increases the overall atomic mismatch and decreases the valence electron concentration (VEC) value, facilitating the formation of a BCC phase. In contrast, a phase transition from BCC to FCC with increasing N content was observed in MoNbTaVW, VAlTiCrMo and TiZrNbAlYCr nitride coatings (Pogrebnjak et al. 2018, Xia et al. 2020, Chen et al. 2020a). These coatings contain a mainly refractory elements, such as V, Cr, Nb, Ta, W, and Mo, which exhibit a BCC structure, therefore, without adding N, the BCC structure was observed. Since binary nitrides, such as TiN, ZrN, VN, and CrN, all exhibit FCC structures, the transformation from the BCC to FCC phase can be observed through the addition of N, so the nitride coatings exhibit an FCC structure.

Another important process parameter that can be tuned is substrate bias. Substrate bias significantly changes the microstructure and properties of nitride coatings, and in rare cases, it also affects their composition. Compared with coatings deposited without substrate bias, the grain sizes are smaller when a substrate bias is applied (Lo et al. 2020). With increasing bias voltage, ion bombardment promotes stronger resputtering effect, and generates a range of defects that can act as nucleation sites (Håkansson et al. 1987). The grain size is thus decreased owing to the increased number of nucleation sites. A distinct grain refinement from 70 to 5 nm was observed in $(Al_{1.5}CrNb_{0.5}Si_{0.5}Ti)$ N coatings when the applied substrate bias increased from 0 to 200 V (Shen et al. 2012). At the same time, the structure of the coatings evolved from a typical coarse columnar structure to a dense and featureless structure. In addition, a transition from a (111) to a (200) dominated texture was also observed with increasing applied bias due to the enhanced ion channeling effect (Shen et al. 2012). Such changes in grain size and texture are commonly seen in reactively sputtered nitride coatings (Huang and Yeh 2009b, Chang et al. 2008, Patsalas et al. 2004, Schell et al. 2002, Lai et al. 2006b). Changes in substrate bias generally do not lead to significant variations in composition. However, the Al content in the (AlCrTaTiZr)N coating was found to be lowered by increased bias voltage due to the higher sputtering yield of Al relative to the other elements (Lai et al. 2006b). The N content was found to be nearly independent of the substrate bias.

In addition to substrate bias, substrate temperature also can affect the structure of the high-entropy nitride coatings (Huang and Yeh 2009c, Chang et al. 2019, Liang et al. 2011a, Lai et al. 2007). Generally, grain size increases as the substrate temperature increases attributed to enhanced adatom mobility and surface diffusion (Huang and Yeh 2009c). Nevertheless, the opposite trend was observed in (AlCrTaTiZr)N coatings (Lai et al. 2007). In addition, the N content was observed to decrease with increasing substrate temperature due to the higher desorption rate of entrapped N atoms from non-equilibrium sites at the coatings surface at a higher substrate temperature (Ong et al. 1996).

6.3.3.2 High-entropy Oxide Coatings

In contrast to the large number of studies on high-entropy nitride coatings, high-entropy oxide coatings have received much less attention. The structure of high-entropy oxide coatings differs across various alloy compositions. $(ZnSnCuTiNb)_{1-x}O_x$ coatings with different O contents all exhibit an amorphous phase structure (Yu et al. 2011). Single-phase crystalline $(AlCrNbTaTi)O_2$ coatings were synthesized at 400 °C by magnetron sputtering (Kirnbauer et al. 2019). All coatings prepared showed a rutile-type solid solution phase, regardless of the relative O_2 flow-rate ratio used (between 30% and 80%). Similar to the effect of adding N, the addition of O also promotes a phase transformation from amorphous to a crystalline structure. For example, sputtered $AlCoCrCu_{0.5}NiFe$ coating showed an amorphous structure, while a HCP phase was formed when the O content in the working gas was between 10% and 50% (Huang et al. 2007). Similarly, crystallization was also observed when the $Ti_xFeCoNi$, $TiFeCoNiCu_x$, and $Al_xCrFeCoNiCu$ HEA coatings were oxidized and annealed (Tsau et al. 2012).

6.3.3.3 High-Entropy Carbide Coatings

(TiAlCrNbY)C films were deposited by reactive sputtering in $(Ar+CH_4)$ reactive atmospheres with different $CH_4/(CH_4 + Ar)$ flow rate ratios (Braic et al. 2010). The metals in high-entropy carbide coatings were found in an almost equiatomic ratio, whereas the C content varied from about 46 at.% to 82 at.%. A single FCC carbide phase of NaCl type with a (111) preferred orientation was observed for compositions close to stoichiometry, that is, C content ranging from 46 at.% and 55 at.%. However, an amorphous structure was found in coatings with higher C concentrations (69–82 at.%) due to the excessive C concentration, which exceeded the standard stoichiometric ratio.

6.3.4 Microstructure of HEA Coatings

Besides HEC coatings, HEA coatings have also been designed and fabricated to overcome the inherently negative properties of ceramic coatings such as brittleness, limited plasticity and chipping

TABLE 6.2

Selected studies on HEA coatings, sorted according to fabrication methods and hardness values, showing compositions, fabrication methods, structures, harnesses (H), and studied properties in addition to nanoindentation

Composition	Method	Structure	H (GPa)	Properties investigated	Reference
NbTaMoW	Ion beam sputtering	BCC	22.8	Nanoscratching	(Tunes and Vishnyakov 2019)
FeAlCuCrCoMn	Magnetron sputtering	FCC	17.5	Corrosion	(Li et al. 2016)
$Al_{2.5}CoCrCuFeNi$	Magnetron sputtering	BCC	15.4	-	(Wu et al. 2014)
$CoCrFeNiAl_{0.3}$	Magnetron sputtering	FCC + ordered BCC	12.3	-	(Liao et al. 2017)
CrCoNi	Magnetron sputtering	FCC + HCP	10	-	(Chen et al. 2018)
$CoCrFeNiCuAl_{2.5}$	Magnetron sputtering	BCC	9.2	Creep	(Ma et al. 2016)
FeMnNiCoCr	Magnetron sputtering	FCC + HCP	9.1	-	(Sha et al. 2020c)
CoCrFeNiCu	Magnetron sputtering	FCC	7.3	Creep	(Ma et al. 2016)
TiVCrAlSi	Laser cladding	BCC + $(Ti,V)_5Si_3$	9.5	-	(Huang et al. 2011)
$FeCrNiCoB_{1.25}$	Laser cladding	FCC + $(Fe, Cr)_2B$	7.9–8.7	Corrosion resistance	(Zhang et al. 2016)
6FeNiCoSiCrAlTi	Laser cladding	BCC	7.7	Magnetic properties	(Zhang et al. 2011c)
$CoCrNiFeAl_{1.8}Cu_{0.7}B_{0.3}Si_{0.1}$	Laser cladding	BCC + Cr_2B	7.5	Thermal stability	(Zhang et al. 2017b)
$CoCrFeNiAl_{0.3}Cu_{0.7}Si_{0.1}B_{0.6}$	Laser cladding	FCC	4.9	-	(He et al. 2017)
FeCoNiCrCu (with Si, Mn and Mo additions)	Laser cladding	BCC	4.4	-	(Zhang et al. 2011a)
CoCrCuFeNi	Laser cladding	FCC	4.2	Electric resistivity, Thermal stability	(Zhang et al. 2014a)
CoCrFeMnNbNi	Tungsten inert gas cladding	FCC + Nb-rich Laves phase	4.6–4.9	Wear	(Huo et al. 2015)

tendency. As can be seen from Table 6.2, due to the high entropy effect, HEA coatings tend to form simple FCC, BCC, FCC+BCC, and FCC+HCP solid solution structures, but intermetallic compounds may also form in some laser clad HEA coatings.

6.3.4.1 Effect of Alloying

Similar to bulk HEAs, most HEA coatings are composed of transition metal elements, among them, Co, Cr, Cu, Fe, Mn, Ni, Ti, and V are the most frequently used elements (Miracle and Senkov 2017). Elements of similar atomic size, such as Cr, Fe, Co, Ni, and Cu, can be referred to as base elements, since they tend to form simple FCC or BCC solid solution structures (Yan et al. 2018). Based on the desired properties for intended applications, functional elements can be added into these base elements. For example, refractory HEA coatings usually contain the refractory elements with high melting temperatures and excellent thermal stability, for example, Cr, Hf, Mo, Nb, Ta, Ti, V, W, and Zr, plus non-refractory elements such as Al and Si for property enhancement and density reduction.

Al is a common alloying addition to HEA coatings as it can enhance many properties. Increasing Al content in HEA coatings promotes the FCC to BCC phase transition due to the increased lattice distortion and the lower packing fraction of the BCC lattice required to accommodate the larger Al atoms. For example, it was found that the equiatomic CoCrCuFeNi coating without Al exhibits an FCC structure, but the addition of Al, for example the $Al_{33.35}Co_{13.33}Cr_{13.33}Cu_{13.33}Fe_{13.33}Ni_{13.33}$ coating, presents with a BCC structure (Wu et al. 2014). Nb is another common element used to improve the mechanical properties of HEA coatings. It was reported that the plasma cladded $FeCoCrNiNb_x$ coatings were composed of FCC and BCC phases. The relative content of BCC and FCC phases varied with the Nb concentration (Fang et al. 2018).

Besides the addition of elements with large atomic radii, such as Al and Nb, the addition of small interstitial elements, such as B and C, can also affect the structure and properties of HEA coatings. With the addition of B into HEA coatings, a boride phase will form along with simple solid solution FCC and BCC phases (Zhang et al. 2016, Liu et al. 2020). However, it was found that the CoCrNiF $eAl_{1.8}Cu_{0.7}B_{0.3}Si_{0.1}$ coating prepared by laser cladding was mainly composed of a BCC solid solution phase, suggesting that the boride formation can be effectively inhibited by laser cladding (Zhang et al. 2017b). Different from the FCC structure formed in the FeMnCoCr HEA, the addition of C in FeMnCoCrC coatings promotes a novel α-Mn crystal structure with highly refined columnar grains (Sha et al. 2020a).

The existing phase-formation rules are mainly based on conventional alloys with one or two principal components. Recently, in order to predict the formation of the HEA phase, phase-formation rules for HEAs were proposed. It was found that the atomic-radius mismatch (δ) and enthalpy of mixing (ΔH_{mix}) could be used to predict the formation of solid solutions, that is, without the formation of compounds in HEAs. Solid solutions tend to form when $-15 KJ/mol \leq \Delta H_{mix} \leq 5 KJ/mol$ and $1\% \leq \delta \leq 6\%$, while compounds were more likely to form when ΔH_{mix} is more negative and δ is larger (Guo et al. 2011, Zhang et al. 2008). The FCC structure is stable at VEC ≥ 8.0 and the BCC structure is stable at VEC < 6.87 and mixed FCC and BCC phases will co-exist at 6.87 ≤ VEC < 8.0 (Figure 6.6) (Guo et al. 2011). However, it was suggested that the VEC-defined FCC and BCC phase boundary was less satisfactory in predicting structure for Mn-containing HEA systems. It is important to note that these phase-formation rules were originally formulated for bulk HEAs. Considering the differences between the phase structure and stability of HEA coatings and their bulk counterparts, as discussed in Section 6.3.2, these phase-formation rules may not be valid for all the HEA coatings.

6.3.4.2 Effect of Processing Parameters

The phase structure in HEA coatings can also be affected by process parameters during deposition. It was found that in dual phase (FCC and HCP) FeMnNiCoCr coatings, the fraction of the HCP phase was found to be greater in coatings deposited at a more negative bias voltage (Sha et al. 2020c), because the low stacking fault energy and the enhanced resputtering effects at higher

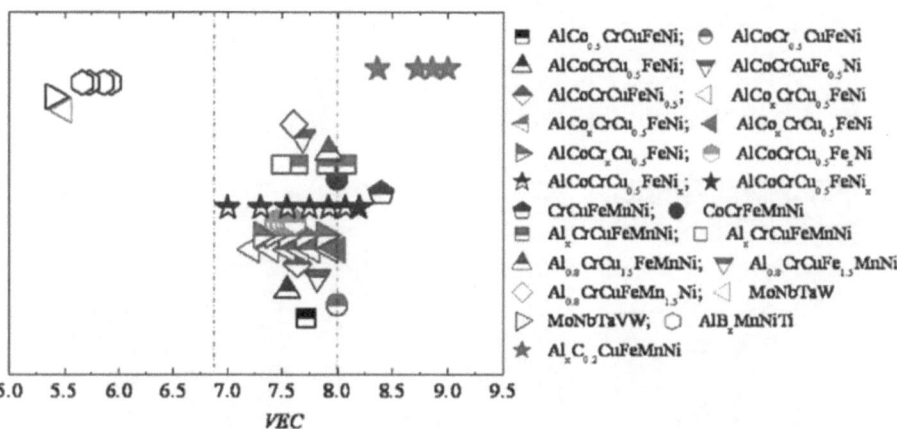

FIGURE 6.6 The dependence of the FCC and BCC phase stability on the VEC values for representative HEA systems (Guo et al. 2011). Fully closed symbols are for single FCC phases; fully open symbols are for single BCC phase; half-closed symbols are for mixed FCC and BCC phases.

bias voltages promote the formation of stacking faults. Another phase transformation from BCC to FCC was revealed in equiatomic AlFeCoNiCuCrV HEA coatings produced by magnetron sputtering when the substrate bias changed from −100 V to −200 V (Shaginyan et al. 2016).

6.3.4.3 Composite Structures

Apart from the above-mentioned HEA coatings, HEA composite coatings have also emerged to further enhance the properties and broaden the applications of HEA coatings. For example, as can be seen in Figure 6.7, a multilayered CoCrNi/Ti composite (Cao et al. 2018b) was deposited via magnetron sputtering. Interestingly, different crystal structures were identified for the different CoCrNi layers. The outermost CoCrNi layer exhibited an FCC structure, while in contrast, both the middle and bottom CoCrNi layers exhibited a BCC structure (Figure 6.7c-e). In addition, the vertically aligned columnar CoCrNi grains contained a high density of nanotwin boundaries. The BCC CoCrNi phase, observed in both the middle and bottom layers, may result from an FCC to BCC phase transformation caused by compressive stress, which is absent in the outmost CrCoNi layer since there is no subsequent ion bombardment effect.

Inspired by the design strategy of Si-containing nitride coatings with nanocomposite structures, crystal–glass HEA nanocomposites (Wu et al. 2020) were designed and synthesized by magnetron sputtering. The crystal–glass HEA nanocomposites were obtained by doping a glass-forming Fe–Si–B system into a crystalline CrCoNi base alloy to form CrCoNi–Fe–Si–B nanocomposites. The nanocomposite exhibited an FCC structured nanocolumnar structure, together with an amorphous phase with an average thickness of ~1 nm formed between the FCC nanograins. The amorphous phase was enriched in Cr. The high content of Cr and B as well as C, O, and Si dopants act synergistically to enhance the glass forming ability (Lu et al. 2000) of the coatings, resulting in the formation of an amorphous phase.

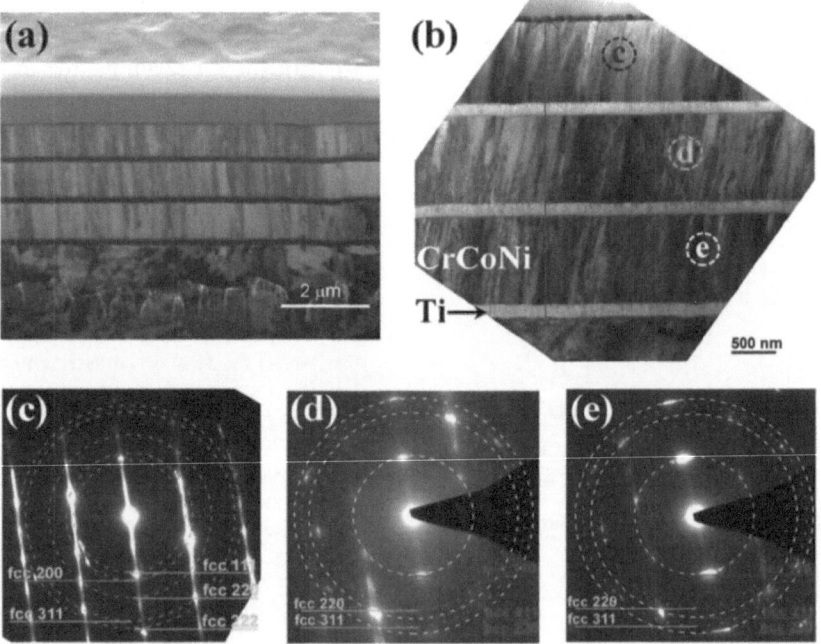

FIGURE 6.7 (a) Cross-sectional FIB image and (b) cross-sectional TEM image of the multilayer CrCoNi/Ti coating; (c–e) SAED patterns for each CrCoNi layer from the circled areas in (b) from top to bottom (Cao et al. 2018b).

6.4 MECHANICAL PROPERTIES

Compared with traditional alloy films, HEA-based coatings exhibit excellent mechanical properties, such as high hardness and strength, good damage tolerance and wear resistance, which are among the most important mechanical properties required for protective coatings.

6.4.1 HARDNESS

At a fundamental level, hardness is the capability of a material to resist plastic deformation, usually measured by indentation methods, in particular for thin films by nanoindentation. Hardness is the most important, and most commonly studied, property of protective coatings. It has been demonstrated that the hardness of some high-entropy nitride coatings can reach the superhard grade, that is, greater than 40 GPa (Cheng et al. 2011a, Huang and Yeh 2009a, b, Sheng et al. 2016, Firstov et al. 2014), higher than those of ternary nitrides and even comparable with multilayered or nanocomposite films, making high-entropy nitrides become a promising alternative to superhard materials. Among the existing high-entropy coatings, the highest hardness value of 66 GPa was determined from (TiVZrNbHf)N coatings which exhibited a simple FCC solid-solution phase (Firstov et al. 2014).

As can be seen from Table 6.1, the hardness values of high-entropy nitride coatings ranging from ~10 to 66 GPa. The properties of the high-entropy nitride coatings presented in Table 6.1 are arranged according to their hardness values, with the hardest coatings listed on top. Clearly, the hardness values of these high-entropy nitrides are closely related to their structure. Generally, high-entropy nitride coatings with single FCC structures show higher hardness ($\geq 30\,\text{GPa}$), while the coatings with other structures, including amorphous, BCC, and FCC + BCC structures, present with comparatively lower hardness ($\leq 20\,\text{GPa}$). As discussed in Section 6.3, an amorphous phase tends to form in high-entropy nitride coatings that contain a majority of weak-nitride-forming elements, while high-entropy nitride coatings composed of the strong nitride elements tend to form a solid solution of binary nitrides with the NaCl-type FCC structure. Therefore, high-entropy nitrides consisting of strong nitride forming elements, can more readily achieve higher hardness values (Firstov et al. 2014, Pogrebnjak et al. 2014, Huang and Yeh 2009a, Cheng et al. 2011a). The low hardness of the high-entropy nitrides can be attributed to the incorporation of weaker nitride formers (Lai et al. 2006a, Lai et al. 2007), leading to a reduced number of strong metal–nitrogen bonds and the formation of a softer amorphous structure. Therefore, the design of high-entropy nitrides for high hardness should avoid the inclusion of weak nitride-forming elements.

The hardness of high-entropy nitride coatings is also dependent on the N content. As can be seen from Figure 6.8, with increasing N_2 flow rate, the hardness of the $(AlCrTiZrHf)N_x$ coatings increases from 17.9 to 33.1 GPa (Cui et al. 2020a). This increase in hardness is mainly attributed to the formation of saturated metal nitride phases as well as solid solution strengthening. However, there exists a maximum hardness at an optimal N_2 flow rate ($N_2:Ar = 5:4$), beyond which further increases in N_2 flow rate will result in a hardness drop. Therefore, optimizing the N_2 flow rate is critical to achieve desired mechanical properties.

The attainment of both high hardness and high toughness is a vital requirement for protective coatings. Ceramic coatings possess ultrahigh hardness compared to metallic counterparts; however, they are often brittle and susceptible to catastrophic failure, and thus unsuited for safety-critical applications. HEA coatings with innovative compositional and microstructural designs are emerging as a new class of coating materials with potential to unite hardness and toughness. For example, the NbTaMoW refractory HEA coating can retain a high hardness of ~22.8 GPa, comparable with most conventional ceramics, and at the same time exhibit impressive mechanical damage tolerance (Tunes and Vishnyakov 2019). As shown in Table 6.2, high hardness values, ranging from ~4 GPa to 22.8 GPa, have been achieved in HEA coatings prepared by different techniques.

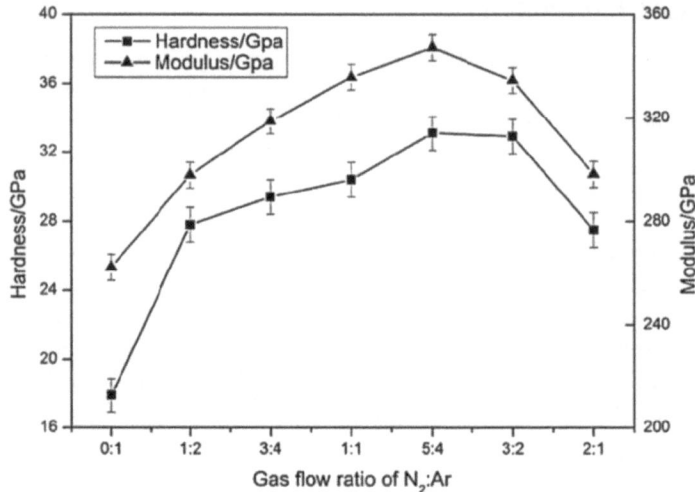

FIGURE 6.8 The dependence of the hardness and elastic modulus on the N_2 flow rate of $(AlCrTiZrHf)N_x$ high-entropy nitride coatings (Cui et al. 2020a).

6.4.2 Hardening Mechanisms

Several major hardening mechanisms are active in HEA-based coatings: (1) solid solution hardening, (2) grain refinement, (3) grain boundary reinforcement, (4) planar defect hardening, and (5) precipitation hardening.

6.4.2.1 Solid Solution Hardening

Solid solution hardening occurs by inserting atoms of an alloying element into the interstitial or substitutional positions of the solvent atoms, forming a solid solution. Such alloying can be used to improve strength, since it can impede the motion of dislocations due to a local elastic strain fields around the solute. In HEA-based coatings, this principle works in the same way. For example, substitutional solid solution hardening was realized in an $Al_xCo_{1.5}CrFeNi_{1.5}Ti_y$ coating by the addition of elements with large atomic radii, such as Al and Ti (Chuang et al. 2011). For $Al_xCoCrCuFeNi$ coatings, the hardness increased from 6.3 GPa (Al-0) to 15.4 GPa (Al-2.5) with the addition of Al (Wu et al. 2014). In addition to substitutional solid solution hardening, an FCC to BCC phase transformation occurred with the increase in Al content, which also contributed to hardening. However, it was suggested that the increased lattice strain and defects caused by the addition of elements into HEA-based coatings with large atomic radii can often lead to high brittleness and reduced ductility and toughness (Yang et al. 2015).

Recently, some studies reported that interstitial solid solution hardening, by the addition of small elements such as B or C, can also be used to improve the mechanical properties of HEA-based coatings. For example, a high hardness of ~12.5 GPa was achieved in $FeMnCoCrC_{0.5}$ coatings (Sha et al. 2020a). The hardness of a $FeCoCrNiB_x$ coating was enhanced with increasing B content, and the maximum hardness reached around 10 GPa (Liu et al. 2020). With the addition of B, an alloy boride phase may form along with simple FCC or BCC solid solutions (Zhang et al. 2016, Liu et al. 2020). However, it was found that, during the laser cladding process, some B atoms formed borides while other B atoms acted as interstitials and became dissolved in the FCC solid solution. The rapid solidification conditions during laser cladding increased the limit of solid solubility of the B atoms in the solid solution (Liu et al. 2020). Thus, more B atoms were retained in the solid solution, making the interstitial solid solution strengthening more effective in HEA-based coatings than in bulk HEAs.

6.4.2.2 Hardening Via Grain Refinement

According to the Hall–Petch relationship, the strength and hardness values of a material are inversely proportional to its grain size (Chawla and Meyers 1999). The higher hardness achieved in HEA coatings, relative to their bulk counterparts, can be partially attributed to their small grain sizes, because the grain boundaries can impede dislocation motion. For example, an equiatomic FeMnNiCoCr coating with a nanograined microstructure prepared by magnetron sputtering exhibited a high hardness of ~6.8 GPa (Dang et al. 2018), much higher than ~2.8 GPa measured in its single-phase bulk counterpart (Yang et al. 2019). The hardness of $(Al_{1.5}CrNb_{0.5}Si_{0.5}Ti)N$ coatings (Shen et al. 2012) increased from 12 to 36 GPa as the grain size decreased from ~70 to ~5 nm, caused by increased substrate bias from 0 to −100 V. It was revealed that the grain boundary strengthening efficiency of CrCoNi-based HEAs is significantly larger than conventional FCC alloys (Zhu et al. 2018), thus, grain refinement is an effective strategy to enhance the hardness of HEA coatings.

6.4.2.3 Hardening via Grain Boundary Reinforcement

As mentioned above, the Hall–Petch effect governs hardening via grain refinement. However, an inverse Hall–Petch effect has been observed for nanocrystalline materials (Schiøtz et al. 1998). This effect implies that nanocrystalline materials get softer as grain size is reduced below a critical value, mainly due to the onset of grain boundary sliding (Schiøtz et al. 1998). To overcome this size limitation in hardening, a novel nanocomposite structure of nanocrystallites separated by thin amorphous layers was proposed, since the amorphous layer eliminates the grain boundary plasticity of the nanocrystallites. Si-containing ternary or quaternary nitride coatings, such as TiSiN and TiSiCN, with nanocomposite structures have gained considerable attention (Ma et al. 2007, Kong et al. 2007).

Recently, Si-containing high-entropy nitride coatings were also synthesized with an expectation of acquiring a nanocomposite structure with enhanced mechanical properties. The nanocomposite structure was successfully formed in $(AlCrTaTiZr)Si_xN$ coatings produced by magnetron sputtering (Cheng et al. 2011b, Lin et al. 2014). However, unlike the ternary Si-containing nitrides, the hardness of the $(AlCrTaTiZr)Si_xN$ coatings showed less dependence on Si content. Moreover, the formation of SiN_x amorphous phase did not result in any hardening effects. It worth noting that, due to the high entropy effect, the Si_3N_4 phase in the high-entropy nitride coatings exhibits a relatively high solubility, so they do not appear as sharp interfaces between the crystalline phases in the $TiN/a-Si_3N_4$ systems (Kong et al. 2007). Thus, there is no extra strengthening benefit from the formation of amorphous SiN_x. In contrast to the success in $(AlCrTaTiZr)Si_xN$ coatings, no nanocomposite structures were observed in (NbTiAlSi)N, (NbTiAlSiW)N, and (AlCrNbSiTiV)N coatings (Sheng et al. 2016, Huang and Yeh 2009a), because of the extended solubility of Si caused by the high entropy effect.

Although the results of hardening via grain boundary reinforcement in high-entropy nitride coatings were disappointing, a crystal–glass CrCoNi–Fe–Si–B HEA nanocomposite coating (Wu et al. 2020) with near theoretical compressive strength and large deformability was successfully designed and synthesized by magnetron sputtering. The high strength was mainly attributed to a synergistic strengthening mechanism including the crystal–glass composite structure, interface-dislocation interactions, and the nanoscale dimensions of the crystalline and amorphous phases.

6.4.2.4 Hardening via Planar Defects

The strengthening effect of planar defects, including twin boundaries and stacking faults, on the mechanical properties of materials has been widely investigated and reported (Lu et al. 2009a, Chen et al. 2016). It has been shown that when a high density of twin boundaries was introduced into ultrafine grains, metals can be significantly strengthened while retaining considerable plasticity and exhibiting strong work hardening (Lu et al. 2009b). These results imply that interface boundary engineering can lead to robust mechanical performance for HEA coatings. The presence of nanotwins is a common feature in HEA coatings with low stacking fault energies (Chen et al. 2018, Cao et al. 2018a, b, Tsianikas et al. 2020, Feng et al. 2018). It has been shown that FeMnNiCoCr coatings with nanotwinned structures achieve exceptional hardness values between 8.5 and 9.25 GPa

(Sha et al. 2020c, Wang et al. 2020), which is much higher than that of the equiaxed nanograined CoCrFeMnNi coatings (6.8 GPa) (Dang et al. 2018). The hardness of these coatings was affected by twin spacing. By decreasing the twin spacing from 2.2–5.6 nm to 1.2–2.5 nm, the hardness of CoCrFeMnNi coatings reduced from 9.25 to 8.5 GPa (Wang et al. 2020), because the ultrathin nanotwins with a thickness of ~2 nm undergo detwinning during plastic deformation, resulting in a weakening effect (Feng et al. 2018, Chen et al. 2020b). In addition to twin boundaries, a high density of stacking faults was also observed in these coatings when deposited by magnetron sputtering (Chen et al. 2018, Feng et al. 2018). Stacking faults are as effective as twin boundaries in obstructing dislocation motion and causing hardening, but stacking faults are more stable relative to the observed detwinning of ultrathin nanotwins (Feng et al. 2018).

6.4.2.5 Precipitation Hardening

HEA-based coatings can also be hardened by the formation of fine precipitates, which impede the movement of dislocations, the dominant carriers of plasticity, by particle shearing or Orowan bypassing mechanisms. Precipitations are usually observed in HEA-based coatings prepared by laser cladding and spraying. A high hardness of ~12.3 GPa was achieved in the $Al_{2.0}CoCrFeNiSi$ HEA coating, mainly attributed to the formation of hard intermetallic Cr_3Si precipitates (Jin et al. 2020). The microhardness of the laser clad refractory TiZrNbWMo coating increased remarkably from 6.9 GPa (as-clad coating) to about 12.8 GPa after heat treatment at 800 °C due to the formation of hard β-Ti_xW_{1-x} precipitations (Zhang et al. 2017c). Recently, elements in the CrFeCoNiPd HEA tend to show periodic aggregation patterns with a wavelength of incipient concentration waves as small as 1–3 nm (Ding et al. 2019). Such a concentration wave results in nanoscale alternating tensile and compressive strain fields, which can effectively resist dislocation movement, resembling precipitation hardening. This unique concentration wave strengthening in HEAs has not yet been studied in HEA-based coatings. Introducing the concentration wave to HEA-based coatings is expected to further enhance the mechanical strength.

6.4.3 Toughness

Toughness is a critical property for protective coatings (Wang and Zhang 2014). However, due to the limitations of specimen size and instrumentation, it is difficult to quantitatively measure the toughness of thin films and coatings, thus, in some studies, excellent damage tolerance was related to a high toughness. The toughness of a (AlTaTiVZrSi)N (Hahn et al. 2019) coating was assessed by bending tests of a notched single cantilever beam using micro- or nanomechanical testing methods (Hahn et al. 2019). It was found that with the addition of 4.9 at.% Si to a (AlTaTiVZr)N coating (Hahn et al. 2019), the hardness and fracture toughness remain unchanged, but the modulus was significantly decreased, leading to ~30% higher strain-to-failure and increased absorbed energy during the generation of cracks. By using a similar method, the toughness of the NbMoTaW coatings (Xiao et al. 2019), deposited by magnetron sputtering and Ar+-ion beam assisted deposition, was measured to be 3.3 $MPa\sqrt{m}$ and 2.9 $MPa\sqrt{m}$, respectively, which are values higher than that of TiN coatings (Bartosik et al. 2017) and nanolayered TiN/CrN thin films (Hahn et al. 2016). The HEA-based coatings with a combination of smart microstructural and compositional design can achieve both high hardness and damage tolerance at the same time, allowing them to be considered as a hard coating candidate for future applications in extreme environments.

6.4.4 Toughening Mechanisms

For protective coatings, toughness is as important as, if not more than, hardness. However, it is conventionally accepted that hardness and toughness are often mutually exclusive in metallic systems. Design of tough HEA-based coatings can be realized via (1) introducing a toughening agent, (2) inducing compressive stresses, (3) phase transformation, and (4) bio-inspired toughening.

6.4.4.1 Toughening via Introducing a Toughening Agent

Incorporating a ductile phase into a coating is the most straightforward route in toughening hard ceramic coatings. The crack resistance of (AlCrNbYZr)N (Fieandt et al. 2020) coatings was enhanced by increasing the N content (Figure 6.9). Nitride coatings with 50–51 at.% N not only show a high hardness, but also a high resistance to cracking since no evidence of crack formation could be observed (Figure 6.9d), which can be related to a high toughness. Such a toughening effect is not generally expected for transition metal nitride materials, and was attributed to the formation of a ductile metallic phase as a toughening agent in the columnar boundaries caused by partial elemental segregation mainly of Y.

In addition to the incorporation of ductile metallic phases in ceramic coatings, a HEA coating was toughened by introducing inert-gas Ar bubbles. Recently, a high hardness ~22.8 GPa was achieved in refractory NbTaMoW thin coatings with simple BCC structures (Tunes and Vishnyakov 2019). As shown in Figure 6.10, the cracking of the NbMoTaW coating was less catastrophic than that observed for TiN coatings, which are prone to crack in an uncontrolled manner during nano-scratching tests, indicating the superior crack resistance of the NbMoTaW coating. Ar bubbles, which can be viewed as clusters of vacancies and Ar atoms, with diameters around of 1.3 ± 0.4 nm were observed in this HEA coating. The origins of such high damage tolerance were attributed to dislocation pinning by Ar bubbles in combination with well-known HEA core effects.

6.4.4.2 Compressive Stress Toughening

Open surfaces of cracks are generally initiated by the presence of tensile stresses. Therefore, toughness will be likely increased if a compressive residual stress is induced in the coatings. A compressive

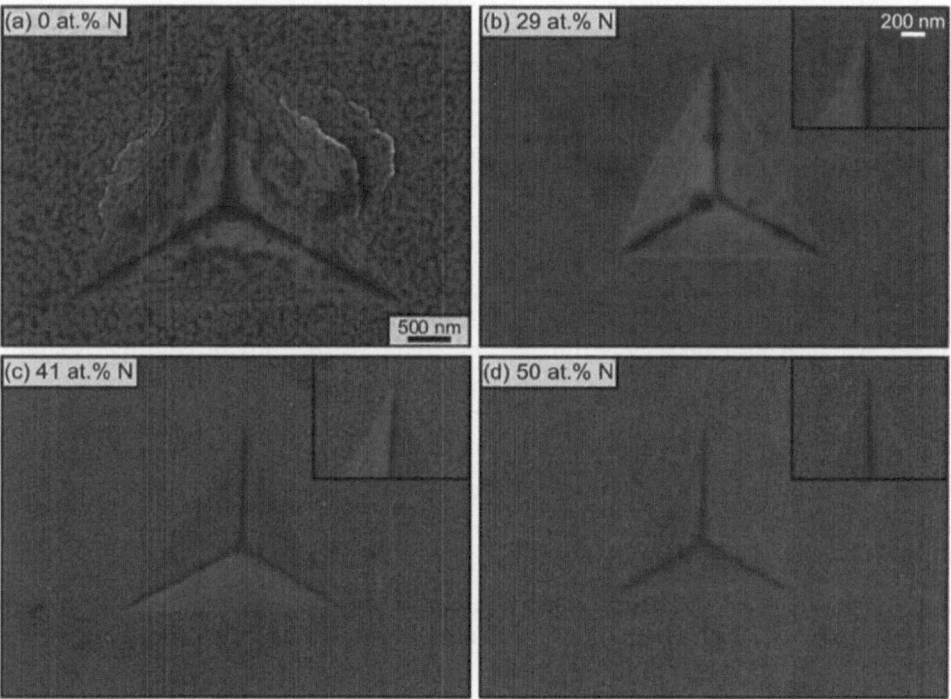

FIGURE 6.9 Scanning electron microscopy (SEM) plan view images of indents from a Berkovich indenter under a load of 50 mN for AlCrNbYZrN coatings with different N content: (a) 0 at.% N, (b) 29 at.% N, (c) 41 at.% N, and (d) 50 at.% N. Scale bars in (a) and in the inset in (b) are valid for all the images and insets (Fieandt et al. 2020).

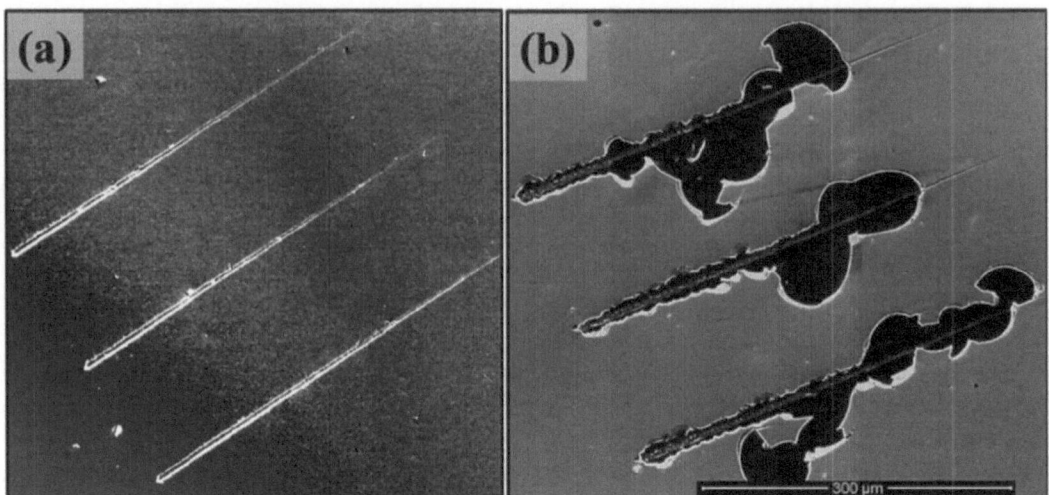

FIGURE 6.10 SEM images of the nanoscratching traces obtained by an old Berkovich indenter under linearly ramped loads up to 500 mN for the as-deposited (a) NbMoTaW and (b) TiN coatings. The scale bar in (b) also applies to (a) (Tunes and Vishnyakov 2019).

residual stress in HEA-based coatings can be introduced by applying a substrate bias during deposition. The toughness of high-entropy nitride coatings was found to be improved by the application of a substrate bias. Compared with the (AlCrTaTiZr)N coating deposited at a substrate bias of 0 V, the coating deposited at −150 V showed insignificant crack formation under indentation loading, indicating a greatly improved toughness. This is because the high residual compressive stress that can effectively inhibit crack initiation (Lai et al. 2006b). These results indicate that high-entropy nitride coatings can exhibit high toughness or damage tolerance, which can be further improved by strategies such as alloying or applying a substrate bias.

6.4.4.3 Phase Transformation Toughening

Phase transformations consume a large amount of energy due to the dimensional variations between crystalline structures, so it can be considered to be an effective toughening mechanism. High hardness and damage tolerance was achieved in dual phase (HCP and FCC) nanotwinned CrCoNi coatings with a low stacking fault energy through combining the benefits of both metastability and boundary engineering (Chen et al. 2020b). Besides the contributions from nanotwins, high damage tolerance originated from a deformation-induced phase transformation from HCP to FCC, through which local stress peaks could be effectively relaxed, inhibiting crack initiation and resulting in extra deformability.

6.4.4.4 Bio-Inspired Toughening

Biological materials have evolved over millions of years through natural selection and optimization. Nature uses its own magical ways to produce strong, yet tough materials with complex, hierarchical microstructure. Mankind has much to learn from nature on how to toughen engineering materials. Inspired by the unique brick-and-mortar structure of nacre, which combines hard inorganic phases with softer organic matrix to achieve high toughness, a multilayered CoCrNi/Ti composite coating was deposited via magnetron sputtering (Cao et al. 2018b). Exceptional damage tolerance was observed in this multilayered coating due to the synergetic deformation of hard CrCoNi layers and soft Ti layers. Toughening via bio-inspired structures has not been extensively studied in HEA-based coatings. Coating deposition methods, such as magnetron sputtering has proved to be effective in the precision control of microstructural features over multiple length scales. With the ability to fine-tune the microstructure development, bio-inspired toughening could be realized in HEA-based coatings and is thus worthy being further researched.

6.4.5 Wear Resistance

For industrial machinery, wear damage can dramatically increase maintenance and replacement costs, so wear resistance is critical for protective coatings. The wear resistance (W) of a hard coating has a strong scaling relationship with its toughness (K_{IC}), generally following the relation $W \sim K_{IC}^4$ (Miserez et al. 2008). Even a moderate increase of toughness can significantly improve wear resistance. As mentioned above, HEA-based coatings can achieve high hardness and toughness, so excellent wear resistance can be expected. The wear resistance of HEA-based coatings can be influenced by composition, substrate bias and substrate temperature.

The wear rate of coatings is inversely proportional to hardness, so increasing the N content in high-entropy nitride coatings can result in higher wear resistance due to the higher coating hardness. For example, with more N present in $(FeMnNiCoCr)N_x$ coatings, the coatings become harder, stiffer, and more wear resistant (Sha et al. 2020b). Moreover, as the N_2 flow rate increases, the friction coefficient of (AlCrTiZrHf)N coatings decreases, mainly due to the formation of the nitride phase, which has a low friction coefficient (Cui et al. 2020a). The addition of Al can improve the wear resistance of HEA-based coatings at high temperature. The wear losses of the cladded $FeCoCrNiMnAl_x$ coatings were found to be significantly decreased by increasing Al content (Cui et al. 2020b). The $FeCoCrNiMnAl_{0.75}$ coating shows the lowest wear loss, only ~17% that of the steel substrate The inclusion of Mo can also reduce the coefficient of friction and the wear rate of HEA-based coatings due to the formation of the lubricious MoO_3 during the sliding process (Cheng et al. 2011a). Compared with conventional coatings, the inclusion of elements in HEA-based coatings that can enhance tribological properties due to their increased solid solubility, can lead to excellent wear resistance.

The tribological properties of HEA-based coatings can also be optimized by controlling residual stress via substrate bias tuning. It was found that the wear rates of (AlCrTaTiZr)N and (AlCrNbSiTiMo)N coatings reduced with increasing the substrate bias (Lai et al. 2008, Lo et al. 2020). However, it should be noted that the highest wear resistance does not correspond to the highest applied bias voltage. The (AlCrNbSiTiMo)N coating deposited at −100 V exhibited the lowest wear rate, while the −150 V and −200 V coatings exhibited a relatively higher wear rate (Figure 6.11). Applying the appropriate substrate bias voltage during deposition can induce denser structures, larger compressive residual stresses and a higher hardness value, thus resulting in enhanced wear resistance. However, it is argued that if the applied bias voltage is too high, excessive residual stresses will promote a higher wear rate. During the wear test spallation readily induced by the applied load on coatings with high compressive stresses coatings can result in deep valleys in the wear track and thus a high wear rate. Therefore, the compressive residual stress needs to be controlled to enhance mechanical strength without promoting spallation.

Along with substrate bias, the temperature of the substrate also effects the tribological properties of HEA-based coatings. By increasing the substrate temperature above room temperature, the wear resistance of $(Cr_{0.35}Al_{0.25}Nb_{0.12}Si_{0.08}V_{0.20})N$ coatings was increased due to both the increased hardness and the elimination of vacancies and/or voids from the coating (Lin et al. 2020). Wear resistance can be further improved by depositing the coating with an applied substrate bias at high substrate temperatures. The highest wear resistance was revealed when a coating was deposited at a substrate temperature of 300 °C and a bias voltage of −150 V due to the combined effect of substrate bias and temperature due to both increased densification and compressive residual stress level.

Compared with traditional nitride coatings, high-entropy nitride coatings have excellent tribological properties. Low wear rates at room temperature were obtained for various high-entropy nitride coatings, ranging from 2×10^{-6} to 3.65×10^{-6} mm^{3}N^{-1}m^{-1} (Cheng et al. 2011a, Lai et al.

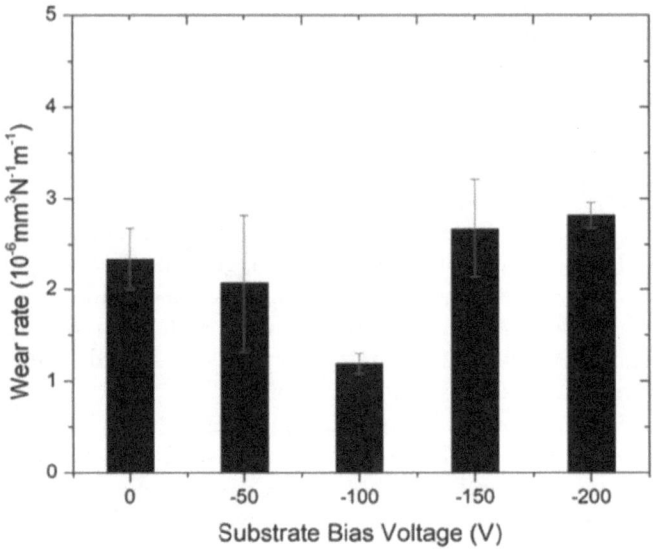

FIGURE 6.11 The relationship between the wear rate and the substrate bias voltages of the AlCrNbSiTiMoN coatings (Lo et al. 2020).

2008, Lin et al. 2020), values much lower than that of TiN and TiCuN (Öztürk et al. 2008). In addition, a $(Cr_{0.35}Al_{0.25}Nb_{0.12}Si_{0.08}V_{0.20})N$ coating possessed a wear rate of 5.4×10^{-6} $mm^3N^{-1}m^{-1}$ under high-temperature testing (600 °C), lower than that for traditional binary and ternary nitrides, and even quaternary nitrides of CrAlSiN and TiAlSiN (Lin et al. 2020). In a wear test at 700 °C, a (AlCrNbSiTiMo)N coating deposited under a substrate bias of -100 V exhibited an even lower wear rate $\sim 1.2 \times 10^{-6}$ $mm^3N^{-1}m^{-1}$. The high hardness, high H^3/E^2 ratio (a reliable indicator to evaluate its anti-wear capability), lower friction coefficient, and good oxidation resistance were the reasons for the relatively low wear rate.

6.5 CORROSION RESISTANCE

HEA-based coatings not only feature excellent mechanical properties, but also can show good corrosion resistance in hostile environments. HEA-based coatings have outstanding resistance to corrosion in saltwater (3.5 wt.% NaCl) compared with most corrosion-resistant alloys and even their bulk HEA counterparts (Shi et al. 2017). The excellent corrosion resistance of HEA-based coatings can be mostly attributed to their single solid solution or amorphous phase structures with highly uniform compositions. The inclusion of elements that enhance corrosion resistance, such as Al and Si, can also lead to further improvements in corrosion performance.

As mentioned in Section 6.3, a single solid-solution phase or amorphous phase tends to form in HEA-based coatings. Amorphous structures are well known to be more corrosion resistant than the crystalline structures, because grain boundaries are chemically unstable. The formation of a single solid solution phase and homogeneous elemental distribution are also important for achieving excellent corrosion resistance. FeAlCoCuNiV and FeAlCuCrCoMn coatings both exhibit dense single phase FCC solid solution structures with minimal elemental segregation, leading to improved corrosion-resistant ability than 201 stainless steel (Dou et al. 2016, Li et al. 2016). An equimolar CrMnFeCoNi coating prepared by laser cladding exhibited a good corrosion resistance. Local elemental segregation led to the formation of Cr-depleted interdendrites that acted as a starting point for corrosion reactions (Ye et al. 2017). The formation of passive films, mainly composed of Cr^{3+} oxide or hydroxide, plays an important role in enhancing the corrosion resistance of the Cr-rich phase. Due to the high stability of Cr^{3+} hydroxide in aqueous solution, the higher Cr content phase, the greater corrosion resistance the alloy would achieve (Yen et al. 2019). $Al_xFeCoNiCuCr$ coatings exhibit better corrosion resistance than that of the bulk alloy, which can be attributed to the relatively homogeneous elemental distribution in these coatings (Ye et al. 2011).

Alloying additions influence the corrosion resistance of HEA-based coatings. Elements that promote corrosion resistance such as Cr, Ni, Co, Cu, Al, and Ti are commonly included in the HEA-based coatings, and this promotes the formation of stable protective surface films resulting in good corrosion resistance (Ye et al. 2011, Qiu et al. 2014). Increases in N concentration in the high-entropy nitride coatings also improve corrosion resistance due to the presence of chemically inert metal–nitrogen bonds (Hsueh et al. 2012). Generally, HEA-based coatings exhibit superior corrosion resistance, so they may find application as wear-resistant protective coatings in environments where corrosion occurs, such as in the petrochemical or marine industries.

6.6 OXIDATION RESISTANCE AND THERMAL STABILITY

Oxidation resistance and thermal stability are critical properties of high-temperature coatings. For example, high-speed cutting without use of a cutting fluid can lead to excessive heat at the tool-chip interface. Thus, protective hard coatings with excellent oxidation resistance and thermal stability are beneficial to improving performance.

Conventionally, alloying additions is the most common strategy used to achieve improvements in oxidation resistance. HEA-based coatings have shown great potential in this regard due to their large solid solubility and compositional freedom. Generally, excellent oxidation resistance has been observed in HEA-based coatings containing elements such as Al, Si, Cr, Ta, and Ti, that can promote the formation dense oxide layers on the coating surface. Such coatings can effectively hinder the inward diffusion of oxygen and, thus, significantly enhance oxidation resistance. It has been found that the addition of Al could lead to the formation of a dense oxide layer on the surface of laser clad $FeCoCrNiAl_x$ coatings (Cai et al. 2019), which significantly improved resistance to high-temperature oxidation. Moreover, oxidation rates decreased with increasing Al concentration. In addition to the formation of the Al- and Cr-rich oxide protective layers, the high oxidation resistance of a $(AlCrTaTiZr)Si_{7.9}N_{50.9}$ coating (Cheng et al. 2011b) was also attributed to the large amount of amorphous SiN_x that had segregated to the grain boundaries. Similarly, it was found that the addition of Si into (AlCrMoTaTi)N coatings significantly improved oxidation resistance (Tsai et al. 2015). The thickness of the surface oxide layer on (AlCrNbSiTi)N high-entropy nitride coatings (Hsieh et al. 2013) was only 80–100 nm after annealing at 900 °C for 2 h in air, which surpasses the performance of many conventional coatings. For example, the thickness of the oxide layer on CrN annealed at 800 °C for 2 h is 330 nm (Milošev et al. 1997). The oxide scale on $Al_{0.52}Ti_{0.43}Cr_{0.03}Y_{0.02}N$ coatings was 400 nm thick after annealing at 950 °C for 1 h (Donohue et al. 1997). It is also suggested that porosity can enhance the diffusion of both oxygen and reactive elements thus reducing the oxidation resistance (Hsu et al. 2017). Therefore, it is critically important to regulate and tailor the microstructure of coatings through control of processing parameters to inhibit the formation of defects, such as pores and cracks in order to enhance oxidation resistance.

HEA-based coatings exhibit excellent thermal stability and they can maintain high strength and phase stability even at high temperatures. This high thermal stability can be attributed to the high mixing entropy, severe lattice distortion, and sluggish diffusion during annealing. For example, a laser clad CoCrCuFeNi coating with an FCC structure possessed excellent thermal stability such that no phase transformations occurred at temperatures up to 1000 °C (Zhang et al. 2014a), and the dendritic morphology of the as-solidified microstructure could be retained at temperatures higher than 750 °C. The hardness of a CoCrCuFeNi coating displayed almost no changes after annealing at 500 °C for 5 h, but a moderate decrease of the hardness was observed after annealing at 750 °C for 5 h due to limited lattice distortion. The (AlCrNbSiTiV)N coating exhibited good thermal stability in both its nanostructure and hardness and retained a simple NaCl-type FCC structure even when annealed at 1000 °C for 5 h (Huang and Yeh 2010). Moreover, grain coarsening was inhibited due to severe lattice distortion. Excellent phase stability and softening resistance at elevated temperatures were also reported for FeCoNiCrCu, 6FeNiCoCrAlTiSi, and (AlCrMoTaTiZr)N coatings (Zhang et al. 2011a, Zhang et al. 2011b, Cheng et al. 2011a). Surprisingly, the microhardness of the laser clad refractory TiZrNbWMo coating increased remarkably from 6.9 GPa (as-clad coating) to about 12.8 GPa after heat treatment at 800 °C due to the enhanced fraction of β-Ti_xW_{1-x} second phase (Zhang et al. 2017c). The excellent oxidation resistance and thermal stability indicate that HEA-based coatings have great potential to be used as protective coatings in harsh environments, such as high temperatures and oxidizing environments.

In order to inhibit the interdiffusion between Si and Cu, robust diffusion barriers are closely pursued. Most of the traditional diffusion barriers composed of two or three elements failed to block the diffusion between Cu and Si at ~600 °C. As a result of their excellent thermal stability and diffusion resistance, HEA-based coatings are considered to be promising candidates for the next generation diffusion barriers. A study on the diffusion resistance of six alloy barriers against the interdiffusion of Cu and Si revealed that with more elements incorporated, the failure temperature of the barriers increased from 550 °C (unitary Ti) to 900 °C (senary TiTaCrZrAlRu HEA) (Chang et al. 2014). The improved diffusion resistance of barriers of the HEA coatings was attributed to their severe lattice distortion strains and a high packing density, as well as a high cohesive energy, which hinder the movements of atoms. Multilayered HEA-based films also show good resistance to the Cu and Si interdiffusion at a high temperature of ~900 °C (Jiang et al. 2020, Chang et al. 2012). In addition to the mechanisms revealed for single-layered diffusion barriers of HEA-based coatings, the interface lattice mismatch in the multilayered structure increases diffusion distance, further enhancing the diffusion resistance.

6.7 SUMMARY AND OUTLOOK

The emergence of HEAs composed of multiple principal elements represents a significant leap from conventional alloys with one, or rarely, two base elements. As such, this development has revolutionized the design of advanced alloys, in both bulk form as well as coatings. During the past 15 years, many alloy systems have been explored to create HEA-based coatings with improved property combinations, leading to extraordinary growth of this field. This chapter provides a comprehensive overview on the processing technologies, microstructures, and properties of HEA-based coatings, including HEC and HEA coatings. HEA-based coatings can be fabricated by a variety of deposition methods, among which magnetron sputtering and laser cladding are the most commonly used and mature techniques. Besides the chemical compositions of HEA-based coating materials, deposition techniques and the variations in processing parameters strongly affect the final microstructure, phase constitution and properties of these coatings. HEA-based coatings have been found to have enhanced or novel properties, such as superior hardness and toughness, excellent resistance against wear, corrosion and oxidation, and exceptional thermal stability; therefore, they exhibit great potential for surface protection. The mechanisms and design criteria for the formation of crystal structures and microstructure, and for achieving the remarkable properties, were also presented in this chapter. In view of these promising findings and technical developments, it is believed that high-performing HEA-based coatings can be developed that rival their conventional counterparts for applications in aerospace, chemical processing and power generation.

Although the available data in the literature are increasing exponentially, it remains unclear whether an exhaustive exploration in a nearly infinite compositional space could uncover new mechanisms that enable the drastically enhanced performance of HEA-based coatings. Bearing this question in mind, this chapter shows that the hardening and toughening mechanisms reported in the HEA-based coatings are broadly similar to those of conventional coatings. However, the present HEA endeavor remains valuable, because the vastly increased compositional space will allow the addition of functional alloying elements, which, assisted by computer simulations, may lead to the creation of new HEA-based coatings with properties that are significantly superior to those of conventional coatings. Despite the recent advancements, our understanding of HEA-based coatings is still limited. A deep understanding of the nanoscale deformation mechanisms governing the superior mechanical properties of HEA-based coatings is desperately needed and may yet lead to the discovery of new hardening and toughening mechanisms. This may be achieved by characterizing dislocation behavior and the initiation and propagation of cracks in HEA-based coatings through combining in-situ nanomechanical characterizations with theoretical simulations.

Comparing with extensive research on the traditional HEAs and emerging HECs, the study of high-entropy intermetallics (HEIs) has just sprung up recently. The HEIs are mostly metallic but exhibit structures resembling that of HECs, in which multiple elements of nearly equal fractions are randomly mixed in only one sublattice while the other sublattice has little mixing. HEIs with multiple non-noble elements and an unusual periodically ordered structure are gaining significant interests due to their potential in electrocatalytic applications, such as hydrogen evolution. To date, few studies have been done on HEI coatings, but it is a rich mine worthy being explored.

It is clear that not all HEA-based coatings exhibit superior properties, therefore, there is a need to develop a fundamental understanding of design criteria for these materials, from which the compositions can identified that can provide superior performance. Of particular concern is that, to date, there is still no complete phase formation rule to support the study of HEA-based coatings. Due to the exceedingly high number of possible compositions for HEA-based coatings, the traditional "trial and error" method has become difficult and even impossible. The design and selection of compositions could be more efficient with high-throughput methods assisted by theoretical calculations. Multitarget co-deposition is arguably the most suitable high-throughput method for the study of HEA-based coatings. Due to the spatial gradient between each target and the substrate, variations in composition can be obtained, that is, a region of the substrate closer to a target exhibits higher concentration of atoms from this specific target. Therefore, a coating with a continuous compositional gradient can be formed on the substrate, achieving rapid alloy screening of HEA-based coatings.

REFERENCES

Bartosik, M., C. Rumeau, R. Hahn, Z. L. Zhang, and P. H. Mayrhofer. 2017. "Fracture toughness and structural evolution in the TiAlN system upon annealing." *Sci. Rep.* 7 (1):1–9.

Braic, M., V. Braic, M. Balaceanu, et al. 2010. "Characteristics of (TiAlCrNbY) C films deposited by reactive magnetron sputtering." *Surf. Coat. Technol.* 204 (12–13):2010–2014.

Braic, V., A. Vladescu, M. Balaceanu, C. R. Luculescu, and M. Braic. 2012. "Nanostructured multi-element (TiZrNbHfTa) N and (TiZrNbHfTa) C hard coatings." *Surf. Coat. Technol.* 211:117–121.

Cai, Y., L. Zhu, Y. Cui, et al. 2019. "High-temperature oxidation behavior of FeCoCrNiAlx high-entropy alloy coatings." *Mater. Res. Express* 6 (12):126552.

Cantor, B., I. T. H. Chang, P. Knight, and A. J. B. Vincent. 2004. "Microstructural development in equiatomic multicomponent alloys." *Mater. Sci. Eng. A* 375:213–218.

Cao, F. Y., P. Munroe, Z. F. Zhou, and Z. H. Xie. 2018a. "Medium entropy alloy CoCrNi coatings: Enhancing hardness and damage-tolerance through a nanotwinned structuring." *Surf. Coat. Technol.* 335:257–264.

Cao, F. Y., P. Munroe, Z. F. Zhou, and Z. H. Xie. 2018b. "Microstructure and mechanical properties of a multi-layered CoCrNi/Ti coating with varying crystal structure." *Surf. Coat. Technol.* 350:596–602.

Chang, H.-W., P.-K. Huang, J.-W. Yeh, et al. 2008. "Influence of substrate bias, deposition temperature and post-deposition annealing on the structure and properties of multi-principal-component (AlCrMoSiTi) N coatings." *Surf. Coat. Technol.* 202 (14):3360–3366.

Chang, Z.-C., S.-C. Liang, S. Han, Y.-K. Chen, and F.-S. Shieu. 2010. "Characteristics of TiVCrAlZr multi-element nitride films prepared by reactive sputtering." *Nucl. Instrum. Methods Phys. Res., B* 268 (16):2504–2509.

Chang, S.-Y., C.-E. Li, S.-C. Chiang, and Y.-C. Huang. 2012. "4-nm thick multilayer structure of multi-component (AlCrRuTaTiZr) Nx as robust diffusion barrier for Cu interconnects." *J. Alloys Compd.* 515:4–7.

Chang, S.-Y., C.-E. Li, Y.-C. Huang, et al. 2014. "Structural and thermodynamic factors of suppressed interdiffusion kinetics in multi-component high-entropy materials." *Sci. Rep.* 4 (1):1–8.

Chang, C. H., P. W. Li, Q. Q. Wu, et al. 2019. "Nanostructured and mechanical properties of high-entropy alloy nitride films prepared by magnetron sputtering at different substrate temperatures." *Adv. Mater. Technol.* 34 (6):343–349.

Chawla, K. K., and M. A. Meyers. 1999. *Mechanical behavior of materials.* Second ed. Cambridge: Prentice Hall.

Chen, T. K., T. T. Shun, J. W. Yeh, and M. S. Wong. 2004. "Nanostructured nitride films of multi-element high-entropy alloys by reactive DC sputtering." *Surf. Coat. Technol.* 188:193–200.

Chen, J.-H., P.-H. Hua, P.-N. Chen, et al. 2008. "Characteristics of multi-element alloy cladding produced by TIG process." *Mater. Lett.* 62 (16):2490–2492.

Chen, Y. J., T. Burgess, X. H. An, et al. 2016. "Effect of a high density of stacking faults on the Young's modulus of GaAs nanowires." *Nano Lett.* 16 (3):1911–1916.

Chen, Y. J., Z. F. Zhou, P. Munroe, and Z. H. Xie. 2018. "Hierarchical nanostructure of CrCoNi film underlying its remarkable mechanical strength." *Appl. Phys. Lett.* 113 (8):081905.

Chen, R., Z. Cai, J. Pu, et al. 2020a. "Effects of nitriding on the microstructure and properties of VAlTiCrMo high-entropy alloy coatings by sputtering technique." *J. Alloys Compd.* 827:153836.

Chen, Y. J., X. H. An, Z. F. Zhou, et al. 2020b. "Size-dependent deformation behavior of dual-phase, nanostructured CrCoNi medium-entropy alloy." *Sci. China Mater.* 64:209–222.

Cheng, K.-H., C.-H. Weng, C.-H. Lai, and S.-J. Lin. 2009. "Study on adhesion and wear resistance of multi-element (AlCrTaTiZr) N coatings." *Thin Solid Films* 517 (17):4989–4993.

Cheng, K.-H., C.-H. Lai, S.-J. Lin, and J.-W. Yeh. 2011a. "Structural and mechanical properties of multi-element (AlCrMoTaTiZr) Nx coatings by reactive magnetron sputtering." *Thin Solid Films* 519 (10):3185–3190.

Cheng, K.-H., C.-W. Tsai, S.-J. Lin, and J.-W. Yeh. 2011b. "Effects of silicon content on the structure and mechanical properties of (AlCrTaTiZr)–Six–N coatings by reactive RF magnetron sputtering." *J. Phys. D: Appl. Phys* 44 (20):205405.

Chuang, M.-H., M.-H. Tsai, W.-R. Wang, S.-J. Lin, and J.-W. Yeh. 2011. "Microstructure and wear behavior of AlxCo1. 5CrFeNi1. 5Tiy high-entropy alloys." *Acta Mater.* 59 (16):6308–6317.

Cui, P. P., W. Li, P. Liu, et al. 2020a. "Effects of nitrogen content on microstructures and mechanical properties of (AlCrTiZrHf) N high-entropy alloy nitride films." *J. Alloys Compd.* 834:155063.

Cui, Y., J. Shen, S. M. Manladan, K. Geng, and S. Hu. 2020b. "Wear resistance of FeCoCrNiMnAlx high-entropy alloy coatings at high temperature." *Appl. Surf. Sci.* 512:145736.

Dang, C. Q., J. U. Surjadi, L. B. Gao, and Y. Lu. 2018. "Mechanical properties of nanostructured CoCrFeNiMn high-entropy alloy (HEA) coating." *Front. Mater.* 5:41.

Dedoncker, R., P. Djemia, G. Radnóczi, et al. 2018. "Reactive sputter deposition of CoCrCuFeNi in nitrogen/argon mixtures." *J. Alloys Compd.* 769:881–888.

Ding, Q., Y. Zhang, X. Chen, et al. 2019. "Tuning element distribution, structure and properties by composition in high-entropy alloys." *Nature* 574 (7777):223–227.

Donohue, L. A., I. J. Smith, W.-D. Münz, I. Petrov, and J. E. Greene. 1997. "Microstructure and oxidation-resistance of Ti1– x– y– zAlxCryYzN layers grown by combined steered-arc/unbalanced-magnetron-sputter deposition." *Surf. Coat. Technol.* 94:226–231.

Dou, D., X. C. Li, Z. Y. Zheng, and J. C. Li. 2016. "Coatings of FeAlCoCuNiV high entropy alloy." *Surf. Eng.* 32 (10):766–770.

Fang, Q., Y. Chen, J. Li, Y. Liu, and Y. Liu. 2018. "Microstructure and mechanical properties of FeCoCrNiNbX high-entropy alloy coatings." *Physica B Condens. Matter* 550:112–116.

Feng, X., G. Tang, X. Ma, M. Sun, and L. Wang. 2013. "Characteristics of multi-element (ZrTaNbTiW) N films prepared by magnetron sputtering and plasma based ion implantation." *Nucl. Instrum. Methods Phys. Res., B* 301:29–35.

Feng, X. B., J. Y. Zhang, K. Wu, et al. 2018. "Ultrastrong Al 0.1 CoCrFeNi high-entropy alloys at small scales: effects of stacking faults vs. nanotwins." *Nanoscale* 10 (28):13329–13334.

Fieandt, K. V., L. Riekehr, B. Osinger, S. Fritze, and E. Lewin. 2020. "Influence of N content on structure and mechanical properties of multi-component Al-Cr-Nb-Y-Zr based thin films by reactive magnetron sputtering." *Surf. Coat. Technol.* 389:125614.

Firstov, S. A., N. I. Danilenko, M. V. Karpets, A. A. Andreev, and E. S. Makarenko. 2014. "Thermal stability of superhard nitride coatings from high-entropy multicomponent Ti–V–Zr–Nb–Hf alloy." *Powder Metall. Met. Ceram.* 52 (9–10):560–566.

Gludovatz, B., A. Hohenwarter, D. Catoor, et al. 2014. "A fracture-resistant high-entropy alloy for cryogenic applications." *Science* 345 (6201):1153–1158.

Guo, S., C. Ng, J. Lu, and C. T. Liu. 2011. "Effect of valence electron concentration on stability of fcc or bcc phase in high entropy alloys." *J. Appl. Phys.* 109 (10):103505.

Hahn, R., M. Bartosik, R. Soler, et al. 2016. "Superlattice effect for enhanced fracture toughness of hard coatings." *Scr. Mater.* 124:67–70.

Hahn, R., A. Kirnbauer, M. Bartosik, S. Kolozsvári, and P. H. Mayrhofer. 2019. "Toughness of Si alloyed high-entropy nitride coatings." *Mater. Lett.* 251:238–240.

Håkansson, G., J.-E. Sundgren, D. McIntyre, J. E. Greene, and W.-D. Münz. 1987. "Microstructure and physical properties of polycrystalline metastable Ti0. 5Al0. 5N alloys grown by dc magnetron sputter deposition." *Thin Solid Films* 153 (1–3):55–65.

He, Y., J. Zhang, H. Zhang, and G. Song. 2017. "Effects of different levels of boron on microstructure and hardness of CoCrFeNiAlxCu0. 7Si0. 1By high-entropy alloy coatings by laser cladding." *Coatings* 7 (1):7.

Hsieh, M.-H., M.-H. Tsai, W.-J. Shen, and J.-W. Yeh. 2013. "Structure and properties of two Al–Cr–Nb–Si–Ti high-entropy nitride coatings." *Surf. Coat. Technol.* 221:118–123.

Hsu, W.-L., Y.-C. Yang, C.-Y. Chen, and J.-W. Yeh. 2017. "Thermal sprayed high-entropy NiCo0. 6Fe0. 2Cr1. 5SiAlTi0. 2 coating with improved mechanical properties and oxidation resistance." *Intermetallics* 89:105–110.

Hsueh, H.-T., W.-J. Shen, M.-H. Tsai, and J.-W. Yeh. 2012. "Effect of nitrogen content and substrate bias on mechanical and corrosion properties of high-entropy films (AlCrSiTiZr) 100– xNx." *Surf. Coat. Technol.* 206 (19–20):4106–4112.

Huang, P.-K., and J.-W. Yeh. 2009a. "Effects of nitrogen content on structure and mechanical properties of multi-element (AlCrNbSiTiV) N coating." *Surf. Coat. Technol.* 203 (13):1891–1896.

Huang, P.-K., and J.-W. Yeh. 2009b. "Effects of substrate bias on structure and mechanical properties of (AlCrNbSiTiV) N coatings." *J. Phys. D: Appl. Phys* 42 (11):115401.

Huang, P.-K., and J.-W. Yeh. 2009c. "Effects of substrate temperature and post-annealing on microstructure and properties of (AlCrNbSiTiV) N coatings." *Thin Solid Films* 518 (1):180–184.

Huang, P.-K., and J.-W. Yeh. 2010. "Inhibition of grain coarsening up to 1000 C in (AlCrNbSiTiV) N superhard coatings." *Scr. Mater.* 62 (2):105–108.

Huang, Y.-S., L. Chen, H.-W. Lui, M.-H. Cai, and J.-W. Yeh. 2007. "Microstructure, hardness, resistivity and thermal stability of sputtered oxide films of AlCoCrCu0. 5NiFe high-entropy alloy." *Mater. Sci. Eng. A* 457 (1–2):77–83.

Huang, C., Y. Z. Zhang, and R. Vilar. 2011. "Microstructure characterization of laser clad TiVCrAlSi high entropy alloy coating on Ti-6Al-4V substrate." *Advanced Materials Research* 154:621–625.

Huo, W.-Y., H.-F. Shi, X. Ren, and J.-Y. Zhang. 2015. "Microstructure and wear behavior of CoCrFeMnNbNi high-entropy alloy coating by TIG cladding." *Adv. Mater. Sci. Eng.* 2015:1–5.

Jansson, U., and E. Lewin. 2013. "Sputter deposition of transition-metal carbide films—A critical review from a chemical perspective." *Thin Solid Films* 536:1–24.

Jiang, C., R. Li, X. Wang, et al. 2020. "Diffusion barrier performance of AlCrTaTiZr/AlCrTaTiZr-N high-entropy alloy films for Cu/Si connect system." *Entropy* 22 (2):234.

Jin, B., N. Zhang, H. Yu, D. Hao, and Y. Ma. 2020. "AlxCoCrFeNiSi high entropy alloy coatings with high microhardness and improved wear resistance." *Surf. Coat. Technol.* 402:126328.

Kirnbauer, A., C. Spadt, C. M. Koller, S. Kolozsvári, and P. H. Mayrhofer. 2019. "High-entropy oxide thin films based on Al–Cr–Nb–Ta–Ti." *Vacuum* 168:108850.

Kong, M., W. J. Zhao, L. Wei, and G. Y. Li. 2007. "Investigations on the microstructure and hardening mechanism of TiN/Si3N4 nanocomposite coatings." *J. Phys. D: Appl. Phys* 40 (9):2858.

Lai, C.-H., S.-J. Lin, J.-W. Yeh, and S.-Y. Chang. 2006a. "Preparation and characterization of AlCrTaTiZr multi-element nitride coatings." *Surf. Coat. Technol.* 201 (6):3275–3280.

Lai, C.-H., S.-J. Lin, J.-W. Yeh, and A. Davison. 2006b. "Effect of substrate bias on the structure and properties of multi-element (AlCrTaTiZr) N coatings." *J. Phys. D: Appl. Phys* 39 (21):4628.

Lai, C.-H., M.-H. Tsai, S.-J. Lin, and J.-W. Yeh. 2007. "Influence of substrate temperature on structure and mechanical, properties of multi-element (AlCrTaTiZr) N coatings." *Surf. Coat. Technol.* 201 (16–17):6993–6998.

Lai, C.-H., K.-H. Cheng, S.-J. Lin, and J.-W. Yeh. 2008. "Mechanical and tribological properties of multi-element (AlCrTaTiZr) N coatings." *Surf. Coat. Technol.* 202 (15):3732–3738.

Lewin, E. 2020. "Multi-component and high-entropy nitride coatings—A promising field in need of a novel approach." *J. Appl. Phys.* 127 (16):160901.

Li, X., Z. Zheng, D. Dou, and J. Li. 2016. "Microstructure and properties of coating of FeAlCuCrCoMn high entropy alloy deposited by direct current magnetron sputtering." *Mater. Res.* 19 (4):802–806.

Li, W., P. Liu, and P. K. Liaw. 2018. "Microstructures and properties of high-entropy alloy films and coatings: a review." *Mater. Res. Lett.* 6 (4):199–229.

Li, J., Y. Huang, X. Meng, and Y. Xie. 2019. "A review on high entropy alloys coatings: Fabrication processes and property assessment." *Adv. Eng. Mater.* 21 (8):1900343.

Liang, S.-C., Z.-C. Chang, D.-C. Tsai, et al. 2011a. "Effects of substrate temperature on the structure and mechanical properties of (TiVCrZrHf) N coatings." *Appl. Surf. Sci.* 257 (17):7709–7713.

Liang, S.-C., D.-C. Tsai, Z.-C. Chang, et al. 2011b. "Structural and mechanical properties of multi-element (TiVCrZrHf) N coatings by reactive magnetron sputtering." *Appl. Surf. Sci.* 258 (1):399–403.

Liao, W. B., S. Lan, L. B. Gao, et al. 2017. "Nanocrystalline high-entropy alloy (CoCrFeNiAl0. 3) thin-film coating by magnetron sputtering." *Thin Solid Films* 638:383–388.

Lin, S.-Y., S.-Y. Chang, C.-J. Chang, and Y.-C. Huang. 2014. "Nanomechanical properties and deformation behaviors of multi-component (AlCrTaTiZr) NxSiy high-entropy coatings." *Entropy* 16 (1):405–417.

Lin, Q. Y., X. H. An, H. W. Liu, et al. 2017. "In-situ high-resolution transmission electron microscopy investigation of grain boundary dislocation activities in a nanocrystalline CrMnFeCoNi high-entropy alloy." *J. Alloys Compd.* 709:802–807.

Lin, Y.-C., S.-Y. Hsu, R.-W. Song, et al. 2020. "Improving the hardness of high entropy nitride ($Cr_{0.35}Al_{0.25}Nb_{0.12}Si_{0.08}V_{0.20}$) N coatings via tuning substrate temperature and bias for anti-wear applications." *Surf. Coat. Technol.* 403:126417.

Liu, L., J. B. Zhu, C. Hou, J. C. Li, and Q. Jiang. 2013. "Dense and smooth amorphous films of multicomponent FeCoNiCuVZrAl high-entropy alloy deposited by direct current magnetron sputtering." *Mater. Des.* 46:675–679.

Liu, D., J. Zhao, Y. Li, W. Zhu, and L. Lin. 2020. "Effects of boron content on microstructure and wear properties of FeCoCrNiBx high-entropy alloy coating by laser cladding." *Appl. Sci.* 10 (1):49.

Lo, W.-L., S.-Y. Hsu, Y.-C. Lin, et al. 2020. "Improvement of high entropy alloy nitride coatings (AlCrNbSiTiMo) N on mechanical and high temperature tribological properties by tuning substrate bias." *Surf. Coat. Technol.* 401:126247.

Lu, Z. P., H. Tan, S. C. Ng, and Y. Li. 2000. "The correlation between reduced glass transition temperature and glass forming ability of bulk metallic glasses." *Scr. Mater.* 42 (7):667–673.

Lu, K., L. Lu, and S. Suresh. 2009a. "Strengthening materials by engineering coherent internal boundaries at the nanoscale." *Science* 324 (5925):349–352.

Lu, L., X. Chen, X. X. Huang, and K. Lu. 2009b. "Revealing the maximum strength in nanotwinned copper." *Science* 323 (5914):607–610.

Ma, S. L., D. Y. Ma, Y. Guo, et al. 2007. "Synthesis and characterization of super hard, self-lubricating Ti–Si–C–N nanocomposite coatings." *Acta Mater.* 55 (18):6350–6355.

Ma, D. C., B. Grabowski, F. Körmann, J. Neugebauer, and D. Raabe. 2015. "Ab initio thermodynamics of the CoCrFeMnNi high entropy alloy: Importance of entropy contributions beyond the configurational one." *Acta Mater.* 100:90–97.

Ma, Y., Y. H. Feng, T. T. Debela, G. J. Peng, and T. H. Zhang. 2016. "Nanoindentation study on the creep characteristics of high-entropy alloy films: fcc versus bcc structures." *Int. J. Refract. Hard Met.* 54:395–400.

Maurya, D. K., A. Sardarinejad, and K. Alameh. 2014. "Recent developments in RF Magnetron sputtered thin films for pH sensing applications—an overview." *Coatings* 4 (4):756–771.

Miao, J. S., C. E. Slone, T. M. Smith, et al. 2017. "The evolution of the deformation substructure in a Ni-Co-Cr equiatomic solid solution alloy." *Acta Mater.* 132:35–48.

Milošev, I., H.-H. Strehblow, and B. Navinšek. 1997. "Comparison of TiN, ZrN and CrN hard nitride coatings: Electrochemical and thermal oxidation." *Thin Solid Films* 303 (1–2):246–254.

Miracle, D. B., and O. N. Senkov. 2017. "A critical review of high entropy alloys and related concepts." *Acta Mater.* 122:448–511.

Miserez, A., J. C. Weaver, P. J. Thurner, et al. 2008. "Effects of laminate architecture on fracture resistance of sponge biosilica: Lessons from nature." *Adv. Funct. Mater.* 18 (8):1241–1248.

Münz, W. D. 1986. "Titanium aluminum nitride films: A new alternative to TiN coatings." *J. Vac. Sci. Technol. A* 4 (6):2717–2725.

Ong, C. W., X.-A. Zhao, Y. C. Tsang, C. L. Choy, and P. W. Chan. 1996. "Effects of substrate temperature on the structure and properties of reactive pulsed laser deposited CNx films." *Thin Solid Films* 280 (1–2):1–4.

Otto, F., Y. Yang, H. B. Bei, and E. P. George. 2013. "Relative effects of enthalpy and entropy on the phase stability of equiatomic high-entropy alloys." *Acta Mater.* 61 (7):2628–2638.

Owen, E. A., and D. M. Jones. 1954. "Effect of grain size on the crystal structure of cobalt." *Proc. Phys. Soc., B* 67 (6):456.

Öztürk, A., K. V. Ezirmik, K. Kazmanlı, et al. 2008. "Comparative tribological behaviors of TiN, CrN and MoNCu nanocomposite coatings." *Tribol. Int.* 41 (1):49–59.

Panjan, P., B. Navinšek, A. Cvelbar, A. Zalar, and I. Milošev. 1996. "Oxidation of TiN, ZrN, TiZrN, CrN, TiCrN and TiN/CrN multilayer hard coatings reactively sputtered at low temperature." *Thin Solid Films* 281:298–301.

Patsalas, P., C. Gravalidis, and S. Logothetidis. 2004. "Surface kinetics and subplantation phenomena affecting the texture, morphology, stress, and growth evolution of titanium nitride films." *J. Appl. Phys.* 96 (11):6234–6246.

Pogrebnjak, A. D., I. V. Yakushchenko, A. A. Bagdasaryan, et al. 2014. "Microstructure, physical and chemical properties of nanostructured (Ti–Hf–Zr–V–Nb) N coatings under different deposition conditions." *Mater. Chem. Phys.* 147 (3):1079–1091.

Pogrebnjak, A. D., I. V. Yakushchenko, O. V. Bondar, et al. 2016. "Irradiation resistance, microstructure and mechanical properties of nanostructured (TiZrHfVNbTa) N coatings." *J. Alloys Compd.* 679:155–163.

Pogrebnjak, A. D., V. M. Beresnev, K. V. Smyrnova, et al. 2018. "The influence of nitrogen pressure on the fabrication of the two-phase superhard nanocomposite (TiZrNbAlYCr) N coatings." *Mater. Lett.* 211:316–318.

Qiu, X. W., Y. P. Zhang, and C. G. Liu. 2014. "Effect of Ti content on structure and properties of Al2CrFeNiCoCuTix high-entropy alloy coatings." *J. Alloys Compd.* 585:282–286.

Ren, B., Z. G. Shen, and Z. X. Liu. 2013. "Structure and mechanical properties of multi-element (AlCrMnMoNiZr) Nx coatings by reactive magnetron sputtering." *J. Alloys Compd.* 560:171–176.

Schell, N., W. Matz, J. Bøttiger, J. Chevallier, and P. Kringhøj. 2002. "Development of texture in TiN films by use of in situ synchrotron x-ray scattering." *J. Appl. Phys.* 91 (4):2037–2044.

Schiøtz, J., F. D. Di Tolla, and K. W. Jacobsen. 1998. "Softening of nanocrystalline metals at very small grain sizes." *Nature* 391 (6667):561–563.

Sha, C. H., Z. F. Zhou, Z. H. Xie, and P. Munroe. 2020a. "Extremely hard, α-Mn type high entropy alloy coatings." *Scr. Mater.* 178:477–482.

Sha, C. H., Z. F. Zhou, Z. H. Xie, and P. Munroe. 2020b. "FeMnNiCoCr-based high entropy alloy coatings: Effect of nitrogen additions on microstructural development, mechanical properties and tribological performance." *Appl. Surf. Sci.* 507:145101.

Sha, C. H., Z. F. Zhou, Z. H. Xie, and P. Munroe. 2020c. "High entropy alloy FeMnNiCoCr coatings: Enhanced hardness and damage-tolerance through a dual-phase structure and nanotwins." *Surf. Coat. Technol.* 385:125435.

Shaginyan, L. R., N. A. Krapivka, S. A. Firstov, and I. F. Kopylov. 2016. "Properties of coatings of the Al–Cr–Fe–Co–Ni–Cu–V high entropy alloy produced by the magnetron sputtering." *J. Superhard Mater.* 38 (1):25–33.

Shen, W.-J., M.-H. Tsai, Y.-S. Chang, and J.-W. Yeh. 2012. "Effects of substrate bias on the structure and mechanical properties of (Al1. 5CrNb0. 5Si0. 5Ti) Nx coatings." *Thin Solid Films* 520 (19):6183–6188.

Sheng, G., and C. T. Liu. 2011. "Phase stability in high entropy alloys: formation of solid-solution phase or amorphous phase." *Prog. Nat. Sci.* 21 (6):433–446.

Sheng, W. J., X. Yang, C. Wang, and Y. Zhang. 2016. "Nano-crystallization of high-entropy amorphous NbTiAlSiWxNy films prepared by magnetron sputtering." *Entropy* 18 (6):226.

Shi, Y. Z., B. Yang, and P. K. Liaw. 2017. "Corrosion-resistant high-entropy alloys: A review." *Metals* 7 (2):43.

Tsai, C.-W., S.-W. Lai, K.-H. Cheng, et al. 2012. "Strong amorphization of high-entropy AlBCrSiTi nitride film." *Thin Solid Films* 520 (7):2613–2618.

Tsai, D.-C., M.-J. Deng, Z.-C. Chang, et al. 2015. "Oxidation resistance and characterization of (AlCrMoTaTi)-Six-N coating deposited via magnetron sputtering." *J. Alloys Compd.* 647:179–188.

Tsau, C.-H., Y.-C. Yang, C.-C. Lee, L.-Y. Wu, and H.-J. Huang. 2012. "The low electrical resistivity of the high-entropy alloy oxide thin films." *Procedia Eng.* 36:246–252.

Tsianikas, S. J., Y. J. Chen, and Z. H. Xie. 2020. "Deciphering deformation mechanisms of hierarchical dual-phase CrCoNi coatings." *J. Mater. Sci. Technol.* 39:7–13.

Tunes, M. A., and V. M. Vishnyakov. 2019. "Microstructural origins of the high mechanical damage tolerance of NbTaMoW refractory high-entropy alloy thin films." *Mater. Des.* 170:107692.

Wang, Y. X., and S. Zhang. 2014. "Toward hard yet tough ceramic coatings." *Surf. Coat. Technol.* 258:1–16.

Wang, Z., C. Wang, Y.-L. Zhao, et al. 2020. "High hardness and fatigue resistance of CoCrFeMnNi high entropy alloy films with ultrahigh-density nanotwins." *Int. J. Plast.* 131:102726.

Wu, Z. F., X. D. Wang, Q. P. Cao, et al. 2014. "Microstructure characterization of AlxCo1Cr1Cu1Fe1Ni1 (x= 0 and 2.5) high-entropy alloy films." *J. Alloys Compd.* 609:137–142.

Wu, G., S. Balachandran, B. Gault, et al. 2020. "Crystal–Glass High-Entropy Nanocomposites with Near Theoretical Compressive Strength and Large Deformability." *Adv. Mater.* 32 (34):2002619.

Xia, A., R. Dedoncker, O. Glushko, et al. 2020. "Influence of the nitrogen content on the structure and proper-
    ties of MoNbTaVW high entropy alloy thin films." *J. Alloys Compd.* 850:156740.
Xiao, Y., Y. Zou, H. Ma, et al. 2019. "Nanostructured NbMoTaW high entropy alloy thin films: High strength
    and enhanced fracture toughness." *Scr. Mater.* 168:51–55.
Yan, X. H., J. S. Li, W. R. Zhang, and Y. Zhang. 2018. "A brief review of high-entropy films." *Mater. Chem.
    Phys.* 210:12–19.
Yang, X., and Y. Zhang. 2012. "Prediction of high-entropy stabilized solid-solution in multi-component alloys."
    Mater. Chem. Phys. 132 (2–3):233–238.
Yang, T., S. Xia, S. Liu, et al. 2015. "Effects of Al addition on microstructure and mechanical properties of
    AlxCoCrFeNi High-entropy alloy." *Mater. Sci. Eng. A* 648:15–22.
Yang, L. X., H. L. Ge, J. Zhang, et al. 2019. "High He-ion irradiation resistance of CrMnFeCoNi high-entropy
    alloy revealed by comparison study with Ni and 304SS." *J. Mater. Sci. Technol.* 35 (3):300–305.
Ye, X. Y., M. X. Ma, Y. Cao, et al. 2011. "The property research on high-entropy alloy AlxFeCoNiCuCr coating
    by laser cladding." *Phys. Procedia* 12:303–312.
Ye, Y. F., Q. Wang, J. T. Lu, C. T. Liu, and Y. C. Yang. 2016. "High-entropy alloy: challenges and prospects."
    Mater. Today 19 (6):349–362.
Ye, Q. F., K. Feng, Z. G. Li, et al. 2017. "Microstructure and corrosion properties of CrMnFeCoNi high entropy
    alloy coating." *Appl. Surf. Sci.* 396:1420–1426.
Yeh, J.-W. 2006. "Recent progress in high entropy alloys." *Ann. Chim. Sci. Mat.* 31 (6):633–648.
Yeh, J. W., S. K. Chen, S. J. Lin, et al. 2004. "Nanostructured high-entropy alloys with multiple principal ele-
    ments: novel alloy design concepts and outcomes." *Adv. Eng. Mater.* 6 (5):299–303.
Yen, C.-C., H.-N. Lu, M.-H. Tsai, et al. 2019. "Corrosion mechanism of annealed equiatomic AlCoCrFeNi tri-
    phase high-entropy alloy in 0.5 M H2SO4 aerated aqueous solution." *Corrosion Science* 157:462–471.
Yu, R.-S., C.-J. Huang, R.-H. Huang, C.-H. Sun, and F.-S. Shieu. 2011. "Structure and optoelectronic properties
    of multi-element oxide thin film." *Appl. Surf. Sci.* 257 (14):6073–6078.
Zhang, S., D. Sun, Y. Q. Fu, and H. J. Du. 2003. "Recent advances of superhard nanocomposite coatings: a
    review." *Surf. Coat. Technol.* 167 (2–3):113–119.
Zhang, Y., Y. J. Zhou, J. P. Lin, G. L. Chen, and P. K. Liaw. 2008. "Solid-solution phase formation rules for
    multi-component alloys." *Adv. Eng. Mater.* 10 (6):534–538.
Zhang, H., Y. Pan, and Y.-Z. He. 2011a. "Synthesis and characterization of FeCoNiCrCu high-entropy alloy
    coating by laser cladding." *Mater. Des.* 32 (4):1910–1915.
Zhang, H., Y. Pan, and Y. Z. He. 2011b. "Effects of annealing on the microstructure and properties of
    6FeNiCoCrAlTiSi high-entropy alloy coating prepared by laser cladding." *J. Therm. Spray Technol.* 20
    (5):1049–1055.
Zhang, H., Y. Pan, Y. Z. He, and H. S. Jiao. 2011c. "Microstructure and properties of 6FeNiCoSiCrAlTi high-
    entropy alloy coating prepared by laser cladding." *Appl. Surf. Sci.* 257 (6):2259–2263.
Zhang, H., Y.-Z. He, Y. Pan, and S. Guo. 2014a. "Thermally stable laser cladded CoCrCuFeNi high-entropy
    alloy coating with low stacking fault energy." *J. Alloys Compd.* 600:210–214.
Zhang, Y., T. T. Zuo, Z. Tang, et al. 2014b. "Microstructures and properties of high-entropy alloys." *Prog.
    Mater. Sci.* 61:1–93.
Zhang, C., G. J. Chen, and P. Q. Dai. 2016. "Evolution of the microstructure and properties of laser-clad
    FeCrNiCoB x high-entropy alloy coatings." *Mater. Sci. Technol.* 32 (16):1666–1672.
Zhang, F. X., S. J. Zhao, K. Jin, et al. 2017a. "Pressure-induced fcc to hcp phase transition in Ni-based high
    entropy solid solution alloys." *Appl. Phys. Lett.* 110 (1):011902.
Zhang, H., H. Tang, Y. He, et al. 2017b. "Effect of heat treatment on borides precipitation and mechanical
    properties of CoCrFeNiAl 1.8 Cu 0.7 B 0.3 Si 0.1 high-entropy alloy prepared by arc-melting and laser-
    cladding." *JOM* 69 (11):2078–2083.
Zhang, M., X. Zhou, X. Yu, and J. Li. 2017c. "Synthesis and characterization of refractory TiZrNbWMo high-
    entropy alloy coating by laser cladding." *Surf. Coat. Technol.* 311:321–329.
Zhang, W. R., P. K. Liaw, and Y. Zhang. 2018a. "Science and technology in high-entropy alloys." *Sci. China
    Mater.* 61 (1):2–22.
Zhang, Y., X.-H. Yan, W.-B. Liao, and K. Zhao. 2018b. "Effects of nitrogen content on the structure and
    mechanical properties of (Al0. 5CrFeNiTi0. 25) Nx high-entropy films by reactive sputtering." *Entropy*
    20 (9):624.
Zhu, Z. G., Q. B. Nguyen, F. L. Ng, et al. 2018. "Hierarchical microstructure and strengthening mechanisms
    of a CoCrFeNiMn high entropy alloy additively manufactured by selective laser melting." *Scr. Mater.*
    154:20–24.

7 High-Temperature Thin Films and Coatings

Xingang Luan, Xingmin Liu, and Yuchang Qing
Northwestern Polytechnical University, China

CONTENTS

7.1 CORROSION RESISTANT COATINGS

7.1.1 POTENTIAL APPLICATION

Environmental degradation of exposed materials resulting from oxidation, corrosion in high-temperature, and corrosive and other aggressive environments, for example, in gas turbine engines, leads to fast and severe failure of serving materials in the hot-section components [1–3]. To keep long-term stability in harsh serving environment in gas turbines due to the thermo-chemo-mechanical degradation, environmental protections are needed [3–6]. For example, the fiber reinforced ceramic matrix composites (CMCs) are considered as one of the prime candidates for applications in turbine engines [7–11]. However, they suffer from rapid degradation due to the volatilization of silica via reaction with hot H_2O vapor in combustion environments [7, 8]. In addition, corrosion induced by molten salts leads to catastrophic failure through silica dissolution [7], while ingested environmental debris causes severe impact damage [7].

In this circumstance, the most representative thermal protection and/or corrosion resistant coatings, also known as thermal and environmental barrier coatings (T/EBCs) deposited on the exposed materials are playing increasingly important roles in counteracting the degradation of substrate materials and ensuring the durability of high-temperature structural components for gas turbines (Figure 7.1) [12].

In the gas turbine, the aggressive threat results from the products of combustion, including hot H_2O vapor, molten salts (Na_2SO_4 and V_2O_5 calcium–magnesium–alumina–silicate (CMAS)), in the original dusts, and volcanic ashes, sands, and runway debris entering the gas turbine together with the intaking air can all reduce the durability of thermal components [13–15]. TBCs introduced onto the Ni-based superalloy thermalmechanical components can ensure the gas inlet temperatures of turbines increasing up to 1500°C, through which the thrust-to-weight ratio and the fuel efficiency in the gas turbines for aircraft propulsion and power generation are significantly improved [13].

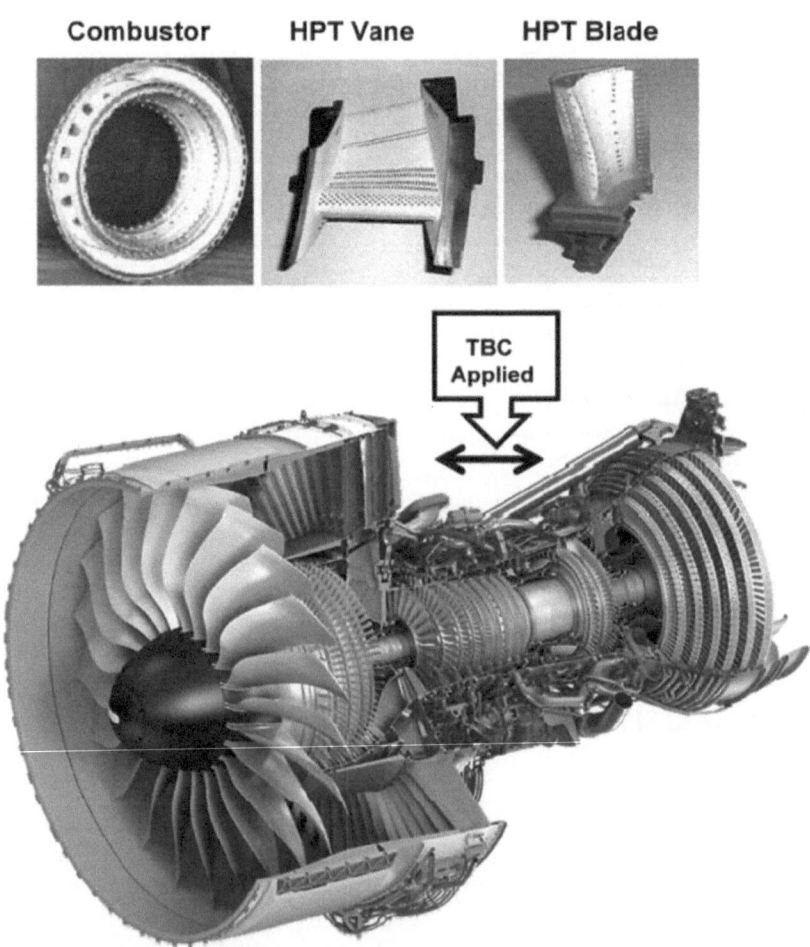

FIGURE 7.1 Examples of T/EBCs application in propulsion engines. Combustor liners, HPT blades, and vanes are coated with T/EBCs [12].

Moreover, the combination of fibers including silicon nitride, carbon, and silicon carbide fibers reinforced CMCs with EBCs is, by now, the most effective strategy to protect CMCs serving in the high-temperature structural components of gas turbines from water vapor and other corrosive gases inside engines and other extreme environments [13, 14].

The intrinsic properties, including environmental durability, phase stability, chemical compatibility, and coefficient of thermal expansion (CTE), are typically considered to be the most important features for the selection of T/EBC materials [12, 16]. T/EBCs are designed for thermal protection from hot gas damage in the engines and have been proved to be a significant functional part in lowering the surface temperature of the substrate components.

7.1.2 PREPARING METHODS

Yttria-stabilized zirconia (YSZ), rare earth zirconate, and rare earth silicate by now are the most commonly reported E/TBC materials [3, 13, 17]. Among them, YSZ coatings can be prepared with a variety of methods including electron beam enhanced physical vapor deposition (EBPVD), atmospheric plasma spraying (APS), suspension plasma spraying (SPS), laser-enhanced chemical vapor deposition (LCVD), and low-pressure plasma spraying (LPPS) [13, 16, 18, 19]. It is worth to note that APS and EB-PVD are two most frequently used methods for the preparation of TBCs coatings [1–5, 13]. Similar with YSZ-based coatings, rare earth zirconate-based TBCs are also prepared with the aforementioned two main methods, due to the unique microstructural characteristics of the coating resulting from APS and EB-PVD [7, 13, 16].

Compared with TBCs, EBCs can be prepared with simpler and lower-cost methods due to the intrinsic thermal stability of ceramic-based materials. In addition to some mainstream preparation processes including APS, EB-PVD, plasma spraying physical vapor deposition (PSPVD), and chemical vapor deposition (CVD), EBCs can also be prepared with dipping, spinning, spin-dipping, painting, and spraying techniques [7, 13].

7.1.3 MICROSTRUCTURE AND PROPERTIES

Typically, E/TBC systems are primarily composed of a two-layer structure: (a) a porous (with porosities of 10–25%) ceramic top coating layer comprised of yttria (Y_2O_3) partially stabilized (about 6–8 wt-%) zirconia (ZrO_2), which was generally referred to as 7YSZ or YSZ; (b) an alumina forming bond coating layer on substrate materials (Ni-based super alloy or CMC-based materials) [7, 20–22].

Generally, the durability and performance of T/EBCs are influenced by four components as shown in Figure 7.2 [12]:

   (i) top coating layer, working as thermal insulation to the substrates;
  (ii) thermally grown oxide (TGO) layer, providing bonding to bond coat layer and decreasing the subsequent oxidation in the bond coat layer;
 (iii) bond coat layer, providing the elements for the formation of TGO in oxidation environment and preventing the substrate from being oxidized;
 (iv) Substrate materials, providing mechanical strength in a harsh environment.

However, YSZ-based TBCs typically suffer from poor phase stability of only up to 1200°C, which seriously limit the high requirement for the high thrust-to-weight ratio and the fuel efficiency.

The early EBCs were mainly composed of Mullite ($3Al_2O_3 \cdot 2SiO_2$), which were used for the protection of molten salt and oxidation corrosion of silicon-based CMCs [20]. However, the SiO_2 suffers from fast volatilization when attacked by hot water vapor. After that, new EBC system

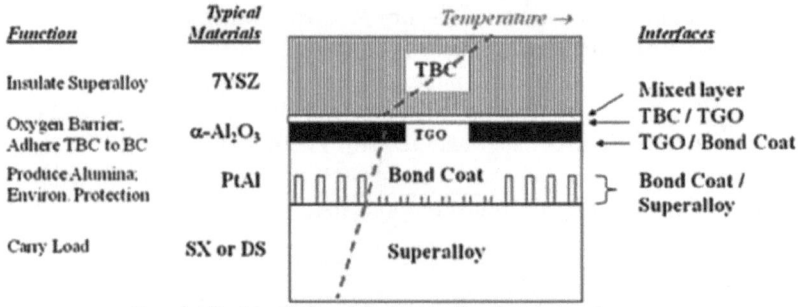

FIGURE 7.2 Thermal barrier coating consists of four main components with unique functions that influence TBC life [12].

possessing a close CTE with the substrates, such as YSZ [21], BaO-SrO-Al$_2$O$_3$-SiO$_2$ (BSAS) [22], rare earth monosilicates (RE$_2$SiO$_5$/REMS), and disilicates (RE$_2$Si$_2$O$_7$/REDS) [23], were developed to counteract the rapid water vapor recession. Figure 7.3 shows the representative scanning electron microscopy (SEM) images of cross sections of different kinds of EBCs including Mullite/SiC, YSZ/Mullite/SiC, BSAS/Mullite/Si/(MI)SiC, and Yb$_2$SiO$_5$/Mullite/Si/(MI)SIC. Moreover, Hafnia [24] and high-entropy materials [13] are further introduced to improve phase stability and service temperature.

7.2 TRIBOLOGICAL COATINGS

7.2.1 POTENTIAL APPLICATION

Tribological characteristics including wear, friction, and lubrication of materials are one of the most vital properties attracting the attention of almost all mechanical engineers in the field of surface engineering, which are the utmost important behaviors when determining the tribological characteristics of critical sliding components [25–28]. Engineering components in the application field of automotive, marine, aerospace industry, power generation, and petrochemical and mining industries, experience wearing when two surfaces of components are in contact or one is moving relative to each other, which could lead to large thermal plastic cyclic strains and worn down during each mission cycle, especially when they are serving under severe and/or harsh conditions of high chamber pressure and heat fluxes [29, 30]. Reports showed that even a reduction of 15–20% in wear/friction can significantly reduce the economic costs in relation to environmental benefits [31]. There are several tribology systems that can effectively reduce the friction coefficient and improve wear resistance [25–30]. Some solid-lubricating materials even can be used up to 1200 °C [32]. Employment of tribological coatings is one of the most effective route in tailoring surface morphology, wear performance, adhesion, and fatigue strength of substrate material without changing properties of the bulk products [28, 33].

In order to fabricate coatings with excellent wear resistance and/or low friction coefficient, coating materials of various systems were developed, including carbon (diamond-like carbon (DLC) [34–36], carbon nanotubes (CNTs) [37–39], graphite-like carbon and graphene) [40–42], (Ti,Al) N-based materials containing (Ti,Al)CN, (Ti,Nb)N, TiN, AlN/TiN, CrN, and Mo [43–47], polymer-based self-lubricating coatings modified with active fillers and solid lubricant composites made of

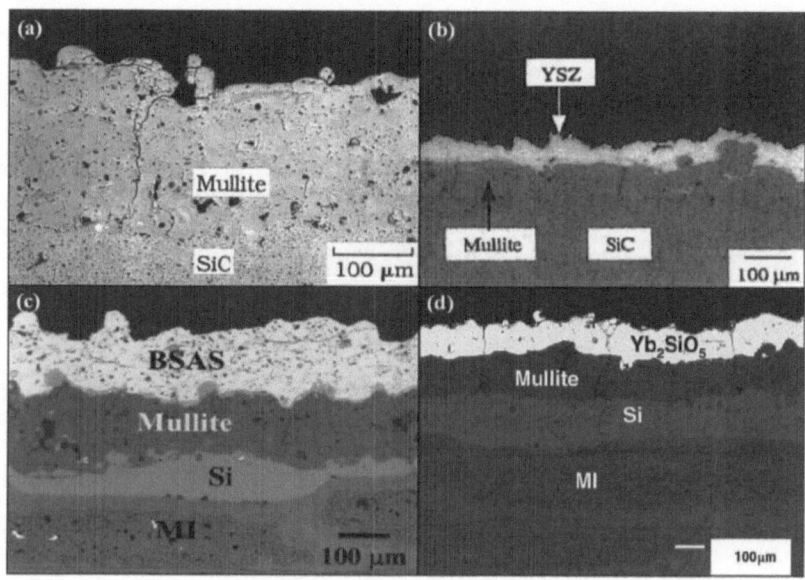

FIGURE 7.3 The development of environmental barrier coating systems for silicate ceramics in gas turbines: (a) Mullite/SiC [20]; (b) YSZ/Mullite/SiC [21]; (c) BSAS/Mullite/Si/(MI)SiC [22]; (d) Yb$_2$SiO$_5$/Mullite/Si/(MI)SIC [23].

oxides [48–50], dichalcogenides, transitional metal nitrides, and/or noble metals (PbO/MoS$_2$, ZnO/MoS$_2$, ZnO/WS$_2$ and Mo$_2$N/MoS$_2$/Ag and YSZ/Ag/Mo/MoS$_2$ [29, 51–60].

7.2.2 Preparing Methods

By now, various surface processing techniques have been developed for the fabrication of self-lubricating and/or wear resistant coatings, including cold spray (CS) and laser-assisted cold spray (LACS), CVD, physical vapor deposition (PVD), electroless coating, thermal and plasma spray processing, electrodeposition, plasma transferred-arc (PTA) surfacing, and laser surface cladding (LSC) [61–70].

CS is a coating deposition method, with which solid powders (1–50 μ in diameter) are accelerated in a supersonic gas jet to velocities up to ca. 1200 m/s and introduced onto the substrate after experiencing the deformation of the particles. Figure 7.4 shows a schematic illustration of (a) cold spray system and (b) material coating sprayed to the substrate with CS technique. Various material systems including metals, polymers, ceramics, composite materials, and nanocrystalline powders can all be deposited using CS [71].

LACS technique is a relatively new and hybrid coating deposition technique which combines the advantages of both CS and laser-softening processes. Different from thermal spray techniques such as plasma spraying and arc spraying, the powders from CS and LACS are not melted during the spraying process [72]. Due to the cold process technique, the initial physical and chemical particle properties of sprayed materials are retained. Moreover, coatings with high hardness and homogeneity are available with CS. Figure 7.5 shows an SEM image of Al-12 wt.% Si coating deposited on the stainless steel substrate prepared with LACS [72]. From the SEM image, we can see a highly dense structure was obtained.

Plasma spray is an important surface engineering technology with benefits of fast deposition rate, minimum heat effect on the substrate, and applicability for various material systems, which makes it widely employed in surface modifications and strengthening of mechanical components [28].

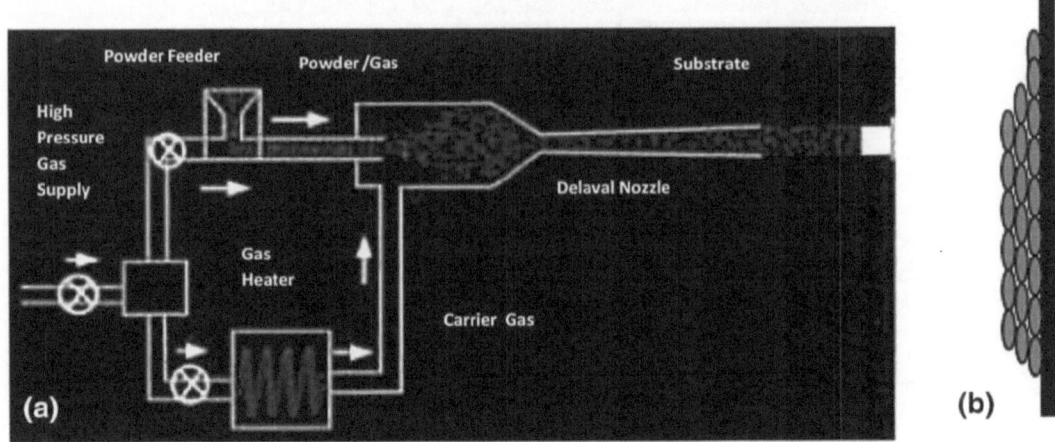

FIGURE 7.4 Schematic illustration of (a) cold spray system and (b) material coating sprayed to the substrate during CS [71].

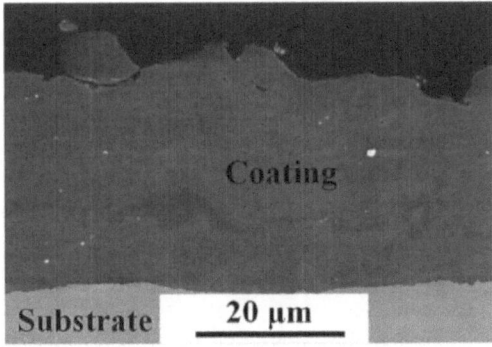

FIGURE 7.5 SEM micrograph of a pore- and crack-free Al-12 wt.%Si coating deposited on the stainless steel substrate prepared with LACS [72].

Plasma spraying possesses advantages including better jet speed, less oxidation, and higher deposition efficiency. Various types of coatings including steel coatings, metal matrix composite (MMC) coatings and ceramic coatings can all be prepared with plasma spray.

In PVD processes, the coatings are deposited to the substrates in vacuum atmosphere by condensation from a flux of neutral or ionized atoms of metals. During the deposition, materials to be deposited were transferred from solid state to vapor phase through heating of an evaporation source or sputtering of a target, and then introduced onto the substrates. The cathodic arc vapor (plasma or arc ion plating) deposition, magnetron sputtering (or sputter ion plating), and combined magnetron and arc processes are the most widely used PVD processes. Several PVD techniques are available for deposition of hard coatings [28].

7.2.3 MICROSTRUCTURE AND PROPERTIES

Hard, wear resistant, and/or self-lubrication coatings are commonly deposited on tools which are used for severe cutting, forming, and casting applications, where the conditions typically result in high temperatures, mechanical loads, and pronounced wear. By now, owing to the low value and desirable mechanical residences, inorganic nanoparticle filler (such as carbon nanotubes and black phosphorus) reinforced polymer-based coatings, Fe-based powders, Al_2O_3-CeO_2 coating, Molybdenum, Al-12Si/SiO_2 composites, Al-12Si/TiB_2/h-BN composites, and ceramic-based coatings are the desired materials for the economic utility of wear protection for diverse substrate substances [73–76]. However, the tribological behavior of coating materials cannot be comprehended just by taking the surface hardness into account only. As hard coatings do not sufficiently reduce friction coefficient and may not provide protection from the adjacent surface asperities. In some cases, the rougher and hard particles in the coatings could be easily peeled off during the wear process, which may lead to aggressive abrasive wear. Therefore, developing of solid lubricant coatings, also known as "chameleon" coatings that may reduce friction and wear rate, and protect the opposing surface is quite necessary. Under the tribological load, the risk of damage due to inaccessibility of the lubricant can be drastically reduced or eliminated by introducing solid lubricants into the material system. For example, polymer-based solid lubricant reinforced coatings are widely used for sliding parts in non-lubrication (dry condition), water lubrication, vacuum, low temperature, or corrosive atmosphere. Generally, polymer-based lubricant coatings possess good flexible and mechanical properties, which make them a new kind of lubrication and wear-resistant coatings instead of traditional metal-based solid lubricant coatings.

7.3 ELECTRIC CONDUCTIVE COATINGS

7.3.1 POTENTIAL APPLICATION

With ever-increasing demand for lightweight, downsizing, and portable devices, electrically conductive coatings used to cover dielectric or less conductive substrates are required in variety of engineering and industrial applications, ranging from gas sensor, touch screens, electrostatic discharge (ESD) and electromagnetic/radio frequency interference (EMI/RFI) shielding in consumer electronics to flat panel and flexible displays, batteries, and fuel cells [77–81]. For example, plastic housings which are used to enclose the electronic equipment are typically electrically nonconductive. However, it is vital to guarantee that the working components of the electronic equipment are effectively shielded from incoming electromagnetic microwave interference signals, as well as preventing leakage of EM waves from the enclosed house. It has become common practice to apply electrically conductive coatings. Besides the engineering application, electrical conductive coatings are also employed in the scientific research. Sputtering of conductive elements from a target onto substrates is used extensively in the preparation of conductive samples for analysis via SEM [82, 83]. Conductive coatings can enhance the image contrast due to higher secondary electron yield reflected by the conductive coating deposited onto the less or non-conductive substrates [82]. The resolution of SEM with field emission may go down to a range of 0.6–3 nm. However, in order to obtain such high resolution from non-conductive substrates, for example, polymer-based samples, deposition of conductive coatings are necessary. Moreover, electrical conductive coatings applied to the TV or monitor screens (or on the glass panels in front of the screen) could conduct away electrostatic charge and prevents the dusts from adhering to the surface of the screens.

7.3.2 PREPARING METHODS

Electrically conductive coatings adhered to the substrates typically can be prepared via standard wet-chemical methods such as spray, roll, drop-casting, drop coating, PVD such as magnetron sputter, ion beam sputtering, penning sputter and electron beam, CVD, and transferring methods [82, 83].

Electrically conductive coatings served in the aforementioned applications are typically prepared by mixing electrically conductive fillers into a polymeric matrices, where the polymers work as binders [84]. Conductive metals (such as silver, gold, and copper) nanofillers (nanoparticles, wires and/or tubes), carbon-based conductive materials (carbon nanotubes, graphene and graphene/carbon nanotubes hybrids), and two-dimensional transitional metal carbides/nitrides (MXene) exhibit great

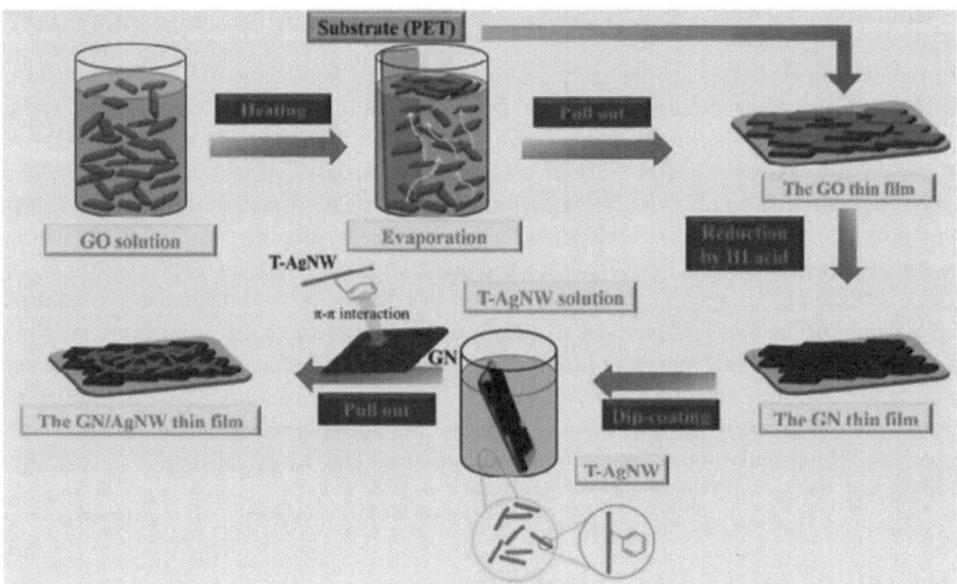

FIGURE 7.6 Schematic depositing process of GO/AgNWs coating on PET substrate [84].

promise as alternatives to conventional conductive materials [84]. Through incorporating the conductive nanomaterials into coating, a combination of their excellent electrical and optical properties with versatile mechanical characteristics superior to those of conventional conductive materials can be expected. These approaches can prepare coatings with thicknesses varying from nanometers to micrometers. For example, Hsiao reported a uniform, high-quality, and electrical conductive graphene nanosheet and silver nanowires (GN/AgNWs) film deposited on PET substrate with dip-coating method (Figure 7.6) [84]. The GN/AgNW hybrid nanomaterial films exhibited a sheet resistance of $71\Omega/cm$ and 85% light transmittance.

PVD method can not only deposit metals but also ceramics. Usually, metals such as gold, palladium, chromium, platinum, silver, and copper which can be used lower than 1000°C, and carbon and their mixtures are employed for the deposition of conductive coatings via the following PVD approaches. a) magnetron sputter is the most popular way of applying conductive coating on nonconductive substrates. In this method, the sputter source (e.g., metals) and the sample are located in a common vacuum chamber (Figure 7.7), [85]. The pressure in the chamber is kept at about 10 Pa and inert gas such as argon or xenon is employed to fill in the chamber; b) compared with magnetron sputter, ion beam sputtering requires a much higher vacuum (8×10^{-3} Pa). Through hitting the target material with a directed ion beam generated by an ion source, the ejected atoms from the solid surface of the target material deposited onto the substrate surface; c) penning sputter technique combines the plasma generation and ejection of target material in one piece of equipment. A neutral particle beam with energy comparable to the kinetic energy of an ion beam sputter coater of is emitted from the source. Sample is maintained under high vacuum and must be kept in motion for a continuous coating. The neutral particle source makes thin coatings possible at room temperature; d) electron beam evaporation technique is applied through heating of the target material in a high vacuum atmosphere. The uncharged particle beam hits the substrate material with low kinetic energy, which leads to the deposition of coating.

Besides the typically employed wet-chemical methods and PVD, CVD is also frequently employed approaches for the preparation of electrical conductivity coatings. For example, single- or few-layered graphene coatings were deposited on silicon or silica substrates with CVD method. Moreover, transferring method is employed to transfer graphene layer onto polymer-based substrate (Figure 7.8), [86].

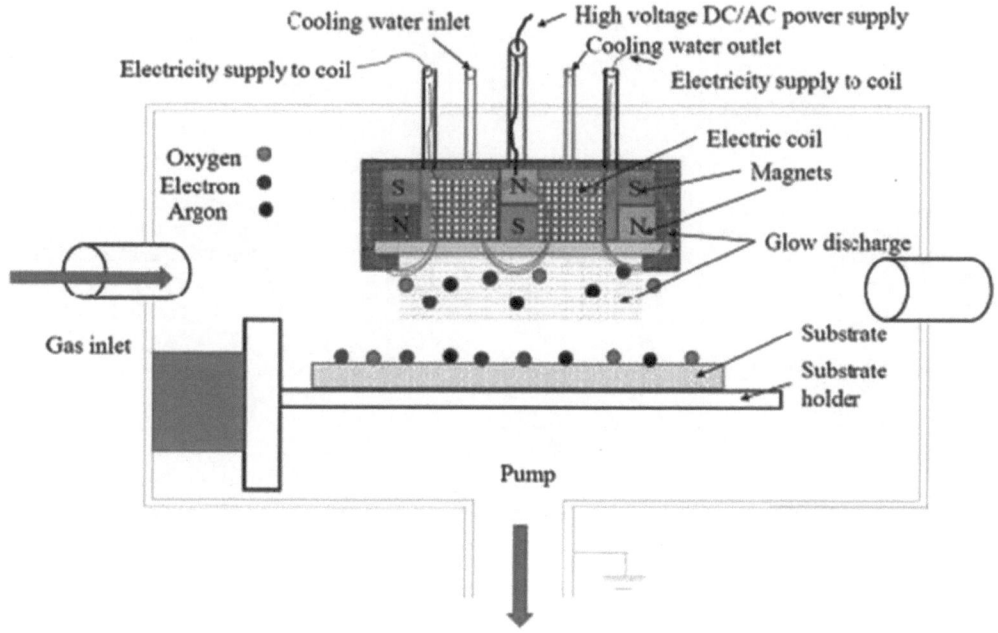

FIGURE 7.7 Schematic illustration of mechanism of magnetron sputter coating system [85].

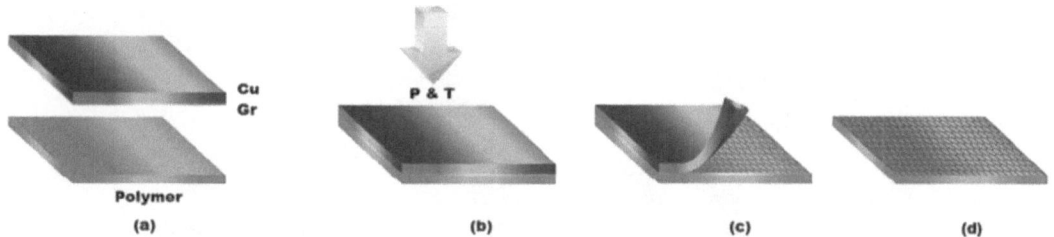

FIGURE 7.8 Schematic of the transfer method and sample after transfer. (a–d) Schematic: (a) graphene/metal and polymer film before transfer. (b) Polymer application step to form the metal/graphene/polymer stack. (c) Peeling of the metal step. (d) Final graphene/polymer stack [86].

7.3.3 Microstructure and Properties

The microstructure and electrical property of resultant coatings vary with the variation of fabrication techniques and materials systems. Due to the excellent electrical property of Ag, various components including particles and nanowire combined with other conductive materials were employed to construct electrical conductive coatings. Sheng-Tsung Hsiao prepared AgNWs modified graphene nanosheet with dip coating method [84]. Graphene oxide thin films with good uniformity and high quality were prepared with a dip-coating method and chemically reduced to highly electrical conductive graphene nanosheet films with hydriodic acid. AgNWs were deposited on the surfaces of the GN film to form high-performance GN/AgNW films. Figure 7.9 shows the SEM image of GN/AgNW films with variation of AgNWs. The contents of AgNWs were varied through adjusting the deposition time. As shown in the SEM image, with increasing the deposition time, the content AgNWs in the scanning scope increases. The increasing content of AgNWs leads to the construction of conductive networks built up by AgNWs.

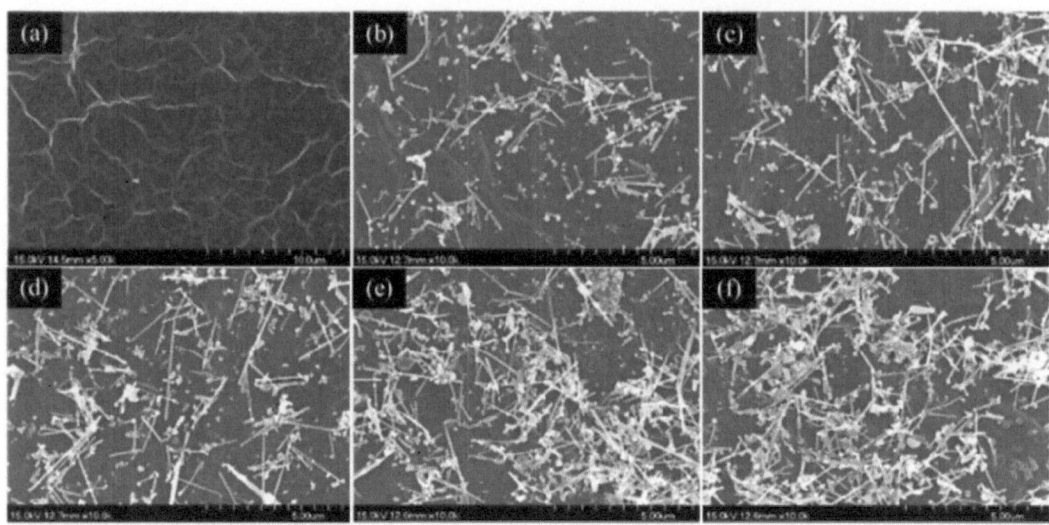

FIGURE 7.9 The FESEM images of (a) GN, (b) GN/AgNW (5min), (c) GN/AgNW (10 min), (d) GN/AgNW (30 min), (e) GN/AgNW (1 h), and (f) GNS/AgNW (3h) TCFs [84].

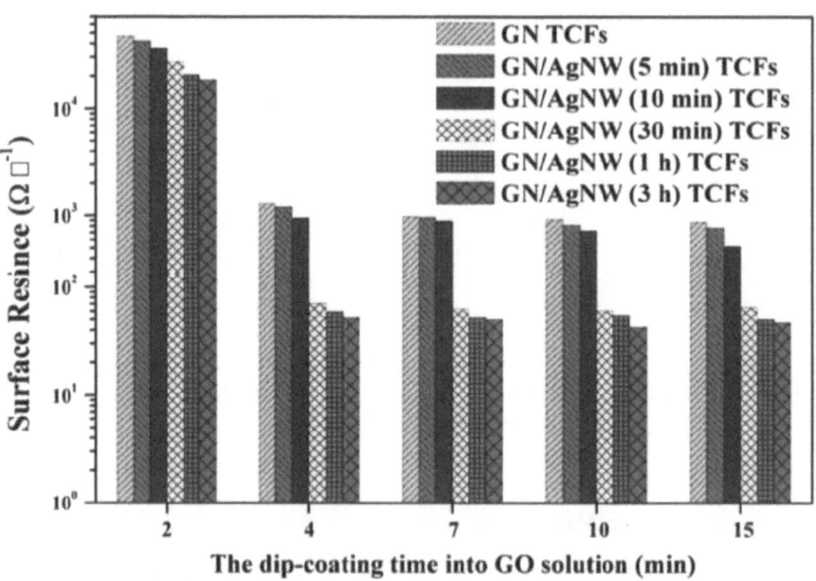

FIGURE 7.10 Surface electrical resistance of TCFs with the various heating time of GO solution and the different dip-coating times in T-AgNW solution [84].

Figure 7.10 shows the surface electrical resistance of resultant GN/AgNW film with various dipping and heating time. With the increase of AgNW content and heating time, the surface electrical resistance of resultant GN/AgNW film exhibits a decreasing tendency, which indicates an increasing electrical conductivity. For example, when the GO solution was heated for 4 min, the GN thin film exhibited a surface electrical resistance of 1.3×10^3 Ω/cm. Table 7.1. summarizes the electrical property of Ag nanowire and nanoparticle coatings and thin films via different deposition methods.

TABLE 7.1

Summary of different deposition methods and properties of the Ag nanowire and nanoparticle coatings and thin films

Materials	Method	Thickness (nm)	Transparency (%)	Electronic Factor	Ref.
Ag NPs	Spin coating	82	Not available	$2.4 \times 10\text{--}5$ W cm	[87]
Ag NPs	Ink-jet printing	530	Not available	$1.6 \times 10\text{--}5$ W cm	[88]
Ag NPs	Ink-jet printing	100	Not available	$3.5 \times 10\text{--}6$ W cm	[89]
Ag NWs	Solution-processed	100	85	10 W/sq.	[90]
Ag NWs	Vacuum filtration	107	85	13 W/sq.	[91]
Ag NPs	Ink-jet printing	300	95	4 ± 0.5 W/sq.	[92]
Ag NWs	Rod-coating	Not available	75	175 W/sq.	[93]
Ag NWs	Spray deposition	Not available	85	33 W/sq.	[94]
Ag NWs	Spray deposition	Not available	80	35 W/sq.	[95]
Ag NWs	Vacuum filtration	Not available	89	69 W/sq.	[96]
Ag NWs	Rod-coating	Not available	91	13 W/sq.	[97]
Ag NWs	Vacuum filtration	Not available	85	10 W/sq.	[98]
Ag NWs	Spray deposition	Not available	90	50 W/sq.	[99]

7.4 OPTICAL COATINGS

7.4.1 POTENTIAL APPLICATION

Optical coatings are one or more thin layers of materials deposited on optical components such as a lens or mirror, which alters the propagation behavior of the light including reflects and transmits [100–102]. The application of optical coating is a way intended to modify the properties of the surface of the substrate in a subscribed way, and therefore the deposition of the coating occurs on the surface of substrate materials [103–107]. Although the coatings introduced onto the surfaces of substrate are expected to employ a series of properties, including environmental, chemical, and thermal, the primary objective is a set of desired spectacular optical properties that alter the quality of the light manipulated by the surface but leaves the direction unchanged. Based on their properties, the optical coating can be divided into two categories: a) anti-reflection coatings, b) high-reflector coatings.

Anti-reflection coatings are employed to reduce unwanted reflection of light from surface of materials, and are commonly applied to spectacle and photographic lenses. High-reflector coatings, typically are used to reflect the light which falls on them.

One of the earliest high-reflector coatings, most probably, was the invention of mirrors during the Rome Empire by introducing molten lead onto glass. However, the hot temperature of the molten metal could destroy the glass into small pieces. After that, improved technique coatings, including tin amalgam coating in 15th century and chemical silvering in 19th century were developed as alternatives. Aluminum, copper, and silver can be employed as the coating materials. Nowadays, the wide application of optical systems in numerous conventional and high technological fields has significantly stimulated the advancement of physics and technology of optical coatings includes transparent dielectric coatings for optical filters (in a broad sense: devices selecting a portion of the transmitted or reflected light, such as anti-reflective (AR) coatings, band pass filters, edge filters, hot/cold mirrors and others, and optical waveguides [108–111]. For example, close to 70% of glass production worldwide, flat glass – for example, window glass, picture glass, laminated glass, motor vehicle windshields and windows, skylight glass, and shaped glass – for example, lenses for precision instruments and for ophthalmic applications, are provided with AR coat. Moreover, the rapidly evolving area of advanced applications includes very narrow band filters (<1 nm) for wavelength division multiplexing and pigments for anti-forgery devices, low laser damage filters, chirped mirrors for ultrashort laser pulse compression, integrated optics for optical signal processing in optical communication, optical computing and optical sensors, and others [112–116].

Transparent conductive oxides (TCOs) and transparent conductive sulfides (TCSs) that possess desirable conductivity and optical permeability have a very important place in optoelectronic industry. The oxides and sulfides of Sn, Zn, In, Cu, Ni, Cd, and their alloys have been extensively studied for their use in semiconductor device technology [117–119]. For example, CdO as an n-type semiconductor has many attractive properties, such as large energy band gap, high transmission coefficient in visible spectral domain, and exceptional luminescence characteristics, which give it good optical conductivity and transmission in the range of visible light. Moreover, ZnSe, ZnTe, CdSe, and CdTe were also employed as optical and conductive coating materials owning to their fascinating optical properties [120–122].

7.4.2 PREPARING METHODS

By now, plenty of fabrication techniques have been developed by the researchers, including ion assisted deposition, ion plating, sputtering, cathodic arc evaporation, pulsed laser deposition, CVD, atomic layer deposition, sol–gel processes, etching and other techniques. Depending on the intrinsic property of coating material system, corresponding preparation techniques are employed. For example, NiO-based optical coating were prepared with electrodeposition, a wet chemical method, RF sputtering, cathodic deposition (electrochemical deposition), DC magnetron sputtering, and successive ionic layer absorption and reaction (SILAR) [123–127]. CVD has been employed to prepare BN and SiBCN coatings on different kinds of optical fibers, aiming to be used up to 1800°C [128–130]. Figure 7.11 exhibits the cross-section morphology of a light-leakage-proof BN coating of silica optical fibers. The coating is prepared at 650 °C by CVD process and the gas flow ratio of

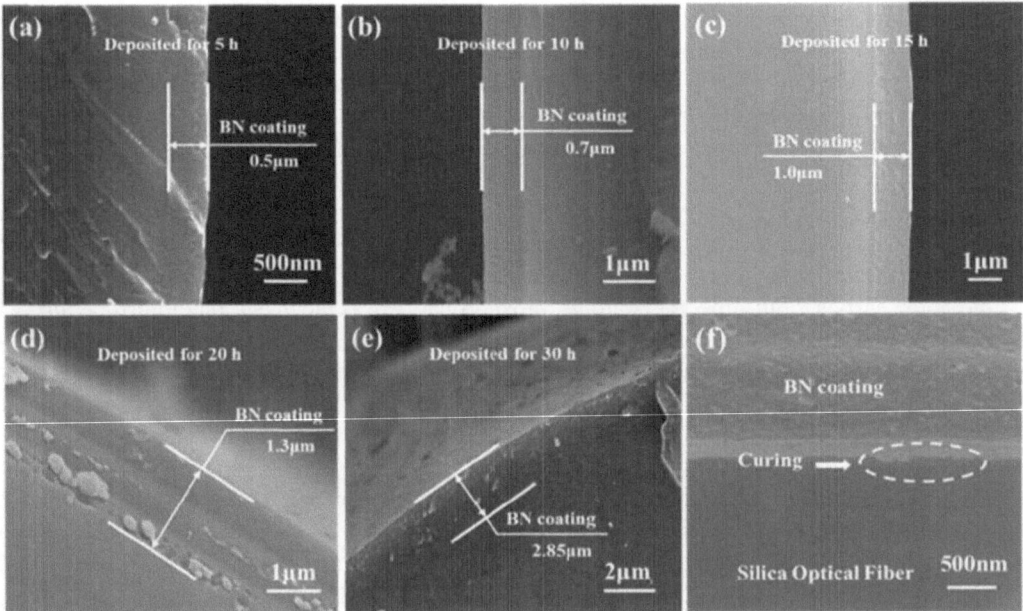

FIGURE 7.11 Cross-section morphology of BN coated silica optical fibers with different deposition time (a–e) and microstructure of the interfacial area between silica optical fiber and BN coating (f) [128].

BCl_3: NH_3: H_2: Ar is 3: 5: 12: 12 [128]. The thickness of BN coating increases as the deposition time rises. The coating would become loose and uneven when the deposition time is too long. This phenomenon relates to two major factors. Firstly, the coating tends to crack when it is too thick because the crystallization degree of BN deposited at 650 °C is quite low (i.e., the BN coating has a high lattice defect density) [131]. Secondly, the axial compressive stress of silica optical fiber would gradually increase as the thickness of BN coating increases, resulting in the crack of the coating. Healing of defects could be found on silica optical fiber in Figure 7.11(f), which is beneficial to the mechanical properties of the silica optical fiber.

SiBCN coating of silica optical can be prepared by polymer derived ceramics (PDC) method [132]. The precursor solution was prepared at a mass ratio of PSNB to THF of 1: 2 and 1: 3. Silica optical fiber is immersed in the prepared precursor for 1 h. The samples were cured at 120 °C in air for 2 h. Finally, the samples were pyrolyzed at 700 °C in Ar for 2 h. The parameters for preparing PDC SiBCN coating are shown in Table 7.2. Figure 7.12 shows the surface morphology of PDC SiBCN coated silica optical fiber. A dense flat SiBCN coating can be obtained when the mass ratio of PSNB to THF is 1: 2. And the coating prepared with three cycles of immersing–curing process is flatter than that with five cycles. As shown in Figures 7.12(b) and (d) that the coating would be flatter as the heating rate of pyrolysis decreases. This phenomenon is related to the volatilization of small-molecule gases such as NH_3, H_2, H_2O, CH_2, and CO_2.

TABLE 7.2

Comparison of parameters for preparing PDC SiBCN coatings [132]

PSNB: THF	Cycles of immersing–curing process	Coating thickness	Heating rate of pyrolysis	Morphology
1:2	3	Not measured	3 °C/min	Figure 7.4(a)
1:3	3	3.0 μm	3 °C/min	Figure 7.4(b)
1:3	5	3.5 μm	3 °C/min	Figure 7.4(c)
1:3	3	0.5 μm	1 °C/min	Figure 7.4(d)

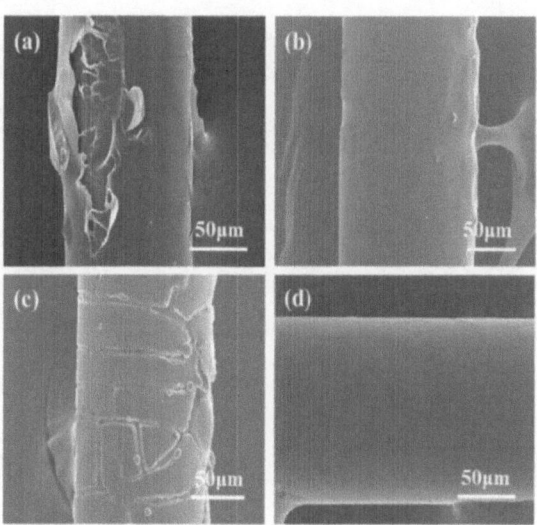

FIGURE 7.12 Surface morphology of PDC SiBCN coated silica optical fiber with different precursor concentration, curing, and pyrolysis methods [132].

Based on the preparing methods of CVD BN and PDC SiBCN coating mentioned above, BN/SiBCN light-leakage-proof coatings of silica optical fiber are prepared [132]. Surface and cross-section morphologies of BN/SiBCN coated silica optical fiber are shown in Figure 7.13. As shown in Figure 7.13, the thickness of BN layer is 0.5 μm. It could be found in Figure 7.13(a) that the coating is flat, just with a few defects on the surface. The small-molecule gases generated during the pyrolysis process is hard to diffuse out when the coating is too thick. Therefore, the coating is porous and rough when SiBCN coating thickness is 2 and 2.5 μm.

Figure 7.14 shows a schematic illustration of SILAR technique for the preparation of Cu doped NiO thin films, which can be described as: 1) adhesion of precursor in the substrates; 2) formation of the $Cu_xNi_{1-x}O$ layer with water bath; 3) drying of substrate in the air, and 4) removing of excessive species from the substrate with deionized water [133]. It is worthy to mention that SILAR method has several advantages such as repeatability, low cost, and deposition temperature. Moreover, thickness, particle size, and morphology of the films can be easily controlled by changing the deposition

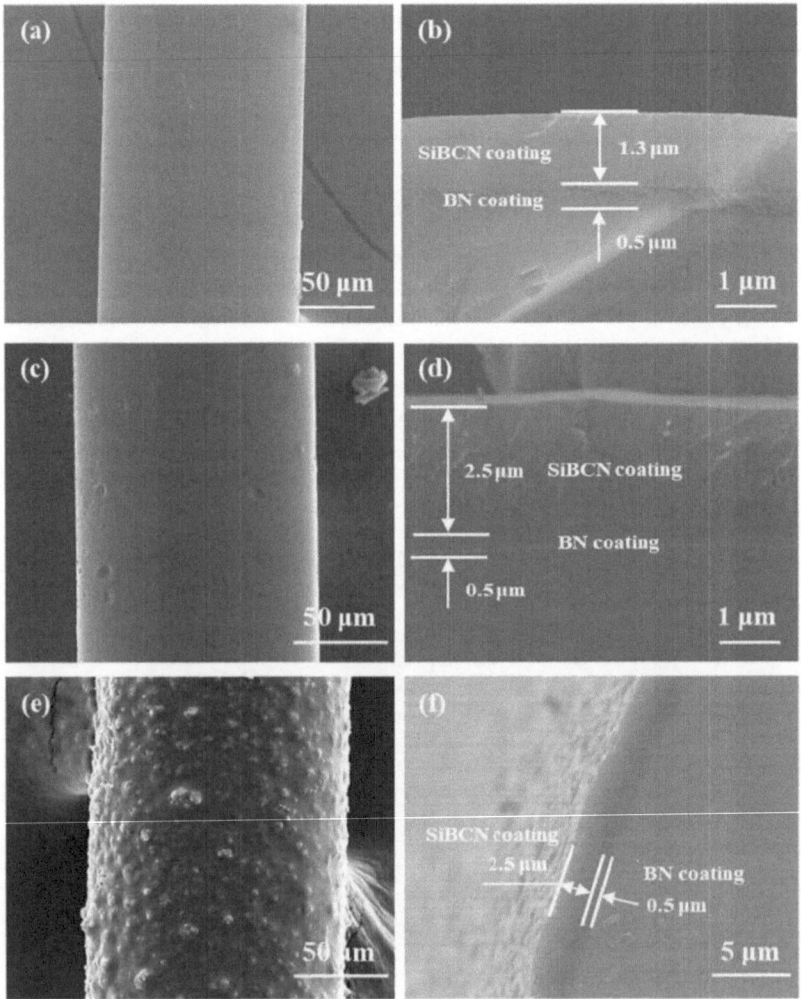

FIGURE 7.13 Surface morphology (a, c, and e) and cross-section morphology (b, d, and f) of BN/SiBCN coated silica optical fiber with different SiBCN thickness (1.3 μm for (a) and (b), 2 μm for (c) and (d), and 2.5 μm for (e) and (f)) [132].

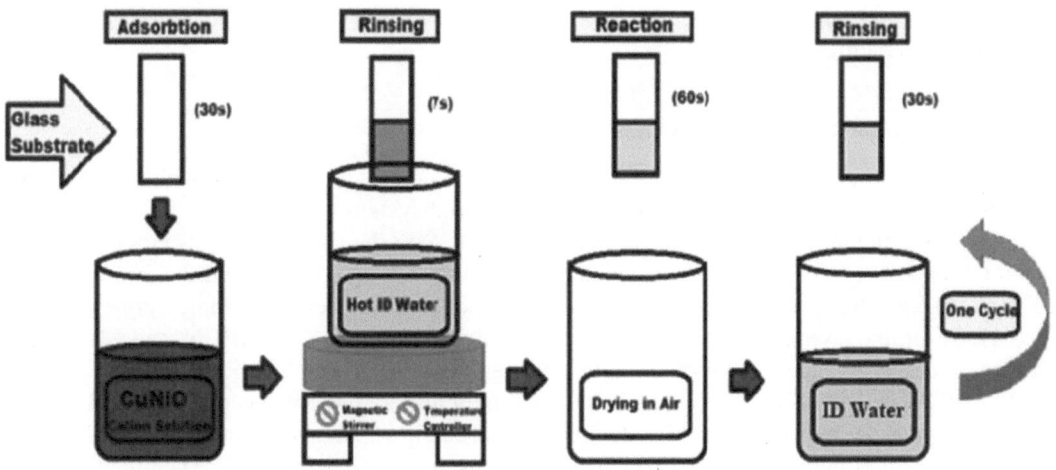

FIGURE 7.14 Schematic representation of SILAR method [133].

cycle, bath temperature, pH, and concentration of the solution in this method. Regardless of substrate type, shape, and surface profile, any substrate can be used for deposition in SILAR.

DC magnetron sputtering, spray pyrolysis, chemical bath deposition (CBD), SILAR, activated reactive evaporation, metal organic chemical vapor deposition (MOCVD), pulsed laser deposition, and sol–gel dip coating were employed to deposit optical thin films [120, 133–136]. According to the comparison of all of these methods, spray pyrolysis is more beneficial than all other methods. The ultrasonic spray pyrolysis has many benefits: (a) it is altogether facile in structure, (b) the necessary arrangement cost is less valuable and ductile for the procedure of alternations, and (c) capability of large area coatings.

ZnO thin films with a melting point of 1950°C have been prepared by many techniques such as molecular beam epitaxy, atomic layer epitaxy, MOCVD, spray pyrolysis, sol-gel method, pulsed laser deposition, magnetron sputtering, and electron beam evaporation [137]. Among these techniques, sol-gel method attracts much attention due to some distinct advantages including low cost, simple deposition equipment, ease of adjustment of composition, ability to carry out doping at molecular level, and ease of fabrication of large-area films. Figure 7.15 shows the schematic sketch of the growth mechanism of ZnO films fabricated with sol-gel method, where ZnO thin films with different thicknesses were prepared by sol-gel method on glass substrate [137].

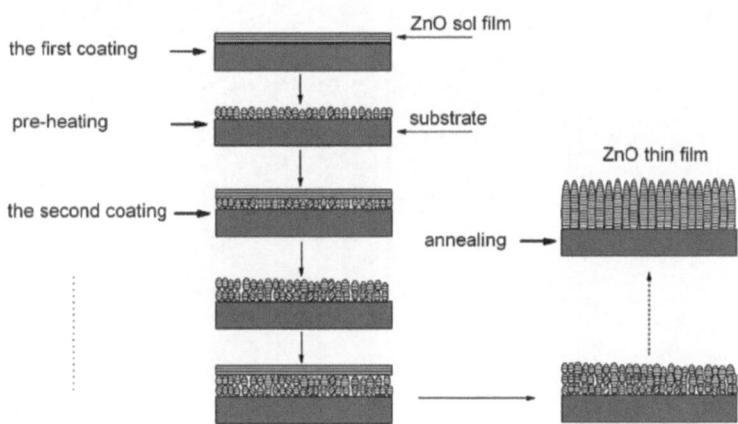

FIGURE 7.15 The schematic sketch of the growth mechanism of self-template process [137].

Moreover, some other techniques such as plasma-enhanced CVD and CBD was employed to prepare optical coatings.

7.4.3 MICROSTRUCTURE AND PROPERTIES

Here, the properties, especially the transparency of optical coatings is discussed with respect to the (micro)structure, coating thickness, frequency, and material system.

According to the light transmittance performance, BN, SiBCN, and BN/SiBCN coatings can be ideal total reflection solutions to protect silica optical fiber from light leakage [128, 130]. As shown in Figure 7.16(a), the laser spreads out from around silica optical fiber, indicating a poor transmittance. However, the laser can go through the silica optical fibers with BN, SiBCN, and BN/SiBCN coatings successfully and stably without light leakage [128, 130].

BN, SiBCN, and BN/SiBCN coated silica optical fiber also shows enhanced mechanical properties at both room temperature and high temperature [128, 130]. As shown in Figure 7.17(a), BN coated fibers exhibit higher strength and tensile modulus than the original fiber. The highest tensile strength of 385 MPa and highest elastic modulus of 61 GPa are achieved when the thickness of BN coating is 1 μm. In addition, the improvement in strength and elastic modulus of BN coated fiber is higher at 700°C (169% for tensile strength and 79% for elastic modulus) than at room temperature (136% for tensile strength and 19% for elastic modulus). Additionally, as shown in Figure 7.17(b), BN coated silica optical fiber keeps a linear deformation behavior at 700 °C, which is beneficial to its application as a strain sensor.

As for the silica optical fibers with BN/SiBCN coating, their tensile strength exhibits an increase of 90% and 72% at room temperature and 700 °C, respectively. It is found that the improvement of the strength of BN/SiBCN coated silica optical fiber is related to the healing of defects and residual compressive stress in the fibers. It could be found in Figure 7.18(b) that BN/SiBCN coated silica

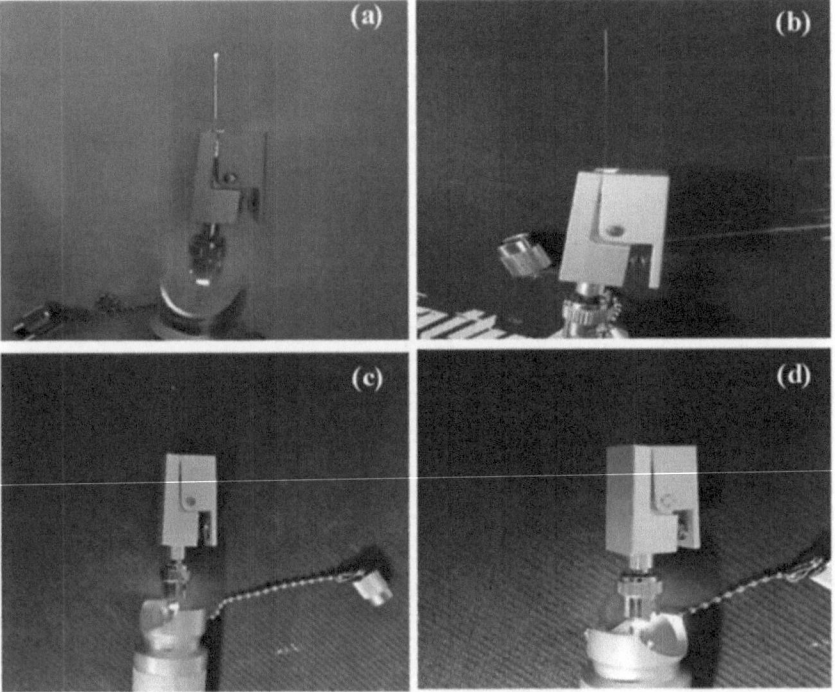

FIGURE 7.16 Transmittance analysis of silica fibers (a) without coating and with (b) BN, (c) SiBCN, and (d) BN/SiBCN coating [128, 132].

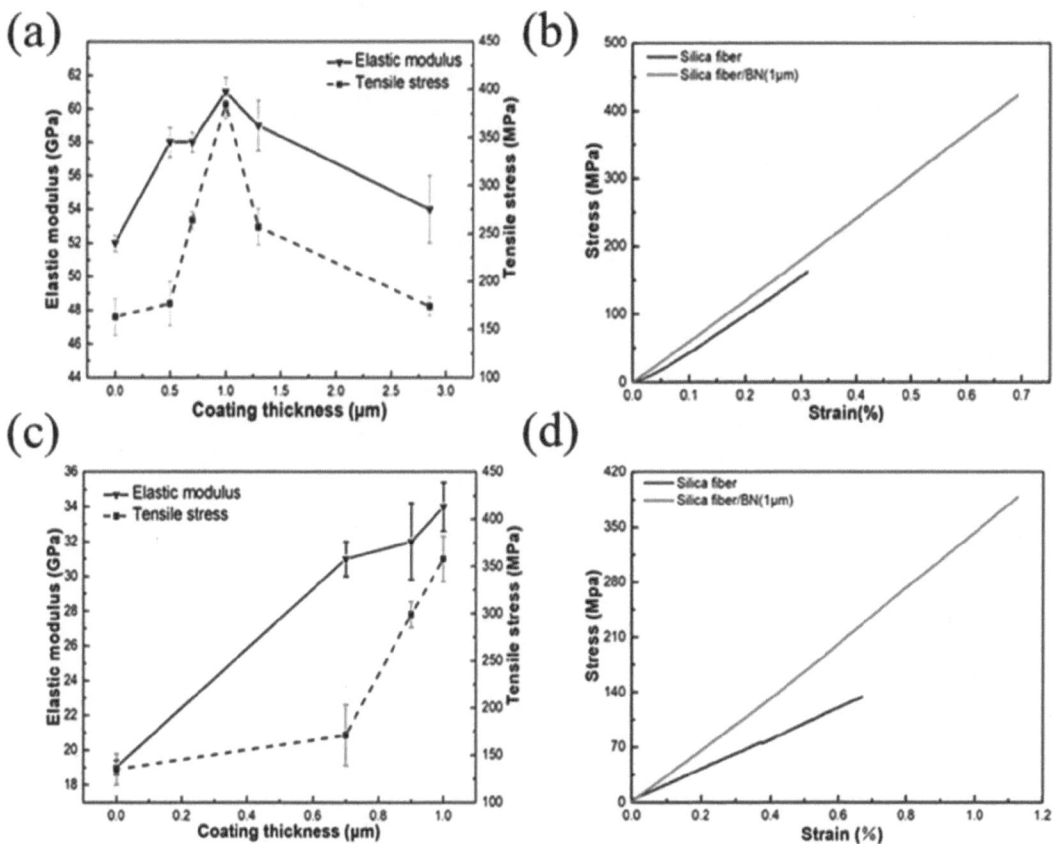

FIGURE 7.17 Tensile strength and elastic modulus of BN coated silica optical fibers with different BN coating thickness (at (a) room temperature and (c) 700 °C); typical tensile stress–strain curves of silica optical fiber and BN coated silica optical fiber with BN coating thickness of 1 μm (at (b) room temperature and (d) 700 °C) [128].

optical fiber exhibits a linear deformation behavior, which is benefit to its application as a strain sensor. Interestingly, an inflection point could be found on the stress–strain curve (see the black circle in Figure 7.18(d)). The point represents the boundary where the crack begins to expand in silica optical fiber and the stress level corresponding to this point is the in-situ strength of silica optical fiber at 700 °C.

Figure 7.19 shows the SEM image of the spin-coated ZnO coatings deposited on silica glass substrates after heat treated at various temperatures [138]. As shown in Figure 19, the ZnO coatings exhibit a homogeneous structure with grain size of about 50 μm. From Figure 19(e–g), we can see grain size of ZnO with the second-heat treatment increased. Moreover, in Figure 19(g), we can see the resultant material exhibits a dense structure. In Figure 7.20 [138], the transmittance of the film with the second-heat treatment in the visible range can reach as high as 90%, which is about 5% higher than the others at wavelengths of 400–800 nm. Surface morphology of optical coatings has strong influence on the optical properties of the films. The increase in transmittance of the second-heat treated coatings may be resulting from decreasing optical scattering caused by the densification of coating, followed by grain growth and the reduction of grain boundary density (Figures 7.19(c) and (e)).

Besides surface morphology, the material species, coating thickness, deposition temperature, and time can also influence the transmittance of the optical coatings. Table 7.3. reviews the preparative

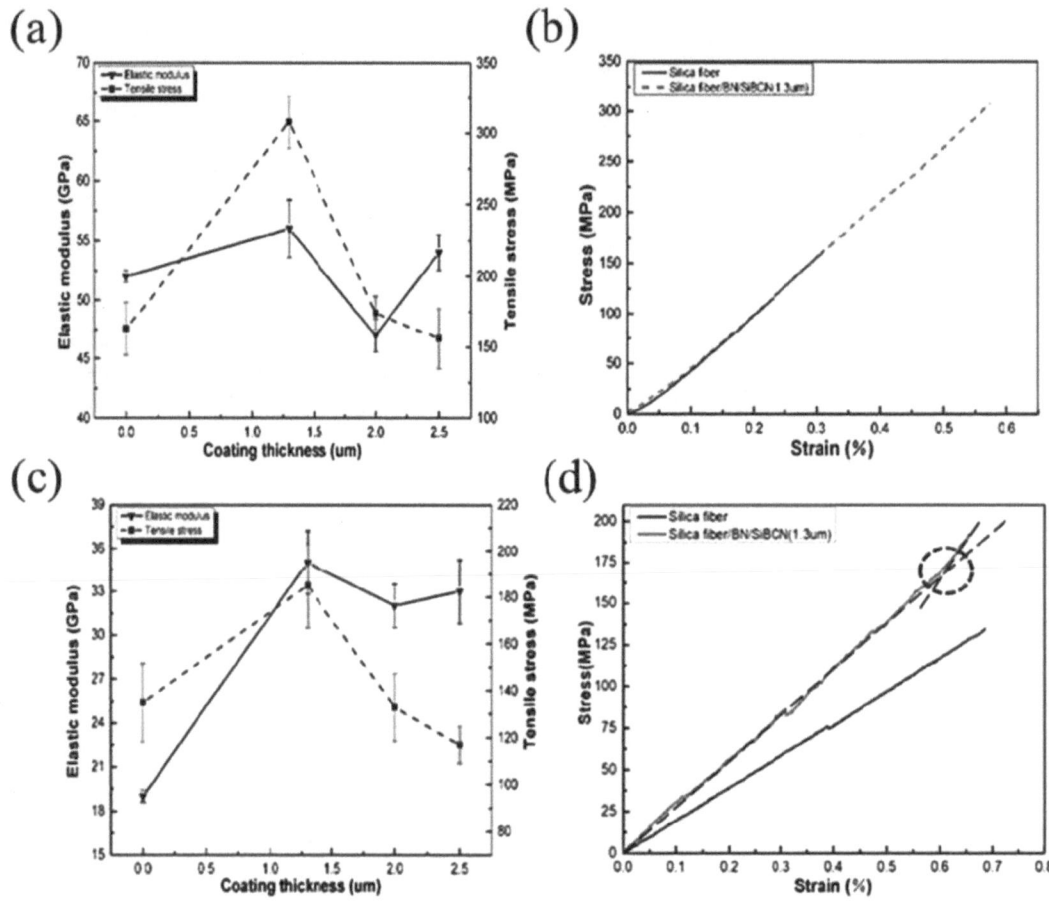

FIGURE 7.18 Tensile strength and elastic modulus of BN/SiBCN coated silica optical fiber with different SiBCN coating thicknesses (at (a) room temperature and (c) 700 °C) and tensile stress–strain curves of the original and BN/SiBCN coated silica optical fiber (at (b) room temperature and (d) 700 °C) [132].

conditions and properties of CBD-ZnS optical coatings [139]. As shown in Table 7.3., the transmittance of the ZnS optical coatings varies with the coating thickness, preparation temperature, and deposition time.

7.5 MAGNETIC COATINGS

7.5.1 POTENTIAL APPLICATION

Magnetic materials have been known with the finding of magnetite (Fe_3O_4) for more than 2500 years [140]. Starting with the invention of compass, they have been a crucial part of human civilization and allowed explorers to travel around the world and explore the lands unknown. Nowadays, with the developments in technology, magnetic materials can be found everywhere from magnetic storage technology, credit cards, magnetic brakes in hybrid vehicles, power transformers, to even biomedical applications. Typically, properties of magnetic materials varies significantly with reduced dimensions. By now, magnetic coatings represent the foundation of most device applications.

With the fast development of electronic devices, coatings exhibiting magnetic functionality have found wide application in various fields of electronics, such as information storage and record, electromagnetic interference compatibility, magnetic field sensors, wireless communication devices,

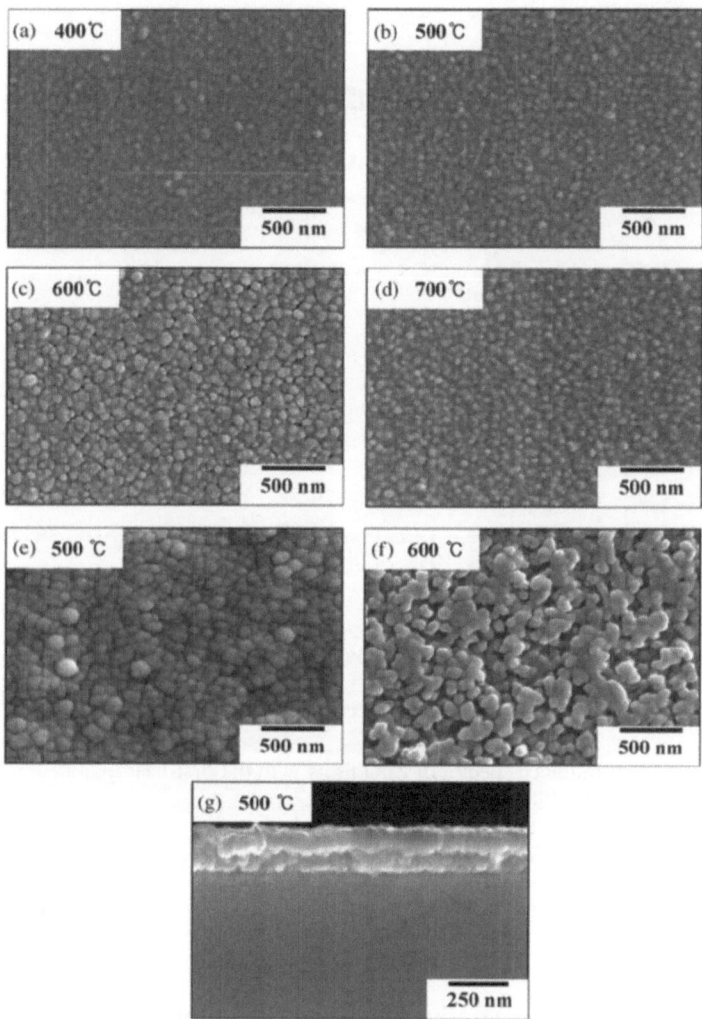

FIGURE 7.19 (a–d) SEM images of the surface morphology and cross section of ZnO coatings heat treated at various temperatures (400–700ºC) ; (e–g): the second-round heat treatment of resultant coatings [138].

spin-based electronic devices, and electrode materials for batteries [141–143]. For different applications, magnetic thin films possessing different properties are required. Among all the parameters, the complex permeability of a magnetic thin film is the most important factor that determines the workability of a magnetic thin film in engineering application and the performances of the devices made from it.

7.5.2 Preparing Methods

Principally, deposition techniques of magnetic coatings can be divided into four categories, which are denoted as electroplating (also known as electro-chemical deposition or electrode-position), electroless (chemical) deposition, CVD, and PVD. For physical deposition techniques, sputtering [6], evaporation [7], and molecular beam epitaxy [8] are three frequently used methods to prepare magnetic thin films.

With the advantages of simplicity, cost-effectiveness, and controllable patterning, electrode-position technique was the first approach that was widely employed for the preparation of magnetic films. Through controlling the deposition parameters including temperature, current, size, and

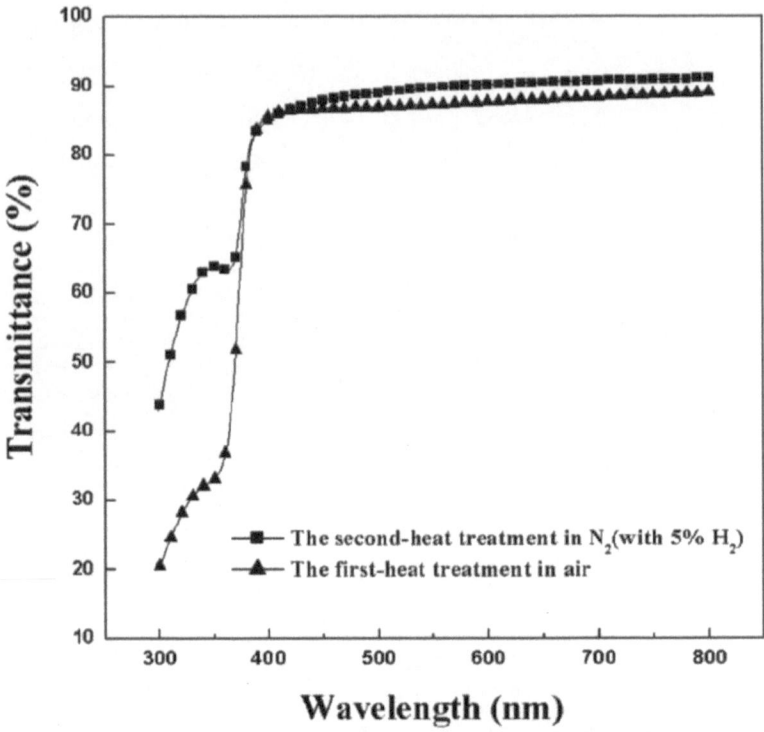

FIGURE 7.20 Optical transmittance spectra of ZnO films with the first-heat treatment at 600°C and with the second-heat treatment at 500°C [138].

TABLE 7.3
Details of recent reports on preparative conditions and properties of CBD-ZnS films [139]

pH value	Thickness (nm)	Temperature (°C)	Deposition time	Substrate	Transparency (%)
10–11.5	-	90	3 h	Glass	70
10	50–110	85	3 h	Glass	80
10–11.5	-	90	3 h	SnO$_2$ coated glass	70
-	120–140	80, 90	20–80 min	Glass	90
-	-	81	40, 60 min	Glass	70–88
10	37–135	80	2 h	Glass	70–80
10	100	80	4 h	ITO coated Glass	60–70
10	-	80	4 h	Glass	85
10	70–140	80	4 h	Glass	75–85
-	80	70	2 h	Soda lime Glass	70
5.9–6.1	140–175	85	4 h	Glass	90
-	-	50–90	1.5–2.5 h	Soda lime Glass	-
9–11	54–122	80	1 h	Glass	75
9.7	90–110	80	20–120 min	Glass	77
10	-	80	2 h	Glass	50–87
10.5	-	75	4–16 h	Soda lime glass	80

composition of electrode, the structure of deposited films can be facilely controlled. However, the electrode position techniques also suffer from their intrinsic disadvantage. Owning to the limited control of layer properties on the nanoscale, the thickness of the deposited films typically vary in the micrometer regime. Moreover, electroplating confronts a serious challenge when depositing pure RE metals due to the negative reduction potential.

Electroless deposition typically occurs through the chemical reduction of the targeted deposition material from a precursor without the additional adding of voltage or current. The obvious disadvantage of electroplating is the limited deposition speed due to the difficulty in controlling the chemical reduction. The advantage of electroless deposition is that the final coatings exhibit much smoother profile. The most suitable electroless plating method for ultrathin layers is dip-coating.

Plasma-assisted sputtering deposition (often known as sputtering) is, by now, the most widely employed coating PVD technique since it allows for the facile and fast fabrication of uniform coatings at a large scale and high quality. The rate of deposited coatings can be easily controlled by changing the power and gas pressure.

7.5.3 MICROSTRUCTURE AND PROPERTIES

Figure 7.21 shows the SEM image of FeCoNi alloy nanocrystals magnetic thin films electrodeposited on an indium tin oxide-coated polyethylene terephthalate sheet from sulfate electrolyte in the

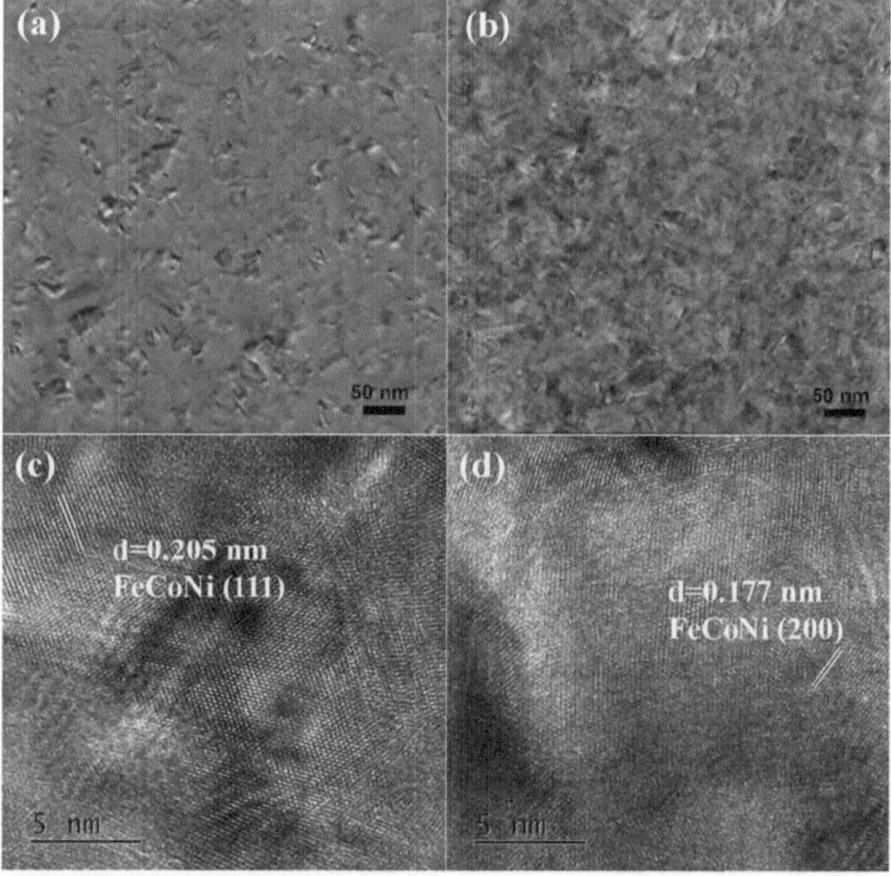

FIGURE 7.21 STEM bright field images of the electrolytes from el1 (a) and el2 (b), STEM bright field images of the FeCoNi electrodeposited from the electrolytes from el1 (c) and el2 (d) [144].

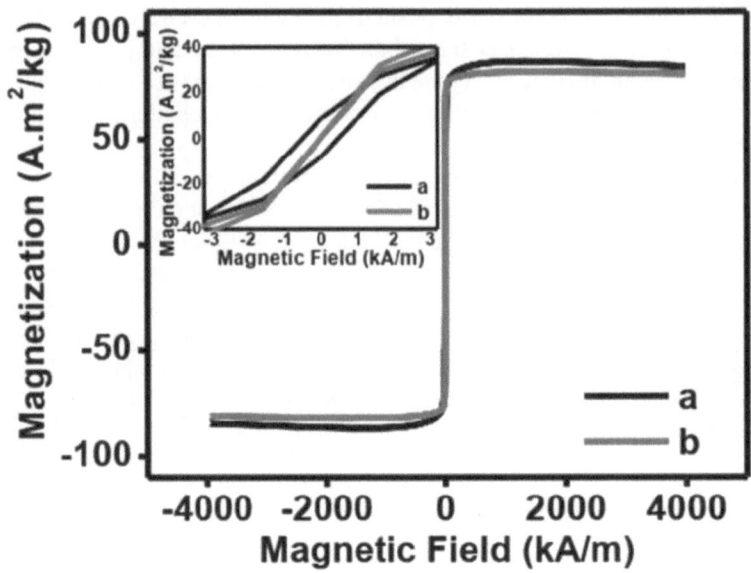

FIGURE 7.22 Magnetization curves of the FeCoNi films electrodeposited from el1 (a) and el2 (b) measured at 300 K. The enlarged curves under low magnetic field are displayed in the inset [144].

presence of saccharin [144]. The film electrodeposited in the presence of saccharin (Figure 7.21b) has more uniform microstructures than the one prepared without saccharin (Figure 7.21a).

The FeCoNi film exhibited excellent soft magnetic behavior at room temperature with a relatively high saturation magnetization (Figure 7.22) [144]. In this case, the decrease in coercivity corresponds to the increase in Ni content and reduction in Fe content. In addition, the coercive field of the crystalline alloy is strongly affected by the material structure and is extremely sensitive to its crystallite size.

7.6 LOW-EMISSIVITY COATINGS

7.6.1 POTENTIAL APPLICATION

With the development of infrared detection technology, infrared stealth becomes more and more important to enhance survivability of military objects effectively [145, 146]. Nowadays, the common method for infrared stealth is decreasing the infrared emissivity of objects according to the Stefan–Boltzmann Law [147]. Therefore, coating low infrared emissivity materials on the object surfaces to weaken infrared characteristic signal of the objects is a good way to achieve the stealthy technology requirement [148]. So far, many kinds of materials with low infrared emissivity have been exploited [149–152]. Among them, Low emissivity films have received more considerable attention due to their light weight and small volume, for instance, indium tin oxide (ITO), PtO_x, TiN_x, Au/Ni, and Pt.

7.6.2 PREPARING METHODS

Low-emissivity thin films were prepared by direct-current (DC) magnetron sputtering method using a metal target. Prior to deposition, the glass substrates were ultrasonically cleaned in deionized water, acetone, and ethanol for 15 min, respectively. The chamber was evacuated to the base pressure of 5.0×10^4 Pa. After the ultimate vacuum was achieved, the target was sputtered in Ar gas (99.999% pure) for 15 min to remove the surface oxide layer [153].

7.6.3 MICROSTRUCTURE AND PROPERTIES

Figure 7.23(a–d) shows SEM images of TiN_x films at various N_2 flow rates. The SEM images reveal that the nanostructured morphology is dense and uniform with trigonal pyramidal shaped grains on the surface of the films. The thickness of TiN_x films measured using SEM is almost the same (1.5 μm). The resistivity of the TiN_x films at different N_2 flow rates is presented in Figure 7.24. The resistivity increases from 305 to 673 μΩ cm as the N_2 flow rate rises from 1 to 8 sccm. In addition, the sputtering pressure is also strongly affected by the material structure. The higher resistivity for higher N_2 flow rate is perhaps contributed to the increment of grain boundary in the film.

Figure 7.25 shows the comparison of the EMI-shielding effectiveness for glass substrate and films deposited on glass. When N_2 flow rate is 1 sccm, the maximum SET of TiN_x films is obtained and greater than 20 dB in 8.2–12.4 GHz range. To achieve a similar EMI-shielding value, the thickness of TiN_x films prepared in this work is only 1.47 μm, which is much smaller than the thickness of the presently known EMI-shielding materials.

SEM micrographs of the ITO films as-deposited and annealed at 200 °C for 1 h are given in Figure 7.26. The as-deposited ITO film shows surface features of densely packed fine particle and neighboring fine particles agglomerating to form a not-obvious large particle shape. After annealing at 200 °C for 1 h, the film is crystalline with regular grain shape and uniform grain size. As we know, film atoms will get more energy at higher temperatures to migrate and crystal nuclei will sufficiently grow up, leading to reconfiguration of grains and formation of particles with perfect crystal structure.

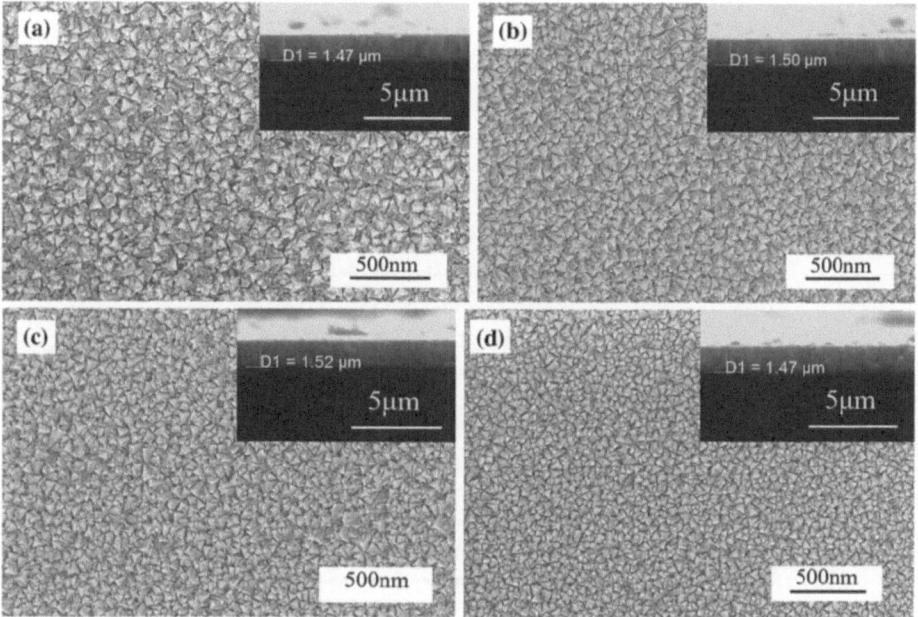

FIGURE 7.23 SEM images of the TiN_x films at various N_2 flow rates: (a) 1 sccm, (b) 2 sccm, (c) 4 sccm, and (d) 8 sccm [154].

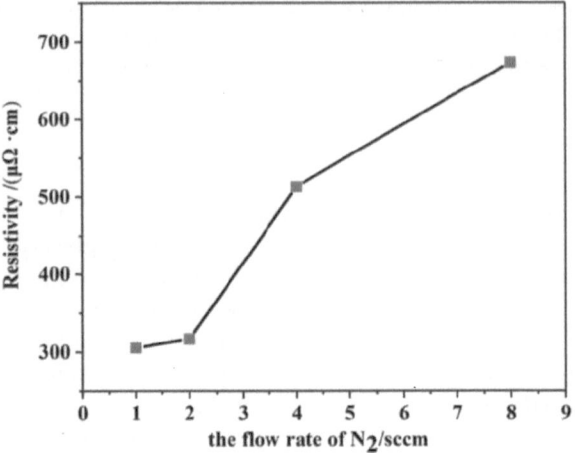

FIGURE 7.24 Resistivity of TiN$_x$ films at various N$_2$ flow rates [154].

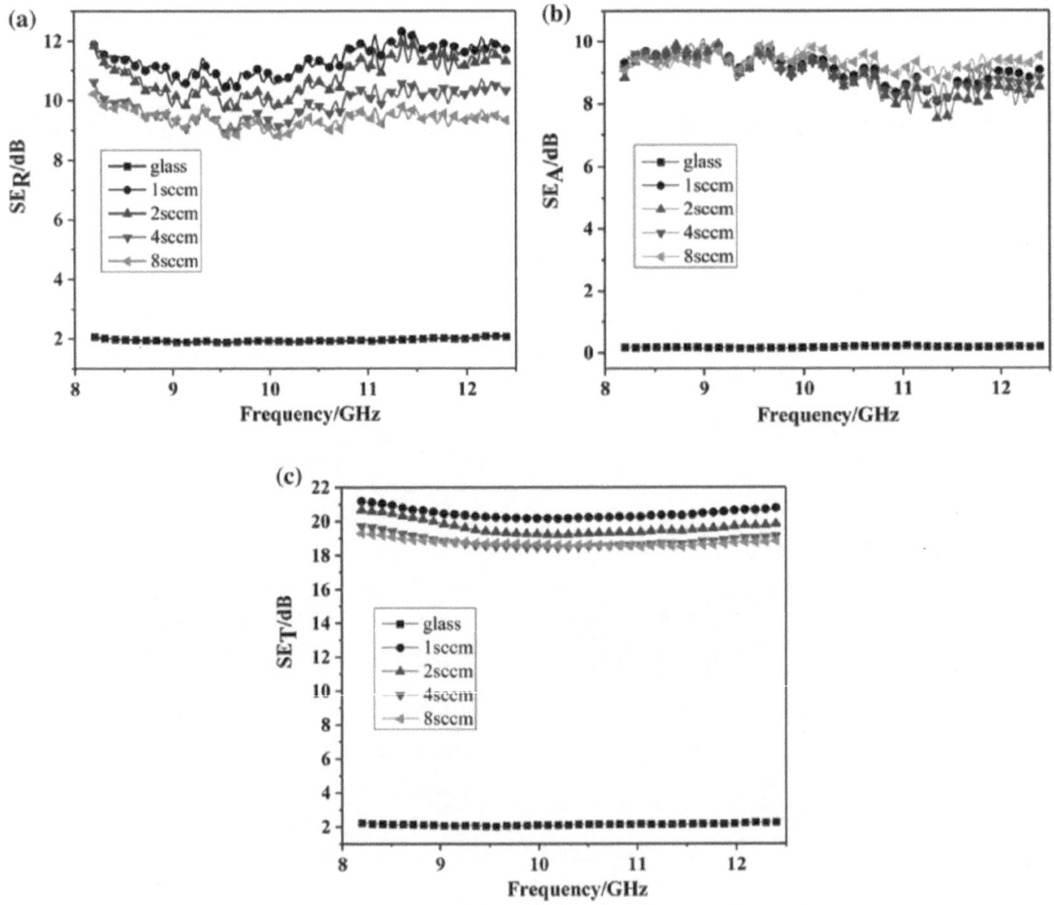

FIGURE 7.25 EMI-shielding effectiveness of the TiN$_x$ films at various N$_2$ flow rates: (a) SER, (b) SEA, and (c) SET [154].

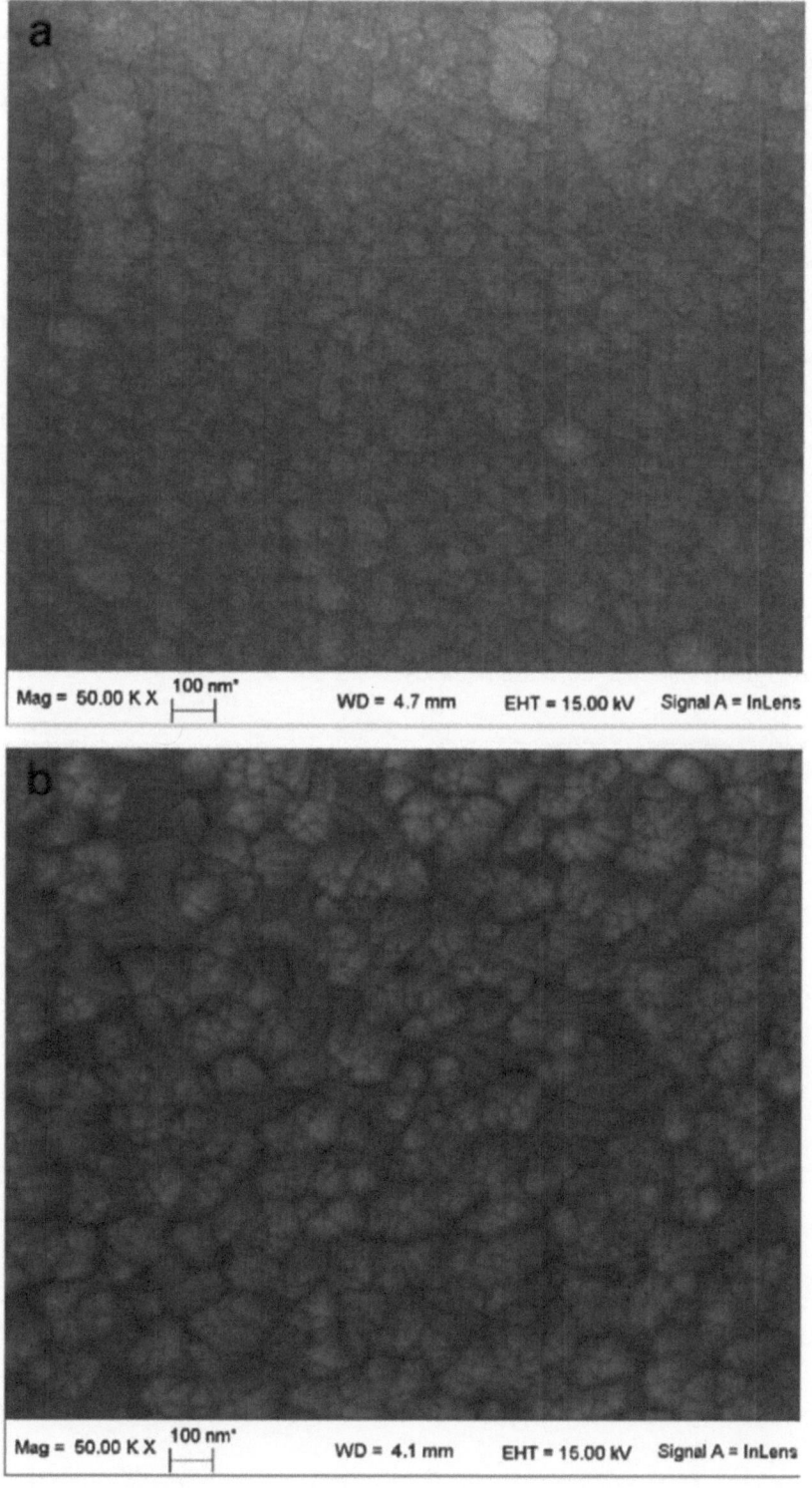

FIGURE 7.26 SEM micrographs of the ITO films as-deposited (a) and annealed at 200 °C for 1 h (b) [155].

The infrared emissivity of all films increases with the increasing temperature and decreases with the decreasing temperature. After the whole process of heating and cooling between room temperature and 350 °C, the infrared emissivity of the films not-annealed and annealed at 100 and 200 °C increases but that of the films annealed at 300 and 400 °C is almost unchanged. The emissivity of ITO films will increase with the increasing absorption, or in other words, with the increasing temperature. So the infrared emissivity of all films increases in the heating process and decreases in the cooling process.

Figure 7.28 shows the SEM images of PtO_x films with different annealing treatments revealing that the post-deposition annealing process would lead to an appreciable change in the surface morphology. It could be seen that the as-deposited thin films were smooth and consisted of nanoparticles. For the films annealed at 400 °C, the surface morphology was very similar with that of the as-deposited. When the films were annealed at 500 °C, their surface became smoother and finer particles distributed uniformly probably due to the slight decomposition of PtO_x films. For the films annealed at 550 °C, fine particles grew and a thin Pt layer formed on the surface, which could be seen from the cross-sectional morphology. After annealing at 600 °C, due to the complete decomposition of PtO_x films and the rapid release of O_2, many nanopores dispersed on the films and the thickness of the films decreased from about 600 nm to about 300 nm.

Figure 7.29. shows the average infrared emissivity of as-deposited PtO_x films and of the ones annealed at different temperatures. The average infrared emissivity of the as-deposited PtO_x films and annealed at 400, 500, 550, and 600 °C is 0.605, 0.641, 0.688, 0.178, and 0.056, respectively. It could be seen that, with the increase of the annealing temperature, the infrared emissivity of the films increased firstly and then decreased quickly.

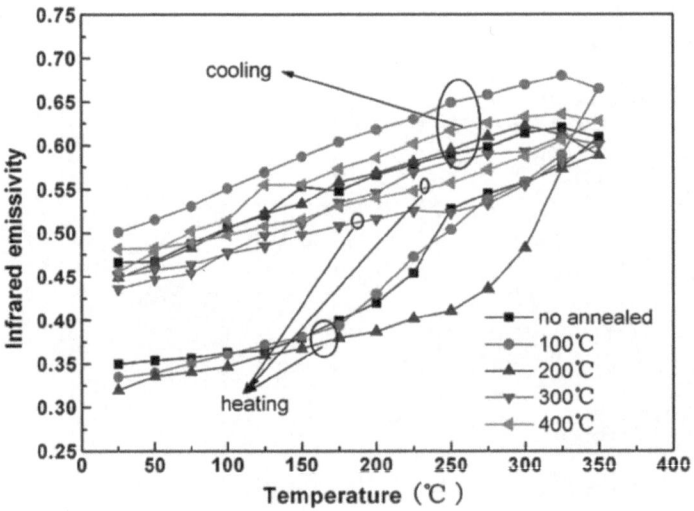

FIGURE 7.27. Temperature dependence of the mean infrared emissivity for not-annealed and annealed films [155].

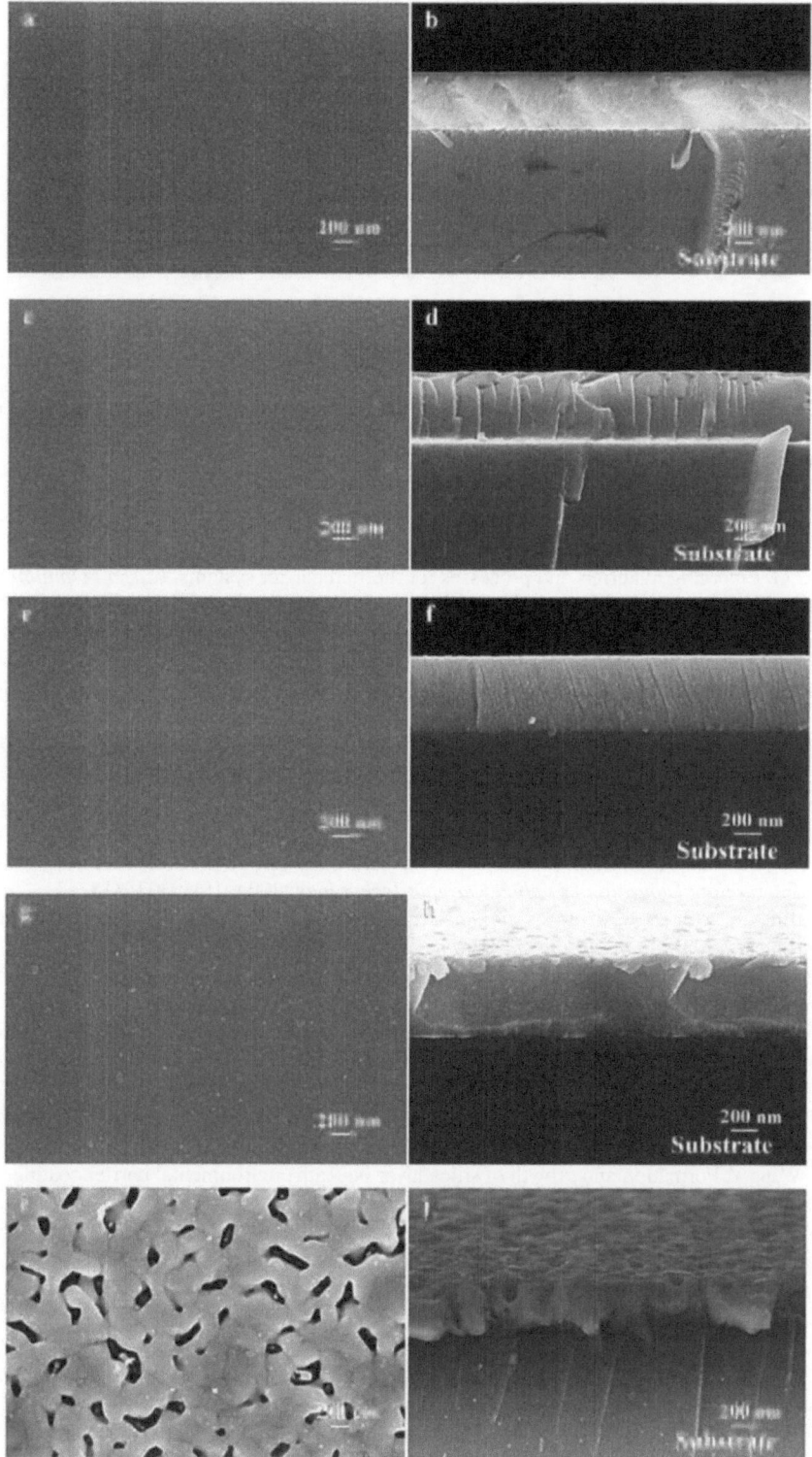

FIGURE 7.28 Surface and cross-sectional morphology of PtO$_x$ films annealed at different temperatures: (a) and (b) as-deposited, (c) and (d) 400 °C, (e) and (f) 500 °C, (g) and (h) 550 °C, (i) and (j) 600 ºC [156].

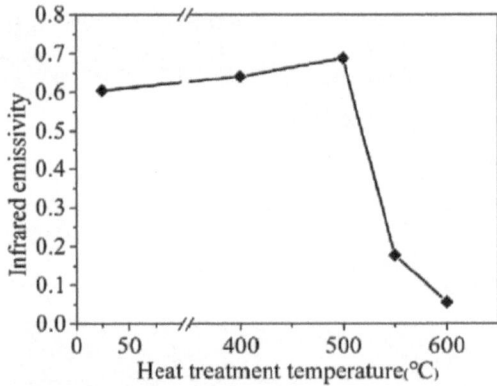

FIGURE 7.29. The infrared emissivity of as-deposited PtO$_x$ films and annealed ones at different temperatures [156].

REFERENCES

[1] Levi C G. Emerging materials and processes for thermal barrier systems. *Current Opinion in Solid State and Materials Science*, 2004, 8(1): 77-91.

[2] Clarke D R, Levi C G. Materials design for the next generation thermal barrier coatings. *Annual Review of Materials Research*, 2003, 33(1): 383–417.

[3] Padture N P, Gell M, Jordan E H. Thermal barrier coatings for gas-turbine engine applications. *Science*, 2002, 296(5566): 280–284.

[4] Li S L, Qi H Y, Yang X G. Oxidation-induced damage of an uncoated and coated nickel-based superalloy under simulated gas environmen. *Rare Metals*, 2018, 37(3): 204–209.

[5] Chen H, Liu Y, Gao Y, et al. Design, preparation, and characterization of graded YSZ/La2Zr2O7 thermal barrier coatings. *Journal of the American Ceramic Society*, 2010, 93(6): 1732–1740.

[6] Schulz U, Saruhan B, Fritscher K, et al. Review on advanced EB-PVD ceramic topcoats for TBC applications. *International Journal of Applied Ceramic Technology*, 2004, 1(4): 302–315.

[7] Lv B, Jin X, Cao J, et al. Advances in numerical modeling of environmental barrier coating systems for gas turbines. *Journal of the European Ceramic Society*, 2020, 40(9): 3363–3379.

[8] Lee K N. Key durability issues with mullite-based environmental barrier coatings for Si-based ceramics. *Journal of Engineering for Gas Turbines and Power*, 2000, 122(4): 632–636.

[9] Liu J, Zhang L, Liu Q, et al. Polymer-Derived SiOC–barium–strontium aluminosilicate Coatings as an Environmental Barrier for C/SiC Composites. *Journal of the American Ceramic Society*, 2010, 93(12): 4148–4152.

[10] Liu J, Zhang L, Hu F, et al. Polymer-derived yttrium silicate coatings on 2D C/SiC composites. *Journal of the European Ceramic Society*, 2013, 33(2): 433–439.

[11] Lu Y, Wang Y. Formation and growth of silica layer beneath environmental barrier coatings under water-vapor environment. *Journal of Alloys and Compounds*, 2018, 739: 817–826.

[12] Darolia R. Thermal barrier coatings technology: critical review, progress update, remaining challenges and prospects. *International Materials Reviews*, 2013, 58(6): 315–348.

[13] Dong Y, Ren K, Lu Y, et al. High-entropy environmental barrier coating for the ceramic matrix composites. *Journal of the European Ceramic Society*, 2019, 39(7): 2574–2579.

[14] Padture N P. Advanced structural ceramics in aerospace propulsion. *Nature Materials*, 2016, 15(8): 804.

[15] Gäumann M, Bezencon C, Canalis P, et al. Single-crystal laser deposition of superalloys: processing–microstructure maps. *Acta Materialia*, 2001, 49(6): 1051–1062.

[16] Chen H F, Zhang C, Liu Y C, et al. Recent progress in thermal/environmental barrier coatings and their corrosion resistance. *Rare Metals*, 2020, 39(5): 498–512.

[17] Hardwicke C U, Lau Y C. Advances in thermal spray coatings for gas turbines and energy generation: a review. *Journal of Thermal Spray Technology*, 2013, 22(5): 564–576.

[18] Dahotre N B, Hampikian J M, Morral J E. Elevated temperature coatings: science and technology IV. *TiÜS*, 2001: 109.

[19] Liu Z, Gong Y, Zhou W, et al. Ultrathin high-temperature oxidation-resistant coatings of hexagonal boron nitride. *Nature Communications*, 2013, 4(1): 1–8.

[20] Lee K N, Miller R A, Jacobson N S. New generation of plasma-sprayed mullite coatings on silicon carbide. *Journal of the American Ceramic Society*, 1995, 78(3): 705–710.

[21] Lee K N, Miller R A. Development and environmental durability of mullite and mullite/YSZ dual layer coatings for SiC and Si3N4 ceramics. *Surface and Coatings Technology*, 1996, 86: 142–148.

[22] Lee K N, Fox D S, Eldridge J I, et al. Upper temperature limit of environmental barrier coatings based on mullite and BSAS. *Journal of the American Ceramic Society*, 2003, 86(8): 1299–1306.

[23] Lee K N, Fox D S, Bansal N P. Rare earth silicate environmental barrier coatings for SiC/SiC composites and Si3N4 ceramics. *Journal of the European Ceramic Society*, 2005, 25(10): 1705–1715.

[24] Poerschke D L, Van Sluytman J S, Wong K B, et al. Thermochemical compatibility of ytterbia–(hafnia/silica) multilayers for environmental barrier coatings. *Acta Materialia*, 2013, 61(18): 6743–6755.

[25] Aouadi S M, Singh D P, Stone D S, et al. Adaptive VN/Ag nanocomposite coatings with lubricious behavior from 25 to 1000° C. *Acta Materialia*, 2010, 58(16): 5326–5331.

[26] Balani K, Harimkar S P, Keshri A, et al. Multiscale wear of plasma-sprayed carbon-nanotube-reinforced aluminum oxide nanocomposite coating. *Acta Materialia*, 2008, 56(20): 5984–5994.

[27] Rajeshshyam R, Venkatraman R, Raghuraman S. The wear resistant nano composite coating on aluminum alloys by plasma spraying technique–A review[C]//AIP Conference Proceedings. *AIP Publishing LLC*, 2020, 2247(1): 050007.

[28] Tyagi A, Walia R S, Murtaza Q, et al. A critical review of diamond like carbon coating for wear resistance applications. *International Journal of Refractory Metals and Hard Materials*, 2019, 78: 107–122.

[29] Voevodin A A, Zabinski J S. Supertough wear-resistant coatings with 'chameleon'surface adaptation. *Thin Solid Films*, 2000, 370(1-2): 223–231.

[30] Gershman I, Gershman E I, Mironov A E, et al. Application of the self-organization phenomenon in the development of wear resistant materials—A Review. *Entropy*, 2016, 18(11): 385

[31] Szeri A Z, McGraw H. Tribology (Friction, Wear, Lubrication). Hemisphere pab., Washington, *Sayfa*, 1980, 1: 30–75.

[32] Shengyu Zhu, Jun Cheng, Zhuhui Qiao, Jun Yang. High temperature solid-lubricating materials: A review. *Tribology International*, 2019,133: 206–223.

[33] Wang J, Ma J, Huang W, et al. The investigation of the structures and tribological properties of F-DLC coatings deposited on Ti-6Al-4V alloys. *Surface and Coatings Technology*, 2017, 316: 22–29.

[34] Hentour K, Marsal A, Turq V, et al. Carbon nanotube/alumina and graphite/alumina composite coatings on stainless steel for tribological applications. *Materials Today Communications*, 2016, 8: 118–126.

[35] Rübig B, Heim D, Forsich C, et al. Tribological behavior of thick DLC coatings under lubricated conditions. *Surface and Coatings Technology*, 2017, 314: 13–17.

[36] Arslan A, Masjuki H H, Kalam M A, et al. Investigation of laser texture density and diameter on the tribological behavior of hydrogenated DLC coating with line contact configuration. *Surface and Coatings Technology*, 2017, 322: 31–37.

[37] Ryu B H, Barthel A J, Kim H J, et al. Tribological properties of carbon nanotube–polyethylene oxide composite coatings. *Composites Science and Technology*, 2014, 101: 102–109.

[38] Moghadam A D, Omrani E, Menezes P L, et al. Mechanical and tribological properties of self-lubricating metal matrix nanocomposites reinforced by carbon nanotubes (CNTs) and graphene–a review. *Composites Part B: Engineering*, 2015, 77: 402–420.

[39] Al-Kawaz A, Rubin A, Badi N, et al. Tribological and mechanical investigation of acrylic-based nanocomposite coatings reinforced with PMMA-grafted-MWCNT. *Materials Chemistry and Physics*, 2016, 175: 206–214.

[40] Niu R, Li J, Wang Y, et al. Structure and tribological behavior of GLCH/nitride coupled coatings on Ti6Al4V by nitriding and magnetron sputtering. *Diamond and Related Materials*, 2016, 64: 70–79.

[41] Wan S, Pu J, Li D, et al. Tribological performance of CrN and CrN/GLC coated components for automotive engine applications. *Journal of Alloys and Compounds*, 2017, 695: 433–442.

[42] Nakahigashi T, Tanaka Y, Miyake K, et al. Properties of flexible DLC film deposited by amplitude-modulated RF P-CVD. *Tribology International*, 2004, 37(11–12): 907–912.

[43] PalDey S, Deevi S C. Single layer and multilayer wear resistant coatings of (Ti, Al) N: a review. *Materials Science and Engineering A*, 2003, 342(1–2): 58–79.

[44] Bars J P, Etchessahar E, Debuigne J. Étude cinétique, diffusionnelle et morphologique de la nitruration du titane par l'azote à haute température: Propriétés mécaniques et structurales des solutions solides Tiα-azote. *Journal of the Less-Common Metals*, 1977, 52(1): 51–76.

[45] Han J G, Yoon J S, Kim H J, et al. High temperature wear resistance of (TiAl) N films synthesized by cathodic arc plasma deposition. *Surface and Coatings Technology*, 1996, 86: 82–87.

[46] Lin K L, Hwang M Y, Wu C D. The deposition and wear properties of cathodic arc plasma deposition TiAIN deposits. *Materials Chemistry and Physics*, 1996, 46(1): 77–83.

[47] Björk T, Westergård R, Hogmark S, et al. Physical vapour deposition duplex coatings for aluminium extrusion dies. *Wear*, 1999, 225: 1123–1130.

[48] Peng S, Guo Y, Xie G, et al. Tribological behavior of polytetrafluoroethylene coating reinforced with black phosphorus nanoparticles. *Applied Surface Science*, 2018, 441: 670–677.

[49] Feng Q, Zou S, Li H, et al. Review of Polymer Self-lubricating Coatings[C]//IOP Conference Series: Earth and Environmental Science. *IOP Publishing*, 2020, 526(1): 012077.

[50] Rudresh B M, Kumar B N R. Influence of experimental parameters on friction and wear mechanisms of PA66/PTFE blend reinforced with glass fiber. *Transactions of the Indian Institute of Metals*, 2018, 71(2): 339–349.

[51] Zabinski J S, Donley M S, Dyhouse V J, et al. Chemical and tribological characterization of PbO-MoS2 films grown by pulsed laser deposition. *Thin Solid Films*, 1992, 214(2): 156–163.

[52] Zabinski J S, Donley M S, McDevitt N T. Mechanistic study of the synergism between Sb2O3 and MoS2 lubricant systems using Raman spectroscopy. *Wear*, 1993, 165(1): 103–108.

[53] Zabinski J S, Prasad S V, McDevitt N T. *Tribology for Aerospace System[C]//Proceedings of NATO Advisory Group of Aerospace Research and Development (AGARD) Conference on Tribology for Aerospace Systems*, Sesimbra, Portugal, May. 1996: 6–7.

[54] Vepřek S. Conventional and new approaches towards the design of novel superhard materials. *Surface and Coatings Technology*, 1997, 97(1–3): 15–22.

[55] Mitterer C, Mayrhofer P H, Beschliesser M, et al. Microstructure and properties of nanocomposite Ti–B–N and Ti–B–C coatings. *Surface and Coatings Technology*, 1999, 120: 405–411.

[56] Mitterer C, Losbichler P, Hofer F, et al. Nanocrystalline hard coatings within the quasi-binary system TiN–TiB2. *Vacuum*, 1998, 50(3–4): 313–318.

[57] Irie M, Ohara H, Nakayama A, et al. Deposition of Ni TiN nano-composite films by cathodic arc ion-plating. *Nuclear Instruments and Methods in Physics Research Section B: Beam Interactions with Materials and Atoms*, 1997, 121(1–4): 133–136.

[58] Musil J, Zeman P, Hrubý H, et al. ZrN/Cu nanocomposite film—a novel superhard material. *Surface and Coatings Technology*, 1999, 120: 179–183.

[59] Musil J, Polakova H, Šuna J, et al. Effect of ion bombardment on properties of hard reactively sputtered Ti (Fe) Nx films. *Surface and Coatings Technology*, 2004, 177: 289–298.

[60] Musil J, Vlček J. Magnetron sputtering of films with controlled texture and grain size. *Materials Chemistry and Physics*, 1998, 54(1–3): 116–122.

[61] Olakanmi E O, Doyoyo M. Laser-assisted cold-sprayed corrosion-and wear-resistant coatings: a review. *Journal of Thermal Spray Technology*, 2014, 23(5): 765–785.

[62] Lupoi R, Cockburn A, Bryan C, et al. Hardfacing steel with nanostructured coatings of Stellite-6 by supersonic laser deposition. *Light: Science & Applications*, 2012, 1: e10.

[63] Bray M, Cockburn A, O'Neill W. The laser-assisted cold spray process and deposit characterisation. *Surface and Coatings Technology*, 2009, 203(19): 2851–2857.

[64] Musil J, Hrubý H. Superhard nanocomposite Ti1– xAlxN films prepared by magnetron sputtering. *Thin Solid Films*, 2000, 365(1): 104–109.

[65] Li Z, Guan X, Wang Y, et al. Comparative study on the load carrying capacities of DLC, GLC and CrN coatings under sliding-friction condition in different environments. *Surface and Coatings Technology*, 2017, 321: 350–357.

[66] Yuwei Ye, Chunting Wang, Yongxin Wang, Wenjie Zhao, Jinlong Li, Yirong Yao. A novel strategy to enhance the tribological properties of Cr / GLC films in seawater by surface texturing. *Surface and Coatings Technology*, 2015, 280: 338–346.

[67] Shum P W, Zhou Z F, Li K Y. Investigation of the tribological properties of the different textured DLC coatings under reciprocating lubricated conditions. *Tribology International*, 2013, 65: 259–264.

[68] Praveen B M, Venkatesha T V, Naik Y A, et al. Corrosion studies of carbon nanotubes–Zn composite coating. *Surface and Coatings Technology*, 2007, 201(12): 5836–5842.

[69] Bai W Q, Xie Y J, Li L L, et al. Tribological and corrosion behaviors of Zr-doped graphite-like carbon nanostructured coatings on Ti6Al4V alloy. *Surface and Coatings Technology*, 2017, 320: 235–239.

[70] Chunting Wang, Yuwei Ye, Xiaoyan Guan, Jianmin Hu, Yongxin Wang, Jinlong Li. An analysis of tribological performance on Cr/GLC film coupling with Si3N4, SiC, WC, Al2O3 and ZrO2 in seawater. *Tribology International*, 2016, 96: 77–86.

[71] Alkhimov A P, Papyrin A N, Kosarev V F, et al. Gas-dynamic spraying method for applying a coating: U.S. Patent 5,302,414[P]. 1994-4–12.

[72] Olakanmi E O, Tlotleng M, Meacock C, et al. Deposition mechanism and microstructure of laser-assisted cold-sprayed (LACS) Al-12 wt.% Si coatings: effects of laser power. *JOM*, 2013, 65(6): 776–783.

[73] Jiang C, Xing Y, Zhang F, et al. Wear resistance and bond strength of plasma sprayed Fe/Mo amorphous coatings. *Journal of Iron and Steel Research International*, 2014, 21(10): 969–974.

[74] Vencl A. Tribological behavior of ferrous-based APS coatings under dry sliding conditions. *Journal of Thermal Spray Technology*, 2015, 24(4): 671–682.

[75] Banerji A, Lukitsch M J, McClory B, et al. Effect of iron oxides on sliding friction of thermally sprayed 1010 steel coated cylinder bores. *Wear*, 2017, 376: 858–868.

[76] Xing Y Z, Liu Z, Wang G, et al. Improvement of interfacial bonding between plasma-sprayed cast iron splat and aluminum substrate through preheating substrate. *Surface and Coatings Technology*, 2017, 316: 190–198.

[77] Kim L A, Anikeeva P O, Coe-Sullivan S A, et al. Contact printing of quantum dot light-emitting devices. *Nano Letters*, 2008, 8(12): 4513–4517.

[78] Bae W K, Kwak J, Lim J, et al. Multicolored light-emitting diodes based on all-quantum-dot multilayer films using layer-by-layer assembly method. *Nano Letters*, 2010, 10(7): 2368–2373.

[79] Shallcross R C, Chawla G S, Marikkar F S, et al. Efficient CdSe nanocrystal diffraction gratings prepared by microcontact molding. *ACS Nano*, 2009, 3(11): 3629–3637.

[80] Pattani V P, Li C, Desai T A, et al. Microcontact printing of quantum dot bioconjugate arrays for localized capture and detection of biomolecules. *Biomedical Microdevices*, 2008, 10(3): 367–374.

[81] Curri M L, Comparelli R, Striccoli M, et al. Emerging methods for fabricating functional structures by patterning and assembling engineered nanocrystals. *Physical Chemistry Chemical Physics*, 2010, 12(37): 11197–11207.

[82] Hilal, Nidal, et al., eds. *Membrane characterization*. Springer, New York, 2017.

[83] Alazemi M, Dutta I, Wang F, et al. Electrically conductive thin films prepared from layer-by-layer assembly of graphite platelets. *Advanced Functional Materials*, 2009, 19(7): 1118–1129.

[84] Hsiao S T, Tien H W, Liao W H, et al. A highly electrically conductive graphene–silver nanowire hybrid nanomaterial for transparent conductive films. *Journal of Materials Chemistry C*, 2014, 2(35): 7284–7291.

[85] Tiwari A. *Handbook of antimicrobial coatings*. Elsevier, Amsterdam, 2017.

[86] Fechine G J M, Martin-Fernandez I, Yiapanis G, et al. Direct dry transfer of chemical vapor deposition graphene to polymeric substrates. *Carbon*, 2015, 83: 224–231.

[87] Chou K S, Huang K C, Lee H H. Fabrication and sintering effect on the morphologies and conductivity of nano-Ag particle films by the spin coating method. *Nanotechnology*, 2005, 16(6): 779.

[88] Lee H H, Chou K S, Huang K C. Inkjet printing of nanosized silver colloids. *Nanotechnology*, 2005, 16(10): 2436.

[89] Kim D, Jeong S, Park B K, et al. Direct writing of silver conductive patterns: Improvement of film morphology and conductance by controlling solvent compositions. *Applied Physics Letters*, 2006, 89(26): 264101.

[90] Lee J Y, Connor S T, Cui Y, et al. Solution-processed metal nanowire mesh transparent electrodes. *Nano Letters*, 2008, 8(2): 689–692.

[91] De S, Higgins T M, Lyons P E, et al. Silver nanowire networks as flexible, transparent, conducting films: extremely high DC to optical conductivity ratios. *ACS Nano*, 2009, 3(7): 1767–1774.

[92] Layani M, Gruchko M, Milo O, et al. Transparent conductive coatings by printing coffee ring arrays obtained at room temperature. *ACS Nano*, 2009, 3(11): 3537–3542.

[93] Liu C H, Yu X. Silver nanowire-based transparent, flexible, and conductive thin film. *Nanoscale Research Letters*, 2011, 6(1): 1–8.

[94] Madaria A R, Kumar A, Zhou C. Large scale, highly conductive and patterned transparent films of silver nanowires on arbitrary substrates and their application in touch screens. *Nanotechnology*, 2011, 22(24): 245201.

[95] Akter T, Kim W S. Reversibly stretchable transparent conductive coatings of spray-deposited silver nanowires. *ACS Applied Materials & Interfaces*, 2012, 4(4): 1855–1859.

[96] Lee J, Lee P, Lee H, et al. Very long Ag nanowire synthesis and its application in a highly transparent, conductive and flexible metal electrode touch panel. *Nanoscale*, 2012, 4(20): 6408–6414.

[97] Zhu S, Gao Y, Hu B, et al. Transferable self-welding silver nanowire network as high performance transparent flexible electrode. *Nanotechnology*, 2013, 24(33): 335202.

[98] Madaria A R, Kumar A, Ishikawa F N, et al. Uniform, highly conductive, and patterned transparent films of a percolating silver nanowire network on rigid and flexible substrates using a dry transfer technique. *Nano Research*, 2010, 3(8): 564–573.

[99] Scardaci V, Coull R, Lyons P E, et al. Spray deposition of highly transparent, low-resistance networks of silver nanowires over large areas. *Small*, 2011, 7(18): 2621–2628.

[100] Cha J H, Kwon S M, Bae J A, et al. Effect of the deposition process of window layers on the performance of CIGS solar cells. *Journal of Alloys and Compounds*, 2017, 708: 562–567.

[101] Kim K, Shafarman W N. Alternative device structures for CIGS-based solar cells with semi-transparent absorbers. *Nano Energy*, 2016, 30: 488–493.

[102] Fang X S, Bando Y, Shen G Z, et al. Ultrafine ZnS nanobelts as field emitters. *Advanced Materials*, 2007, 19(18): 2593–2596.

[103] Piegari A and Flory F. *Optical thin films and coatings: From materials to applications.* Woodhead Publishing, Oxford, 2018.

[104] Uhlmann D R, Boulton J M, Teowee G T, et al. Sol-gel synthesis of optical thin films and coatings[C]// Sol-Gel Optics. *International Society for Optics and Photonics*, 1990, 1328: 270–295.

[105] Kruschwitz J D T, Pawlewicz W T. Optical and durability properties of infrared transmitting thin films. *Applied Optics*, 1997, 36(10): 2157–2159.

[106] Marsh K J, Savage J A. Infrared optical materials for 8–13 μm—current developments and future prospects. *Infrared Physics*, 1974, 14(2): 85–97.

[107] Black P W, Wales J. Materials for use in the fabrication of infrared interference filters. *Infrared Physics*, 1968, 8(3): 209–222.

[108] Willey R R. Estimating the number of layers required and other properties of blocker and dichroic optical thin films. *Applied Optics*, 1996, 35(25): 4982–4986.

[109] Samyn P, Achten M. Present trend and architect's expectation on coated glass. *Journal of Non-Crystalline Solids*, 1997, 218: 1–6.

[110] Kawachi M. Silica waveguides on silicon and their application to integrated-optic components. *Optical and Quantum Electronics*, 1990, 22(5): 391–416.

[111] Mir J M, Agostinelli J A. Optical thin films for waveguide applications. *Journal of Vacuum Science & Technology A: Vacuum, Surfaces, and Films*, 1994, 12(4): 1439–1445.

[112] Reiss S M. Coatings for cars and communications. *Optics & Photonics News*, 1997, 8(10): 31–34.

[113] Dobrowolski J A, Ho F C, Waldorf A. Research on thin film anticounterfeiting coatings at the National Research Council of Canada. *Applied Optics*, 1989, 28(14): 2702–2717.

[114] Phillip R W, Bleikolm A F. Optical coatings for document security. *Applied Optics*, 1996, 35(28): 5529–5534.

[115] Szipöcs R, Ferencz K, Spielmann C, et al. Chirped multilayer coatings for broadband dispersion control in femtosecond lasers. *Optics Letters*, 1994, 19(3): 201–203.

[116] Kärtner F X, Matuschek N, Schibli T, et al. Design and fabrication of double-chirped mirrors. *Optics Letters*, 1997, 22(11): 831–833.

[117] Kose S, Atay F, Bilgin V, et al. In doped CdO films: Electrical, optical, structural and surface properties. *International Journal of Hydrogen Energy*, 2009, 34(12): 5260–5266.

[118] Bhosale C H, Kambale A V, Kokate A V, et al. Structural, optical and electrical properties of chemically sprayed CdO thin films. *Materials Science and Engineering B*, 2005, 122(1): 67–71.

[119] Deokatea R J, Salunkhea S V, Agawanea G L, et al. Structural, optical and electrical properties of chemically sprayed nanosized gallium doped CdO thin films. *Journal of Alloys and Compounds*, 2010, 496: 357–363.

[120] Ullah H, Rahaman R, Mahmud S. Optical properties of cadmium oxide (CdO) thin films. *Indonesian Journal of Electrical Engineering and Computer Science*, 2017, 5(1): 81–84.

[121] Lilhare D, Sinha T, Khare A. Influence of Cu doping on optical properties of (Cd–Zn) S nanocrystalline thin films: a review. *Journal of Materials Science: Materials in Electronics*, 2018, 29(1): 688–713.

[122] Al Kuhaimi S A, Tulbah Z. Structural, Compositional, Optical, and Electrical Properties of Solution-Grown Zn x Cd1– x S Films. *Journal of the Electrochemical Society*, 2000, 147(1): 214.

[123] Firat Y E, Peksoz A. Efficiency enhancement of electrochromic performance in NiO thin film via Cu doping for energy-saving potential. *Electrochimica Acta*, 2019, 295: 645–654.

[124] Varunkumar K, Hussain R, Hegde G, et al. Effect of calcination temperature on Cu doped NiO nanoparticles prepared via wet-chemical method: Structural, optical and morphological studies. *Materials Science in Semiconductor Processing*, 2017, 66: 149–156.

[125] Chen S C, Kuo T Y, Lin Y C, et al. Preparation and properties of p-type transparent conductive Cu-doped NiO films. *Thin Solid Films*, 2011, 519(15): 4944–4947.

[126] Zhao L, Su G, Liu W, et al. Optical and electrochemical properties of Cu-doped NiO films prepared by electrochemical deposition. *Applied Surface Science*, 2011, 257(9): 3974–3979.

[127] Huang A, Lei L, Chen Y, et al. Minimizing the energy loss of perovskite solar cells with Cu+ doped NiOx processed at room temperature. *Solar Energy Materials and Solar Cells*, 2018, 182: 128–135.

[128] Luan X G, Xu X, Li M, et al. Design, preparation, and properties of a boron nitride coating of silica optical fiber for high temperature sensing applications. *Journal of Alloys and Compounds*, 2020, 850(5): 156782.

[129] Chen S, Zhang Q Q, Luan X G, et al. Sapphire optical fiber with SiBCN coating prepared by chemical vapor deposition for high-temperature sensing applications. *Thin Solid Films*, 2020, 709: 138242.

[130] Luan X G, Yu R, Zhang Q Q, et al. Boron nitride coating of sapphire optical fiber for high temperature sensing applications. *Surface and Coatings Technology*, 2019, 363: 203–209.

[131] Cheng Y, Yin X W, Liu Y S, et al. BN coatings prepared by low pressure chemical vapor deposition using boron Trichloride-ammonia-hydrogen-argon mixture gases. *Surface and Coatings Technology*, 2010, 204(16): 2797–2802.

[132] Luan X G, Xu X, Rong Y, et al. BN/SiBCN light-leakage-proof coatings of silica optical fiber for long term sensors at high temperatures. *Chinese Journal of Aeronautics*, 2021, 34(5): 93–102.

[133] Taşdemirci T Ç. Synthesis of copper-doped nickel oxide thin films: Structural and optical studies. *Chemical Physics Letters*, 2020, 738: 136884.

[134] Al-Ogili H K J. Effect of Thickness to the Structure Properties of CdO Thin Films. *Engineering and Technology Journal*, 2011, 29(8): 1536–1544.

[135] Balu A R, Nagarethinam V S, Suganya M, et al. Effect of solution concentration on the structural, optical and electrical properties of SILAR deposited CdO thin films. *J. Electron Devices*, 2012, 12: 739–749.

[136] Ziabari A A, Ghodsi F E, Kiriakidis G. Correlation between morphology and electro-optical properties of nanostructured CdO thin films: Influence of Al doping. *Surface and Coatings Technology*, 2012, 213: 15–20.

[137] Xu L, Li X, Chen Y, et al. Structural and optical properties of ZnO thin films prepared by sol–gel method with different thickness. *Applied Surface Science*, 2011, 257(9): 4031–4037.

[138] Jin-Hong Lee, Kyung-Hee Ko, Byung-Ok Park. Electrical and optical properties of ZnO transparent conducting films by the sol–gel method. *Journal of Crystal Growth*, 2003, 247(1–2): 119–125

[139] Sinha T, Lilhare D, Khare A. Effects of various parameters on structural and optical properties of CBD-grown ZnS thin films: a review. *Journal of Electronic Materials*, 2018, 47(2): 1730–1751.

[140] Rasic G.. *Advanced nano deposition methods*. Wiley, Weinheim, 2016.

[141] Kodama R H. Magnetic nanoparticles. *Journal of Magnetism and Magnetic Materials*, 1999, 200(1–3): 359–372.

[142] Reiss G, Hütten A. Applications beyond data storage. *Nature Materials*, 2005, 4(10): 725–726.

[143] Wu W, Wu Z, Yu T, et al. Recent progress on magnetic iron oxide nanoparticles: synthesis, surface functional strategies and biomedical applications. *Science and technology of advanced materials*, 2015, 16(2): 023501.

[144] Setia Budi, Budhy Kurniawan, Derrick M Mott, Shinya Maenosono, Akrajas Ali Umar, Azwar Manaf. Comparative trial of saccharin-added electrolyte for improving the structure of an electrodeposited magnetic FeCoNi thin film. *Thin Solid Films*, 2017, 642: 51–57.

[145] L. Yuan, X. Weng, L. Deng, Influence of binder viscosity on the control of infrared emissivity in low emissivity coating, *Infrared Physics & Technology*, 2013, 56: 25–29.

[146] G. Fang, Z. Liu, K.L. Yao, Fabrication and characterization of electrochromic nanocrystalline WO_3/Si (111) thin films for infrared emittance modulation applications, *Journal of Physics D: Applied Physics,* 2001, 34: 2260.

[147] X. Bu, Y. Zhou, M. He, Z. Chen, T. Zhang, Optically active SiO_2/TiO_2/polyacetylene multilayered nanospheres: preparation, characterization, and application for low infrared emissivity, *Applied Surface Science*, 2014, 288: 444–451.

[148] Y. Zhu, Y. Zhou, T. Zhang, M. He, Y. Wang, X. Yang, Y. Yang, Preparation and characterization of lactate-intercalated Co-Fe layered double hydroxides and exfoliated nanosheet film with low infrared emissivity, *Applied Surface Science*, 2012, 263: 132–138.

[149] Z. Mao, W. Wang, Y. Liu, L. Zhang, H. Xu, Y. Zhong, Infrared stealth property based on semiconductor (M)-to-metallic (R) phase transition characteristics of W-doped VO_2 thin films coated on cotton fabrics, *Thin Solid Films*, 2014, 558: 208–214.

[150] D. Liu, H. Cheng, X. Xing, C. Zhang, W. Zheng, Thermochromic properties of Wdoped VO_2 thin films deposited by aqueous sol-gel method for adaptive infrared stealth application, *Infrared Physics & Technology*, 2016, 77: 339–343.

[151] Q. Fu, W.W. Wang, D.L. Li, J.J. Pan, Research on surface modification and infrared emissivity of In_2O_3: W thin films, *Thin Solid Films*, 2014, 570: 68–74.

[152] H.H. Huang, M.H. Hon, Effect of N2 addition on growth and properties of titanium nitride films obtained by atmospheric pressure chemical vapor deposition, *Thin Solid Films*, 2002, 416: 54–61.

[153] L.L. Lu, F. Luo, Z.B. Huang, W.C. Zhou, D.M. Zhu, Research on optical reflectance and infrared emissivity of TiN_x films depending on sputtering pressure. *Infrared Physics & Technology*, 2018, 91: 63–67.

[154] Lu L L, Luo Fa, et al. Effect of N_2 flow rate on electromagnetic interference shielding effectiveness of TiN_x films. *Applied Physics A*, 2018, 124: 721.

[155] Sun K W, Zhou W C, Effects of air annealing on the structure, resistivity, infrared emissivity and transmission of indium tin oxide films. *Surface & Coatings Technology*, 2012, 206: 4095–4098.

[156] Kang W B, Zhu D M, et al. Effects of annealing temperature on the structure, electrical resistivity and infrared emissivity of PtOx films. *Vacuum*, 2017, 145: 174–178.

8 Roads Toward Surface Protection of Magnesium Alloys

Wenling Xie
Southwest University, China

Bin Liao
Beijing Normal University, China

Sam Zhang
Southwest University, China

CONTENTS

8.1 INTRODUCTION

Magnesium and magnesium alloys are a kind of light alloys with great application prospect in the electronics and aerospace industries [1, 2] because of their lowest density among all the engineering metallic materials, good strength-to-weight ratio, shock absorption, electromagnetic shielding, thermal conductivity, recycling ability, etc. However, poor corrosion and wear resistance caused by high chemical activity and low hardness are the biggest obstacles to their wide applications [3–6]. As corrosion and abrasion always start from the surface of a material (such as steel, magnesium alloy, etc.), surface coating is thus an economic and effective way for material protection [7, 8]. In the last few decades, studies of all kinds of coatings on different substrates are many. These substrates include silicon plates [9–11], stainless steels [12, 13], mild steels [14], tool steels [15], aluminum alloys [16], titanium alloys [17], and magnesium alloys [18, 19]. These studied magnesium alloys consist of Mg-Al alloys (such as AZ31, AZ91, and AM50 magnesium alloys, and the elements of aluminum, zinc, and manganese aim at improving hardness, strength, ductility, corrosion resistance, and castability of magnesium), Mg-Zn-Zr alloys (such as ZK60 magnesium alloy, and the elements of zinc and zirconium are aim to strengthen magnesium) [2], and Mg-Ca alloys (the element of calcium is aim to improve biocompatibility of magnesium). More details about magnesium alloys are given in ref. [2]. Unless specified, all coatings discussed in this chapter are coatings on magnesium and magnesium alloys. Coatings are easy to obtain by all kinds of methods, such as electroplating [20, 21], microarc oxidation (MAO) [22, 23], dip coating [24, 25], chemical vapor deposition (CVD) [26], physical vapor deposition (PVD) [27, 28], and their combination processes [29–31].

To date, no comprehensive review of protection coatings on magnesium and magnesium alloys are available. This chapter aims to give a comprehensive understanding of roads toward surface protection of magnesium alloys. We discuss improvement of wear and corrosion resistance of surface protection coatings on magnesium alloys by nanofiller blended single-layer composite coating, the main coating with support coating and/or top coating, and multi-interface coating.

8.2 NANOFILLER BLENDED SINGLE-LAYER COMPOSITE COATING

Currently, many coatings have already been applied to protect magnesium alloys. Single aluminum, chromium, or nickel coating has good corrosion resistance but low hardness [27, 32]. Organic and organic/inorganic hybrid coatings [33, 34] are not stable under thermal exposure or mechanical impact, or their applications are environmentally untenable [35, 36]. Nitrides and carbides ceramic coatings have excellent corrosion and at the same time, high hardness, and thus superb wear resistance. But the huge differences in physical properties (hardness, elastic modulus, and thermal expansion coefficient) between these ceramic coatings and the magnesium alloys result in tremendous internal stress in the coating which, in turn, gives rise to low interfacial bonding and weak bonding strength [37, 38]. For example, diamond-like carbon (DLC) coating partially cracked and peeled off from the AZ31 magnesium alloy [37]. To take advantage of each material and avoid its disadvantages, composite coatings at least consisting of two different kinds of phases, become an obvious choice.

Nanocomposite coating, as the name implies, refers to a composite coating with at least one phase being nanometer in at least one dimension. The limit for the size of this range is arbitrarily set at 100 nm. However, in practice, the structure of nanocomposite coatings refers to the size below 10 nm [39]. This chapter discusses the nanocomposite coatings with a size below 100 nm. Nanocomposite coatings could offer multiple good properties simultaneously which otherwise in traditional materials are not possible, for instance, high hardness and toughness [40–44], self-lubrication [38, 45], self-healing ability [46, 47], wettability [48], anti-bacterial property [49, 50], thermal stability [51, 52], biocompatibility [24], good corrosion resistance [53], good wear resistance [54, 55], and so on. As nanomaterials have surface and interface effects, small size effect, quantum size effect, and macroscopic quantum tunneling effect, nanocomposite coatings thus have some unique optical [56], catalytic [57], and electromagnetic properties [58]. Currently, nanocomposite coatings have many industrial applications in high-speed machining, tooling, automotive, aerospace, biomedical sectors, and so on. Unlike bulk materials, coatings with a thickness of micrometers or even nanometers can be prepared layer-by-layer (LBL) [59–63].

According to the dimension of the nanofillers in the composite coatings, D. Martínez-Martínez also calls them 0D, 1D, and 2D-nanocomposites coatings, as shown in Figure 8.1 [39]. 0D, 1D, and 2D nanocomposite coatings indicate that the nanofillers are nanoparticles, nanotubes, and nanolayers, respectively. Different kinds of nanocomposite coatings have various performances. Nitrides [6, 8], oxides [36], DLC [37, 38] ceramic matrix nanocomposite coatings have high hardness and high chemical inertness. Metal [5, 27], organic [47], and polymer matrix nanocomposite coatings [46] have good corrosion resistance. In addition, special properties can be gained in composite coatings such as biocompatibility, anti-bacterial, and self-lubricating properties. For example, hydroxyapatite (HA) nanocomposite coatings [43], chitosan (CS) nanocomposite coatings [50], and poly(lactic-co-glycolic acid) or PLGA nanocomposite coatings [63] have good bioactivity, biocompatibility, and anti-bacterial properties. MoS_2 nanocomposite coatings [45] and graphene oxide coatings [31] have self-lubricating properties.

8.2.1 BLENDING OF NANOFILLER FOR NANOCOMPOSITE COATING

Nanofillers are materials of nanoscale in at least one dimension, such as nanoparticles, nanowhiskers/nanotubes, or nanolayers (graphene and graphene oxide). These are typical nanofillers: SiC [20, 64, 65], TiO_2 [66, 67], WC [53], W [68], CaP [29], B_4C [69, 70], Fe_3O_4 [50], SiO_2 [71, 72], ZrO_2 [73], TiN [74], carbon nanotubes (CNTs) [21, 75, 76], graphene [23, 77], and graphene oxide (GO) [78–80]. Nanofillers are blend in coatings to form "nanocomposite" coatings. Figures 8.2–8.4 show the morphology of the nanoparticles (three-dimensional nanosize), CNTs (two-dimensional nanosize), and nanolayers (one-dimensional nanosize) in composite coatings, respectively. Nanofiller blended composite coatings have been prepared on the magnesium alloy substrates by various methods, such as MAO/plasma electrolytic oxidation (PEO) [64, 65], electrodeposition [20, 43, 81],

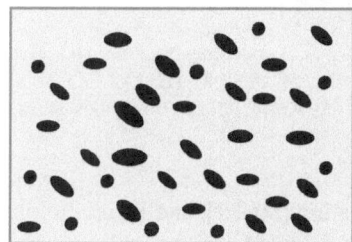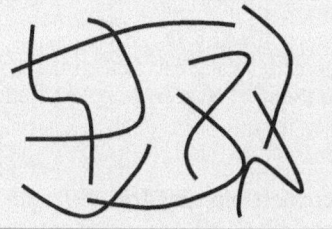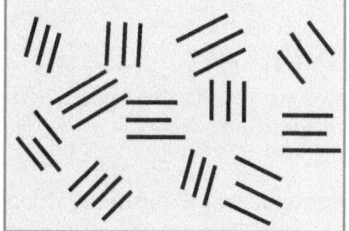

FIGURE 8.1 Scheme of the different types of nanocomposite coatings. (a) 0D, formed by nanoparticles. (b) 1D, formed by nanotubes. (c) 2D, formed by nanolayers. The matrix is represented by a gray background (Based on Martínez-Martínez, D. 2013).

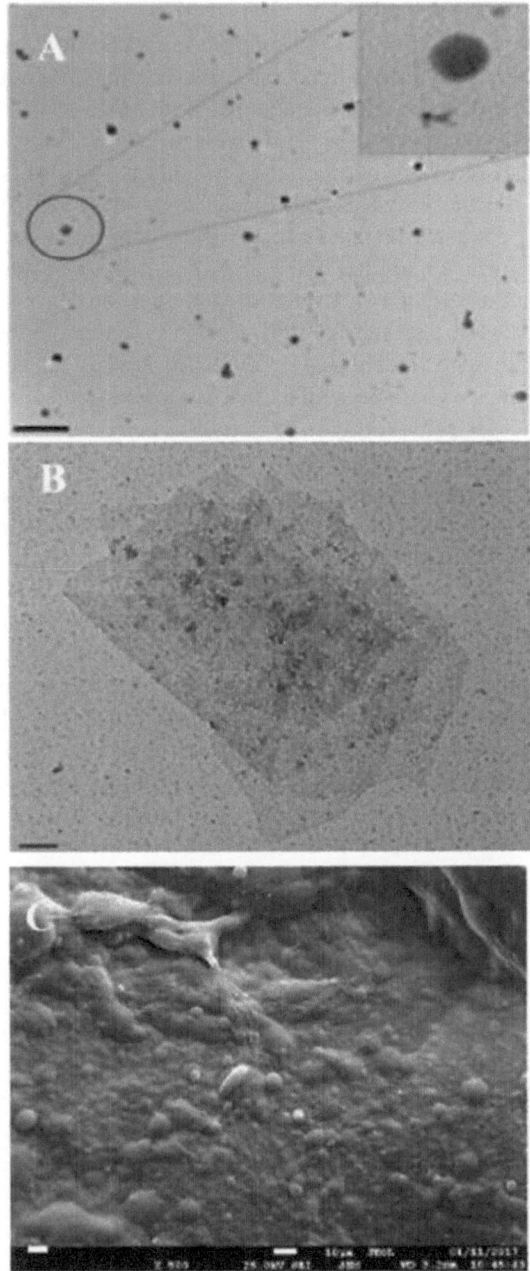

FIGURE 8.2 Electron microscopy images of (A) TiO$_2$ NPs prepared at 60 °C (scale bar 50 nm). (B) The poly(BMA)-GO-TiO$_2$ nanocomposite blend (scale bar 200 nm). (C) SEM of the poly(BMA)-GO-TiO$_2$ nanocomposite blend (scale bar 10 μm) (Source: Nazeer, A. A., Al-Hetlani, E., Amin, M. O., Quiñones-Ruiz, T., Lednev, I. K. 2019).

electrophoretic deposition [50], electroless planting [68, 69], spin coating [80, 82], anodic oxidation coating [83], dip coating [84], plasma chemical vapor deposition (PCVD) [85], electropolymeriza-tion [86], and PVD [37, 38].

For electroplating, anodic oxidation, electroless plating, and MAO processes, nanofillers can be directly added and suspend in the electrolyte solution, then move in the electrolyte by mechanical

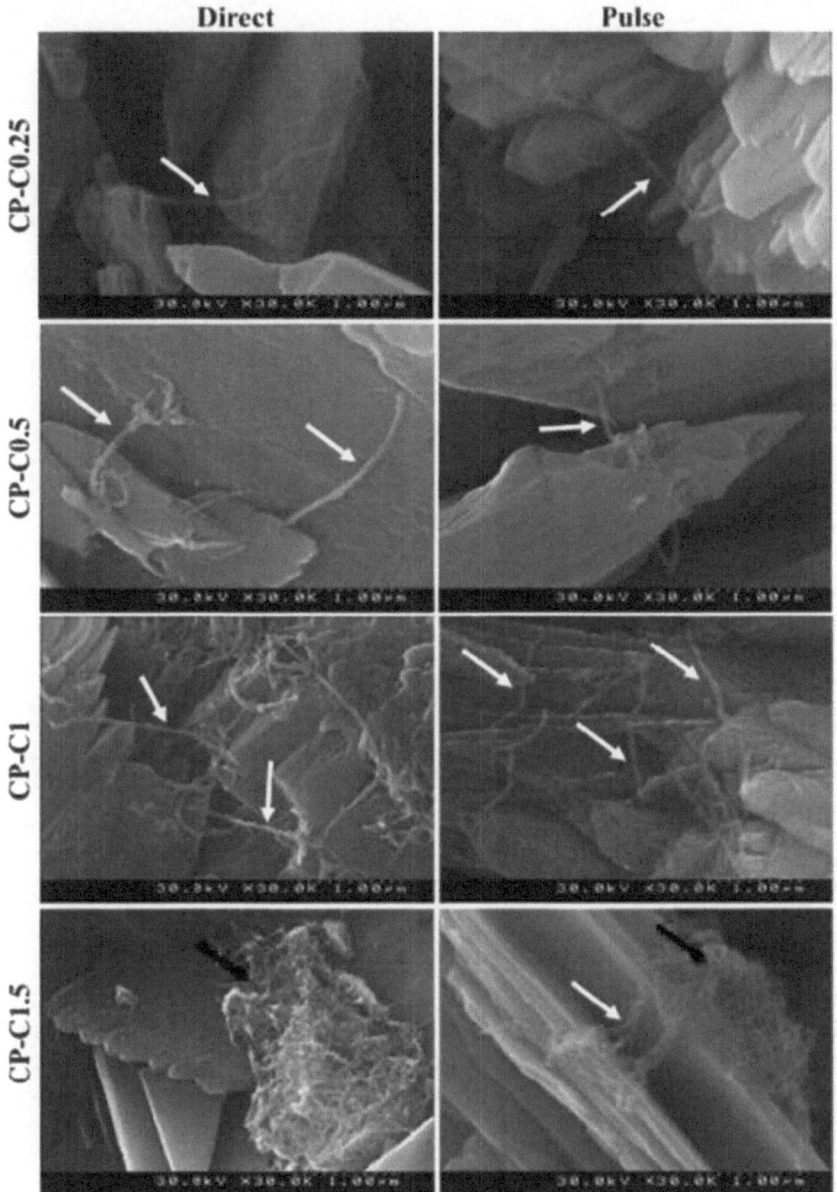

FIGURE 8.3 FESEM images of HA composite coatings blended with different amounts of CNTs using direct and pulse voltage at high magnification (CNTs is marked by the white arrow and aggregation of CNTs is marked by the red arrow) (Source: Khazeni, D., Saremi, M., Soltani, R. 2019).

stirring, and reach the outer surface of the substrate, and finally are absorbed in the coating to form a nanocomposite coating [87–89]. Sankara Narayanan et al. have demonstrated in detail, the movement of nanoparticles in the electrolyte solution and their absorption in the MAO coating [90]. The blending mechanism of nanoparticles is controlled by the following factors: transport particles to the anode–electrolyte interface and the oxide growth front, trap the particles in the growing oxide matrix, and retain the embedded particles during the subsequent growth of the oxide layer [91, 92]. Nanofiller blended coatings prepared by spin coating and dip coating first need to mix and disperse nanoparticles in a solution, and then directly plate the solution on the magnesium alloy substrates.

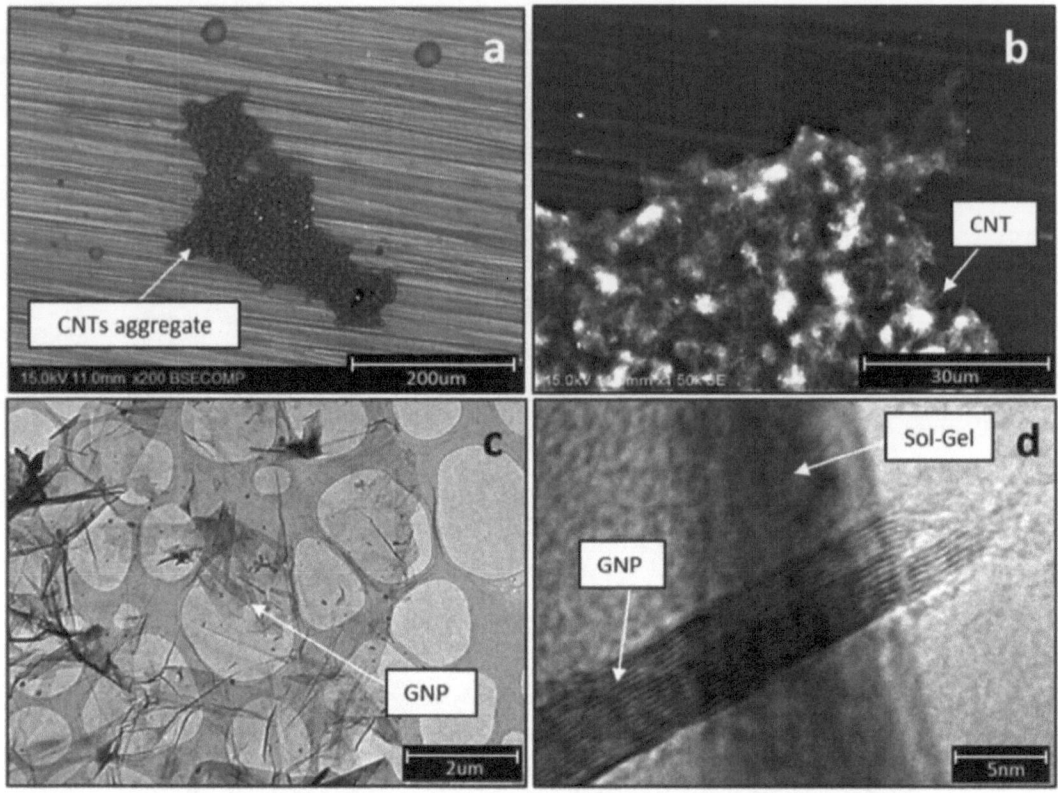

FIGURE 8.4 Nanocharge dispersions. (a) MWCNTs aggregate embedded in the sol-gel coating. (b) Detail of MWCNTs aggregate. (c) TEM micrograph of GNPs. (d) GNPs embedded in the sol-gel coating (Source: Fernández-Hernán, J. P., López, A. J., Torres, B., Rams, J. 2020).

Nanofiller blended coatings prepared by PVD usually have nanometer-size grain, controllable properties, which can be prepared by multiple targets or alloy target, and do not need additional nanofiller dispersion process [80, 82, 93] and complicate pretreatment [20, 32, 50], such as alkali washing, pickling, and activation. Cr blended DLC coatings [37], Ti and N co-blended DLC ((Ti: N)-DLC) coatings [38], Cu blended TiN coatings [12, 94], Au blended AlN coatings [56], and TiN blended SiN_x (nc-TiN/a-SiN_x) coatings [95–97] are the classic nanofiller blended composite coatings.

8.2.2 PERFORMANCE OF NANOFILLER BLENDED COMPOSITE COATING

Nanofiller blended coatings can significantly improve the properties of magnesium alloys due to complex phases formed. They usually show good corrosion resistance, high hardness, high toughness, low friction, high adhesion, good load-bearing capacity, etc. Apart from the nanoscale size effect, nanofillers have some special physical and chemical properties. Nitride [87], oxide [50, 59, 98], and carbide [69, 99] nanoparticles (NPs) have high hardness and chemical inertness. Multiwalled carbon nanotubes (MWCNTs) have many surface groups, such as carboxyl groups. Literatures [29, 100, 101] revealed that the halloysite nanotubes (HNT) and CNTs are one of the most favorable reinforcement materials, especially in nanocomposites because of their unique structural features of high aspect ratio and excellent mechanical properties. It is worth noting that CNTs may react with other organic compounds, such as HA during the coating process [43]. Graphene is a unique two-dimensional carbon material consisting of a single layer of carbon atoms joined by sp^2 covalent bonds, has excellent electrical properties, highest thermal conductivity, and extraordinary mechanical properties [102–104]. The graphene layer itself can play a role in preventing corrosion [78, 105, 106] and can be used as a moisture-proof layer and gas layer of organic composite materials [73]. GO plates reject corrosive anions due to their negatively charged functional groups [78]. Graphene and GO have been proven to be effective in controlling corrosion in MAO coatings [77, 78], anodic oxidation coatings [83], and electropolymerization coatings [86] to provide good wear resistance and self-lubricity. Table 8.1 summarizes the performance improvement of magnesium alloys by nanocomposite coatings blended with these NPs, CNTs, and nanolayers (graphene and GO). The optimized concentration of the nanofillers, stirring mode, and stirring rate are also listed in Table 8.1. The performance improvement of hardness, wear resistance, corrosion resistance, and adhesion strength is attributed to the following functions of nanofillers in coatings:

8.2.2.1 Strengthening and Toughening of the Second Phase

Based on Orowan's mechanism [107], due to the second phase strengthening, the hard nanoparticles blended in the coating suppress the plastic deformation of the matrix under load. High-hardness SiC (~3000 HV) NPs are embedded in the PEO coating and distribute on the surface of the oxide coating, which will significantly reduce the friction and shear stress between the coating and the counter grinding ball [64]. Therefore, the hardness of the coated AZ31 Mg alloy exceeded 550 HV, eight times higher than that of the uncoated sample (64 HV). Reference [20] electrodeposited Ni coatings blended with 0–15 g/L SiC NPs on the AZ91 magnesium alloy and increased the microhardness from the uncoated sample of 74 HV and Ni coated sample of 286 HV to 523 HV. The increase of hardness comes from fine grain strengthening and dispersion strengthening (the second phase strengthens). Generally, the wear rate decreases with the increase of hardness. The wear rate of Ni-SiC nanocomposite coating with 15 g/L SiC concentration was 0.6×10^3 mm^3.m^{-1}, which is only 12% of that of the uncoated AZ91 (5×10^3 mm^3.m^{-1}) and nearly 1/4 of that of the Ni coated sample (2.3×10^3 mm^3.m^{-1}).

Compared with NPs, CNTs and nanolayers are superior in improving the mechanical properties of the coating, and even can obtain some special properties. Khazeni et al. prepared HA coatings blended CNTs with a concentration of 0.25–1.5 wt.% on AZ31 magnesium alloy substrates by electrodeposition process [43, 76]. The HA-1 wt.% CNTs composite coating using pulse deposition mode shows the best mechanical properties among all coatings, it had the highest elastic modulus of 135.4 ± 6.2 GPa, hardness of 540 ± 3 HV, toughness of 1.96 ± 0.72 MPa.m$^{-1/2}$, and adhesion strength of 19.6 ± 1.1 MPa. The hardness improvement is due to the strengthening effect of the stiffer CNTs fillers in the HA matrix. The enhancement in elastic modulus is mainly attributed to the following three factors: (i) high elastic modulus of CNTs, equivalent to diamond, (ii) CNTs uniform dispersion and bridging between HA grains, as shown in Figure 8.3, (iii) the carboxyl group of functionalized CNTs reaction with HA produces chemical bonds (such as -COO-Ca-OOC- [108]) and (IV) bridging effect (will be discussed in detail later), the interfaces between CNTs and HA have a good load transfer efficiency [109–111], thereby increasing mechanical interlocking of CNTs and the HA matrix. Vatan et al. [77] studied the PEO coatings (PEO coating for 5–20 min) with and without 5 wt.% graphene incorporation prepared on the AZ31 magnesium alloy substrates. After PEO coating for 20 min, the maximum hardness of the graphene-PEO composite coatings was up to 547 HV. The wear rate of the graphene-PEO composite coating was reduced to 1/4 of that of the PEO coating due to higher hardness. In addition, since graphene can be used as a solid lubricant in coating, the blending of graphene reduces the friction coefficient of the coating. Han et al. prepared PEO coatings with graphene (0–250 mg/L concentration) incorporation on the AZ91 Mg alloys [23]. With the increase of the graphene concentration, the hardness of GO-PEO coatings increased and peaked at 808 HV (with 250 mg/L concentration of GO blended). This is 16 times as much as that of the bare AZ91 Mg alloy (52 HV) and twice that of the PEO coating (350 HV). Zhang et al. found

TABLE 8.1

Summary of performance improvement by nanocomposite coatings blended with three kinds of fillers and their optimized parameters

Category of the Nanofillers	Preparation method	Nanofillers	Matrix	Coating thickness (µm)	Substrate	Improvement of performance	Optimized parameters	Optimal parameters	Year	Ref.
Nanoparticles	PEO	Al_2O_3	PEO	~21.5	AZ31	Anti-corrosion↑ Anti-wear↑	Concentration of fillers: 10–40 g/L Stirring rate: 50–700 rpm	Concentration of fillers: 30 g/L Stirring rate:100 rpm	2017	[59]
	Electrophoretic deposition (EPD)	Bioglass (BG)-Fe_3O_4	Chitosan (CS)	10 ± 0.2	AZ91	Anti-corrosion↑	Concentration of fillers (Fe_3O_4): 1, 3, 5 wt.%	1 wt.%	2020	[50]
	Electroless deposition	W-P-B_4C	Ni	50	AZ91D	Hardness↑ Anti-wear↓	Concentration of fillers: 4 g/L		2013	[69]
	PEO	TiO_2	PEO	~5	Mg-1% Ca	Anti-corrosion↑ Anti-wear↑ Hydrophobicity↑	Concentration of fillers: 0, 4, 8 g/L	4 g/L	2016	[98]
	Electrodeposition	SiC	Ni	35 ± 5	AZ91	Anti-corrosion↑ Hardness↑ Anti-wear↑	Concentration of fillers: 0, 5, 10, 15 g/L	15 g/L	2013	[20]
	MAO	SiC	MAO	13.06	AZ31	Anti-corrosion↑ Anti-wear↑	Coating time: 5, 10, 15, and 20 min	Anti-corrosion:20 min Anti-wear:5 min	2016	[65]
	PVD	TiN	DLC	0.5	AZ80	Anti-wear↑ Internal stress↓	Gas flow rate (C_2H_2/N_2): 8 sccm/7 sccm		2013	[38]
	PVD	Cr	DLC	0.662	AZ31	Internal stress↓ Adhesion strength↑	Concentration of fillers: 2.34, 12.1, 31.5 at.%	2.34 at.%	2010	[37]
Nanotubes	Electrodeposition	CNTs	HA	6.5~7.8	AZ31	Anti-corrosion↑ Hardness↑ Toughness↑ Adhesion strength↑	Concentration of fillers: 0.25, 0.5, 1, 1.5 wt.%	1 wt.%	2019, 2019	[43, 76]
	EPD	CNTs-CaP	Chitosan (CS)	420	AZ91D	Cell compatibility↑	Concentration of fillers: 0.05~0.4 g/L	0.1~0.25 g/L	2016	[29]

Group										
	Sol-gel	CNTs	SiO_2	3 ± 0.4	WE54	Anti-wear↑	Stirring mode: mechanical mixing (MM) and ultrasonic probe mixing	MM	2011	[25]
Nanolayers	PEO	BTA-HNT	PEO	36.2 ± 2.91	AM50	Adhesion strength↑ Anti-wear↓	Concentration of fillers: 10 g/L		2016	[100]
	PEO	Graphene	PEO	8.19	AZ31	Hardness↑ Friction coefficient↓	Coating time: 5, 10, 15, 20 min	20 min	2018	[77]
	Electrochemically polymerize	GO	PEDOT	-	Mg	Anti-corrosion↑	Concentration of fillers: 10 mg/ml		2017	[86]
	PEO	GO	PEO	34.3	AZ31	Anti-corrosion↑	Concentration of fillers: 1, 2, 3 g/L	2g/L	2016	[78]
Hybrid	Electrophoresis and electrodeposition	GO-CNTs	Ni	~3.4	AZ91D	Hardness↑ Anti-wear↑	Concentration of fillers: 0~0.06g/L	0.04 g/L	2018	[21]
	Spin coat	$GO-TiO_2$	BMA	70 ± 5	AZ31	Anti-corrosion↑ Hydrophobicity↑	Concentration of fillers: 0.3 wt.%		2019	[73]

that the GO-PEO coatings blended with 20 and 40 ml/L of GO both showed a high hardness of ~573 HV, which is higher than that of PEO coatings without and with 5 and 10 ml/L GO incorporation [22].

Dong et al. prepared Ni-GO-CNTs composite coatings blended with 0.2 g/L GO and 0–0.06 g/L CNTs on AZ91D magnesium alloy substrates by combining electrophoresis and electrodeposition methods [21]. Both GO and CNTs have high strength and toughness and large surface area. During the friction and wear process, GO and CNTs will be exposed and form a lubricating film. This lubricating film will restrict direct contact between friction pairs. CNTs also play a load-supporting role in friction procedure. They regard that the cross-linking effect of CNTs and GO can further enhance the physicochemical properties of the coating. However, the experiment results were not as good as expected. Compared with the hardness below 110 HV and wear loss above 0.55×10^{-6} kg/m for the Ni-GO composite coating, the hardness slightly higher than 130 HV and little lower wear loss of ~0.45×10^{-6} kg/m were gained for Ni-GO-CNTs composite coating with 0.04 g/L CNTs added. Fernández-Hernán et al. revealed that the functionalized graphene nanoplatelets (COOH-GNPs) blending can improve the toughness and anti-corrosion behavior of the sol-gel coated magnesium alloy because the presence of COOH-GNPs hinders chloride ions from reaching the surface of the substrate, which delays the starting of the corrosive process [36].

8.2.2.2 Fine Grain Strengthening

Figure 8.5 shows the hardness of a material as a function of the grain size. Obviously, according to the "Hall–Petch" relationship, as the grain size decreases, the hardness of the material increases as the diffusion and migration of dislocations are hindered [49]. Addition of nanoparticles refines grains of the coatings [20, 59]. In the XRD spectra of the Ni-P-B_4C and Ni-W-P-B_4C coatings, three Ni peaks appeared at $2\theta = 44.7°$, $57.5°$, and $76.7°$, respectively, as shown in Figure 8.6. Compared with the Ni-P coating, the Ni peaks in the Ni-P-B_4C and Ni-W-P-B_4C coatings significantly broadened. The calculated average grain sizes of the Ni-P, Ni-P-B_4C, and Ni-W-P-B_4C coatings were 2.7, 1.9, and 1.5 nm, respectively, which indicates that NPs blending refined the grains of the coatings. After Ni-P, Ni-P-B_4C, and Ni-W-P-B_4C coatings, the surface hardness of AZ91D magnesium alloys increased from 100 MPa to 700, 1000, and 1290 MPa, respectively [69].

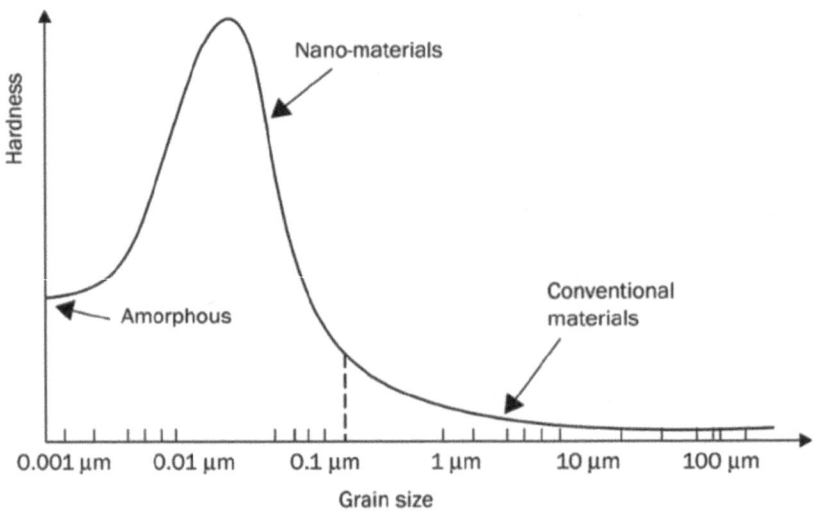

FIGURE 8.5 Hardness of a material as a function of the grain size (Source: Santo, L., Davim, J. P. 2012).

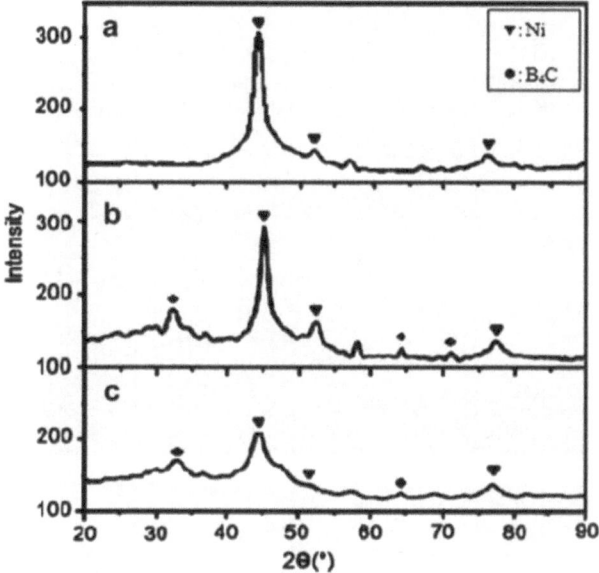

FIGURE 8.6 XRD pattern of the electroless (a) Ni-P. (b) Ni-P-B₄C. (c) Ni-W-P-B₄C deposition on the AZ91D magnesium alloy (Source: Araghi, A., Paydar, M. H. 2013).

8.2.2.3 Bridging Effect

Nanowhiskers and nanotubes can bridge particles in coating, thereby enhancing the toughness of the coating. Figures 8.7 and 8.3 show the SEM morphologies of the AlN nanowhiskers prepared by the direct nitridation method [14] and the CNTs blended in coating [43], respectively. Lahiri et al. took a HA coating blended with CNTs as an example to explain the bridging process in detail [110]. The bridging process starts such that the HA matrix first deforms under stress due to its lower elastic modulus, and then, due to the good interface bonding between the CNTs and the HA matrix, the stress is effectively transferred from the HA matrix to the CNTs reinforced component. As a result, CNTs absorb more stress than the HA matrix, causing the elastic modulus of the HA-CNTs coating increases. The bridging of CNTs is an important reason for the toughness improvement of HA-CNTs composite coatings [111–113]. Khazeni et al. prepared the HA coatings blended with 0–1.5 wt.% CNTs on the AZ31 magnesium alloy [43, 76]. The highest elastic modulus of 135.4 ± 6.2 GPa was obtained with 1 wt.% CNTs blending. López et al. confirmed that the incorporation of CNTs into the SiO_2 coating causes a slight decrease in the friction coefficient under both mechanical mixing (MM) and ultrasonic probe mixing (UM) system [25]. The minimum wear rate under MM mode was 0.9 mm³/N. m, reduced to 39% and 51% of the untreated and heat-treated substrates, respectively. The crack bridging and pulling out of CNTs are the mechanism to improve the wear resistance of SiO_2-CNTs composite coatings. Zhang et al. used CNTs to adsorb gentamicin, and then the positively charged CNTs with gentamicin molecules and other components were moved in the electrophoresis solution to be deposited on the AZ91D substrate under an electric field [29]. Through this method, with 0.025–0.0625 g/L CNTs blending, there was a strong interaction between the gentamicin molecule and the CaP/chitosan coatings. The addition of CNTs increased the loading of gentamicin in the CaP/chitosan coatings, improved the cell compatibility of magnesium-based CaP/chitosan, and reduced the release rate of gentamicin. This results in a positive impact on the biological properties of the coating.

8.2.2.4 Decreasing Internal Stress in the Coating

Due to ion bombardment in the PVD process, unavoidable internal stress in the coating will increase with the increase of the deposition time. Some metal doping reduces the internal stress of the coating,

FIGURE 8.7 SEM morphology of AlN nanowhiskers prepared by the direct nitridation method (Source: Aal, A. A., Bahgat, M., Radwan, M. 2006).

increases the bonding strength between the coating and the substrate, will prevent coating failure or peeling during the corrosion process [37, 38]. Dai et al. revealed that the DLC film deposited by a hybrid beams deposition system partially cracked and peeled off from the AZ31 substrate due to the poor adhesion strength. While all Cr doped DLC films exhibited no localized delamination and good adhesion to the AZ31 substrate [37]. This is due to the combined effects of two aspects: the increase of the Cr content reduces the internal stress and the formation of the CrC crystalline phases in the amorphous DLC improves the interface bonding. The DLC coating with 2.34 at.% Cr incorporation gained the highest critical load of ~8 N. The DLC coatings with 2.34 at.% and 12.1 at.% Cr incorporation showed lower stable wear coefficient and more shallow wear tracks than that of the uncoated, DLC coated, and 31.5 at.% Cr-DLC coated samples. The results of ref. [38] also showed that the significant improvement in wear and corrosion resistance of the DLC coating imbedded with TiN crystal on the MAO coated AZ80 magnesium alloy comes mainly from the improvement of the bonding strength, lubricating characteristics, and the I_D/I_G ratio.

8.2.2.5 Sealing of the Holes

Except for good adhesion, in general, an ideal anti-wear protective coating should have a dense structure without pores or cracks to prevent corrosive media from penetrating the substrate. Studies

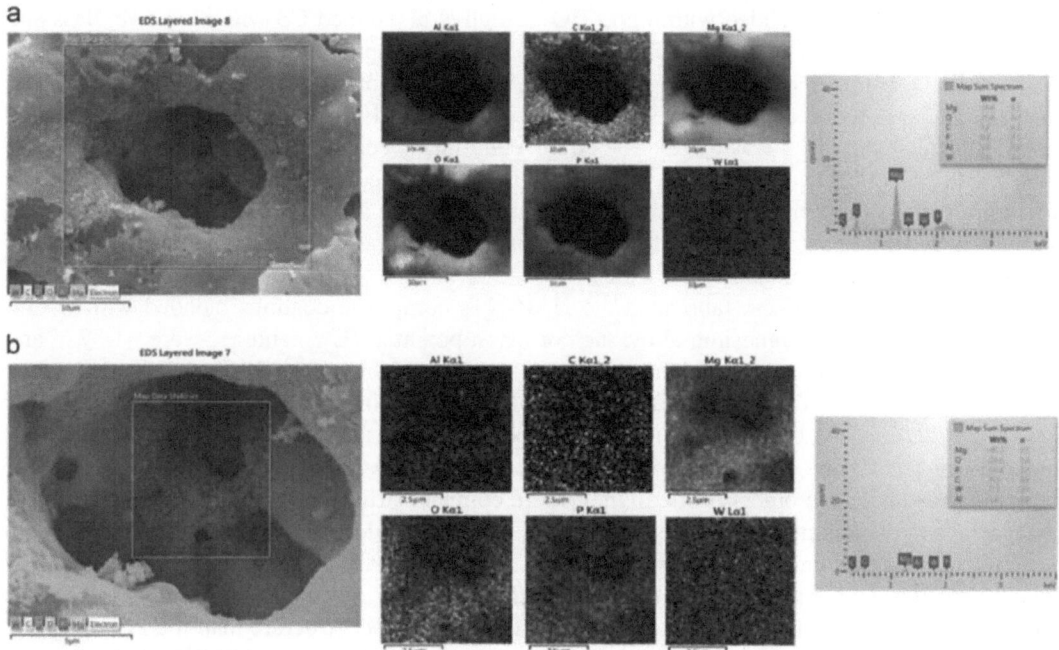

FIGURE 8.8 EDS map obtained from SEM image with high magnification from a single pore at free surface of sample NP15 (with WC blended and coating for 15 min): (a) outside and (b) bottom of pore (Source: Vatan, H. N., Ebrahimi-kahrizsangi, R., Kasiri-asgarani, M. 2016).

have shown that nanofillers can easily seal the coating holes formed during the coating growth process owing to the nanoscale size of the nanofillers [64]. Figure 8.8 verifies that the WC nanoparticles are absorbed and trapped inside of pores in PEO coating [53]. As the holes-sealing by nanofillers, the porosity of the coating reduced [59], the submicron defects of the coating reduced, and a denser coating was obtained [50, 91, 114]. Cui found that the pure DLC film had an obvious columnar structure. After Cr doping and Cr/H co-doping, the columnar microstructure of the DLC coating became indistinct, indicating that the NPs blending makes DLC coating more compact [115]. Due to the denser structure of coating and the holes in the coating can be partially sealed by nanofillers, the passage of corrosive media into the coating surface will be restricted, resulting in the improvement of corrosion resistance of coated the magnesium alloys [65, 116, 117]. In addition, compact coatings have appropriate mechanical properties [118].

The concentration of nanofillers play an important role in the performance of the composite coatings. For the PEO coatings blended with Al_2O_3 NPs on the AZ31 magnesium alloy substrates, with the concentration of Al_2O_3 NPs increased from 0 to 40 g/L, the surface porosity of the Al_2O_3-PEO nanocomposite coatings decreased [59]. Corrosion current density decreased from 3.10×10^{-5} A/cm^2 (the uncoated sample) to 6.52×10^{-7} A/cm^2 (Al_2O_3-PEO coated sample) with 30 g/L Al_2O_3 added. Due to the sealing holes effect and a coaxial morphology with a fine distribution of the SiC particles, the corrosion current density (i_{corr}) of Ni-15 g/L SiC coated AZ91 magnesium alloy is nearly 3 and 4 orders of magnitude lower than that of the Ni coated and uncoated magnesium alloy samples, respectively [20]. Daroonparvar et al. found that the porosity of 4 g/L TiO_2-PEO was lower than the PEO coating and 8 g/L TiO_2-PEO coating, the minimum i_{corr} of TiO_2-PEO coatings was 1.5×10^{-7} A/cm^2 with 4 g/L TiO_2 incorporation. This is lower than that of the uncoated (2.555×10^{-4} A/cm^2) and PEO coated (6.8×10^{-6} A/cm^2) Mg-1%Ca alloys [98]. In addition, the relative amount of rutile TiO_2 phase in the PEO coating blended with 4 g/L TiO_2 was more than that in the PEO coating blended with 8 g/L TiO_2. The rutile TiO_2 phase is stable in the corrosive medium [119], which is another reason why the corrosion resistance of the PEO coating blended with 4 g/L TiO_2 was better than that of the PEO coating blended with 8 g/L TiO_2. Singh et al. studied CS with bioactive glass and 1, 3, and 5 wt.% of Fe_3O_4 nanoparticles blending by the co-precipitation method [50]. The bioglass (BG)-1% Fe_3O_4 nanocomposite coating (i_{corr} was $3.74 \times 10^{-7} A/cm^2$) has a superior corrosion resistance as compared to the uncoated AZ91 alloy (i_{corr} was $4.87 \times 10^{-4} A/cm^2$). It was also better than the BG coated sample (i_{corr} was $5.38 \times 10^{-7} A/cm^2$) and Fe_3O_4 coated sample (i_{corr} was $6.28 \times 10^{-7} A/cm^2$). BG-$Fe_3O_4$-CS nanocomposite coating can be widely used as a suitable material for orthopedic applications to improve the hemocompatibility of the Mg alloys. Its uniform and crack-free structure is the dominant reason for the better protection for the magnesium alloys than the BG coated and Fe_3O_4 coated magnesium alloys.

Khazeni and his coworkers fabricated the HA-CNTs composite coatings blended with CNTs. For the uncoated AZ31 magnesium alloy, the corrosion potential (E_{corr}) and i_{corr} were -1.57 V and $44.25 \times 10^{-6} A/cm^2$, respectively. Using direct and pulse cathodic electrodeposition modes, when blended with 1 wt.% CNTs, the HA-CNTs nanocomposite coatings obtained the minimum corrosion current density of $1.07 \times 10^{-6} A/cm^2$ and $0.72 \times 10^{-6} A/cm^2$, and the relative positive corrosion potential of -1.35 V and -1.31 V, respectively [43, 76]. The HA-1wt.% CNTs coatings had better corrosion resistance than others is due to their more compact structure, which reduces and fills the passageways in the composite coatings and hinders the simulated body fluid (SBF) solution to permeate into the magnesium alloy substrate. Sun et al. studied the PEO coating blended with the HNT loaded corrosion inhibitor benzotriazole (BTA) on the AM50 magnesium alloy substrate [100]. The HNT-PEO and BTA-HNT-PEO coatings both had a more compact structure than the PEO coating. The porosity decreased from 31.9% for the PEO coating to 13.3% for the HNT-PEO coating and 15.2% for the BTA-HNT-PEO coating. The BTA-HNT-PEO coating exhibited the minimal change in impedance during the 12 h of immersion and provided the best protection for the magnesium alloy substrate.

Han et al. added 0–250 mg/L graphene into the electrolyte during the PEO procedure [23]. With 250 mg/L graphene incorporation, the corrosion current density of the PEO coatings gained the minimum value of $1.15 \times 10^{-7} A/cm^2$, which is 3 orders and 1 order of magnitude lower than that of the uncoated and PEO coated AZ91 alloys, respectively. The improvement of corrosion resistance is attributed to the thicker inner layer and denser structure of the PEO blended with graphene. In addition, the graphene sheets embedded in the inner layer of the PEO coating provide better protection for the magnesium alloy by preventing the penetration of the aggressive medium. Wen et al. found that the MAO coating process decreased i_{corr} from $2.124 \times 10^{-3} A/cm^2$ of the uncoated AZ31 magnesium alloy to $1.221 \times 10^{-4} A/cm^2$, and increased E_{corr} from -1.708 V of the uncoated sample to -1.598 V. With HA and GO co-incorporation, the HA-GO-MAO coating had a much lower i_{corr} of $3.643 \times 10^{-5} A/cm^2$ and more positive E_{corr} of -1.472 V [79]. The biodegradable HA-GO-MAO coating provided better protection for magnesium alloy than the MAO coating because it had a lower porosity of 0.2%, which is only 10% of the MAO coating. This indicates that HA and GO can effectively seal pores in the MAO coating. The results of Zhao et al. showed that the PEO coatings blended with 1–3 g/L GO can significantly improve the corrosion resistance of the AZ31 magnesium alloys. It is worth noting that the PEO-2g/L GO coating had the lowest corrosion current density of $3.29 \times 10^{-8} A/cm^2$, indicating that it provided the best protection for the AZ31 magnesium alloy [78]. This is due to its densest microstructure and lowest porosity, as shown in Figure 8.9. In the PEO-GO composite coatings, the nanosheets can act as a barrier to hinder the aggressive electrolyte to penetrate the AZ31 magnesium alloy surface and increase the tortuosity for the electrolyte diffusion pathway.

8.2.2.6 Barrier Function in Restrain Corrosion Medium

Most nitrides, carbides, oxides, CNTs, and GO nanofillers have good corrosion resistance. Uniform distribution of these nanofillers in coatings will effectively reduce the coating area in contact with corrosive media [20]. Without TiO_2 nanofiller blending, corrosive media easily penetrates to the magnesium alloy substrate along with the pores in the PEO coating, as shown in Figure 8.10. When

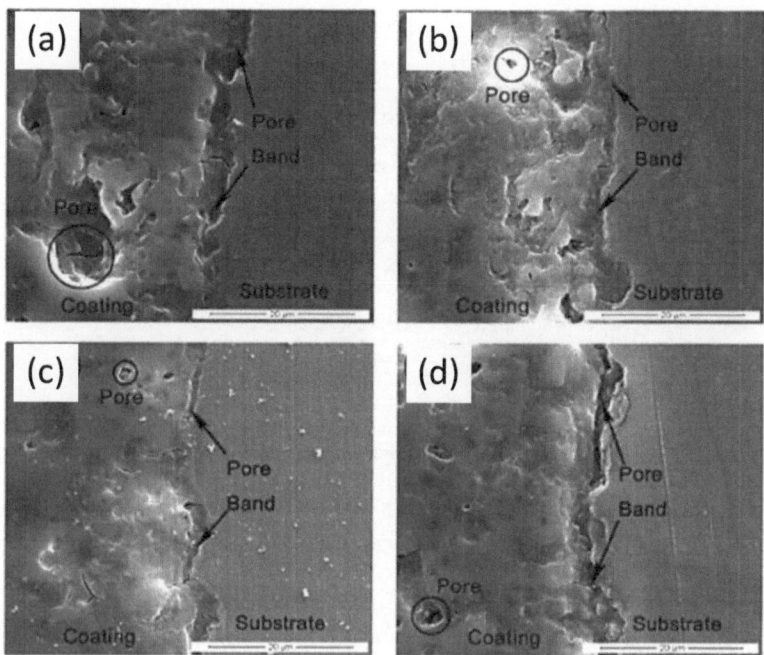

FIGURE 8.9 Cross-section morphologies of the PEO coatings of the four samples. (a) MAO coating. (b) MAO coating blended with 1 g/L GO. (c) MAO coating blended with 2 g/L GO. (d) MAO coating blended with 3 g/L GO (Source: Zhao, J. M., Xie, X., Zhang, C. 2017).

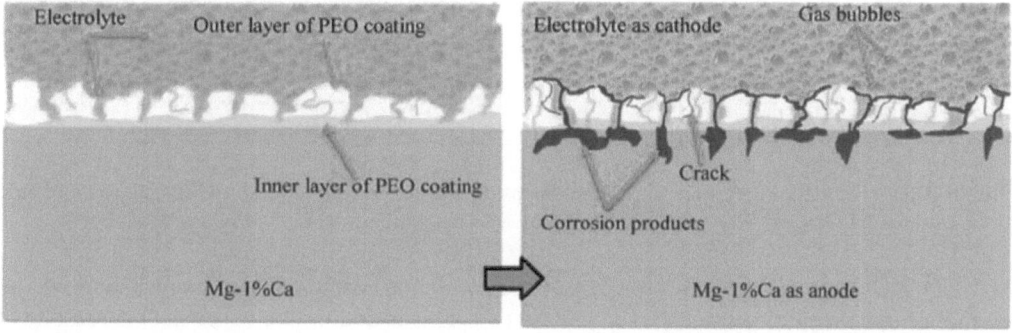

FIGURE 8.10 Schematic illustration of the corrosion mechanism of the PEO coated Mg alloy in 3.5 wt.% NaCl solution (Source: Daroonparvar, M., Yajid, M. A. M., Yusof, N. M. 2016).

TiO_2 nanofillers are blended in PEO coating, these nanofillers can act as a barrier in the PEO coating, change the corrosion path, and block the corrosion diffusion pathway by making them longer and more encapsulated, and even prevent its progress [73, 98], as demonstrated in Figure 8.11. CNTs and GO nanofillers are endowed with better corrosion resistance due to CNTs maybe react with other organic compounds, such as hydroxyapatite during the coating process [43], and GO sheets can reject anions that are usually corrosive due to negatively charged functional groups [86], as previously stated.

Nazeer et al. investigated the poly(BMA)-GO-TiO_2 nanocomposite coating with 0.3% TiO_2 NPs and GO nanolayers co-blending on the AZ31 magnesium alloy substrate by spin coating [80]. After

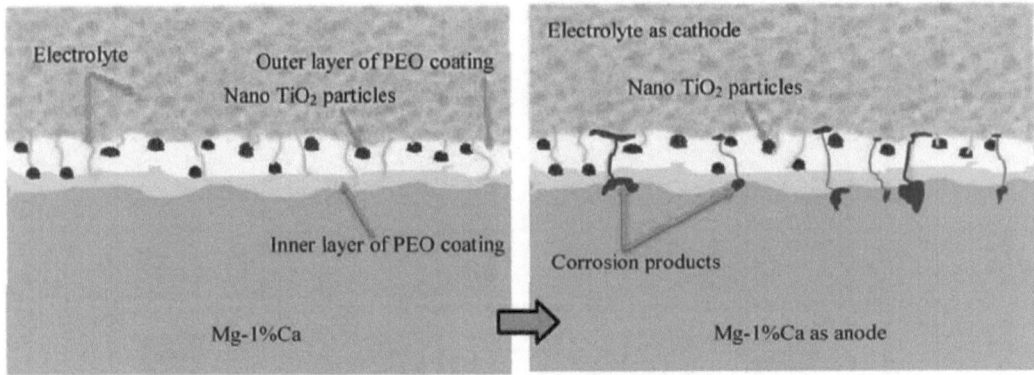

FIGURE 8.11 Schematic illustration of the corrosion mechanism of the PEO-TiO$_2$ coated Mg alloy in 3.5 wt.% NaCl solution (Source: Daroonparvar, M., Yajid, M. A. M., Yusof, N. M., Bakhsheshi-Rad, H. R. 2016).

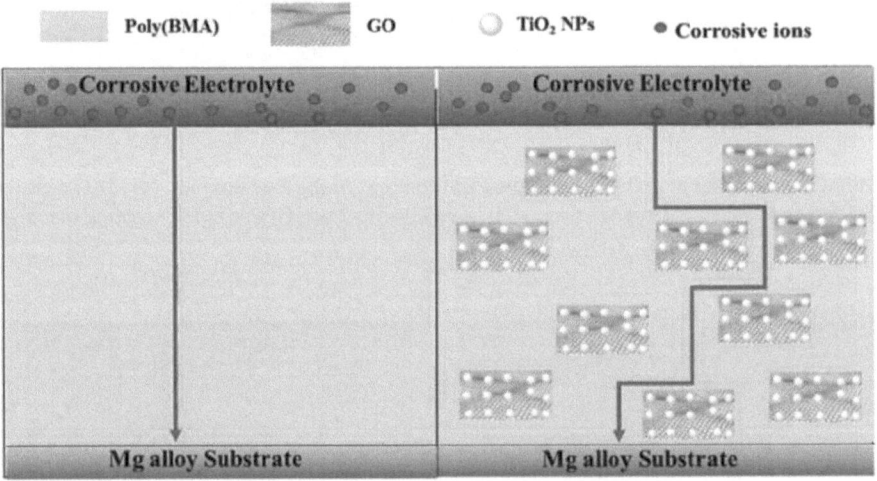

FIGURE 8.12 Schematic representation of the corrosion protection of poly(BMA)-GO-TiO$_2$ nanocomposite coating (Source: Nazeer, A. A., Al-Hetlani, E., Amin, M. O., Quiñones-Ruiz, T., Lednev, I. K. 2019).

BMA-GO-TiO$_2$ coating, i$_{corr}$ was 8.3×10^{-7} A/cm^2, which is nearly four orders of magnitude lower than that of the uncoated sample (7.4×10^{-3} A/cm^2). This is due to two reasons. One reason is that GO and TiO$_2$ act as physical barriers and increase the tortuosity of the diffusion path of electrolyte ions into the magnesium alloy substrate. Another reason is that the TiO$_2$ nanoparticles can improve the adhesion of nanocomposites to the metal surface by forming a strong binding with the metal surface (Ti-O-metal). Although there are many hydroxyl groups on the surface of TiO$_2$ nanoparticles, the authors regard that the inclusion and dispersion of TiO$_2$ nanoparticles in polymer composites are good, as shown in Figure 8.12, which leads to the decrease of surface energy and the increase of contact angle to 122.8° and hence increase of its hydrophobicity.

8.2.2.7 Others

After blending with nanofillers with special properties, the composite coatings will have some special anti-bacterial [120, 121], biological [50], self-lubricity [38, 64], hydrophobicity [73], and other properties. For example, MoS$_2$ and graphite have great advantages in friction reduction as their self-lubricating. CNTs have a lubricating effect and their own negatively charged carboxyl group, can induce apatite formation during biomineralization [29]. Due to negatively charged functional groups, GO sheets can be used as a corrosion inhibitor to repel anions that normally facilitate corrosion [86, 122]. Dong et al. regard that the cross-linking effect of CNTs and GO can enhance the physicochemical properties of the CNTs and GO [21].

The friction coefficient of the MAO coated, DLC-MAO coated, and TiN blended DLC-MAO coated samples were around 0.5, 0.2, and 0.1, respectively. The authors regard that the much lower friction coefficient of the TiN blended DLC-MAO coated sample is due to the self-lubricating mechanism and the high binding strength [38]. Vatan et al. described that SiC nanoparticles pulled out on the PEO coatings' surface to produce a "rolling effect" and act as lubricants during the sliding test. Therefore, the SiC-PEO nanocomposite coated AZ31 magnesium alloy had a low wear rate of 7 mg/N m and a low frictional coefficient of ~0.2 [64].

Catt et al. have developed a conducting polymer 3,4-ethylenedioxythiphene (PEDOT)-GO composite coatings on the Mg substrate in ethanol media by electropolymerization [86]. The PEDOT-GO coating showed a more positive E_{corr} of -1.55 V than that of the uncoated Mg substrate (-1.705 V). And the i_{corr} of the PEDOT-GO coated sample was reduced to 12.3 µA, which is 25% of that of the uncoated sample (49.6 µA). Its barrier effect is one of the reasons for the PEDOT-GO coated sample had better corrosion resistance. Meanwhile, the reaction between the MgO of the magnesium surface and PEDOT-GO coating may further improve the corrosion resistance of the PEDOT-GO coated magnesium sample. In addition, the large GO sheets are trapped in the polymer film, resulting in a net negative charge produced in the film, thereby increasing the barrier to the penetration of corrosion negative ions such as Cl$^-$ [86, 122]. Zhang et al. found that the GO-PEO coating with 20 ml/L GO incorporation displayed the lowest wear rate of 1.34×10^{-4} mm^3/N m, which is only 20% of that of the uncoated sample. This is ascribed to the lower surface roughness, higher surface hardness, and self-lubricating property of the 20 ml/L GO-PEO coating, which alleviates the abrasive and adhesive wear [22].

8.2.3 Concentration Optimization of Nanofiller Blended in Composite Coating

A low concentration of nanofiller blending has little effect on the performance enhancement of coatings, while a high concentration of nanofillers will lead to agglomeration and worsening the performance. Figure 8.4 (a and b) shows the agglomeration of CNTs in the sol-gel coating [36]. Please note that for micro arc microarc oxidation, plating, and chemical plating coating preparation processes, the concentration of nanofillers in the coating is not equal to the concentration in the electrolyte solution. The concentration of nanofillers in these coatings relates to the concentration of nanofillers in electrolyte, deposition method, stirring method, and stirring speed. For spin coating, spraying, and laser cladding, the concentration of nanoparticles almost has no change before and after coating in case of uniform distribution of the nanofillers in the prepared solutions or powders. In the PVD deposition process, the concentration of nanofillers is a combined effect of a customized target, gas flow rate, and other process parameters. Table 8.1 shows some examples of concentration optimization of nanofillers in coatings.

Khazeni and his coworkers [43, 76] fabricated HA-CNTs composite coatings blended with 0–1.5 wt.% CNTs. The HA-CNTs nanocomposite coating blended with 1 wt.% CNTs had the minimum of the corrosion current density, which indicates that CNTs blending concentration has an optimal value. Comparing the pulse deposition and direct deposition modes, the corrosion resistance of coatings using pulse deposition was better than that of coating using direct deposition. This means that the deposition mode affects the performance of the HA-CNTs nanocomposite coatings. In the study of GO-CNTs co-blended Ni coatings, with the concentration of CNTs increased from 0 to 0.06 g/L, the hardness of coatings increased and reached the maximum above 130 HV and the minimum wear loss of $\sim 0.45 \times 10^{-6}$ kg/m with 0.04 g/L of CNT incorporation [21]. With the CNTs concentration exceeding 0.04 g/L, the agglomeration of CNTs, abundant holes, and cracks defects generated. These agglomerated particles would fall off during friction and wear, resulting in serious wear, so the wear loss increased. Zhang et al. revealed that, with the increase of CNTs concentration from 0.05 g/L to 0.25 g/L, the cell compatibility of the CaP/chitosan coatings firstly increased. When the CNTs content exceeded 0.25 g/L, CNTs accumulated and interlaced, the growth of HA retarded, and the cell compatibility of the CaP/chitosan coatings decreased [29]. The GO-PEO coatings with 20 and 40 ml/L GO incorporation both showed high microhardness of around 573 HV; this is higher than that of coatings without and with 5 and 10 ml/L GO incorporation [22]. The lowest wear rate of 1.34×10^{-4} mm^3/N was obtained for GO-PEO coating with 20 ml/L GO incorporation, which is only 20% of that of the uncoated sample. The best anti-corrosion and anti-wear performance were gained when Al_2O_3 NPs blended with 30 g/L concentration [59]. Under the same conditions, the highest absorption of nanoparticles in coatings was at a stirring speed of 100 rpm.

Dai et al. found that in the Cr-DLC nanocomposite coatings by a hybrid beams deposition system, the concentration of Cr affected the crystalline structure and morphology of coatings, the internal stress in coatings, bonding strength to the substrates, and the wear resistance of the coated AZ31 alloy samples. The DLC coating had a classic amorphous structure, with the Cr concentration increased from 2.34 at.% to 31.5 at.% through adjusting the Ar and CH_4 gas flow rate, the CrC carbide was formed and its amorphous phase structure could be transformed into a crystalline structure, the grain size of the coatings increased, and the structure of coatings became to a segregated bigger "flowerlike". In terms of bonding strength, the DLC film partially cracked and peeled off from the AZ31 substrate due to the poor adhesion strength; while all Cr doped DLC films exhibited no localized delamination and good adhesion to the AZ31 magnesium alloy substrates. It can be explained by two aspects competition mechanisms: with the increase of the Cr content, the increase of the internal stress of the coating leads to a decrease of the critical load, while the formation of the CrC crystal phase in the amorphous DLC improves the interface bonding. The DLC film with 2.34 at.% Cr gained the highest critical load around 8 N. The 2.34 at.% Cr-DLC coated sample and 12.1 at.% Cr-DLC coated sample owned lower stable friction coefficient and shallower wear tracks than the uncoated, DLC coated, and 31.5 at.% Cr-DLC coated samples [37].

Table 8.2 summarizes the preparation method, thickness, the hardness ratio (H_c/H_s), the coefficient of friction ratio (COF_c/COF_s), the wear rate ratio (W_c/W_s), and the i_{corr} ratio (i_c/i_s) of the coating to the substrate. Obviously, compared with other nanofillers blended composite coatings, graphene and CNTs blended coatings have much higher hardness [23, 123]. The TiN embedded DLC coatings on the MAO coated the AZ80 magnesium alloy has excellent anti-corrosion performance due not only to the compact structure of the coating but also to the good bonding strength of the coating to the substrate and the barrier function of the MAO support coating [38]. The GO and TiO_2 co-blended poly(BMA) coating shows much better anti-corrosion protection than GO blended BMA coating for magnesium alloy [73], which provides a new way to improve corrosion resistance through co-blending. In addition, the wear rate of the CNTs blended TiO_2 coating with a thickness of 135 μm on the NiCrAlY support coating (thickness of 100 μm) is 1/35 of that of the uncoated Mg-Ca magnesium alloy sample, which means that a thicker composite coating could significantly improve the wear resistance of the magnesium alloy [123].

Based on these analyses, different nanofillers have different degrees of improvement on the performances of the coated magnesium alloys. Note that the improvement mechanisms usually do not work alone, but are often a combination of multiple mechanisms. For example, the microhardness of the NiCrAlY/$nTiO_2$-CNT coating was more than 850 HV, which is higher than that of the NiCrAlY/$nTiO_2$ (around 720 HV) coated and 17 times of the uncoated Mg alloys (49.2 HV) [123]. The remarkable hardness improvement by the NiCrAlY/$nTiO_2$-CNT coating is mainly due to the following reasons: (i) the denser structure, (ii) the bridging effect of the CNTs, and (iii) the second phase strengthening of CNTs [124]. Besides, the NiCrAlY/$nTiO_2$-CNT composite coating exhibited higher bonding strength. This is due to the presence of CNTs in the coating as a reinforcing fiber to fix the coating matrix together, thereby improving the bonding strength to the substrate [125]. The incorporation of CNTs into the NiCrAlY/$nTiO_2$-CNT nanocomposite coating also significantly enhanced the tribological behavior of the coated Mg alloy due to the bridging effect and the lubrication of the CNTs, and the low porosity of the nanocomposite coating [126].

TABLE 8.2

Summary of preparation method, thickness, hardness ratio of the coating to the substrate (H_c/H_s), coefficient of friction ratio of the coating to the substrate (COF_c/COF_s), wear rate ratio of the coating to the substrate (W_c/W_s), and corrosion current density ratio (i_c/i_s) of the coating to the substrate of the nanofiller blended single-layer composite coatings

Preparation method	Fillers	Matrix	Coating thickness (μm)	Substrate	H_c/H_s	COF_c/COF_s	W_c/W_s	i_c/i_s	Year	Ref.
MAO/PEO	TiO$_2$	PEO	~5	Mg-Ca	5.16	-	-	1/1703.3	2016	[98]
	Al$_2$O$_3$	PEO	~21.5	AZ31				6.521×10^{-7} (A/cm^2)/-	2017	[59]
	SiC	PEO	~10	AZ31	8.59	0.15/0.4		1/251.9	2016	[64]
	SiC	MAO	13.06	AZ31B				9.2×10^{-6}/-	2016	[65]
	GO	PEO	34.3	AZ31				1/327.3	2016	[78]
	Graphene	PEO	~70	AZ91	15.54	-	-	1/1547.8	2018	[23]
Electrodeposition	SiC	Ni	35±5	AZ91	7.07	-	1/8.33	1/747.1	2013	[20]
	CNTs	HA	7.8	AZ31	540 HV/-	-	-		2019	[43]
	BG-Fe$_3$O$_4$	Chitosan	10±0.2	AZ91				1/1302.1	2020	[50]
	CNTs	HA	<7.8	AZ31				1/61.5	2019	[76]
	GO-CNTs	Ni	~3.4	AZ91D	130 HV/-		-		2018	[21]
Electroless plating	W-P-B$_4$C	Ni	50	AZ91D	1.84	-	-	1/18.9 (Ni-P coating)	2013	[69]
	P-B$_4$C	Ni	50	AZ91D	1.84	1.6/1.0	1/12.86	1/4.2 (Ni-P coating)	2010	[99]
	W-P	Ni	24	AZ91D	6	-	-	1/52.7	2007	[68]
Atmospheric plasma spraying	CNT	TiO$_2$	100 (NiCrAlY)/135 (TiO$_2$-CNT, top layer)	Mg-Ca	17.28	0.13/0.44	1/35		2017	[123]
Sol-gel	SiO$_2$	CNTs	3±0.4	WE54	-	0.4/0.43	1/4.44		2011	[25]
PVD	TiN	DLC	5 (MAO)/0.5 (PVD, top layer)	AZ80	-	0.1/0.2	-	1/9366.4	2013	[38]
	Cr	DLC	0.662	AZ31	-	0.3/0.3	-	1/0.34	2010	[37]
	Cr	DLC	3.78	AZ91D		0.17/0.35	1/741.7	1/0.41	2019	[115]
	Cr and H	DLC	3.50	AZ91D		0.09/0.35	1/539.48	1/9.86	2019	[115]
Spin coating	GO-TiO$_2$	Poly(butyl methacrylate)	70 ± 5	AZ31				1/8915.7	2019	[73]
	GO							1/1804.9		

8.3 MAIN COATING WITH SUPPORT COATING AND/OR TOP COATING

The previous section has illustrated using the nanofiller blending to create nanocomposite coatings to improve the performance of magnesium alloys. However, based on the results in Table 8.2, although the hardness of the magnesium alloys coated with nanofiller blended nanocomposite coatings has been significantly improved [23, 123], the hardness is still relatively low, and the maximum hardness is less than 1000 HV. Therefore, nanofiller blending nanocomposite coatings can only improve the performance of the magnesium alloys to a certain extent. We notice that [38, 123] with a thick support coating, the GO-TiO$_2$ blended poly coating has excellent corrosion resistance [73]. It is difficult to say their significant performance improvement has nothing to do with the thick support coating or the excellent corrosion resistance of the polymer matrix. As mentioned in our introduction, the polymer coatings have some limitations in application. In addition, we have already mentioned that some hard coatings have excellent anti-wear properties, but the low hardness of magnesium alloys cannot provide load support for the coatings, resulting in coatings failure [127]. In addition, nanofillers can partially seal pores in the coating; however, there are still some holes will be maintained, as shown in Figure 8.9, monolayer coatings cannot efficiently prevent the corrosion liquid from invading the magnesium alloys [78]. To improve the performances of the coated magnesium alloys, the proposed coating structure design is shown in Figure 8.13, where the coating system consists of the main coating and support and/or top coating. The support coating acts as a hardness transition for hard coatings, thus provides load support for the softer substrate. This layer also effectively serves as a bonding layer between the main coating and the substrate. Because this layer is also immediately on top of the underneath magnesium alloy, it also becomes the last defense or barrier against inward diffuse of any corrosion medium. The top coating seals holes in the coating and gives coating hydrophobicity/superhydrophobicity properties. More examples of the top coating and the support coating are described later. In this section, we focus on the roles and mechanisms of the support coating and top coating in improving the performance of coating.

The support coatings in use today include MAO coating (such as TiN-DLC/MAO layer/AZ80 [38], TiO2/MAO layer/AZ80 [128]), nitrogen ions implanted transition layer (Ti coating/N-implanted layer/Mg-Ca-Zn) [129], physical vapor deposited SiO$_2$ (thus GO coating/SiO$_2$/Mg-Ca-Zn) [130], atomic layer deposited ZrO$_2$ (thus PLGA/ZrO$_2$/AZ31) [63], etc.

Apart from the support coating between the magnesium alloys surface and the main coating, top coatings have also been used to seal surface holes in the coatings, and/or give the coating self-lubricity, making the coating hydrophobic and super-hydrophobic, etc. Absorption created top layers on the PEO coated Mg-Mn-Ce alloy substrates include super-dispersed polytetrafluoroethylene (SPTFE) [131], hydrophobic (HP), and superhydrophobic (SHP) nanocomposite layers [132]. The poly-lactic acid (PLA) by dip-coating was used as the top sublayer on the MAO coated Mg-Ca alloy substrate [133]. The graphene oxide layers by LBL self-assemble and electrodeposition were used as the

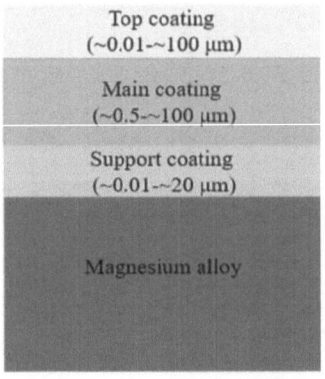

FIGURE 8.13 Schematic diagram of the main coating with support coating and/or top coating.

top sublayer on the PEO coated ZK60 and AZ91 magnesium alloy substrates [30, 31]. Also, Al/AlCr bilayer coatings using PVD consisting of a linear ion source and a magnetron-sputtering source were prepared on the Mg-Zn-Ce-La alloy substrate [134] and the NiCrAlY/nTiO$_2$-CNT bilayer composite coating by an atmospheric plasma spraying [123] was prepared on the Mg-Ca alloy substrate.

8.3.1 Main Coating with Support Coating

Support coating improves the performance of the coatings owing to the following three reasons: (i) the support coating forms hardness transition between the magnesium alloy and the coating. For example, a thick MAO coating (up to 20 microns) has a higher hardness than magnesium alloys to provide a hardness transition and load-supporting for the magnesium alloy and the coating [128, 135]. (ii) The support coating diminishes the thermal expansion coefficient mismatch between the ceramic top coating and the magnesium alloy substrate, resulting in enhancement of the bonding strength of the coating to the substrate. Good bonding can relax and adjust the stress distribution, form a supporting layer [41], and prevent coatings from peeling off from the magnesium substrates, (iii) the support coating acts as a barrier to effectively inhibit the penetration of corrosive media into the magnesium alloy [38, 63, 130].

The hardness transition, load-supporting, and good bonding functions provided by the support coatings help the coating to improve the hardness and wear resistance of the magnesium alloys. An MAO support coating has metallurgical bonding with the magnesium alloy substrate [38]. TiO$_2$/MAO coated magnesium alloy with ~19 μm thick MAO support coating has reported a hardness of 700 HV, 10 times higher than that of the uncoated sample (67 HV) [128]. MCrAlY (M = Ni, Co) support coating prevent the atmospheric plasma sprayed (APS) coating debonding [126]. Bakhsheshi-Rad et al. found that the APS thick NiCrAlY layer creates good mechanical bonding between the magnesium alloy substrate and the TiO$_2$ coating [123]: Compare to the hardness of 49.2 HV for the uncoated sample, high hardness of 850 HV for the NiCrAlY/nTiO$_2$-CNTs bilayer coated Mg-Ca alloy achieved. For the uncoated Mg-Ca alloy, the wear rate was around 0.7 mm^3/N. m, and the coefficient of friction was 0.44. After coating with NiCrAlY/nTiO$_2$-CNTs, the wear rate was reduced to around 0.02 mm^3/N. m and the coefficient of friction was reduced to 0.13.

Gradient multilayer is also used as a support coating. For example, the Al/AlN/CrAlN/CrN/MoS$_2$ gradient coating was prepared on the AM60 magnesium alloy substrate combined with N ion implantation and plasma immersion ion implantation (Al/AlN/CrAlN/CrN layer, the support coating) and magnetron sputtering (MoS$_2$ layer, the main coating) processes [61]. Figure 8.14 (a) shows the clear gradient structure of the Al/AlN/CrAlN/CrN/MoS$_2$ gradient coating. The interfaces between Al/AlN/CrAlN/CrN layers are not very clear, indicating high bonding strength obtained

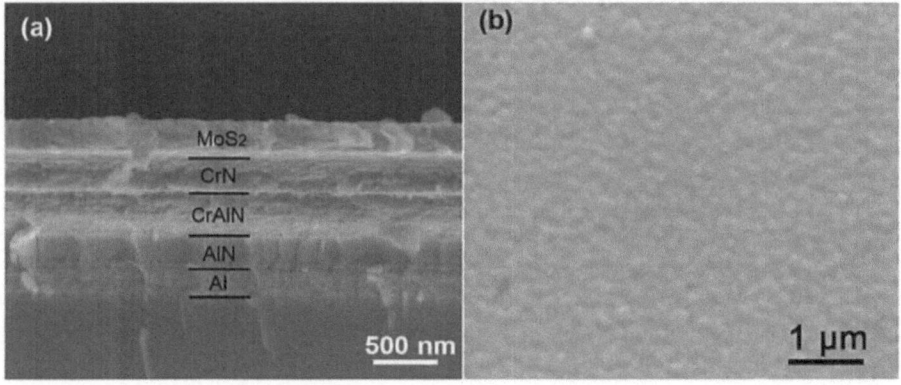

FIGURE 8.14 Cross section (a) and surface images (b) of the gradient duplex coating (Source: Xie, Z., Luo, Z., Yang, Q., Chen, T., Tan, S., Wang, Y., Luo, Y. 2014).

using high-energy ion implantation during deposition. Figure 8.14 (b) shows that the gradient coating possesses a uniform surface with the agglomeration of some nanometer-sized particles. The hardness of the uncoated, MoS_2 coated, and gradient multilayer coated AM60 alloy were 0.93, 1.43, and 19.03 GPa, respectively. In addition, during the friction test, the single MoS_2 coating failed, while the $Al/AlN/CrAlN/CrN/MoS_2$ gradient multilayer coating withstood the whole friction course with a low and stable friction coefficient of ~0.1. Its wear rate of ~2.31×10^{-6} mm^3/N. m is one order of magnitude lower than that of the uncoated magnesium alloy (~3.15×10^{-4} mm^3/N. m) [61]. Its better wear resistance and the low friction coefficient are due to these aspects: (i) the high hardness and smooth surface of the gradient coating decreases wear rate and friction coefficient and (ii) the $Al/AlN/CrAlN/CrN$ gradient multilayer coating provides hardness transition and load support for the soft Mg alloy substrate. As MoS_2 coating has low hardness, its wear life depends on the substrate's load-bearing capacity [136, 137]. (iii) The top MoS_2 coating forms a solid lubrication film on the contact surface during the friction course as its special layered structure [138], which leads to a low friction coefficient [61].

Xie et al. studied nitrogen ion implanted/$AlN/CrAlN/MoS_2$-phenolic resin coating [45]. Nitrogen ions react with Al and CrAl to form AlN and CrAlN gradient layers (the support coating), which improves the bonding strength between MoS_2 blended phenolic resin coating (the main coating) and the magnesium alloy substrate. The gradient hardness of the nitrogen ion implanted/$AlN/CrAlN$ support coating provides sufficient load support for the soft magnesium alloy. Compared with the 0.93 GPa hardness and 34.52 GPa elastic modulus of the uncoated AM60 alloy, the nitrogen ion implanted/AlN/MoS_2-phenolic resin coated sample had a higher hardness of 16.95 GPa and elastic modulus of 175.12 GPa. The MoS_2-phenolic resin main coating enhances the corrosion resistance of AM60 magnesium alloy due to its excellent anti-corrosion performance. The i_{corr} of the nitrogen ion implanted/AlN/MoS_2-phenolic resin coated magnesium alloy sample was 2.31×10^{-9} A/cm^2 in 1 M NaCl solution, which is 4 and 2 orders of magnitude lower than that of the uncoated and single-layer MoS_2-phenolic resin coated magnesium alloy samples, respectively.

The support coating can improve the bonding strength and act as a barrier, which is beneficial to the corrosion resistance of the coating. Wu et al. used the nitrogen ions implantation layer as a support coating before the titanium coating deposition [129], after 24 h of immersion in a 3.5 wt.% NaCl solution, the corrosion current density of 2.38×10^{-5} A/cm^2 of the uncoated sample declined to 1.01×10^{-5} A/cm^2. This is attributed to: (i) nitrogen ion bombardment increased strong bonding between the substrate and the coating, which limits the delamination of Ti coating. In addition, nitrogen ions can react with Mg-Al alloy to form AlN phase, which improves bonding strength of the coating to the magnesium alloy substrate, (ii) both AlN and the compact original Ti coating have good thermal stability and good corrosion resistance. For the SiO_2/GO coating, as shown in Figure 8.15, the nano-SiO_2 coating prepared by PVD deposition has a uniform surface with a bubble structure (Figure 8.15 (b, e)). After the dip-coating graphene oxide on the SiO_2 coated magnesium alloy, the composite coating has a sheet-like morphology, as shown in Figure 8.15 (c, f). The dense nano-SiO_2 layer can act as a barrier layer (Figure 8.15 h, e) to prevent the penetration of corrosion media to the Mg-Ca-Zn alloy substrate [130]. W. Yang revealed that the corrosion current density s of the uncoated, MAO coated, DLC/MAO coated, and (Ti: N)-DLC/MAO coated samples were 5.159×10^{-5}, 2.002×10^{-6}, 2.135×10^{-7}, and 5.508×10^{-9} A/cm^2, respectively. The authors regard that the formation of a galvanic cell between the MAO support coating and the substrate was largely restrained by the top (Ti: N)-DLC film [38]. While other people consider that there will be no formation of the galvanic cell between the MAO coating and the magnesium alloy because the MAO coating has chemical inertness [128]. Therefore, the (Ti: N)-DLC/MAO coating markedly improved the corrosion resistance of the coated magnesium alloy might due to the MAO support coating acts as the barrier layer to efficiently restrain the infiltration of the corrosion medium to the magnesium alloy substrate.

DLC/AlN/Al coating was prepared on the AZ31 magnesium alloy by a hybrid ion beam deposition system (including a linear ion source and a magnetron sputtering source) [139]. Compared with the uncoated AZ31 alloy, after coating, E_{corr} positively transformed from -1.57 V to -1.48 V, and i_{corr}

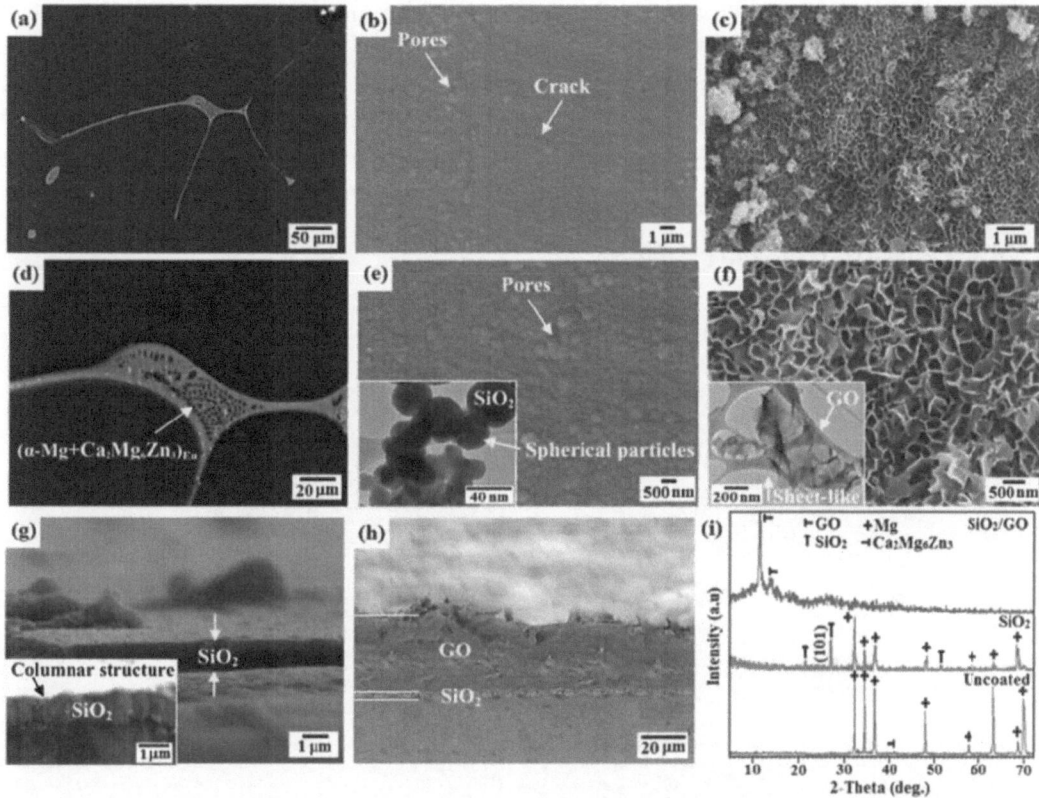

FIGURE 8.15 Surface morphology of (a, d) uncoated Mg alloy. (b, e) single-layer SiO_2 coated, (c, f) bi-layer SiO_2/GO coated and cross-sectional SEM micrographs of (g) single-layer SiO_2 coated and (h) bi-layer SiO_2/GO coated and (g) X-ray diffraction patterns of uncoated and coated Mg alloy samples (Source: Bakhsheshi-Rad, H. R., Hamzah, E., Kasiri-Asgarani, M., Saud, S. N., Yaghoubidoust, F., Akbari, E. 2016).

dropped from 2.25×10^{-5} A/cm² to 1.28×10^{-6} A/cm², respectively. The improvement was not that fantastic because even though most of the surface area of the Al/AlN/DLC coating is very dense, it also shows micropores and inevitable through-thickness defects.

A self-assembled graphene oxide (GO, top layer)/8-hydroxyquinoline (8-Hq)/inorganic (IC, under-layer) multilayer coating was prepared on the AZ31 Mg alloy using PEO and dip-chemical coating [34]. Compared with the corrosion rate of ~70 mpy for the uncoated AZ31 alloy, the corrosion rate of GO/8-Hq/IC multilayer coated sample significantly decreased to 0.0087 mpy.

To notice that a thicker support coating can play a more active role in improving corrosion resistance performance. Liu et al. prepared top PLGA spin coating combining a 2.925/11.7 nm thick ZrO_2 bottom nanofilm by atomic layer deposition and on AZ31 substrate [63]. With the 2.925 and 11.7 nm thick ZrO_2 support coating, the corrosion current density of the PLGA/ZrO_2 composite coatings were 1.038×10^{-6} and 5.601×10^{-9} A/cm², which is 2 and 4 orders of magnitude lower than that of the uncoated sample (5.124×10^{-4} A/cm²), respectively. The PLGA/ZrO_2 composite coating with 11.7 nm thick ZrO_2 support coating had a better corrosion resistance than that with 2.925 nm thick ZrO_2 support coating [140]. Qiu et al. prepared graphene oxide coating on the PEO coated ZK60 magnesium alloys via a self-assembly method. With PEO coating for 120 s and 360 s, the thickness of PEO coatings was 9 µm and 17 µm, respectively. Then the GO coating with a 4–6 µm of thickness was prepared by dip repeating three times. The GO sheets were embedded into the pores of the PEO coatings, which resulted in the porosity reduction of the PEO coatings. The i_{corr} of the uncoated, PEO (120 s) coated, GO/PEO (120 s) coated, PEO (360 s) coated, GO/PEO (360 s) coated samples were 1.05×10^{-4}, 7.28×10^{-7}, 2.26×10^{-7}, 6.72×10^{-8}, and 1.45×10^{-8} A/cm², respectively. Among these coatings, the GO/PEO (360 s) coated sample had the lowest i_{corr} and best corrosion resistance performance [30].

8.3.2 Main Coating with Top Coating

The top coating improves the performance of the coatings owing to the following.

8.3.2.1 Sealing Holes in the Coating

Daroonparvar et al. used an air plasma spraying nano-TiO₂ top coating to seal pores in the MAO coating [128]. As shown in Figure 8.16, the TiO₂/MAO composite coating consists of a TiO₂ sealing

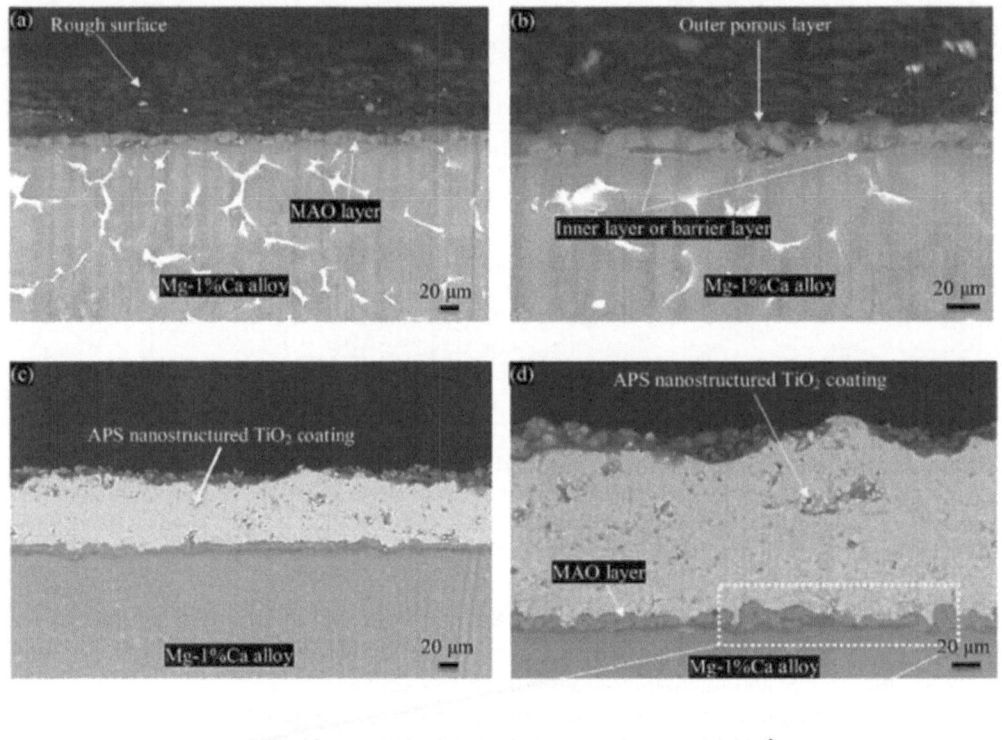

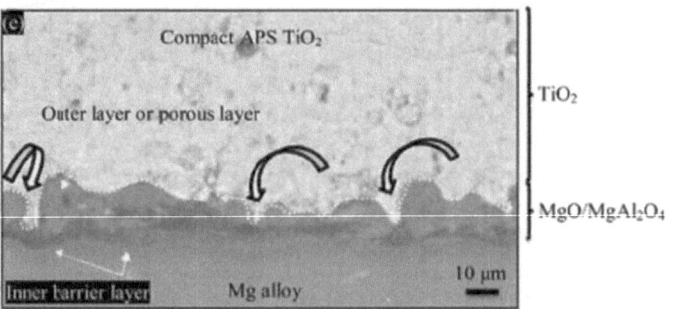

FIGURE 8.16 Cross section of the coatings on the Mg-1%Ca alloy. (a) MAO layer. (b) double-layered structure of the MAO coating on the Mg alloy. (c) Duplex MAO layer/air plasma sprayed nanostructure TiO₂ composite coating on the Mg-1%Ca alloy at low magnification. (d) Duplex MAO layer/air plasma sprayed nanostructure TiO₂ composite coating on the Mg-1%Ca alloy at high magnification. (e) Penetration of the melted TiO₂ particles into the pores and microcracks of the porous layer of the MAO coating (Source: Daroonparvar, M., Yajid, M. A. M., Yusof, N. M., Bakhsheshi-Rad, H. R., Hamzah, E., Mardanikivi, T. 2015).

layer, the outer porous layer of the MAO coating, and an inner barrier layer of the MAO coating. There are no obvious microcracks or micropores at the bonding interface (Figure 8.16 d and e). The TiO_2 coating consists of both fully melted and partially melted regions within splats, is integrated with the MAO layer by physically interlocking. This is due to the melted TiO_2 particles penetrate the pores and microcracks of the MAO coating. Shang et al. prepared GO top coating to cover the surface of the MAO film and seal the micropores in the MAO film [31]. Qiu et al. found that the graphene oxide solution (the top coating) can be embedded into the pores in the PEO coating and the porosity of the PEO coating decreases [30].

For the Mg-Ca alloy, the i_{corr} was 255.5 $\mu A/cm^2$, which was reduced to 8.5 $\mu A/cm^2$ after MAO coating [128]. After the nanostructured TiO_2 top coating on the MAO coated sample, the i_{corr} was 0.07 $\mu A/cm^2$, 2 and 4 orders of magnitude lower than the MAO coated and uncoated samples, respectively. The reasons are: (i) the inert MAO support coating acts as a barrier layer. (ii) TiO_2 top coating seals pores in the MAO coating as shown in Figure 8.16. (iii) The MAO support coating provides good bonding strength of TiO_2 coating and the Mg alloy substrate. (iv) The 100 ± 20 μm thick TiO_2 coating effectively prevents the corrosion media from penetrating the magnesium alloy substrate. Bakhsheshi-Rad et al. dipped a GO top coating on the Mg-1 wt.% Ca-6 wt.% Zn alloy coated with nano-SiO_2 coating by PVD deposition [130]. The corrosion current density of the SiO_2/GO coated magnesium alloy was 5.95×10^{-6} A/cm^2, which is lower than that of the uncoated sample (2.397×10^{-4} A/cm^2) and the SiO_2 coated sample (3.38×10^{-5} A/cm^2). This is due to the barrier function of the dense nano-SiO_2 bottom layer, the high corrosion resistance of the GO coating, and the hole-sealing effect of the GO coating. Shang et al. electrodeposited a GO top coating on the MAO coated AZ91 magnesium alloy [31]. The GO top coating covers the surface of the MAO film and seals the micropores of the MAO film. And the GO/MAO composite coatings are smoother than the single MAO film. The i_{corr} of the uncoated, MAO coated, and GO/MAO coated samples was 3.902×10^{-5}, 2.136×10^{-6}, and 8.115×10^{-8} A/cm^2, respectively.

Ning et al. prepared DLC film and DLC films blended with H, Cr, and Si on the MAO coated AZ31 magnesium alloy substrates and found that compared with single MAO coating, the duplex coatings showed pore as well as pore-diameter reduction, which indicates the DLC films (the top coating) have a certain sealing holes effect. The H-DLC (the top coating)/MAO (the main coating) coated magnesium alloy samples had the lowest i_{corr} of 3.26×10^{-7} A/cm^2, which is two orders of magnitude lower than the uncoated sample [141]. Furthermore, the wear rate of bilayer coated AZ31 magnesium alloys was 3.4–4.3×10^{-6} mm^3/ N. m, which is much lower than that of the MAO coated sample with a wear rate of 10.5×10^{-6} mm^3/ N. m.

Lu et al. [4] adopted an Mg-rich primer (MRP) top coating to seal the holes in the anodic oxidation (AO) film and found that the AO/MRP system provides excellent protection for the AZ91D alloy. The i_{corr} of bare AZ91D substrate was 7.83×10^{-6} A/cm^2, it decreased to 2.87×10^{-7} A/cm^2 after the AO/MRP coating. They believe (i) the epoxy resin penetrates the pores of the AO film to play a role in sealing holes. (ii) The AO coating and the substrate have a good metallurgical bond, and the penetration of epoxy resin into the pores of the AO film enhances the bonding strength between the MRP film and the AO film, so the substrate/AO/MRP system has good bond strength. (iii) The AO film will hinder the electrical contact between the Mg particles and the AZ91D substrate, thereby preventing the consumption of these particles in the MRP.

Fernández-Hernán et al. [36] found that the sol-gel/functionalized graphene nanoplatelets (COOH-GNPs) blended sol-gel bilayer coating showed the best corrosion protection for the AZ31 alloy because the COOH-GNPs top coating makes it is difficult for the chloride ions to penetrate the surface of the substrate and postpones the start of the corrosive process.

8.3.2.2 Making the Coating Hydrophobic and Super-Hydrophobic

Bakhsheshi-Rad et al. have pointed out that a single MAO coating with a thickness of 8–10 μm cannot provide sufficient protection for the biomedical applications of Mg-Ca alloys in a SBF physiological environment, even though its i_{corr} has changed from 241.5 mA/cm^2 of the uncoated samples to 7.3 mA/cm². The hydrophobic PLA top coating with a thickness of 37–40 μm has a good sealing holes effect on the MAO coating, and the i_{corr} was further reduced to 0.03 mA/cm², which greatly improves the corrosion resistance of magnesium alloy in SBF environment [133]. Gnedenkov et al. also prepared the HP and SHP nanocomposite top coatings to further improve the corrosion resistance of the PEO coated magnesium alloy [131, 132].

8.3.2.3 Giving the Coating Self-Lubricity

Yang et al. prepared the TiN blended DLC top coating on the MAO coated AZ80 magnesium alloy substrate [38]. The wear coefficient of the MAO coated sample was about 0.5, that of the DLC/ MAO coated and the (Ti: N)-DLC/MAO coated samples were reduced to ~0.2 and 0.1, respectively. This is due to the low friction coefficient of the DLC coating, the self-lubricating mechanism of TiN blending.

8.3.2.4 Improving the Coating Biocompatibility Property

Liu et al. spun the PLGA top coating on the 2.925 and 11.7 nm thick ZrO_2 coated AZ31 substrate [63]. After ZrO_2 coating, the Young's modulus of the coated magnesium alloy increased from 25 GPa (uncoated sample) to 38 GPa, and the surface hardness increased from approximately 0.95 GPa (uncoated sample) to 1.15 GPa. After ZrO_2/PLGA coating, the Young's modulus and hardness of the coated samples decreased to 16.7 GPa and 0.25 GPa, respectively, which are within the range of the natural bone's Young's modulus of 10–30 GPa and hardness of 0.234–0.76 GPa [142, 143]. The decrease of the surface hardness and Young's modulus is beneficial to the biomedical applications of magnesium-based alloys.

However, the main coating with supporting coating and/or top coating does not always improve the corrosion resistance of the coating. Daroonparvar et al. prepared the Al monolayer and Al/AlCr bilayer coatings on the Mg-Zn-Ce-La substrate using PVD consisting of a linear ion source and a magnetron-sputtering source [128]. The result shows that the Al/AlCr bilayer coating (AlCr film as the top coating) has larger surface roughness, coarse grains, and a looser structure than the Al coating. As a result, although the Al/AlCr bilayer coating (around 1.741 μm) was much thicker than the Al monolayer coating (0.762 μm), its corrosion current density of 105.5 μA/cm² is higher than that of the Al/AlCr bilayer coating (2.09 μA/cm²). The microcracks and pinholes in the Al/AlCr coating became a location for establishing an electrochemical cell between the Al and AlCr layer at the defect site (at the beginning of the corrosion), which will accelerate the corrosion process. It infers that the dense structure of coating is more important than the coating thickness to hinder the start of the corrosion process.

Table 8.3 summarizes the coating preparation method, thickness, and hardness ratio of the coating to the substrate (H_c/H_s), the corrosion current density ratio of the coating to the substrate (i_c/i_s), and the wear rate ratio of the coating to the substrate (W_c/W_s) of the main coating with support coating and/or top coating.

In terms of improving the corrosion resistance of the magnesium alloy surface, compared with the single-layer composite coatings blended with nanofillers (Table 8.2), the main coating with support coating and/or top coating has a more excellent effect, especially the ZrO_2/PLGA coating [63]. However, the improvement of hardness and wear resistance by the main coating with support coating and/or top coating needs further study.

8.4 MULTI-INTERFACE COATING

8.4.1 Multi-Interface

A single protective coating is insufficient to meet the requirements of both wear and corrosion resistance all at once. Holleck et al. figured out that multilayer coatings allow the introduction of metastable and multicomponent materials in a gradient multilayer arrangement. With this, the different

TABLE 8.3

Summary of the coating preparation method, thickness, and hardness ratio of the coating to the substrate (H_c/H_s), the corrosion current density ratio of the coating to the substrate (i_c/i_s), and the wear rate ratio of the coating to the substrate (W_c/W_s) of the main coating with support coating and/or top coating

Coating structure	Preparation method	Coating (under layer/top layer)	Coating thickness (under lay/top layer, μm)	Substrate	H_c/H_s	W_c/W_s	i_c/i_s	Year	Ref.
Main coating with support coating	MAO and APS	MAO/TiO$_2$	19/100 ± 20	Mg-1% Ca	10.44	-	1/3650	2015	[128]
	MAO and PVD	MAO/(Ti: N)-DLC	5/0.5	AZ80	-	-	1/9380	2013	[38]
		MAO/DLC	5/0.48				1/241.64		
	PVD (ion implantation and magnetron sputtering)	N ion implantation/Ti	-/2	AZ31	-	-	~1/2.35	2009	[129]
	PVD and dip coating	SiO$_2$/GO	1/30	Mg-Ca-Zn	-	-	1/40.28	2016	[130]
	ALD and spin-coating	ZrO$_2$/PLGA	0.0117/-	AZ31	1/3.8	-	1/91483	2018	[63]
	PEO and self-assemble	PEO/GO	17/4-6	ZK60	-	-	1/7241.37	2015	[30]
	APS	NiCrAlY/nTiO$_2$-CNT	100/135	Mg-1.1% Ca	~17.27	~1/35	-	2017	[123]
	PVD	Al/AlCr	0.762/0.98	Mg-Zn-Ce-La	-	-	1/1.85	2014	[134]
Main coating with top coating	MAO and electrodeposition	MAO/GO	~12/~3.4	AZ91	-	-	1/480.83	2020	[31]
	PEO and triboelectric	PEO/superdispersed polytetrafluoroethylene	60/-	Mg-Mn-Ce	-	-	1/370.37	2014	[131]
	PEO and absorption	PEO/superhydrophobic coating	~4/1-3	Mg-Mn-Ce	-	-	1/4400	2013	[132]
	MAO and dip coating	MAO/hydrophobic poly-lactic acid (PLA)	8-10/30-35	Mg-Ca	-	-	1/8050	2016	[133]
	Dip coating	sol-gel/sol-gel	1.5/0.7	AZ31B	-	-	1/31.81	2020	[36]
	MAO and PVD	sol-gel/sol-gel blended with MWCNTs	5/2	AZ31B	-	-	1/120.68	2020	[141]
		sol-gel/sol-gel blended with COOH-GNPs					1/14		
		MAO/DLC					1/56.9		
		MAO/H-DLC					1/907.98		
		MAO/Si-DLC					1/56.27		
		MAO/Cr-DLC					1/3.59		

layers, functions, and structural designs are realized simultaneously. Considering the material selection of each layer, the adjustment of the interface volume and structure, and the optimization of the sublayer sequence and thickness, the performance of coatings can be tailored [144]. Multi-interface is easily formed in gradient multilayer and alternating multilayer coatings. Figure 8.17 shows the schematic diagram of the multi-interface in alternating multilayer. The alternating multilayer coatings are formed by alternating deposition of layers A and B. The sum of the thickness of layers A and B is a period, which is represented by the symbol Λ. For instance, metal/nitride [62, 145], metal/metal [27], and nitride/nitride [6, 28] alternating multilayer coatings are readily deposited by PVD.

In this chapter, the alternating multilayers obtained by alternating layers A and B for n times are represented by A/B n-layer. With the multi-interface, it is possible to group good properties into one, such as high hardness, hard yet tough, excellent corrosion resistance, and good wear resistance [146–149]. Voevodin et al. elaborated on the methods and mechanisms of multilayer nanocomposite coating to improve hardness, toughness, and low friction [150]. Tian et al. verified that the TiN/CrN multilayer coating had a hardness of 26 GPa, which is 1.7 times high than that of the rule-of-mixtures (15 GPa) [28]. In this section, we discuss the gradient multilayer and alternating multilayer coatings in detail.

Most alternating multilayer coatings are prepared through PVD deposition on various magnesium alloy substrates. For example, the TiN/AlN 21-layer alternating multilayer and superlattice NbN/CrN nano-multilayers with a period of 2–8 nm were prepared on the AZ31 magnesium alloy substrates by PVD deposition using laboratory and industrial-scale coating devices [6]. The Hf/Si$_3$N$_4$ n-layer (n = 2, 10, 20) alternating multilayers were prepared on the AZ91D magnesium alloy substrates by magnetron sputtering [145]. The TiN/CrN 6-layer and TiN/CrN 12-layer alternating multilayer coatings were prepared on the AZ91D magnesium alloy substrates via a pulsed bias cathodic arc and arc-glow plasma PVD processes [28, 151], respectively. The Hf/HfN 6-layer alternating multilayer was prepared on the AZ91D magnesium alloy by magnetron sputtering [62]. The Ti/Cr 6-layer alternating multilayer was prepared on the AZ91D magnesium alloy substrate by arc-added glow plasma depositing [27]. Some other processes also have been used in multilayer coating preparation on magnesium alloy substrates. For instance, the PA/Ce n-layer (n = 1, 2, 4, 6) alternating multilayers were prepared on the AZ31B magnesium alloys by immersion process [3]. The electroless Ni-P/electroplated Ni/electroless Ni-P triple-layer was prepared on the AZ31 magnesium alloy substrate [152]. The Chi/heparinized graphene oxide (HGO) 5-layer alternating multilayer was prepared on the AZ31B magnesium alloy substrate through the electrostatic adsorption method [24]. After the cerium-based conversion layer, branched poly(ethylene imine, PEI) layer, and GO coating through the immersion process, the PEI/poly(acrylic acid, PAA) 10-layer alternating multilayer coating was prepared on the AZ31 magnesium alloy substrate via LBL [46].

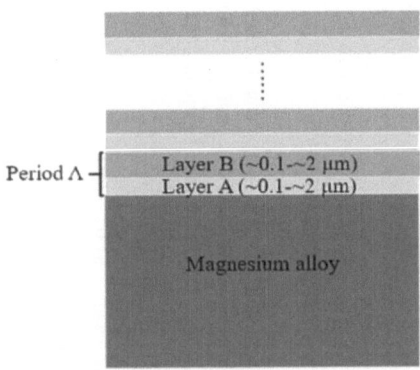

FIGURE 8.17 Schematic diagram of the multi-interface coating.

8.4.2 Improvement and Mechanism of Multi-Interface Effect on Coating Properties

Many studies have shown that the multi-interface coatings significantly improve the surface properties of magnesium alloys due to the following mechanisms of the multi-interface effect.

8.4.2.1 Improving Hardness and Wear Resistance through Blocking Dislocation Movement

The strategy of a multi-interface structure is a promising method to improve hardness and toughness. With this structure, the interface between the sublayers can deflect cracks and provide obstacles for dislocation movement [10, 153]. High hardness and toughness of coating are beneficial to improve the wear resistance of magnesium alloys [154].

Tian et al. prepared the TiN/CrN 6-layer alternating multilayer coating on the AZ91D magnesium alloy via pulsed bias cathodic arc PVD process [28]. The hardness of the CrN single layer, TiN single layer, and TiN/CrN multilayer coating are 12.8, 17.2, and 26 GPa, respectively, which are much higher than that of the magnesium alloy substrate (~1 GPa). This relates to the deformation mechanism of the support coating interface, the inhibition of columnar growth, and the dense coating structure. For the TiN/CrN multilayer coated AZ91D magnesium alloy, the wear friction coefficient of 0.08 is much lower than that of the uncoated sample (~0.28), owing to the higher hardness of TiN/CrN multilayer. The wear rate of the TiN/CrN multilayer coated AZ91D magnesium alloy was 0.026 $mm^3/N \cdot m$, only 1/75 (~1.958 $mm^3/N \cdot m$) of that of the uncoated sample, which is due to the higher fatigue and fracture strength of multiple layers lead to the reduction of surface deformation and damage during friction.

8.4.2.2 Improving the Adhesion of the Coating and the Load Support for the Substrate

As previously discussed in Section 8.3.1, the gradient multilayers can form a hardness transition between the substrate and coating, which is beneficial to improve the performance of coatings [45, 61, 139]. On the one hand, the hardness transition can improve the bonding force between the coating and substrate, as previously mentioned, good bonding force can prevent the coating peeling off from the substrate. On the other hand, the hardness transition can provide load support for the coating.

For Ti/Cr 6-layer alternating multilayer coating on the AZ91magnesium alloy by arc-added glow plasma deposition [27], after the thermal cycling experiment (repeating heating to 300 °C, and then quenching in water for three times), it remained intact and no bubbling or peeling had occurred, which means Ti/Cr multilayer coating has good bonding to the magnesium alloy substrate.

8.4.2.3 Serving as a Barrier Against Corrosion Penetration

The microstructure mismatch between the sublayers will interrupt the columnar growth of each sublayer [28, 60, 62], and cracks and holes in the sublayers will also be deflected at the interface. So, the Cl$^-$ corrosion channel will be deflected at these irregular interfaces and boundaries. This phenomenon is like the flexural effect of the multilayer toughening mechanism in terms of mechanical properties [60]. Therefore, the multilayer structure effectively blocks the penetration channel of the corrosive medium and can greatly improve the corrosion resistance of the coated magnesium alloys.

Liang et al. utilized plasma etching technology to form multi-interfaces with 100, 200, and 400 nm thicknesses in the Al/TiAl/TiAlN coating [60]. Although the cyclic plasma etching process cannot eliminate the columnar structure, the coating consists of multiple interfaces. As shown in Figure 8.18, the TiAlN film with a 100 nm interface period appears in Figure 8.18(c). Tafazoly et al. studied the electroless Ni-P/electroplated Ni/electroless Ni-P triple coating on the AZ31 alloy substrate. After Ni-P (20 μm thick)/Ni (4 μm thick)/Ni-P (20 μm thick) coating, the i_{corr} of the AZ31 alloy significantly decreased from 8.0×10^{-4} A/cm^2 to 6.0×10^{-7} A/cm^2, owing to the following three reasons: (i) the thin electroplated Ni layer between two electroless layers inhibits the reaction of magnesium with the plating bath during the electroless process, (ii) more compact structure and lower porosity in the Ni-P/Ni/Ni-P coating and (iii) multilayer structure prevents the penetration of electrolyte through the coating [152]. The use of hydrophobic coatings further improves the corrosion resistance of the coating, such as HGO coating. After Chi monolayer coating and Chi/HGO 5-layer alternating multilayer coating [24], the i_{corr} of the uncoated AZ31 alloy sample dropped from 8.863×10^{-5} A/cm^2 to 2.295×10^{-6} A/cm^2 and 7.483×10^{-7} A/cm^2, respectively. The Chi/HGO multilayer coating significantly improved the corrosion resistance of the magnesium alloy is attributed to: (i) negatively charged HGO can act as a barrier to anion erosion and (ii) the Chi/HGO multilayer coating was denser and much thicker than the mono-layer chitosan film. This indicates that multilayer coating may be a good barrier to the corrosion medium.

The Ti/Cr 6-layer alternating multilayer coating was prepared on the AZ91D alloy by alternative depositing Ti and Cr sublayers via arc-added glow plasma depositing [27]. Based on the results of the salt spray test, the erosion of the Ti/Cr multilayer coated sample began at a testing time of 48 h; this time is 12 times longer than that of the uncoated sample. Tian et al. prepared the TiN/CrN multilayer coating on the AZ91D magnesium alloy via pulsed bias cathodic arc PVD process [28]. The i_{corr} of the TiN/CrN multilayer coated sample was 4.92×10^{-4} A/cm^2, which is lower than that of the uncoated sample (3.495×10^{-3} A/cm^2). The multi-interface acts as a barrier that hindered the path for corrosive media to penetrate to the substrate.

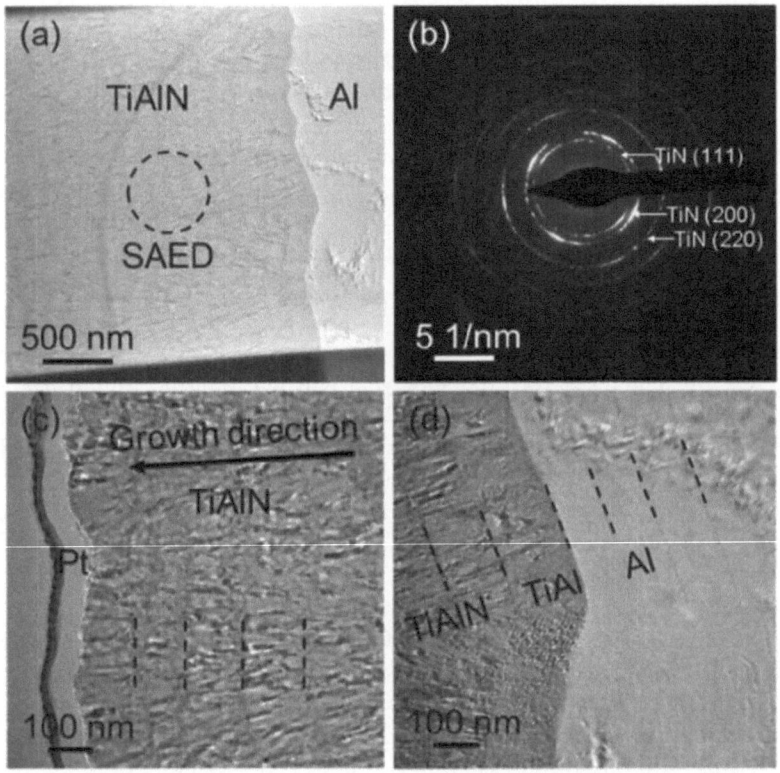

FIGURE 8.18 Cross-sectional TEM images of Al/TiAlN nanocomposite films with 100 nm interface period. (a) Bright-field images, (b) selected area electron diffraction (SAED) patterns, (c) top layer, and (d) transition zone of TiAlN/TiAl/Al layers (Source: Liang, F., Shen, Y., Pei, C., Qiu, B., Lei, J., Sun, D. 2019).

8.4.2.4 Synergizing of Multi-Interface and Top Coating for Total Performance Improvements

There is also a synergy effect between the sublayers of the multilayers. Compared with the corrosion rate of ~70 mpy for the uncoated AZ31 alloy, the corrosion rate of GO/8-Hq/IC multilayer coated sample significantly decreased to 0.0087 mpy [34]. The authors believe that the GO/8-Hq/IC coating significantly improves the corrosion resistance owing to the following: (i) the addition of top GO coating [86, 122], (ii) GO and 8-Hq were adsorbed on the surface of magnesium and acted as mixed-type inhibitors to restrict both cathodic and anodic reactions, and (iii) due to the existence of air pocket and SHP surfaces in the micro flower-like structure and the secondary nanosheets, the corrosive medium can hardly penetrate the magnesium alloy surface. Gao et al. studied Chi/HGO multilayer coating by LBL through electrostatic adsorption method on AZ31 alloy [24]. Chi/HGO multilayer coating greatly improved the corrosion resistance of the magnesium alloy due to the erosion inhibition by the electronegatively HGO layer. For cardiovascular materials, good hydrophilicity is conducive to the improvement of biocompatibility, resulting in reduced platelet adhesion and good affinity for endothelial cells. The Chi/HGO multilayer coating also reduces the hemolysis rate and platelet adhesion, promotes the adhesion and proliferation of endothelial cells, and has good biocompatibility [24]. For the Ce (IV)/PEI/GO/(PEI/PAA)$_{10}$ multilayer coating [46], the graphene oxide layer can act as the corrosion inhibitor because of its relatively high aspect ratio, which creates a more tortuous path for the penetration of corrosive electrolytes. Furthermore, the PEI/PAA multilayer top coating has a rapid self-healing performance. Therefore, the corrosion resistance of Ce (IV)/PEI/GO/(PEI/PAA)$_{10}$ coated AZ31 alloy was significantly improved.

Some organic compounds have special functions in the coating. Ou et al. prepared (PA/Ce) n-layer (n = 1, 2, 4, 6) alternating multilayer coatings by alternative immersion into the aqueous solutions of phytic acid (PA) and CeCl$_3$ for n cycles on the AZ31B magnesium substrates [3]. Before coating, the cleaned Mg alloy was first immersed in boiling water for 20 min, then taken out and dried with nitrogen. The i_{corr} of the Mg-H$_2$O-(PA/Ce)$_2$ coated sample (4.84×10^{-6} A/cm^2) is two orders of magnitude lower than that of the uncoating sample (3.47×10^{-3} A/cm^2). With the hexadecyltrimethoxysilane (HDMS) molecules treatment (the top coating), the i_{corr} of the Mg-H$_2$O-(PA/Ce)$_2$ coated sample reduced to 4.71×10^{-8} A/cm^2, which is probably due to the following aspects: (i) the coating is more interlinked via the coordination bonding of P-O-Ce between PA and Ce^{3+} ions, (ii) an efficient corrosion inhibition of Ce^{3+} ion, and (iii) the physical barrier effect and the SHP modifying of the HDMS layer.

8.4.3 THE INFLUENCE OF SUBLAYER THICKNESS AND NUMBER OF INTERFACES ON THE MULTI-INTERFACE EFFECT

Coatings always have defects, such as pin-holes, pores, and cracks. These defects in the coating will become channels for the corrosive liquid to reach the magnesium alloy substrate, leading to corrosion of the magnesium alloy. Once the corrosive liquid attacks the alloy the galvanic cell forms due to a large potential difference between the coating and the alloy [37] thereby accelerating the corrosion process. Therefore, to provide good protection for the magnesium alloy substrate, a coating should have two characteristics: one is a compact structure that prevents the penetration of the corrosive medium; the other is a reasonable coating structure to reduce or avoid galvanic corrosion. In this section, we will focus on the influence of sublayer/interface thickness and the number of interfaces on the multi-interface effect.

8.4.3.1 Sublayer Thickness

The thickness of each sublayer is an important influence factor of corrosion resistance of coatings. Zhang et al. [62] revealed the Hf coating had a columnar structure with pores. Hf has a hcp crystalline structure and HfN has a fcc crystalline structure. The introduced HfN layer will interrupt the columnar growth of the Hf layer (shown in Figure 8.19), and the porosity of the Hf/HfN multilayer coatings will be reduced. Therefore, the Hf/HfN multilayer coatings greatly improved the corrosion

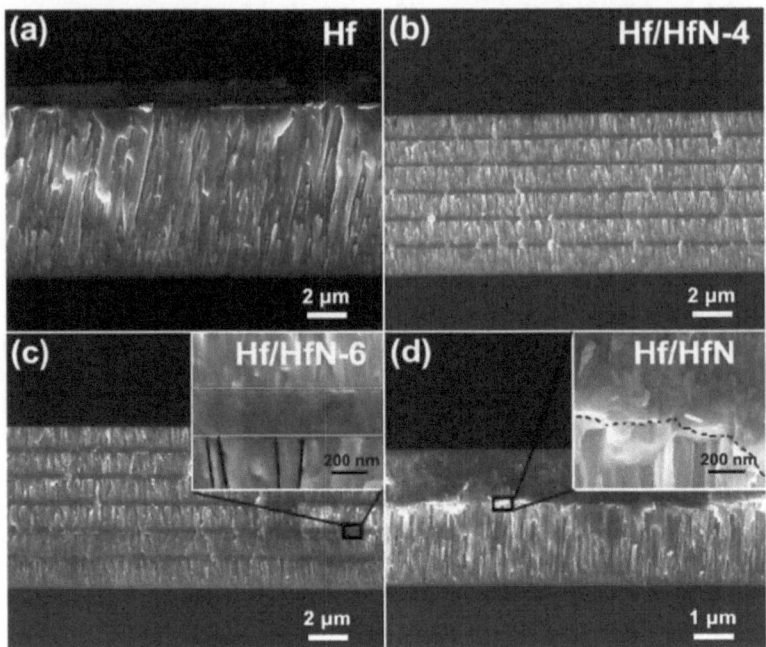

FIGURE 8.19 Cross-sectional SEM morphologies of the coatings. (a) Hf. (b) Hf/HfN-4. (c) Hf/HfN-6. (d) Hf/HfN bilayer coating with a local amplification (inset) (Source: Zhang, D., Qi, Z., Wei, B., Shen, H., Wang, Z. 2018).

resistance of magnesium alloys. In addition, HfN coating not only has a dense structure but also does not cause galvanic corrosion with magnesium alloy due to its chemical inertness. The i_{corr} of Hf/HfN-6 (HfN and Hf sublayers were deposited for 6 min and 14 min, respectively) was lower than that of the Hf/HfN-4 (HfN and Hf sublayers were deposited for 4 min and 16 min). The former was 1×10^{-8} A/cm^2 and the latter was 2×10^{-8} A/cm^2. This indicates that a thicker HfN sublayer is beneficial to the improvement of the corrosion resistance of the coating [62]. Gao et al. support that a thicker Chi/HGO coatings (by increasing the number of cycles of LBL assembly) can further improve the corrosion resistance of magnesium alloys and adjust the biodegradation behavior [24].

However, for the Al/TiAl/TiAlN coating with 100, 200, and 400 nm thicknesses of the multi-interface, the i_{corr} were 9.37×10^{-8}, 1.46×10^{-7}, and 1.06×10^{-6} A/cm^2, respectively. This indicates that the thinnest interface thickness (100 nm) had the highest corrosion resistance improvement for the Al/TiAl/TiAlN multilayer coated magnesium alloy [60].

8.4.3.2 Number of Interfaces

With almost the same overall thickness and different values of n, Hf/Si$_3$N$_4$ n-layer alternating multilayer coatings have different period thickness and the number of interfaces. The results showed that the i_{corr} of these coatings is determined by n, and the larger n means that more sublayers are introduced into the multilayer coating, resulting in a smaller through-holes thickness or pin-holes. With n increased from 2 to 10, the lowest i_{corr} of 2.798×10^{-6} A/cm^2 was gained, which is much lower than that of the uncoated sample (1.627×10^{-4} A/cm^2) and (Hf/Si$_3$N$_4$) 2-layer alternating multilayer (9.729×10^{-6} A/cm^2). However, when n = 20, the i_{corr} increased to 8.590×10^{-6} A/cm^2 due to the increase of porosity in the coating. It infers that much more sublayers increase the risk of corrosion medium permeating through the interfaces of multilayer coatings. This result means that the period thickness of the alternative multilayers has an optimal value [145]. The amorphous structure of the Si$_3$N$_4$ sublayer causes corrosion medium piercing through the interface of multilayer coatings needing higher energy is another reason for the improvement of corrosion resistance of multilayer coatings.

Table 8.4 summarizes the preparation method, thickness, hardness ratio of the coating to the substrate (H_c/H_s), elastic modulus ratio of the coating to the substrate (E_c/E_s), corrosion current density

TABLE 8.4

Summary of preparation method, thickness, hardness ratio of the coating to the substrate (H_c/H_s), elastic modulus ratio of the coating to the substrate (E_c/E_s), i_{corr} ratio of the coating to the substrate (i_c/i_s), and wear rate ratio of the coating to the substrate (W_c/W_s) of the multi-interface coatings

Preparation method	Coating	Coating thickness (μm)	Substrate	H_c/H_s	E_c/E_s	W_c/W_s	i_c/i_s	Year	Ref.
PVD (ion implantation and magnetron sputtering)	Al/AlN/CrAlN/CrN/MoS$_2$ gradient multilayer	~1.8	AM60	20.46	7.48	1/136.36	1/20.19	2014	[61]
PVD (arc-added glow plasma depositing)	Ti/Cr alternative multilayer	2	AZ91	430 HK$_{0.01}$/-	-	-	-	2007	[27]
PVD (arc-glow plasma depositing)	(TiN/CrN) 12 alternative multilayers	2	AZ91D	1433 HK$_{0.01}$/-		1/75.31	-	2006	[151]
PVD (ion implantation and mechanical spraying)	Ion implantation/AlN/MoS$_2$-phenolic resin gradient multilayer	12.57	AM60	18.22	5.07	-	1/38484	2015	[45]
PVD (cathodic arc deposition)	TiN/CrN alternative multilayer	2	AZ91D	~26		1/75.31	1/7.10	2009	[28]
PVD (hybrid ion beam deposition system)	DLC/AlN/Al gradient multilayer	1	AZ31	~6	>2.17	-	1/17.57	2010	[139]
PVD (magnetron sputtering)	(Hf/Si$_3$N$_4$) n alternative multilayer, n = 2, 10, 20	10 ± 0.2	AZ91D	-			1/58.15	2017	[145]
PVD (magnetron sputtering)	(Hf/HfN) 6 alternative multilayers	10 ± 0.1	AZ91D	-			1/16270	2018	[62]
Immersion	(PA/Ce) n alternative multilayer and hexadecyltrimethoxysilane (HDMS) modifying, (n = 1, 2, 4, 6) treatment	0.76 ((PA/Ce)2)	AZ31B	-			1/73673	2019	[3]
PVD (ion source hybrid magnetron sputtering)	Al/TiAl/TiAlN gradient multilayer	5.15	AZ91D	31.3 GPa/-			1/140.87	2019	[60]
Electroless and electroplating	Ni-P/Ni/Ni-P multilayer	44	AZ31	-			1/1333.33	2012	[152]
Plasma electrolytic oxidation and dip-chemical coating	GO/8-Hq/IC(PEO) multilayer	-	AZ31	-			~1/8009 (corrosion rate)	2019	[34]
Immersion	Chitosan (Chi)/heparinized graphene oxide (HGO) alternative multilayer	0.755	AZ31B	-			1/118.44	2019	[24]
Spin-assisted layer-by-layer (LBL)	(PEI/PSS/8HQ/PSS)$_{25}$	-	AZ91D	-			1/8.33	2011	[155]

TABLE 8.5

Summary of preparation method, representative coating, and performance improvement mechanism of the nanofiller blended single-layer composite coating, main coating with support coating and/or top coating, and multi-interface coating

Performance improvement method	Preparation method	Representative coating	Performance improvement mechanism	Ref.
Nanofiller blended single-layer composite coating	PEO (MAO), electrodeposition, electroless planting, spin coating, dip coating, PVD, sol-gel	SiC-PEO, graphene-PEO, SiC-Ni, CNTs-HA, (BG-Fe$_3$O$_4$)-chitosan, (P-B$_4$C)-Ni, CNTs-TiO$_2$, TiN-DLC, Cr-DLC, (GO-TiO$_2$)-butyl methacrylate, CNTs-SiO$_2$	The second phase strengthens, sealing holes, bridging, fine grain strengthens, increase of adhesion strength, barrier effect, etc.	[25, 38, 43, 59, 64, 73, 76]
Main coating with support coating and/or top coating	MAO combined APS, MAO combined PVD, ALD combined spin coating, PEO combined self-assemble, MAO combined dip coating, APS, PVD, dip coating, PVD combined dip coating, MAO combined electrodeposition, PEO combined triboelectric method	MAO/TiO$_2$, MAO/(Ti: N)-DLC, ZrO$_2$/PLGA, PEO/GO, MAO/PLA, NiCrAlY/nTiO$_2$-CNT, Al/AlCr, sol-gel/sol-gel blended with COOH-GNPs	Hardness transition, increase of adhesion strength, and barrier layer of support coating, sealing hole, hydrophobicity and superhydrophobicity, self-lubricity, biocompatibility top coating	[30, 36, 63, 123, 133, 134]
Multi-interface coating	PVD, ion implantation combined mechanical spraying, immersion, plasma electrolytic oxidation combined dip-chemical coating, electroless combined electroplating, molten salt bath	Al/AlN/CrAlN/CrN/MoS$_2$, Ion implantation/AlN/MoS$_2$-phenolic resin, TiN/CrN multilayer, magnetron sputtering, (Hf/HfN)$_6$, (PA/Ce)$_n$, Al/TiAl/TiAlN, GO/8-Hq/IC(PEO)	Barrier layer, good bonding strength, synergy effect, hardness transition, the obstruction of multi-interface to dislocation movement	[3, 28, 34, 45, 60–62, 151]

ratio of the coating to the substrate (i_c/i_s), and wear rate ratio of the coating to the substrate (W_c/W_s) of multi-interface coatings. Compared with the data in Tables 8.2 and 8.3, the performance improvement in hardness [28, 60], wear resistance [28, 61, 151], and corrosion resistance [34, 45, 62] by multi-interface coating are much better than that by nanofiller blended composite coating and the main coating with support coating and/or top coating. Table 8.5 summarizes the coating preparation method, representative coating, and performance improvement mechanism.

8.5 SUMMARY

This chapter gave a comprehensive review on roads toward strengthening of coatings for wear and corrosion protection on magnesium alloys. We discussed nanofiller blended single-layer composite coatings, main coating with support coating and/or top coating (support coating, main coating, and the top coating), and multi-interface coatings.

In nanofiller blended single-layer composite coatings, the coatings have nanofillers blended in, where nanofillers dispersed in the coating matrix thus giving rise to dispersion strengthening (second phase strength), bridging, and self-lubricating. The improvement in corrosion resistance comes from pores sealing. The choice of special properties of nanofillers gives rise to special properties such as biocompatibility and anti-bacterial properties. Nanoparticles (nitrides, carbides, oxides), nanotubes, and nanolayers (graphene and graphene oxide) can be blended into coatings by micro-arc oxidation, electrodeposition, electrophoretic deposition, electroless planting, spin coating, anodic oxidation coating, and dip coating, etc. Nanoparticles of metals, nitrides, and carbides can blend in ceramic or diamond-like carbon coatings by physical vapor deposition.

The main coating with support coating and/or top coating further improves the performance of the coating. The support coating functions as a hardness transition for the nanocomposite main coating and improves adhesion on the alloy surface. The support coating also provides load support and acts as a barrier against penetration of corrosion media. MAO and atmospheric plasma spraying coatings are frequent examples of the support coatings. The top coating seals holes thus blocking corrosion medium from entering the coating. People also design special top coatings for hydrophobicity, super-hydrophobicity, self-lubrication, and biocompatibility. For example, graphene oxide, TiO_2, poly-lactic acid, poly, and Mg-rich primer coatings.

Multiple interface coatings achieve improvement of coating performance through inhibiting the movement of dislocations, formation of hardness transition, preventing crack propagation, interrupting columnar growth of the coating, refinement of the coating, and barrier layer to block inward penetration of corrosion media. Compared with nanofiller blended single-layer composite coatings, multi-interface coatings provide better protection against corrosion and wear. Metal/nitride, metal/metal, and nitride/nitride alternating multilayer coatings are readily deposited by PVD. The thickness of sublayer A or B and the number of interfaces will strongly affect the performance of the coating.

ACKNOWLEDGEMENTS

This work was supported by Fundamental Research Funds for the Central Universities: SWU118105.

REFERENCES

[1] Mordike, B. L., Ebert, T., Magnesium: Properties- applications- potential. *Materials Science and Engineering A* 2001, A302: 37–45.

[2] Esmaily, M., Svensson, J. E., Fajardo, S., Birbilis, N., Frankel, G. S., Virtanen, S., Arrabal, R., Thomas, S., Johansson, L. G., Fundamentals and advances in magnesium alloy corrosion. *Progress in Materials Science* 2017, 89, 92–193.

[3] Ou, J., Chen, X., Corrosion resistance of phytic acid/Ce (III) nanocomposite coating with superhydrophobicity on magnesium. *Journal of Alloys and Compounds* 2019, 787: 145–151.

[4] Lu, X. Y., Feng, X. G., Zuo, Y., Zhang, P., Zheng, C. B., Improvement of protection performance of Mg-rich epoxy coating on AZ91D magnesium alloy by DC anodic oxidation. *Progress in Organic Coatings* 2017, 104: 188–198.

[5] Yang, L. Q., Li, Z. Y., Zhang, Y. Q., Wei, S. Z., Liu, F. Q., Al-TiC in situ composite coating fabricated by low power pulsed laser cladding on AZ91D magnesium alloy. *Applied Surface Science* 2018, 435: 1187–1198.

[6] Hollstein, F., Wiedemann, R., Scholz, J., Characteristics of PVD-coatings on AZ31hp magnesium alloys. *Surface & Coatings Technology* 2003, 162: 261–268.

[7] Tokunaga, T., Ohno, M., Matsuura, K., Coatings on Mg alloys and their mechanical properties: A review. *Journal of Materials Science & Technology* 2018, 34: 1119–1126.

[8] Altun, H., Sen, S., The effect of PVD coatings on the corrosion behaviour of AZ91 magnesium alloy. *Materials & Design* 2006, 27 (10): 1174–1179.

[9] Andreasen, K. P., Jensen, T., Petersen, J. H., Chevallier, J., Bøttiger, J., Schell, N., The structure and the corresponding mechanical properties of magnetron sputtered TiN-Cu nanocomposites. *Surface and Coatings Technology* 2004, 182: 268–275.

[10] Guo, Z., Ma, D., Zhang, X., Li, J., Feng, J., Preparation and toughening of a-CuZr/c-ZrN nano-multilayer hard coatings. *Applied Surface Science* 2019, 483: 432–441.

[11] Abadias, G., Tse, Y. Y., Michel, A., Jaouen, C., Jaouen, M., Nanoscaled composite TiN/Cu multilayer thin films deposited by dual ion beam sputtering: growth and structural characterisation. *Thin Solid Films* 2003, 433 (1–2):166–173.

[12] Balashabadi, P., Larijani, M. M., Jafari-Khamse, E., Seyedi, H., The role of Cu content on the structural properties and hardness of TiN-Cu nanocomposite film. *Journal of Alloys and Compounds* 2017, 728: 863–871.

[13] Voevodin, A. A., Walck, S. D., Zabinski, J. S., Architecture of multilayer nanocomposite coatings with super-hard diamond-like carbon layers for wear protection at high contact loads. *Wear* 1997, 203–204: 516–527.

[14] Aal, A. A., Bahgat, M., Radwan, M., Nanostructured Ni-AlN composite coatings. *Surface & Coatings Technology* 2006, 201 (6): 2910–2918.

[15] Kumar, V., Sinha, S. K., Agarwal, A. K., Tribological studies of dual-coating (intermediate hard with top epoxy-graphene-base oil composite layers) on tool steel in dry and lubricated conditions. *Tribology International* 2018, 127: 10–23.

[16] Aal, A. A., Hard and corrosion resistant nanocomposite coating for Al alloy. *Materials Science and Engineering A* 2008, 474 (1–2): 181–187.

[17] Li, H., Xie, Y., Li, K., Huang, L., Huang, S., Zhao, B., Zheng, X., Microstructure and wear behavior of graphene nanosheets-reinforced zirconia coating. *Ceramics International* 2014, 40: 12821–12829.

[18] Liu, C., Liu, Y., Wang, Q., Liu, X. W., Bao, Y., Wu, G., Lu, J., Nano-dual-phase metallic glass film enhances strength and ductility of a gradient nanograined magnesium alloy. *Advanced Science* 2020: 2001480.

[19] Wu, G., Zeng, X., Li, G., Yao, S., Wang, X., Preparation and characterization of ceramic/metal duplex coatings deposited on AZ31 magnesium alloy by multi-magnetron sputtering. *Materials Letters* 2006, 60: 674–678.

[20] Fini, M. H., Amadeh, A., Improvement of wear and corrosion resistance of AZ91 magnesium alloy by applying Ni-SiC nanocomposite coating via pulse electrodeposition. *Transactions of Nonferrous Metals Society of China* 2013, 23: 2914–2922.

[21] Dong, Y.-r., Sun, W. C., Liu, X.-J., Jia, Z.-J., Guo, F., Ma, M., Ruan, Y.-Y., Effect of CNTs concentration on the microstructure and friction behavior of Ni-GO-CNTs composite coatings. *Surface & Coatings Technology* 2018, 359: 141–149.

[22] Zhang, Y. L., Chen, F., Zhang, Y., Du, C. W., Influence of graphene oxide additive on the tribological and electrochemical corrosion properties of a PEO coating prepared on AZ31 magnesium alloy. *Tribology International* 2020, 146: 106135.

[23] Han, B. J., Yang, Y., Li, J., Deng, H., Yang, C. B., Effects of the graphene additive on the corrosion resistance of the plasma electrolytic oxidation (PEO) coating on the AZ91 magnesium alloy. *International Journal of Electrochemical Science* 2018, 13: 9166–9182.

[24] Gao, F., Hu, Y. D., Gong, Z. H., Liu, T., Gong, T., Liu, S., Zhang, C., Quan, L., Kaveendran, B., Pan, C. J., Fabrication of chitosan/heparinized graphene oxide multilayer coating to improve corrosion resistance and biocompatibility of magnesium alloys. *Materials Science & Engineering C* 2019, 104: 109947.

[25] López, A. J., Ureña, A., Rams, J., Wear resistant coatings: silica sol-gel reinforced with carbon nanotubes. *Thin Solid Films* 2011, 519 (22): 7904–7910.

[26] Perez-Mariano, J., Lau, K. H., Sanjurjo, A., Caro, J., Casellas, D., Colominas, C. J. S., Technology, C., TiSiN nanocomposite coatings by chemical vapor deposition in a fluidized bed reactor at atmospheric pressure (AP/FBR-CVD). *Surface & Coatings Technology* 2006, 201 (6): 2217–2225.

[27] Cui, C. E., Miao, Q., Pan, J. D., Ti/Cr multi-layer coating on magnesium alloy AZ91 by arc-added glow plasma depositing technique. *Surface & Coatings Technology* 2007, 201 (9): 5400–5403.

[28] Tian, L. H., Liu, E. Q., Fan, A. L., Qin, L., Liu, D. X., Tang, B., Pan, J. D. In effect of TiN/CrN multilayer coating by cathodic arc deposition on wear and corrosion behaviours of AZ91D magnesium alloy, *Materials Science Forum*, 2009, 610–613: 870–873.

[29] Zhang, J., Wen, Z. H., Zhao, M., Lig, G. Z., Dai, C. S., Effect of the addition CNTs on performance of CaP/chitosan/coating deposited on magnesium alloy by electrophoretic deposition. *Materials Science and Engineering C* 2016, 58: 992–1000.

[30] Qiu, Z. Z., Wang, R., Wu, J. Z., Zhang, Y. S., Qu, Y. F., Wu, X. H., Graphene oxide as a corrosion-inhibitive coating on magnesium alloys. *RSC Advances* 2015, 5: 44149–44159.

[31] Shang, W., Wu, F., Wang, Y. Y., Baboukani, A. R., Wen, Y. Q., Jiang, J. Q., Corrosion resistance of micro-arc oxidation/graphene oxide composite coatings on magnesium alloys. *ACS Omega* 2020, 5: 7262–7270.

[32] Fini, M. H., Amadeh, A., Improvement of wear and corrosion resistance of AZ91 by magnesium alloy applying Ni-SiC nanocomposite coating via pulse electrodeposition. *Transactions of Nonferrous Metals Society of China* 2013, 23: 2914–2922.

[33] Pourhashem, S., Saba, F., Duan, J., Rashidi, A., Hou, B., Polymer/inorganic nanocomposite coatings with superior corrosion protection performance: A review. *Journal of Industrial and Engineering Chemistry* 2020, doi:10.1016/j.jiec.2020.04.029.

[34] Zoubi, W. A., Kim, M. J., Kim, Y. G., Ko, Y. G., Progress in organic coatings fabrication of graphene oxide/8-hydroxyquinolin/inorganic coating on the magnesium surface for extraordinary corrosion protection. *Progress in Organic Coatings* 2019, 137: 105314.

[35] Feil, F., Fürbeth, W., Schütze, M., Purely inorganic coatings based on nanoparticles for magnesium alloys. *Electrochimica Acta* 2009, 54 (9): 2478–2486.

[36] Fernández-Hernán, J. P., López, A. J., Torres, B., Rams, J., Silicon oxide multilayer coatings doped with carbon nanotubes and graphene nanoplatelets for corrosion protection of AZ31B magnesium alloy. *Progress in Organic Coatings* 2020, 148: 105836.

[37] Dai, W., Wu, G. S., Wang, A. Y., Preparation, characterization and properties of Cr-incorporated DLC films on magnesium alloy. *Diamond and Related Materials* 2010, 19 (10): 1307–1315.

[38] Yang, W., Ke, P. L., Fang, Y., Zheng, H., Wang, A. Y., Microstructure and properties of duplex (Ti:N)-DLC/MAO coating on magnesium alloy. *Applied Surface Science* 2013, 270: 519–525.

[39] Martínez-Martínez, D., *Nanocomposite coatings. Springer US*: 2013.

[40] Zhang, S., Sun, D., Fu, Y. Q., Du, H., Zhang, Q., Effect of sputtering target power on preferred orientation in nc-TiN/a-SiNx nanocomposite thin films. *Journal of Metastable and Nanocrystalline Materials* 2005, 23 (23): 175–178.

[41] Zhang, S., Wang, H. L., Ong, S. E., Sun, D., Bui, X. L., Hard yet tough nanocomposite coatings-present status and future trends. *Plasma Processes and Polymers* 2010, 4 (3): 219–228.

[42] Chen, J., Ma, B., Liu, G., Song, H., Wu, J., Cui, L., Zheng, Z., Wear and corrosion properties of 316L-SiC composite coating deposited by cold spray on magnesium alloy. *Journal of Thermal Spray Technology* 2017, 26: 1381–1392.

[43] Khazeni, D., Saremi, M., Soltani], R., Development of HA-CNTs composite coating on AZ31 magnesium alloy by cathodic electrodeposition. Part 1: Microstructural and mechanical characterization. *Ceramics International* 2019, 45: 11174–11185.

[44] Zhang, S., Ali, N., *Nanocomposite thin films and coatings*. Imperial College Press: London, 2007.

[45] Xie, Z. W., Chen, Q., Chen, T., Gao, X., Yu, X. G., Song, H., Feng, Y. J., Microstructure and properties of nitrogen ion implantation/AlN/CrAlN/MoS₂-phenolic resin duplex coatings on magnesium alloys. *Materials Chemistry & Physics* 2015, 160 (4): 212–220.

[46] Fan, F., Zhou, C., Wang, X., Szpunar, J., Layer-by-Layer assembly of a self-healing anticorrosion coating on magnesium alloys. *ACS Applied Materials & Interfaces* 2015, 7: 27271–27278.

[47] Kartsonakis, I. A., Balaskas, A. C., Koumoulos, E. P., Charitidis, C. A., Kordas, G., Evaluation of corrosion resistance of magnesium alloy ZK10 coated with hybrid organic–inorganic film including containers. *Corrosion Science* 2012, 65: 481–493.

[48] Gheytani, M., Aliofkhazraei, M., Bagheri, H. R., Masiha, H. R., Sabour Rouhaghdam, A., Wettability and corrosion of alumina embedded nanocomposite MAO coating on nanocrystalline AZ31B magnesium alloy. *Journal of Alloys and Compounds* 2015, 649: 666–673.

[49] Santo, L., Davim, J. P., Materials and Surface Engineering ‖ Nanocomposite coatings: A review. *Materials and surface engineering* 2012: 97–120.

[50] Singh, S., Singh, G., Bala, N., Aggarwal, K., Characterization and preparation of Fe₃O₄ nanoparticles loaded bioglass-chitosan nanocomposite coating on Mg alloy and in vitro bioactivity assessment. *International Journal of Biological Macromolecules* 2020, 151: 519–528.

[51] Moszner, F., Cancellieri, C., Chiodi, M., Yoon, S., Ariosa, D., Janczak-Rusch, J., Jeurgens, L. P. H., Thermal stability of Cu/W nano-multilayers. *Acta Materialia* 2016, 107: 345–353.

[52] Liu, Z. K., Qiu, S., Ying, S., Jiao, Y., Chen, X. Q., Wang, H. R., Jiang, Z. H., High-emissivity composite-oxide fillers for high temperature stable aluminum-chromium phosphate coating. *Surface & Coatings Technology* 2018, 349: 885–893.

[53] Vatan, H. N., Ebrahimi-kahrizsangi, R., Kasiri-asgarani, M., Tribological performance of PEO-WC nanocomposite coating on Mg alloys deposited by plasma electrolytic oxidation. *Tribology International* 2016, 98: 253–260.

[54] Ma, F. L., Li, J. L., Zeng, Z. X., Gao, Y. M., Structural, mechanical and tribocorrosion behaviour in artificial seawater of CrN/AlN nano-multilayer coatings on F690 steel substrates. *Applied Surface Science* 2018, 428: 404–414.

[55] Amadeh, A., Rahimi, A., Farshchian, B., Moradi, H., Corrosion behavior of pulse electrodeposited nano-structure Ni-SiC composite coatings. *Journal of Nanoscience and Nanotechnology* 2010, 10: 5383–5388.

[56] Figueiredo, N. M., Vaz, F., Cunha, L., Rodil, S. E., Cavaleiro, A., Structural, chemical, optical and mechanical properties of Au doped AlN sputtered coatings. *Surface & Coatings Technology* 2014, 255 (255): 130–139.

[57] Hosseini, M. G., Abdolmaleki, M., Ashrafpoor, S., Najjar, R., Deposition and corrosion resistance of electroless Ni-PCTFE-P nanocomposite coating. *Surface & Coatings Technology* 2012, 206: 4546–4552.

[58] Li, Q., Fan, S., Han, W., Sun, C., Liang, W., Coating of carbon nanotube with nickel by electroless plating method. *Physica Scripta* 2013, T158: 014015.

[59] Asgari, M., Aliofkhazraei, M., Darband, G. B., Rouhaghdam, A. S., Evaluation of alumina nanoparticles concentration and stirring rate on wear and corrosion behavior of nanocomposite PEO coating on AZ31 magnesium alloy. *Surface and Coatings Technology* 2017, 309: 124–135.

[60] Liang, F., Shen, Y., Pei, C., Qiu, B., Lei, J., Sun, D., Microstructure evolution and corrosion resistance of multi interfaces Al-TiAlN nanocomposite films on AZ91D magnesium alloy. *Surface & Coatings Technology* 2019, 357: 83–92.

[61] Xie, Z., Luo, Z., Yang, Q., Chen, T., Tan, S., Wang, Y., Luo, Y., Improving anti-wear and anti-corrosion properties of AM60 magnesium alloy by ion implantation and Al/AlN/CrAlN/CrN/MoS$_2$ gradient duplex coating. *Vacuum* 2014, 101: 171–176.

[62] Zhang, D., Qi, Z., Wei, B., Shen, H., Wang, Z., Microstructure and corrosion behaviors of conductive Hf/HfN multilayer coatings on magnesium alloys. *Ceramics International* 2018, 44: 9958–9966.

[63] Liu, X., Yang, Q., Li, Z., Yuan, W., Zheng, Y., Cui, Z., Yang, X., Yeung, K. W. K., Wu, S., A combined coating strategy based on atomic layer deposition for enhancement of corrosion resistance of AZ31 magnesium alloy. *Applied Surface Science* 2018, 434: 1101–1111.

[64] Vatan, H. N., Ebrahimi-kahrizsangi, R., Kasiri-asgarani, M., Structural, tribological and electrochemical behavior of SiC nanocomposite oxide coatings fabricated by plasma electrolytic oxidation (PEO) on AZ31 magnesium alloy. *Journal of Alloys and Compounds* 2016, 683: 241–255.

[65] Vatan, H. N., Ebrahimi-kahrizsangi, R., Kasiri-asgarani, M., Growth, corrosion and wear resistance of SiC nanoparticles embedded MAO coatings on AZ31B magnesium alloy. *Protection of Metals Physical Chemistry of Surfaces* 2016, 52 (5): 859–868.

[66] Li, Z. L., Liu, H., Liu, W., Zhou, D. Q., Hou, J. Y., Preparation and properties of nano-ceramic TiO$_2$/Ni composite electroplating on magnesium alloy. *Corrosion & Protection* 2010, 31 (3): 208–211.

[67] Ma, Z., Wang, R., Wang, C., Li, Z. C., Research on characteristics of Ni-Cu-P/nano-TiO$_2$ composite electroless plating of magnesium alloy. *Materials Reports* 2008, 22, 105–108.

[68] Zhang, W. X., Huang, N., He, J. G., Jiang, Z. H., Jiang, Q., Lian, J. S., Electroless deposition of Ni-W-P coating on AZ91D magnesium alloy. *Applied Surface Science* 2007, 253: 5116–5121.

[69] Araghi, A., Paydar, M. H., Electroless deposition of Ni-W-P-B$_4$C nanocomposite coating on AZ91D magnesium alloy and investigation on its properties. *Vacuum* 2013, 89: 67–70.

[70] Pushpanathan, D. P., Alagumurthi, N., Devaneyan, S. P., On the microstructure and tribological properties of pulse electrodeposited Ni-B$_4$C-TiC nano composite coating on AZ80 magnesium alloy. *Surfaces and Interfaces* 2020, 19 (21): 100465.

[71] Li, D., Chen, F., Xie, Z. H., Preparation of high corrosion resistance nanocomposite coatings on Mg alloy by two-step method. *Ordnance Material Science and Engineering* 2017, 40 (1): 28–31.

[72] Yan, L., Si-rong, Y., Jin-dan, L., Zhi-wu, H., Dong-sheng, Y., Microstructure and wear resistance of electrodeposited Ni-SiO$_2$ nano-composite coatings on AZ91 HP magnesium alloy substrate. *Transactions of Nonferrous Metals Society of China* 2011, 21: 483–488.

[73] Liu, Y., Lu, G. L., Liu, J. D., Li, L., Effect of surface crosslinking on absorption properties of super absorbent polymers. *Journal of Functional Materials* 2012, 43 (5): 650–652+656.

[74] Yu, L. H., Huang, W. G., Zhao, X., Study on Ni-P-nano TiN electroless composite coating. *Surface Technology* 2009, 38 (5): 17–19.

[75] Sun, Y., Lu, Y. M., Xie, X. H., Ma, H. J., Liu, S. J., Gu, Y. H., Study of the electrochemical properties of HA/CNTs coating and its preparation on the surface of magnesium alloy. *Journal of Beijing Institute of Petrochemical Technology* 2018, 26 (2): 5–12.

[76] Khazeni, D., Saremi, M., Soltani], R., Development of HA-CNTs composite coating on AZ31 Magnesium alloy by cathodic electrodeposition. Part 2: Electrochemical and in-vitro behavior. *Ceramics International* 2019, 45: 11186–11194.

[77] Vatan, H. N., Adabi, M., Investigation of tribological behavior of ceramic-graphene composite coating produced by plasma electrolytic oxidation. *Transactions of the Indian Institute of Metals* 2018, 71: 1643–1652.

[78] Zhao, J. M., Xie, X., Zhang, C., Effect of the graphene oxide additive on the corrosion resistance of the plasma electrolytic oxidation coating of the AZ31 magnesium alloy. *Corrosion Science* 2017, 114: 146–155.

[79] Wen, C. L., Zhan, X. Z., Huang, X. G., Xu, F., Luo, L. J., Xia, C. S., Characterization and corrosion properties of hydroxyapatite/graphene oxide bio-composite coating on magnesium alloy by one-step micro-arc oxidation method. *Surface & Coatings Technology* 2017, 317: 125–133.

[80] Nazeer, A. A., Al-Hetlani, E., Amin, M. O., Quiñones-Ruiz, T., Lednev, I. K., A poly (butyl methacrylate)/graphene oxide/TiO$_2$ nanocomposite coating with superior corrosion protection for AZ31 alloy in chloride solution. *Chemical Engineering Journal* 2019, 361: 485–498.

[81] Zhang, J., Chi, Y. X., Guo, X. L., Jiao, S. Q., Zhang, X. Y., The effects of carbon nanotubes on drug release behavior of calcium phosphate/chitosan coating on magnesium alloy substrate. *Heilongjiang Medicine and Pharmacy* 2015, 38 (2): 6–7, 3.

82. Chen, L., You, G. Q., Ma, X. L., Ming, Y., Bai, S. L., Performance of organic composite heat dissipation coating of magnesium alloy. *Ordnance Material Science and Engineering* 2015, 39 (1): 89–93.

[83] Han, B. J., Yang, Y., Huang, Z. J., You, L., Huang, H., Wang, K. J., A composite anodic coating containing graphene on AZ31 magnesium alloy. *International Journal of Electrochemical Science* 2017, 12: 9829–9843.

[84] Wu, H. J., Yang, F. Y., Peng, C. Z., Guo, W. M., Wang, X. M., Corrosion resistance of carbon nanotubes/silane composite coatings on AZ91D magnesium alloy. *Journal of Shaoyang University (Natural Science Edition)* 2016, 13 (3): 69–75.

[85] Veprek, S., Reiprich, S., A concept for the design of novel superhard coatings. *Thin Solid Films* 1995, 268: 64–71.

[86] Catt, K., Li, H., Cui, X. T., Poly (3,4-ethylenedioxythiophene) graphene oxide composite coatings for controlling magnesium implant corrosion. *Acta Biomaterialia* 2017, 48: 530–540.

[87] Guo, W. H., Zhang, G. D., Ni-SiC disquisition on sediment mechanism of Ni-SiC composite plating. *Journal of East China University of Metallurgy* 1999, 16 (3): 235–239.

[88] Wang, G. B., Ni-P-Al$_2$O$_3$ Mechanism of composite Ni-P-Al$_2$O$_3$ electroless plating. *Journal of Northeast University of Technology* 1991, 12 (5): 548–552.

[89] Wang, P., Wu, T., Li, J., Jia, X. H., Gong, C. L., Guo, X. Y., Doping mechanism of nano TiO$_2$ in micro-arc oxidation coating on aluminum alloy. *Rare Metal Material and Engineering* 2017, 46 (2): 479–483.

[90] Sankara Narayanan, T. S. N., Park, I. S., Lee, M. H., Strategies to improve the corrosion resistance of microarc oxidation (MAO) coated magnesium alloys for degradable implants: Prospects and challenges. *Progress in Materials Science* 2014, 60: 1–71.

[91] Necula, B. S., Fratila-Apachitei, L. E., Berkani, A., Apachitei, I., Duszczyk, J., Enrichment of anodic MgO layers with Ag nanoparticles for biomedical applications. *Journal of Materials Science Materials in Medicine* 2009, 20 (1): 339–345.

[92] Necula, B. S., Apachitei, I., Tichelaar, F. D., Fratila-Apachitei, L. E., Duszczyk, J., An electron microscopical study on the growth of TiO2-Ag antibacterial coatings on Ti6Al7Nb biomedical alloy. *Acta Biomaterialia* 2011, 7: 2751–2757.

[93] Barati Darband, G., Aliofkhazraei, M., Hamghalam, P., Valizade, N., Plasma electrolytic oxidation of magnesium and its alloys: Mechanism, properties and applications. *Journal of Magnesium and Alloys* 2017, 5 (1): 74–132.

[94] Elyutin, A. V., Blinkov, I. V., Volkhonsky, A. O., Belov, D. S., Properties of nanocrystalline arc PVD TiN-Cu coatings. *Inorganic Materials* 2013, 49 (11): 1106–1112.

[95] Zhang, S., Sun, D., Fu, Y., Du, H., Effect of sputtering target power on microstructure and mechanical properties of nanocomposite nc-TiN/a-SiN$_x$ thin films. *Thin Solid Films* 2004, 447 (3): 462–467.

[96] Patscheider, J., Nanocomposite hard coatings for wear protection. *Mrs Bulletin* 2003, 28: 180–183.

[97] Bendavid, A., Martin, P. J., Cairney, J., Hoffman, M., Fischer-Cripps, A. C., Deposition of nanocomposite TiN-Si$_3$N$_4$ thin films by hybrid cathodic arc and chemical vapor process. *Applied Physics A (Materials Science & Processing)* 2005, 81 (1): 151–158.

[98] Daroonparvar, M., Yajid, M. A. M., Yusof, N. M., Bakhsheshi-Rad, H. R., Preparation and corrosion resistance of a nanocomposite plasma electrolytic oxidation coating on Mg-1% Ca alloy formed in aluminate electrolyte containing titania nano-additives. *Journal of Alloys and Compounds* 2016, 688: 841–857.

[99] Araghi, A., Paydar, M. H., Electroless deposition of Ni-P-B$_4$C composite coating on AZ91D magnesium alloy and investigation on its wear and corrosion resistance. *Materials & Design* 2010, 31 (6): 3095–3099.

[100] Sun, M., Yerokhin, A., Bychkova, M. Y., Shtansky, D. V., Levashov, E. A., Matthews, A., Self-healing plasma electrolytic oxidation coatings doped with benzotriazole loaded halloysite nanotubes on AM50 magnesium alloy. *Corrosion Science* 2016, 111: 753–769.

[101] Li, H., Zhao, Q., Li, B., Kang, J., Yu, Z., Fabrication and properties of carbon nanotube-reinforced hydroxyapatite composites by a double in situ synthesis process. *Carbon* 2016, 101: 159–167.

[102] Sanjinés, R., Abad, M. D., Vaju, C., Smajda, R., Mioni, M., Magrez, A., Electrical properties and applications of carbon based nanocomposite materials: An overview. *Surface & Coatings Technology* 2011, 206 (4): 727–733.

[103] Balandin, A. A., Ghosh, S., Bao, W., Calizo, I., Teweldebrhan, D., Miao, F., Lau, C. N., Superior thermal conductivity of single-layer graphene. *Nano Letters* 2008, 8 (3), 902–907.

[104] Lee, C., We, X., Kysar, J. W., Hone, J., Measurement of the elastic properties and intrinsic strength of monolayer graphene. *Science* 2008, 321: 385–388.

[105] Krishnamurthy, A., Gadhamshetty, V., Mukherjee, R., Chen, Z., Ren, W., Cheng, H. M., Koratkar, N. J. C., Passivation of microbial corrosion using a graphene coating. *Carbon* 2013, 56: 45–49.

[106] Ramezanzadeh, B., Ahmadi, A., Mahdavian, M., Enhancement of the corrosion protection performance and cathodic delamination resistance of epoxy coating through treatment of steel substrate by a novel nanometric sol-gel based silane composite film filled with functionalized graphene oxide nanosheets. *Corrosion Science* 2016, 109: 182–205.

[107] Scattergood, R. O., Koch, C. C., Murty, K. L., Brenner, D., Strengthening mechanisms in nanocrystalline alloys. *Materials Science and Engineering A* 2008, 493: 3–11.

[108] Zhao, X. N., Chen, X. Y., Zhang, L., Liu, Q. Y., Wang, Y., Zhang, W. G., Zheng, J. M., Preparation of nano-hydroxyapatite coated carbon nanotube reinforced hydroxyapatite composites. *Coatings* 2018, 8: 357.

[109] Basirun, W. J., Nasiri-Tabrizi, B., Baradaran, S., Overview of hydroxyapatite-graphene nanoplatelets composite as bone graft substitute: Mechanical behavior and in-vitro biofunctionality. *Critical Reviews in Solid State and Materials Sciences* 2018, 43: 269–269.

[110] Lahiri, D., Singh, V., Keshri, A. K., Seal, S., Agarwal, A., Carbon nanotube toughened hydroxyapatite by spark plasma sintering: Microstructural evolution and multiscale tribological properties. *Carbon* 2010, 48: 3103–3120.

[111] Baradaran, S., Moghaddam, E., Basirun, W. J., Mehrali, M., Sookhakian, M., Hamdi, M., Moghaddam, M. R. N., Alias, Y., Mechanical properties and biomedical applications of a nanotube hydroxyapatite-reduced graphene oxide composite. *Carbon* 2014, 69: 32–45.

[112] Meng, Y. H., Tang, C. Y., Tsui, C. P., Chen, D. Z., Fabrication and characterization of needle-like nano-HA and HA/MWNT composites. *Journal of Materials Science: Materials in Medicine* 2008, 19: 75–81.

[113] Echeberria, J., Rodríguez, N., Vleugels, J., Vanmeensel, K., Reyes-Rojas, A., Garcia-Reyes, A., Domínguez-Rios, C., Aguilar-Elguézabal, A., Bocanegra-Bernal, M. H., Hard and tough carbon nanotube-reinforced zirconia-toughened alumina composites prepared by spark plasma sintering. *Carbon* 2012, 50 (2): 706–717.

[114] Bi, X. Q., Wei, Y. L., Effects of addition of Nano-diamond on structure and properties of Ni-P composite coating of magnesium alloy. *Surface Technology* 2016, 45 (12): 68–72.

[115] Cui, X. J., Ning, C. M., Shang, L. L., Zhang, G. A., Liu, X. Q., Structure and anticorrosion, friction, and wear characteristics of pure diamond-like carbon (DLC), Cr-DLC, and Cr-H-DLC films on AZ91D Mg alloy. *Journal of Materials Engineering and Performance* 2019, 28: 1213–1225.

[116] Aliofkhazraei, M., Rouhaghdam, A. S., Wear and coating removal mechanism of alumina/titania nanocomposite layer fabricated by plasma electrolysis. *Surface & Coatings Technology* 2011, 205: S57–S62.

[117] Lim, T. S., Ryu, H. S., Hong, S. H., Electrochemical corrosion properties of CeO_2-containing coatings on AZ31 magnesium alloys prepared by plasma electrolytic oxidation. *Corrosion Science* 2012, 62: 104–111.

[118] Lima, R. S., Marple, B. R., Thermal spray coatings engineered from nanostructured ceramic agglomerated powders for structural, thermal barrier and biomedical applications: A review. *Journal of Thermal Spray Technology* 2007, 16: 40–63.

[119] Berger-Keller, N., Bertrand, G., Filiatre, C., Meunier, C., Coddet, C., Microstructure of plasma-sprayed titania coatings deposited from spray-dried powder. *Surface and Coatings Technology* 2003, 168: 281–290.

[120] Othman, S. H., Salam, N. R. A., Zainal, N., Basha, R. K., Talib, R. A., Antimicrobial activity of TiO_2 nanoparticle-coated film for potential food packaging applications. *International Journal of Photoenergy* 2014, 2014: 1–6.

[121] Hussein-Al-Ali, S. H., Zowalaty, M. E. E., Kura, A. U., Geilich, B., Fakurazi, S., Webster, T. J., Hussein, M. Z., Antimicrobial and controlled release studies of a novel nystatin conjugated iron oxide nanocomposite. *BioMed Research International* 2014, 2014: 1–13.

[122] Richard Prabakar, S. J., Hwang, Y. H., Bae, E. G., Lee, D. K., Pyo, M., Graphene oxide as a corrosion inhibitor for the aluminum current collector in lithium ion batteries. *Carbon* 2013, 52: 128–136.

[123] Bakhsheshi-Rad, H. R., Hamzah, E., Ismail, A. F., Aziz, M., Daroonparvar, M., Parham, S., Hadisi, Z., Yajid, M. A. M., Titania-carbon nanotubes nanocomposite coating on Mg alloy: microstructural characterisation and mechanical properties. *Materials Science and Technology* 2017, 34: 378–387.

[124] Singh, V., Diaz, R., Balani, K., Agarwal, A., Seal, S., Chromium carbide-CNT nanocomposites with enhanced mechanical properties. *Acta Materialia* 2009, 57 (2), 335–344.

[125] Arai, S., Suzuki, Y., Nakagawa, J., Yamamoto, T., Endo, M., Fabrication of metal coated carbon nanotubes by electroless deposition for improved wettability with molten aluminum. *Surface & Coatings Technology* 2012, 212: 207–213.

[126] Jamali, H., Mozafarinia, R., Shoja-Razavi, R., Ahmadi-Pidani, R., Comparison of hot corrosion behaviors of plasma-sprayed nanostructured and conventional YSZ thermal barrier coatings exposure to molten vanadium pentoxide and sodium sulfate. *Journal of the European Ceramic Society* 2014, 34 (2): 485–492.

[127] Liang, J., Wang, P., Hu, L., Hao, J., Tribological properties of duplex MAO/DLC coatings on magnesium alloy using combined microarc oxidation and filtered cathodic arc deposition. *Materials Science and Engineering A* 2007, 454 (16): 164–169.

[128] Daroonparvar, M., Yajid, M. A. M., Yusof, N. M., Bakhsheshi-Rad, H. R., Hamzah, E., Mardanikivi, T., Deposition of duplex MAO layer/nanostructured titanium dioxide composite coatings on Mg-1%Ca alloy using a combined technique of air plasma spraying and micro arc oxidation. *Journal of Alloys and Compounds* 2015, 649: 591–605.

[129] Wu, G., Ding, K., Zeng, X., Wang, X., Yao, S., Improving corrosion resistance of titanium-coated magnesium alloy by modifying surface characteristics of magnesium alloy prior to titanium coating deposition. *Scripta Materialia* 2009, 61 (3): 269–272.

[130] Bakhsheshi-Rad, H. R., Hamzah, E., Kasiri-Asgarani, M., Saud, S. N., Yaghoubidoust, F., Akbari, E., Structure, corrosion behavior, and antibacterial properties of nano-silica/graphene oxide coating on biodegradable magnesium alloy for biomedical applications. *Vacuum* 2016, 131: 106–110.

[131] Gnedenkov, S. V., Sinebryukhov, S. L., Zavidnaya, A. G., Egorkin, V. S., Puz', A. V., Mashtalyar, D. V., Sergienko, V. I., Yerokhin, A. L., Matthews, A., Composite hydroxyapatite-PTFE coatings on Mg-Mn-Ce alloy for resorbable implant applications via a plasma electrolytic oxidation-based route. *Journal of the Taiwan Institute of Chemical Engineers* 2014, 45 (6): 3104–3109.

[132] Gnedenkov, S. V., Egorkin, V. S., Sinebryukhov, S. L., Vyaliy, I. E., Boinovich, L. B., Formation and electrochemical properties of the superhydrophobic nanocomposite coating on PEO pretreated Mg-Mn-Ce magnesium alloy. *Surface & Coatings Technology* 2013, 232: 240–246.

[133] Bakhsheshi-Rad, H. R., Hamzah, E., Ebrahimi-Kahrizsangi, R., Daroonparvar, M., Medraj, M., Fabrication and characterization of hydrophobic microarc oxidation/poly-lactic acid duplex coating on biodegradable Mg–Ca alloy for corrosion protection. *Vacuum* 2016, 125: 185–188.

[134] Daroonparvar, M., Yajid, M. A. M., Yusof, N. M., Bakhsheshi-Rad, H. R., Hamzah, E., Kamali, H. A., Microstructural characterization and corrosion resistance evaluation of nanostructured Al and Al/AlCr coated Mg-Zn-Ce-La alloy. *Journal of Alloys and Compounds* 2014, 615: 657–671.

[135] Cui, X.-J., Ping, J., Zhang, Y.-J., Jin, Y.-Z., Zhang, G.-A., Structure and properties of newly designed MAO/TiN coating on AZ31B Mg alloy. *Surface & Coatings Technology* 2017, 328: 319–325.

[136] Wang, L., Zhao, S., Xie, Z., Huang, L., Wang, X. J. N. I., Research, M. i. P., MoS$_2$/Ti multilayer deposited on 2Cr13 substrate by PVD. *Nuclear Instruments and Methods in Physics Research* 2008, 266 (5): 730–733.

[137] Kim, S. K., Ahn, Y. H., Kim, K. H., MoS2-Ti composite coatings on tool steel by d.c. magnetron sputtering. *Surface and Coatings Technology* 2003, 169-170: 428–432.

[138] Qin, X., Ke, P., Wang, A., Kim, K. H., Microstructure, mechanical and tribological behaviors of MoS$_2$-Ti composite coatings deposited by a hybrid HIPIMS method. *Surface & Coatings Technology* 2013, 228: 275–281.

[139] Wu, G., Dai, W., Zheng, H., Wang, A., Improving wear resistance and corrosion resistance of AZ31 magnesium alloy by DLC/AlN/Al coating. *Surface & Coatings Technology* 2010, 205: 2067–2073.

[140] Wang, D. P., Wang, S. L., Wang, J. Q., Relationship between amorphous structure and corrosion behaviour in a Zr-Ni metallic glass. *Corrosion Science* 2012, 59: 88–95.

[141] Ning, C. M.; Cui, X. J.; Shang, L. L.; Zhang, Y. J.; Zhang, G. A., Structure and properties of different elements doped diamond-like carbon on micro-arc oxidation coated AZ31B Mg alloy. *Diamond & Related Materials* 2020, 106: 107832.

[142] Zheng, Y. F., *Magnesium Alloys as Degradable Biomaterials*. CRC Press: 2015: 87–142.

[143] Zysset, P. K., Guo, X. E., Hoffler, C. E., Moore, K. E., Goldstein, S. A., Elastic modulus and hardness of cortical and trabecular bone lamellae measured by nanoindentation in the human femur. *Journal of Biomechanics* 1999, 32 (10): 1005–1012.

[144] Holleck, H., Schier, V., Multilayer PVD coatings for wear protection. *Surface and Coatings Technology* 1995, 76–77: 328–336.

[145] Zhang, D., Qi, Z., Wei, B., Wu, Z., Wang, Z., Anticorrosive yet conductive Hf/Si$_3$N$_4$ multilayer coatings on AZ91D magnesium alloy by magnetron sputtering. *Surface & Coatings Technology* 2017, 309: 12–20.

[146] Tsai, S. H., Duh, J. G., Microstructure and mechanical properties of CrAlN/SiN$_x$ nanostructure multilayered coatings. *Thin Solid Films* 2009, 518 (5): 1480–1483.

[147] Chen, W., Yue, L., Jie, Z., Zhang, S., Liu, S., Kwon, S. C., Preparation and characterization of CrAlN/TiAlSiN nano-multilayers by cathodic vacuum arc. *Surface & Coatings Technology* 2015, 265 (4): 205–211.

[148] Ou, Y. X., Lin, J., Tong, S., Che, H. L., Sproul, W. D., Lei, M. K., Wear and corrosion resistance of CrN/TiN superlattice coatings deposited by a combined deep oscillation magnetron sputtering and pulsed dc magnetron sputtering. *Applied Surface Science* 2015, 351 (OCT.1): 332–343.

[149] Ou, Y. X., Ouyang, X. P., Liao, B., Zhang, X., Zhang, S., Hard yet tough CrN/Si$_3$N$_4$ multilayer coatings deposited by the combined deep oscillation magnetron sputtering and pulsed dc magnetron sputtering. *Applied Surface Science* 2020, 502: 144168.1–144168.9.

[150] Voevodin, A. A., Zabinski, J. S., Muratore, C., Recent advances in hard, tough, and low friction nanocomposite coatings. *Tsinghua Science and Technology* 2005, 10 (6): 665–679.

[151] Qiang, M., Cai-e, C., Jun-de, P., Ping-ze, Z., Improving wear resistance of magnesium alloy AZ91D by TiN-CrN multilayer coating. *Transitions of Nonferrous Metals Society of China* 2006, (16): 1802–1805.

[152] Tafazoly, M., Monirvaghefi, M., Salehi, M., Saatchi, A., Tabatabaei, F., Verdian, M. M., Characterisation and corrosion performance of multilayer nano nickel coatings on AZ31 magnesium alloy. *Journal of Nanoscience and Nanotechnology* 2012, 8: 19–26.

[153] Bull, S. J., Jones, A. M., Multilayer coatings for improved performance. *Surface and Coatings Technology* 1996, 78: 173–184.

[154] Leyland, A., Matthews, A., On the significance of the H/E ratio in wear control: a nanocomposite coating approach to optimised tribological behaviour. *Wear* 2000, 246 (1–2): 1–11.

[155] Cai, K., Sui, X., Hu, Y., Li, Z., Lai, M., Luo, Z., Liu, P., Yang, W., Fabrication of anticorrosive multilayer onto magnesium alloy substrates via spin-assisted layer-by-layer technique. *Materials Science and Engineering C* 2011, 31: 1800–1808.

9 Correlation between Coating Properties and Industrial Applications

Yin-Yu Chang
National Formosa University, Taiwan

Heng-Li Huang
China Medical University, Taiwan

Jui-Ting Hsu
China Medical University, Taiwan

Ming-Tzu Tsai
Hungkuang University, Taiwan

CONTENTS

9.1 PRINCIPLES OF FUNCTIONAL COATING DESIGN FOR INDUSTRIAL APPLICATIONS

Design of useful devices and components for structural parts depends upon their specific applications. Scientist and engineers not only deal with the development of materials but also with the processing and synthesis of materials and manufacturing techniques to meet the requirement of applications. It is important to establish the relationships among material composition, properties, performance, and microstructure of the products, and the method which the material to be synthesized or processed. A lot of technical applications need the material having well-defined surface properties to meet specific requirements. For this purpose, surface functionalization of metallic materials and functional coatings have become one of the most lively research areas recently. Table 9.1 shows general requirement of functional coatings for typical tooling and biomedical applications. A functional surface, even a protective surface, involves proper chemical parts and specific morphological characteristics that determine the interaction with the surrounding environment and subsequent effectiveness of materials. The advances in surface engineering and vacuum vapor deposition for the preparation of nanostructured thin films and coatings allow controlling the composition and microstructure of materials down to the nanoscale by organic, inorganic, or hybrid coatings [1–3]. A new generation

TABLE 9.1

General requirement of functional coatings for tooling and biomedical applications

Tooling applications	Biomedical applications
• Hardness(H), young's modulus(E), and H/E	• Antibacterial and antimicrobial properties
• Low friction and wear resistance	• Biocompatibility and osteointegration (e.g., implants)
• Thermal stability (e.g., high-speed and dry cutting)	• Corrosion resistance
• Fracture toughness (e.g., piercing and forming dies)	• Low friction and wear resistance (e.g., artificial joints)
• Oxidation resistance	• Hydrophobic and hydrophilic properties (e.g., stent,
• Antisticking (e.g., plastic injection molds and parts)	biosensors and medical devices)
• Fatigue (e.g., piercing and forming dies)	• High surface area and adsorption capability (e.g., implants)
• Corrosion resistance and chemical inertness	• UV protection (e.g., biosensors and medical devices)

of functional coatings with grain sizes and structural effects in the nanoscale for mechanical applications and used as biomaterials for load-bearing applications has emerged. An example of potential application using the functional and protective coatings is surface functionalization and protection of medical tools. In order to minimize the risk of infections, antimicrobial low-friction and wear resistant protective coatings for metallic medical surfaces have been designed and developed. In the framework of the research results, nanostructured ceramic materials such as oxides, carbides, nitrides, oxygen nitrides, and carbonitrides with additionally added with silver nanoparticles were developed. The doped silver was applied to guarantee adequate antimicrobial properties [4–7].

Before fabricating the mechanical parts of a device or an equipment, the selection and design of materials play a very important role. The selection and design of system materials depend upon the bulk mechanical properties such as hardness, strength, elastic modulus, and fracture toughness, as well as controlled surface characteristics such as surface roughness, coefficient of friction and wear resistance. In addition, cost-effectiveness and product performance play an important role when designing and selecting the materials. Among all the manufacturing methods, combining surface engineering techniques and vacuum vapor deposition for the preparation of nanostructured thin films and coatings is also viable solution because it can be applied uniformly over large area of a bulk material. Coatings can provide multiple benefits such as high hardness, low friction coefficient, hydrophobic/ hydrophilic surface, improved oxidation resistance, wear resistance, and corrosion resistance [8–10]. In the case of mechanical applications, a lot of research activities around the world dealt with either the optimization of existing tribological resistant coatings, or the development of new types of nanostructured coatings which showed excellent mechanical properties and good adhesion to the substrate also at high temperatures. Figure 9.1 presents a generalized structure of a typical coated tribological pair and the design requirement of the coatings. The contact

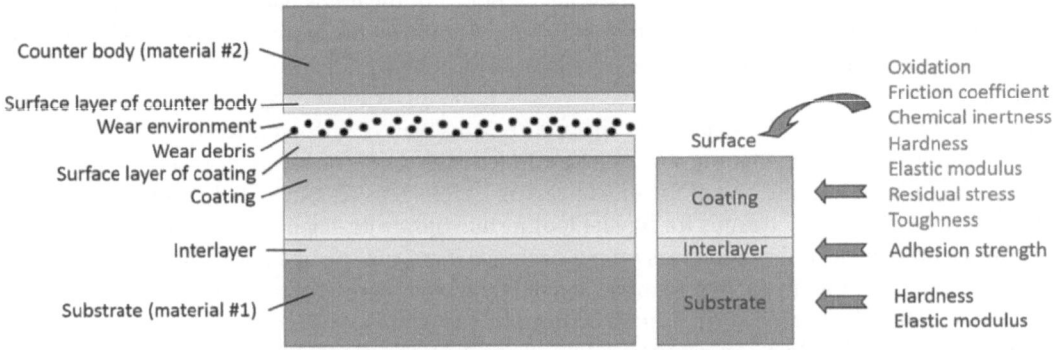

FIGURE 9.1 Schematic structure of a coated tribological pair and the design requirement of the coatings.

environment with and without lubricant (water, oil, or grease) is an important parameter to be considered influencing the tribological behavior. Both the design of a protective coating and the wear performance strongly depend on the operating environment to possess coating effectiveness. Especially, the adhesion strength and mechanical properties of hard ceramic coatings on metallic substrates can be improved by adding an interlayer, which is also known as an adhesion, buffer, or seed layer. The material selection, composition manipulation, and thickness regulation of the interlayer are the key parameters to control the adhesion strength between the coating and substrate and residual stress of the coatings, and therefore influence the mechanical properties and applications [11–14]. In 2020, Chang et al. [11] studied the effect of interlayer design on the mechanical properties of AlTiCrN and multilayered AlTiCrN/TiSiN hard coatings. The AlTiCrN with Ti and Cr interlayers showed high residual stress. The AlTiCrN/TiSiN multilayer coatings with intermediate CrN and transition AlTiN/CrN multilayers under the top AlTiCrN/TiSiN multilayers possessed lower residual stress. The residual stress was changed by modifying interlayer configuration before the top AlTiCrN/TiSiN multilayers. The design of AlTiCrN/TiSiN multilayer coatings with interlayers of CrN and AlTiCrN/CrN can decrease residual stress and possessed good impact fatigue performance, which can be effective to reduce damage and cracking of hard coatings during cyclic impact loading. A large amount of studies has been conducted to pursue high hardness and low friction coefficient of the coatings. However, such applications are investigated for specific contact conditions, such as cutting, sliding, or molding. Therefore, many suppliers of vacuum deposition equipment, such as Oerlikon Balzers (Switzerland), Platit (Switzerland), Kobe steel (Japan), Cemecon (Germany), Ionbond (The Netherlands), Hauzer (The Netherlands), and Surftech (Taiwan), provide a lot of specific coatings for various applications.

In the field of biomaterials applications, an example is nanostructured and nanocomposite coatings on implants and medical devices which enhanced mechanical properties and provide good bioactivity and osteointegration. Recent advances in materials science and surface engineering have brought high-performance, multifunctional materials with bioactive properties [15, 16]. Metallic biomaterials such as titanium and titanium alloys are widely used in implants such as joint prostheses and dental implants because of their excellent mechanical properties, corrosion resistance, and biocompatibility. However, due to the insufficient antibacterial and biological characteristics of titanium, the formation of strong bonds with the soft and hard tissue is usually affected, which influence the application of such materials in the field of biomedical applications. The main driving force for developing biocompatible and antibacterial coatings is the improved performance of functionalized surfaces that cannot be achieved by the typical bulk materials such as Ti and its alloys. Coatings on substrate bulk biomaterials can simultaneously satisfy multiple requirements of good bioactivity in biological environments and mechanical properties, wear resistance, and chemical corrosion resistance. The surface chemistry including hydrophilicity and hydrophobicity of the coating materials play an important role in the attachment of different adhesins to the material surface [17]. Such antibacterial coatings have many applications ranging from prevention of hospital-acquired infection or implants to bathroom equipment. There are major issues focusing on the anti-adhesion and contact-killing surfaces. We must be careful that the surface may become contaminated with materials that attach under a layer of dead cells, resulting in their deactivation. Anti-infective coatings need to be tailored according to the specific clinical application. Innovative technologies are developing new biomaterials and surfaces with antibacterial properties, relying either on hydrophobic, bactericidal, or antibiofilm activities [18].

Over the past decades, a broad range of antibacterial compounds have been developed for release-based systems [15]. For example, silver (Ag) is known to deactivate enzymes by binding to thiol groups and inhibit the respiratory chain, and is a good candidate for antibacterial element [19]. Ag ions have shown to possess strong antimicrobial properties but cause no immediate and serious risk for human health, which led to an extensive use of silver-based products in many applications. However, there is still a toxicity risk of Ag nanoparticles when in widespread use, and it could increase silver release in the environment, which can have negative impacts on ecosystems [20].

Design and measurement of Ag ion fraction of Ag nanoparticles are crucial for toxicity studies [21, 22]. In addition to titanium, tantalum has been adopted to manufacture biomedical implants used for orthopedic, dental, and craniofacial surgeries in clinical practice. Similar to titanium, Ta is highly unreactive and biocompatible in the body. Ta does not exhibit toxicity to surrounding cells, nor does it inhibit local cell growth of surrounding bone. Tantalum pentoxide (Ta_2O_5) possesses good corrosion resistance and biocompatibility. The results of the antibacterial experiments revealed that the Ta_2O_5 coated Ti samples had superior antibacterial properties compared to the uncoated Ti. Cell viability analyses also showed that the Ta_2O_5 films exhibited better cell growth compared to the Ti specimens in human skin fibroblasts and human osteosarcoma cells [23]. Engineering strategies are being developed to tune the properties of the antibacterial materials (concentration, distribution, and grain size), the surface morphologies (porosity and surface roughness), and the overall microstructure of the antibacterial coatings.

Figure 9.2 shows the SEM images illustrating examples of passive strategies to improve the antibacterial and biological characteristics of coatings. TiO_2 is one of the most representative inorganic materials and takes advantages of strong photo-catalytic activity, excellent biocompatibility, good osteo-compatibility, easy sol-gel synthesis route, and low-cost price [24]. Controlling the porous

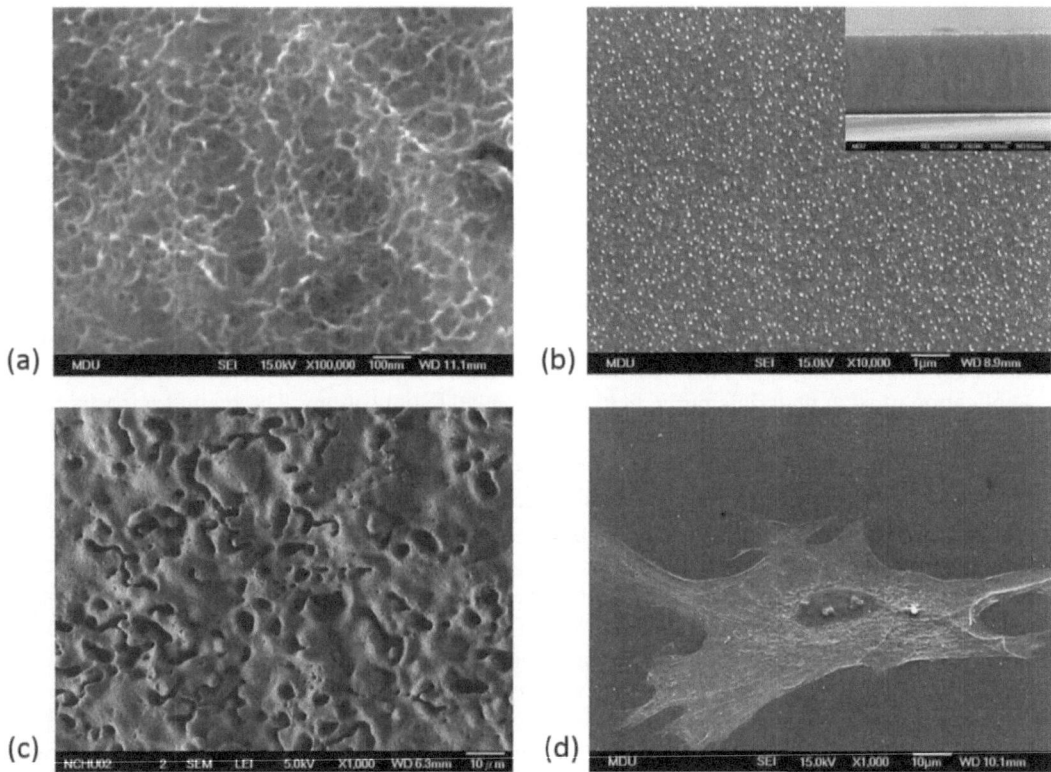

FIGURE 9.2 SEM images illustrating examples of passive strategies to improve the antibacterial and biological characteristics of coatings. (a) Porous TiO_2 coatings on Ti with a crosslinking morphology exhibit cell bioactivity. (b) SEM micrographs of TaN coatings doped with Ag. Small figures on the right top images shows its cross-sectional morphology. The TaN-Ag coatings improve antibacterial performance with compatible biological response. (c) SEM micrographs of tantalum oxide coatings deposited on MAO pre-treated Ti. The presence of micropores is typical of MAO treatment on titanium. The porous Ta_2O_5 film pre-treated with MAO procedure possessed both great biocompatibility and antibacterial abilities. (d) SEM micrograph of human fetal skin fibroblasts cells cultured on Ta-C:H-coated Ti. Ta-C:H coatings exhibited good biocompatibility for human fetal skin fibroblast cells derived from soft tissues and demonstrated the high biological performance.

surface of TiO_2 coatings on Ti with a crosslinking morphology can exhibit good cell bioactivity, as shown in Figure 9.2(a). Figure 9.2(b) shows SEM micrographs of TaN coatings doped with Ag. Small figures on the right top images shows its cross-sectional morphology. The TaN-Ag coatings can improve antibacterial performance with compatible biological response [25]. The versatility of the surface morphology features including surface roughness, pore size and distribution, and coating surface coverage as well as the large range of surface chemistries for both metal oxides and pure noble metals offers wide scope to functionalize nanoparticles and networks of metal substrate materials. Novel efficient techniques have recently been investigated and conducted. However, the main limitations of the current fabrication techniques are the difficulty to process stable and homogeneous surface structure and the control of surface morphology due to the high reactivity of nanostructured metal substrates [26]. Recently, unlike conventional anodizing, microarc oxidation (MAO) method has been developed and this method uses high operating voltages that form a porous oxide layer on the surface of metallic materials, such as titanium and aluminum, which increases the surface roughness and area. For example, the presence of micropores is typical of MAO treatment on titanium. As shown in Figure 9.2(c), the porous Ta_2O_5 film pre-treated with MAO procedure can possess both great biocompatibility and antibacterial abilities [23]. In addition to Ta_2O_5, nanocomposite Ta-C:H coatings are also developed to enhance the biocompatibility of metallic Ti for biomedical applications, as shown in Figure 9.2(d). Ta-C:H coatings exhibited good biocompatibility for human fetal skin fibroblast cells derived from soft tissues and demonstrated the high biological performance [27]. Alternatively, special architectures, which possess their own unique mechanical and chemical properties, can be incorporated in coatings to control antibacterial performance and bioactivity. These include nanotubes (e.g., amorphous carbon nanotubes or TiO_2 nanotubes), nanowires, and nanofibers [28–30].

9.2 MECHANICAL PROPERTIES AND TOOLING APPLICATIONS OF NANOSTRUCTURED HARD COATINGS

Nanomaterials with high hardness and wear resistance in specific applications and severe environment have been developed and available for a long time and are currently used in the fabrication of some critical tools and components in machining, molding, biomedical implants and components, and a wide range of mechanical fields. During the last two decades, multifunctional nanostructured and nanocomposite coatings have become increasingly more popular, mainly because of their impressive mechanical properties and tribological performance characteristics as well as advances of new coating technologies, such as integrated cathodic arc evaporation (CAE) and magnetron sputtering (MS) vacuum coating system using pulsed power combining the advantages of the arc evaporation and sputtering technologies. For the production of nanostructured or nanocomposite coatings, it may be necessary to combine two or more of the deposition methods in a unique system. In addition, the use of high-ionization power sources like CAE and high-power impulse magnetron sputtering (HiPIMS) is important for fabricating nanostructured coatings with strong adhesion strength, compact and dense structure, and superior mechanical properties.

Cutting is an important process in the manufacturing industry and it is the main application field of hard coatings. How to choose the right tool coating for the machining application? The solution of wear problems starts with a detailed examination of the tribological system with all influence factors that are involved. From this, it can be deduced what tribological conditions and wear mechanisms are to be identified. The selection of the proper coated tools depends first on the materials to be machined, such as carbon steel, mold tool steels, nonferrous alloys, glass fiber composite materials, even ceramics. Second, the mechanical properties including hardness and ductility of the material is an important parameter. Generally, a high surface hardness of the coating is one of the best ways to increase tool life. Third, cutting condition and cutting parameters with/without lubricant, even the type of lubricant are influencing the design and selection of coatings. Fourth, the material of the tool, such as high speed steel and WC-Co, is a major to be concerned. There is a vast selection of Physical Vapor Deposition (PVD), Chemical Vapor Deposition (CVD), and alternate surface treatments that are readily available from your manufacturers or coating facilities. Surface friction and wear reduction remain as acute technological challenges from prehistoric times to modern days. For machining applications, coating characteristics including hardness, wear resistance, surface lubricity, oxidation temperature, and anti-seizure play a major role in determining which coating treatment may be the most beneficial for a specific application [31]. Table 9.2 presents typical commercial single-layer coatings and their material characteristics for machining applications. Hard coatings for cutting such as milling, tapping, and drilling all vary and are application-specific [32, 33]. TiN coating is one of

TABLE 9.2
Commercial single-layer coating material characteristics for machining applications

Coating type	Color*	Hardness (GPa)	Coating characteristics	Example of applications
TiN	Gold yellow	23–28	A general purpose coating designed for moderate to high abrasion applications. Reduces friction and prevents adhesive wear and BUE formation.	Milling, hobbing, tapping, and stamping with HSS tools.
TiCN	Gray	27–31	Addition of carbon adds higher hardness and better surface lubricity than TiN. Low friction, high fracture strength, and excellent abrasive wear resistance.	Milling, hobbing, tapping, and stamping with HSS tools.
$Ti_{1-x}Al_xN$	Brown blue	28–35	High thermal hardness, oxidation resistance, and thermal impact resistance.	Universal for drilling, milling, reaming, and turning. Especially suitable for dry machining.
$Ti_{1-x}Al_xCN$	Dark blue	25–30	High thermal hardness and oxidation resistance. Anti-seizure properties to prevent adhesive wear and BUE formation.	Milling, hobbing, tapping, stamping, and punching with WC-Co tools.
CrN	Silver-gray	18–22	Excellent toughness, and improved wear and corrosion resistance. Non-sticky surface against most other materials.	Cutting, piercing, and forming tools.
$Ti_{1-x}Al_xCrN$	Light gray	31–35	High heat and oxidation resistance as well as good toughness.	Milling, hobbing, tapping, stamping, and punching
$Al_{1-x}Cr_xN$	Dark gray	31–34	High wear resistance with excellent hot hardness and thermal shock stability as well as good toughness.	Dry and wet machining at high cutting speed.
DLC	Black	5–40	DLC is hard, amorphous carbon film with a significant fraction of sp3-hybridized carbon atoms with and without hydrogen. Hardness of nonhydrogenated coatings is almost an order of magnitude higher and hence may be aptly termed DLC. Coefficient of friction of DLC is ~0.1, which is comparable to that of Teflon. Chemical inertness to prevent adhesive wear and BUE formation.	Good for cutting graphite, glass fiber composites, aluminum alloy, and many other abrasive materials.

* The color depends on the composition and structure of the coating and it is only a general perception.

the most popular hard coatings, which can effectively increase the life of tools and many other applications. The presence of carbon in TiCN acts as a lubricant to reduce friction and wear. AlTiN is formed by replacing Ti atoms in TiN face-centered cubic with Al atoms. At high temperatures, stable Al_2O_3 are formed on the surface of AlTiN coating, which inhibits oxygen atoms into the coating and improves the high temperature oxidation resistance of the coating. Diamond-like carbon (DLC) coating is a solid lubricant with great commercial application value due to its low friction coefficient and high wear resistance [34–36]. Depending upon the testing environment, the coefficient of friction can be as low as 0.01. However, the tribological properties of DLC are strongly affected by the deposition method. The architecture of the hard coatings has different characteristics in terms of its structure, morphology, composition, gradient, grain size, and defects, which are influenced by the deposition method and parameters, which have an impact on the characteristics of the coating [37].

Figure 9.3 schematically illustrates the development of tribological and mechanical coatings and the typical design architectures that are used for nanostructured and nanocomposite coatings. When specific application conditions in tribological environment become severe, nanostructured coatings with high hardness, toughness, and low friction become high demand for controlling friction and wear. These coatings certainly represent a new class within the very broad field of surface engineering and are the result of breakthrough developments in PVD and plasma enhanced CVD (PECVD) technologies in recent years [38]. Figure 9.4 shows typical application products of the multicomponent hard coatings, such as specific end mills coated with $Ti_xAl_{1-x}SiN$, $AlTiCr_xN$, and $Ti_xAl_{1-x}SiN/TiSiN$; $AlTiCr_xN$ coated gear cutters; $Ti_xAl_{1-x}Si(O)N$ coated screw taps; and $AlTiCr_xN$ coated shaft rods. The important development in the formulation and use of tribological and hard coatings is due to the ability of designed materials to possess low friction and wear resistance under specific conditions including room temperature, high temperature, corrosive or dry environment. During the 1980s, transition metal nitrides, carbides, and oxides attracted the greatest attention for tribological applications. Most of the mechanical components and tools are used to be coated with conventional binary hard coatings, such as TiN with gold color and CrN with metallic gray color. During the last decade, attention has shifted to nanocomposite coatings mainly because of their superior tribological properties and diverse application possibilities in transportation, aerospace, and manufacturing

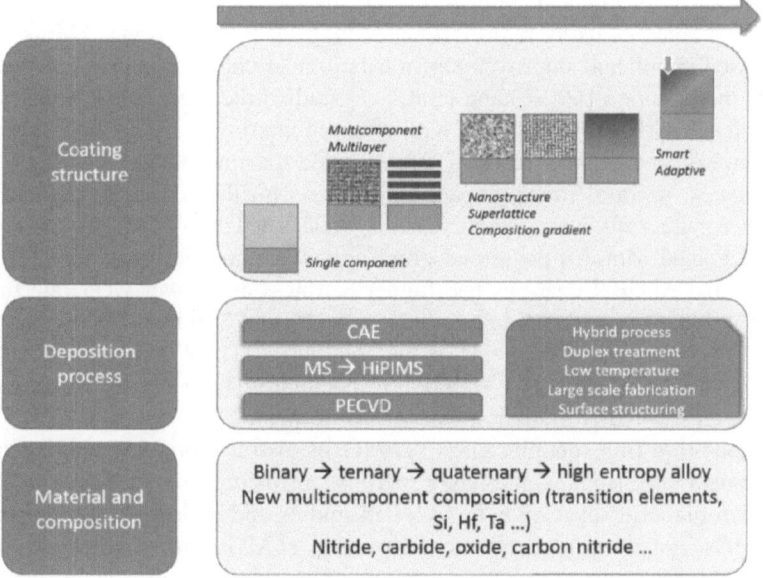

FIGURE 9.3 Schematic illustrations of the development of tribological and mechanical coatings and the typical design architectures that are used for nanostructured and nanocomposite coatings.

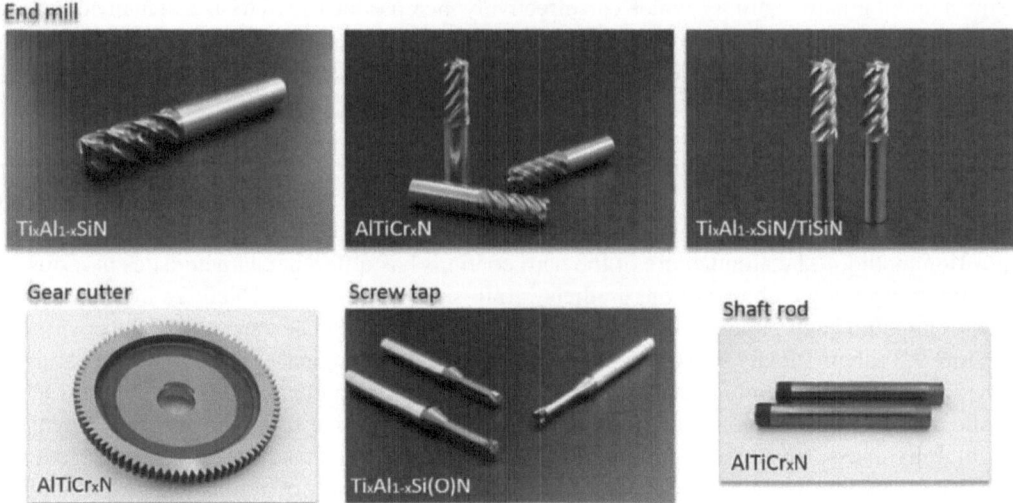

FIGURE 9.4 Typical application products of the multicomponent hard coatings: specific end mills coated with $Ti_xAl_{1-x}SiN$, $AlTiCr_xN$, and $Ti_xAl_{1-x}SiN/TiSiN$; $AlTiCr_xN$ coated gear cutters; $Ti_xAl_{1-x}Si(O)N$ coated screw taps; and $AlTiCr_xN$ coated shaft rods (Courtesy of Surftech Co., Taiwan).

fields. Even the typical TiN, the structure and properties of TiN coatings strongly depends on the degree to which the process parameters can be controlled. The mechanical property and biocompatibility of Ti_xN_y coated Ti alloys strongly depends on nitrogen content of the film [39]. Nowadays, the mechanical and biomedical components and tools are coated with nanostructured or nanocomposite coatings even high-entropy alloy (HEA) coatings with gradient, duplex, multilayered, and nanocomposite architectures because of their improved performance characteristics and longer durability to severe application environment and diverse application possibilities in manufacturing, transportation, aerospace, and biomedical fields. Table 9.3 shows the recent typical nanostructured and nanocomposite coatings for mechanical applications. The research team of Prof. Duh fabricated CrMoN/SiNx multilayered coatings using radio frequency magnetron sputtering. CrMoN/SiNx multilayer coatings possessed significant improvement in tribological characterizations at high temperatures by architecture modification [40]. Chang et al. [11] studied the mechanical properties of AlTiCrN and multilayered AlTiCrN/TiSiN coatings with different interlayers. The single-layer AlTiCrN with Cr interlayer showed the highest residual stress, while the multilayered AlTiCrN/TiSiN coating possessed the lowest residual stress. Impact fatigue tests for the AlTiCrN series and multilayered AlTiCrN/TiSiN coated tools were conducted. The AlTiCrN/TiN, AlTiCrN/CrN, and multilayered AlTiCrN/TiSiN coated samples possessed good impact fatigue performance [11]. By adding Si and B, AlTiSiN and AlTiBN coatings had higher hardness and lower wear rate than AlTiN. The formation of dense and fine columnar structures in the B- and Si-added AlTiSiN and AlTiBN coatings influenced the mechanical behavior of the coatings. AlTiBN and AlTiSiN possessed higher hardnesses of 35 ± 1.2 GPa and 38 ± 1.1 GPa, respectively. The wear resistance was significantly improved for the AlTiBN after oxidation post-treatment [41].

The amorphous thin film metallic glass (TFMG) is also a promising coating material having superior mechanical performance, fatigue properties, corrosion resistance, unique thermal properties, and antimicrobial efficacy. Chu et al. [42] had studied and reviewed both properties and applications of TFMGs. Potential applications of Zr-based TFMGs can be extended to medical tools such as surgical blades and microsurgery scissors. They also use $Zr_{53}Cu_{33}Al_9Ta_5$ TFMG for the coating of syringe needles and compares the results with those obtained using titanium nitride and pure titanium coatings. The $Zr_{53}Cu_{33}Al_9Ta_5$ TFMG coatings possessed non-stick feature reducing

TABLE 9.3

Recent typical nanostructured and nanocomposite hard coatings for mechanical applications

Coating type	Hardness (GPa)	Coating characteristics	Example of applications	Ref.
Multilayered CrMoN/SiNx	23–27	Lowest COF (0.22) and wear rate were obtained under 600 °C, showing noticeable improvement as compared to CrMoN. Significant improvement in tribological characterizations by architecture modification.	Anti-wearing under high temperature condition.	[40]
AlTiSiN, AlTiBN	35–38	Good thermal stability and wear resistance while minimizing abrasion wear.	Metal cutting and forming.	[41]
Multilayered AlTiCrN/TiSiN	34	Low residual stress and good adhesion strength. Possess good impact fatigue performance, which can be effective to reduce damage and cracking of hard coatings during cyclic impact loading.	Metal forming and piercing.	[11]
Zr-based TFMGs	6–11	Good adhesion to steel with better ductility than ceramic hard coating films. Non-stick feature reducing insertion forces by ~66% and retraction forces by ~72% tested using polyurethane rubber block.	Surgical blades, syringe needles and microsurgery scissors.	[42–44]
ta-C	70–90	High chemical inertness, biocompatibility, and wear resistance.	Precision cutting in medical surgery, dental burr, carbon fiber-reinforced composites machining applications. Diverse applications in automotive sector and recording media.	[52]
CrAlSiN	37	Superior friction and wear behavior at high temperatures. The overall better mechanical properties of the CrAlSiN hard coating are manifested in better tribological properties compared to the CrAlN coating.	Turning and milling tools.	[53]
AlCrTiSiN	41	The AlCrTiSiN-coated cutter had a longer service life (increased by approximately 50% over an AlCrN coating).	HSS cutting tools.	[54]
Ti-TiN-(Ti,Cr,Al)N, Zr-ZrN-(Zr,Al,Si)N	32–34	Multilayer composite nanostructured coatings possessed higher tool life by 4–4.5 times than commercial TiN and ZrN for milling Ti alloy.	Cutting tools for Ti alloy.	[55]
TiAlSiN/TiSiN/TiAlN	36–38	Superior anti-adhesive and anti-abrasive behavior when compared to TiAlN layer. Improved tool life for cutting hard AISI 52100 steel.	Ceramic cutting tools.	[56]
TiSiCN	45	Designed and prepared under the guidance of CVD phase diagrams. Superior cutting performance for continuous wet turning of nodular cast iron.	Wet turning and dry milling tools.	[57]

insertion forces by ~66% and retraction forces by ~72% tested using polyurethane rubber block [43]. Nitrogen can also be added into the TFMGs to improve the mechanical properties. Lee et al. [44] developed a Zr-Cu-Al-Ag-N TFMG coating for improved mechanical, corrosion, and anti-microbial property for biomedical applications. With a minor Ag content, the antimicrobial rate is over 99.999%.

HEA coatings in the form of thin or thick films on substrates have been explored early in the 2000s by Yeh et al. [45] even when bulk HEAs were found to be synthesizable and have promising properties. In 2019, Tüten et al. [46] fabricated TiTaHfNbZr HEA films on Ti6Al4V alloy by RF magnetron sputtering. Homogenous surface topography and a fine grained amorphous structure was obtained, and the HEA coatings with good tribological properties can serve as an effective protection against wear and cracking especially for long-term orthopedic implants. Hahn et al. [47] showed that the high-entropy nitride (Al,Ta,Ti,V,Zr)N coating exhibited a hardness of ~30 GPa and a fracture toughness of 2.4 MPa$\sqrt{}$m. Alloying this high-entropy nitride with ~5 at% Si does not influence the hardness and fracture toughness but lowers significantly the elastic response leading to a significantly improved damage tolerance. Kirnbauer et al. [48] had studied the thermal stability of single-phase crystalline (Al,Cr,Nb,Ta,Ti)O$_2$ high-entropy oxide coatings synthesized by reactive magnetron sputtering using Al–Cr–Nb–Ta–Ti-compound target. The coating stayed single-phase crystalline after annealing up to 1200 °C. The HEA coatings can also be designed in a multilayer structure. Typical example is (TiZrNbHfTa)N/WN multilayer coatings deposited by vacuum arc evaporation. Good mechanical properties allow to use (TiZrNbHfTa)N/WN multilayered coatings as a protective materials [49].

Usually, ternary TiAlN coatings attracted considerable industrial interest because of their excellent tribological performance and high oxidation resistance at high temperatures. Recently, multi-component CrAlSiN and TiAlSiN coatings have been developed in order to gain high hardness (>38 GPa) and good thermal stability at temperatures exceeding 800 °C. These coatings are supplied by some coating equipment companies such as Platit, Kobe steel, and Cemecon Co. and they are specially used for high speed and dry cutting of hardened mold steels. The glass-to-metallic mold sticking is a major problem for industrial glass forming processes. Chang et al. [50] exhibited that the CrAlSiN showed a low oxidation rate and a non-wetting characteristic superior to TiAlSiN and AlTiN coatings. To counter the complex stress profile of cutting and forming tools, coating process combinations of thermal spraying and PVD maybe enable to improve coating properties, that is, thermal insulation as well as wear protection, for instance, which cannot be obtained by the sole coating processes. Tillmann et al. [51] applied ceramic Al$_2$O$_3$ thermal barrier coatings on AISI H11 tool steel and subsequently polished to serve as a substrate for TiAlSiN coatings. Titanium interlayers contribute to good adhesion of TiAlSiN on steel and alumina as Ti compensates compressive and tensile stresses of the substrates. The dense thermal barrier coatings surface is crucial for the adhesion and hardness of TiAlSiN.

In addition, DLC (especially the tetragonal amorphous carbon, ta-C) coatings have also maintained a high level interest for numerous industrial applications where efficiency, performance, and reliability are of great importance. The strong covalent bonding and high sp^3 content in ta-C coatings result in high mechanical hardness, stiffness, and chemical and thermal stability that make them well-suited for harsh tribological conditions [52].

In recent years, a broad growing market of coated cutting tools has been developed, and the hard coatings of cutting tools were driven by the demand on the increased usage of difficult-to-cut materials. In the case of cutting difficult-to-cut materials of Inconel super alloy, a nickel-based alloy which is difficult to shape because of its high tribological and thermal properties. Erosion of the cutting edge occurs due to diffusion wear after abrasive wear. Decrease of cutting force and limitation built-up edge (BUE) phenomenon are necessary. AlTiN coating possessing high temperature oxidation resistance shows good tribological behavior and it prevents sticking phenomena on rake face of the cutting tool. Microabrasive blasting can be adopted as a surface modification as well as the AlTiN coating treatment technique [58]. Sanchette et al. [59] showed that a nanolayered TiN/AlTiN coating is a proper coating design for cutting Inconel alloys. This coating induces the lowest cutting force on the tool and limits adhesive BUE phenomenon. This coating is also very effective for avoiding strong abrasive wear due to high hardness and toughness of the coating material.

Adaptive tribological coatings were recently developed as new smart materials that were designed to adjust their surface chemical composition and structure as a function of changes in the working environment to minimize friction coefficient and wear between contact surfaces. Previous studied the design and fabrication of nitride-based smart coatings for tribological applications using two-phase and three-phase materials based on a postulate that the material would demonstrate different adaptation mechanisms as a function of the working temperature. Typically using a hard load-bearing transition metal nitride and a soft lubricious material. The concept is the soft metal inclusions diffuse to the surface when operated and heated to high temperatures, for example, > 250 °C, and provide lubrication to the contact surfaces at high temperatures. The most widely used soft lubricious materials in industry are noble metals (Au, Ag, Pt, and Cu), graphite, boron nitride, and MoS_2. An example of $Mo_2N/MoS_2/Ag$ three-phase coating system was the nitride-based "chameleon" coatings based on a postulate that the material would demonstrate adaptation mechanisms as a function of the working temperature. At high temperatures (e.g., > 500 °C), lubricious oxide phases were expected to form at the surface to improve the tribological performance during severe operating environment [60, 61].

In order to effectively design the coating systems for the specific applications, the chemical, mechanical and tribological properties of the coatings have been widely studied. However, the complicated connections between the tribological and mechanical behaviors could occur during cutting process, such as the cyclic mechanical impact and thermal impact on coated tools under interrupted cutting (e.g., end milling). An engineer tries to find some innovative characterization methods before field application of the tools, and this allowed the adaption of cutting conditions to the coating properties. Zha et al. [62] had studied the correlation of the fatigue impact resistance of TiSiN/TiAlN bilayer coating and TiSiN/TiAlN nanolayered coating with their cutting performance in machining Ti alloy, which is also a difficult-to-cut material. They found that nearly circular shape cracks occurred due to the changed direction of crack propagation in TiSiN/TiAlN nanolayered coating during the cyclic impact tests with the low impact force. This can be contributed to improve stress concentration resistance of the nanolayered TiSiN/TiAlN coating. In the cutting test of Ti alloy, the TiSiN/TiAlN nanolayered coated tool showed excellent resistance to the sliding friction, fatigue impact, and stress concentration under the low feed rate cutting condition. This could be associated with the improved performance to resist fatigue impact and stress concentration at the low impact load. CrN is widely replaced by the ternary compound CrAlN. For these coatings, intrinsic hardening as well as solid solution hardening effect played an important role. The addition of Al and Si into CrN to form CrAlSiN results in coatings with superior properties compared to those of CrAlN and of CrSiN with regard to thermal stability and hardness. Gradient and multilayered coatings composed of different transition nitride layers show superior mechanical strength, such as hardness, adhesion, and wear resistance, as compared to monolayered coatings due to their specific interfaces. TiVN-based coatings are known to show a combination of high hardness and toughness and the possibility to form V_2O_5, a low friction oxide phase, promoting low friction coefficient in dry sliding contacts and consequently these coatings are of interest for tribo-systems. Chang et al. [63] showed that the design of multilayered CrAlSiN/TiVN coating exhibited superior impact fatigue resistance (higher than 4×10^5 impacts) at room temperature and 500 °C as well as good oxidation resistance compared to single-layer TiVN and CrAlSiN. AlCrN/TiAlSiN multilayer coatings have also drawn great attention due to combining both high hardness of TiAlSiN coating and good oxidation resistance of AlCrN coating. This AlCrN/TiAlSiN multilayer coatings showed improved thermal stability and oxidation resistance for high temperature tribological applications [64]. Mo can be added into AlCrN to form AlCrN/AlCrMoN multilayer coatings to improve the mechanical properties with good adhesion strength, which could be attributed to interfacial strengthening induced by the multilayer structure. The AlCrN/AlCrMoN coating can exhibit low friction coefficient and wear rate at room temperature and high temperature due to the optimal structure [65].

When hard coatings are applied to mechanical and tooling applications, a comprehensive investigation of hard coatings deposited by PVD requires a precise knowledge of their plastic behavior. Nanoindentation is a commonly applied method to determine the hardness and the elastic modulus of PVD coatings. Determination of flow curve of PVD thin coatings using a combination of nanoindentation and finite element (FE) simulation is a main subject nowadays. Bobzin et al. [66] studied the plastic behavior of the studied coating systems by combining the simulated flow curves and the results of the analysis of indentation imprints. Their results showed a higher resistance of the nanostructured CrN/AlN-multilayer coating against plastic deformation was obtained compared to single-layer CrN and AlN. For industry, downtime due to tool substitution and recalibration is one of the largest contributors to the final part cost, making an increase in the tool life a priority. The coating and the substrate constitute a system, the application performance of which depends on material parameters, such as hardness, Young's modulus, and the interface material properties between layer and substrate, and yield strength of the substrate. Failure under tribological situations can occur not only through coating wear but also by chipping of the coating from the substrate caused by adhesive failure and high plastic deformation of the substrate. Ojos et al. [67] used an analytical approach for the simple prediction of hard-coating failure for tooling systems which correlated to finite element analyses for ceramic coatings. The failure map was investigated experimentally. The prediction simulation model can be used for hard-coating and substrate property optimization. In order to design a proper coating/tool system for a specific tooling application, based on the model inputs, the predictive simulation method could be useful for design justification and coating optimization, and it is more and more important for coating selection and design.

9.3 BIOCOMPATIBILITIES AND ANTIBACTERIAL PROPERTIES OF NANOSTRUCTURED COATINGS FOR BIOMEDICAL APPLICATIONS

In recent years, more and more materials containing both biocompatibilities and antibacterial properties received attention due to the expansion of the use of medical devices or implants. Generally, biomedical materials with biocompatibilities are not guaranteed to have antibacterial properties. Why is it important for medical devices or implants to have both biocompatibilities and antibacterial properties? When microbial infections occur in medical devices or implants, this effect may lead to possible sepsis infection in patients. If bacteria invade the medical devices or implants, the contaminated surface of the medical devices or implants will form bacterial biofilms, which will seriously damage the function and performance of the implant itself, gather inflammatory cells affecting integration between the surrounding tissues and implant surface. The most serious condition might cause the patient to suffer systemic bacterial infection. It should be mentioned here that once a layer of mature bacterial biofilms is formed on the surface, the usual effect of antibiotic treatment of conventional medical therapies is not significant [68], resulting in the removal of medical implants as the only option to eradicate infections.

Taking dental implants as an example, dental implants usually consist of three different components: (1) fixture, which can stably anchor the implant to the alveolar bone of the upper or lower jaw; (2) abutment, connect the fixture to the artificial crown across the gingival tissue; (3) artificial tooth structure composed of crown, dental bridge, or denture [69]. The oral cavity is an open environment, but also an environment full of bacteria. Tens of millions of dental implants are put into the patient's mouth every year. This means that every day, a large number of people carry dental implants in contact with food and bacteria in the mouth and may face the risk of peri-implantitis [70] which is one of the major peri-implant disease.

The predecessor of peri-implantitis is peri-implant mucositis, and peri-implant mucositis is a type that only involves Inflammatory diseases of soft gum tissue, usually manifested as exudate, swelling, and/or bleeding [70], but if the bacterial inflammation becomes more severe, it can enter the stage of peri-implantitis, leading to a more destructive inflammatory process, resulting in the formation of the pocket around the implant and the loss of the supporting bone around the dental implant [71]. Infections of peri-implantitis are likely to cause bacteremia and the risk of distant transfer of infection, especially in the presence of other implanted devices (such as heart valves, orthopedic implants, and prostheses). Therefore, the impact of infection around the implant is not only local, but also a clinical problem that needs to be addressed seriously.

Coating a nanoscale layer of metal or composite films with both biocompatibilities and antibacterial properties on medical devices or implants is a good idea that can solve the above problems. Coatings with nanostructured characteristics can exploit the different antimicrobial mechanisms of nanomaterials if the synthesis methods are able to tune the composition, morphology, and mechanical properties of the coatings [72]. Over the years, many related studies have explored this topic. In the field of artificial orthopedics and dental implants, titanium (Ti) and titanium alloys have been widely used as metal biomaterials and have become the gold standard among currently available biomaterials due to good mechanical properties and biocompatibility [73]. The biocompatibility of Ti is attributed to the presence of natural or artificial titanium dioxide (TiO_2) surface layer [74]. Although Ti has excellent biomechanical properties, any Ti-based implants inserted into biological tissues may cause bacterial infections on and around the surface of biological materials. It is important to improve the antibacterial properties of the surface of Ti or other biocompatible metal implants and reduce the possibility of infection during and after surgery. Therefore, how to make the surface of Ti or other biocompatible metal implants have antibacterial properties is considered to be one of the key factors that may achieve the osseointegration of the implants. For example, MgO, Graphene Oxide (GO)-Ag, Zn, Ag, and Cu ions have been used to develop antibacterial coating [75]. Introducing GO could increase the mechanical properties while nano-Ag could impart antibacterial function for polymer scaffolds. Nanoscale GO-Ag system showed a synergistic reinforcing effect for the polymer scaffolds. The increasing demand for titanium and its alloys used for implants results in the need for innovative surface treatments that may both increase corrosion resistance and biocompatibility and demonstrate antibacterial protection at no cytotoxicity. In addition to MAO, Ossowska et al. [76] designed two-stage anodization to form oxide coatings on the Ti–13Nb–13Zr alloy. The two-stage electrochemical oxidation could result in the bi-layer oxide coating in which the alloy was subjected to gaseous oxidation and then to the electrochemical oxidation. This technique resulted in an improvement of mechanical properties and corrosion resistance. No significant effects on biologic properties were observed.

Table 9.4 shows the typical nanostructured and nanocomposite coatings for biomedical applications. Figure 9.5 shows typical application products of the multicomponent biomedical coatings for dental implants. Major components of dental implants include implant fixture, abutment, and abutment screw. Antibacterial and cell biocompatible performances are necessary. Adding Ag into a nanoscale layer of Ti or other biocompatible metal coatings has already showed both biocompatibilities and antibacterial properties [6, 7, 25, 77, 78]. Silver ions significantly inhibit the growth of bacteria and are deposited in vacuoles and cell walls in the form of particles [79]. They constrain cell division and extinguish the cell membrane and bacterial content [80], resulting in an increase in the size of bacterial cells, and the cytoplasmic membrane, cytoplasmic content, and outer cell layer of bacterial all showed structural abnormalities. Finally, the interaction of silver ions with nucleic acids leads to the death of bacteria. Ag in the form of nanoparticles in the composite coating can anchor to the bacterial cell wall and consequently infiltrate it. This action will cause physical changes in the bacterial membrane, like the membrane damage, which can lead to cellular contents leakage and bacterial death. In addition, the positive charge confers electrostatic attraction between Ag nanoparticles and negatively charged cell membrane of the microorganisms, thereby facilitating Ag attachment onto cell membranes [81, 82].

In 2002, Chang et al. [6] have embedded the Ag nanoparticles into the TiO_2 film on the Ti-based plates, the elemental contents of Ag of TiO_2 coatings were 1.5, 2.3, 4.2, and 5.9 at.%. In this study, *Streptococcus mutans* (ATCC 31383, Bioresource Collection and Research Centre, Hsinchu City,

TABLE 9.4

Recent typical nanostructured and nanocomposite coatings for biomedical applications

Coating type	Coating characteristics	Example of applications	Ref.
TiO$_2$/Ag	High antibacterial properties, and TiO$_2$/Ag-coated (5.9% Ag) after 72 h of cell adhesion were similar to the uncoated Ti.	Metallic implants / Medical devices.	[6]
TaN-Ag	Nice antibacterial performance and compatible biological response.	Metallic implants.	[25]
Ta$_2$O$_5$-Ag	Exhibit improved antibacterial effects against *S. aureus* and have good skin fibroblast cell cellular biocompatibility.	Metallic implants / Medical devices.	[77]
ZrO$_2$-Ag	Good biological compatibility and antibacterial effects on *S. aureus* and *A. actinomycetemcomitans*.	Metallic implants / Medical devices.	[78]
ZnO	High antibacterial properties and low biocompatibility.	Medical devices.	[6]
Co doped ZnO	Exhibited good antimicrobial activity against colonization and formation of biofilms, and proved to be suitable support for mesenchymal cellular adhesion and proliferation.	Metallic implants.	[83]
Ti-Zn-O	With a lower content of Zn (7.6 ± 1.3 at.%), Ti-Zn-O coating not only provide antibacterial properties, but maintain the biocompatibility to cells.	Metallic implants.	[84]
Ta(Zn)O	High power impulse magnetron sputtering (HiPIMS) was used for the deposition on Ti with rough and porous surface, which was pre-treated by plasma electrolytic oxidation (PEO). The Ta(Zn)O exhibited improved antibacterial abilities against *S. aureus* and *A. actinomycetemcomitans*, and showed good cell viability and low cytotoxicity in HSF cells.	Metallic implants.	[85]
Ti-V-O	Good antibacterial performance but low cell viability.	Medical devices.	[86]
PEO-pre-treated Ta$_2$O$_5$	Both nice biocompatibility and antibacterial abilities.	Metallic implants.	[23]
ZrCN/ amorphous carbon	ZrCN/a-C coatings with carbon content higher than 12.7 at.% promote antibacterial performance with good cell compatibility.	Metallic implants.	[87]
Ti-Zr alloy	Ti-Zr alloy films deposited onto Ti substrates with MAO. Cells onto Ti-Zr films exhibited notable osteogenic genes expression.	Metallic implants.	[88]
ZrO$_2$	Desirable biocompatibility for cell adhesion, proliferation, and viability of the mesenchymal stem cells and osteoblast cells.	Metallic implants.	[89]
Y-doped TiO$_2$	Released Yttrium damaged the bacterial cytoplasmic membrane, with improving implant antibacterial properties.	Bone tissues in dentistry and orthopedics.	[90]
Mn-incorporated TiO$_2$	Mn incorporation enhances cellular response and corrosion resistance of TiO$_2$. The antibacterial capability is mainly ascribed to Mn$_3$O$_4$.	Metallic implants.	[91]
TaC and TaC/a-C	TaC/a-C coating, which contained two metastable phases (TaC and a-C), was more biocompatible with MG-63 cells.	Metallic implants.	[92]
Ta$_2$O$_5$	Good antibacterial performance and cellular biocompatibility.	Metallic implants.	[93]
Ag/Ag-Ta$_2$O$_5$	Ag segregation process, hydrophobicity, adhesion strength, crystallization, and hardness progressively improved after the annealing up to 400 °C.	Metallic implants.	[94]
Diamond-like carbon (DLC)	Possessed good chemical resistance against strong acidity as well as sterilization resistance.	Medical devices.	[95]
Zr-Cu-Al-Ta thin film metallic glass (TFMG)	TFMG reduced surface roughness than Ti coating and reduced the attachment of cancer cells by up to ~87%.	Medical instruments preventing the adhesion of platelet and cancer cells.	[96]

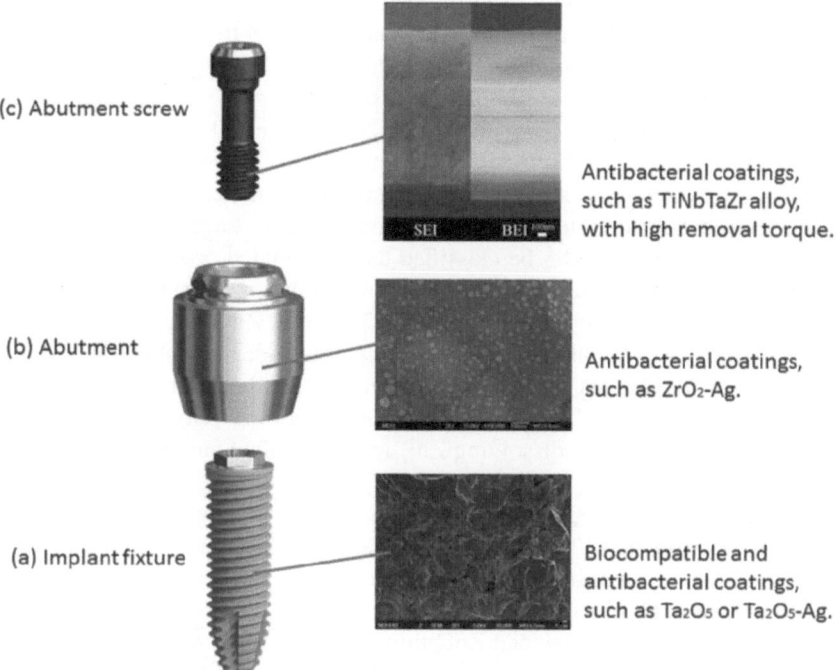

(c) Abutment screw

Antibacterial coatings, such as TiNbTaZr alloy, with high removal torque.

(b) Abutment

Antibacterial coatings, such as ZrO₂-Ag.

(a) Implant fixture

Biocompatible and antibacterial coatings, such as Ta₂O₅ or Ta₂O₅-Ag.

FIGURE 9.5 Major components of a dental implant: (a) implant fixture; (b) abutment; and (c) abutment screw with biocompatible and antibacterial coatings.

Taiwan) was used for antibacterial testing. For the cell viability exam, the human gingival fibroblast (HGF) cells which were obtained from patients during oral surgery extractions, and the process was approved by the Ethics Committee of China Medical University Hospital were exercised. For the results, they found that the amount of Ag in the TiO_2/Ag compound coating showed an outcome on the inhibition of bacterial growth, especially for the 4.2, and 5.9 at.%. of Ag contents of TiO_2/Ag coatings. However, the cell compatibility of all TiO_2/Ag-coated samples after 72 h of HGF cells cultured were similar to the value of the uncoated Ti.

In 2010 and 2014, Huang et al. [25, 77] used different Ag cathode powers to produce TaN-Ag and Ta_2O_5-Ag films containing different levels of Ag nanoparticles. These studies used *Staphylococcus aureus* which are considered to be one of the causes of implant-associated infections and peri-implantitis [97] for antibacterial testing. The biocompatibility of HGF cells or human skin fibroblast cell line (CCD-966SK) was examined with a 3-(4,5-Dimethylthiazol-2-yl)-2,5-diphenyltetrazolium bromide (MTT) assay for assessing cell viability on those coatings. For the results of TaN-Ag films, with the highest Ag content of 21.4 at.% TaN-Ag films can presented the lowest bacterial retention. The antibacterial performances of Ta_2O_5 and Ta_2O_5-Ag films were also superior to that of the uncoated and Ta-coated samples. However, the performance of cell viability of Ta_2O_5-Ag coatings was significantly lower than Ta_2O_5 and showed the obviously cytotoxic for cells. It looks like that high contents of Ag nanoparticles are still harmful to cells. In 2013 the samples of ZrO_2 coating and 3 types of ZrO_2-Ag coatings deposited on pure-Ti plate were produced by Huang et al. [78]. The Ag contents of ZrO_2-Ag films were 3.1, 10.6 and 17.8 at.%. In this study, not only the *S. aureus* (Gram-positive bacteria) but also *Actinobacillus actinomycetemcomitans* (Gram-negative bacteria) were used to examine antibacterial effect of ZrO_2-Ag coatings. Additionally, not only the cytocompatibility and adhesive morphology, but mRNA expression of HGF cells on the coatings were also determined. The results of bacterial retention were lower for ZrO_2-Ag coatings than the uncoated

Ti surfaces. It indicated that because of Ag the ZrO_2-Ag films showed lower bacterial retention than the uncoated Ti surfaces. Strategically adding nanoparticle Ag in to ZrO_2 to form ZrO_2-Ag coatings inhibited both Gram-positive and Gram-negative bacteria significantly. ZrO_2-Ag coatings with Ag at less than 10.6 at. % also provided the comparable results to improve the cell viability and proliferation. For the mRNA expression of HGF cells, it also showed that 3.1% of Ag of ZrO_2-Ag films even maintain the functions of type I and type III collagens in cells.

Zinc oxide (ZnO) is one of the most widely used inorganic materials, commonly used in photocatalysts, optoelectronic devices, textiles, pigments, antibacterial agents, cosmetics, etc. [98]. The antibacterial mechanism of ZnO can be classified as light-related or light-independent. Similar to TiO_2, ZnO has photocatalytic activity under ultraviolet radiation to generate reactive oxygen species (ROS) through photocatalytic water splitting. Therefore, ZnO has been widely studied as an antibacterial agent [99]. For the antibacterial effect of light-independent of ZnO, it was found that the surface of ZnO nanoparticles can generate oxygen free radicals or hydrogen peroxide (H_2O_2) through oxygen defect sites, which may induce the antibacterial activity of ZnO nanoparticles under dark conditions [100]. In addition, dissolving Zn^{2+} ions from ZnO nanoparticles is another possible antibacterial activity mechanism [101].

In 2012, Chang et al. [6] coated ZnO nanorods on Ti plate, and they found that with the increase of the amount of electroplated ZnO nanorods, the antibacterial capacity of ZnO-coated samples was raised. However, biocompatibility was also low on ZnO coated samples than uncoated Ti plate. Voicu et al. [83] doped ZnO with different concentrations of cobalt (Co) and obtained a dense layer of microstructure. This kind of Co doped ZnO thin films deposited by spin coating enhanced antimicrobial abilities against colonization and biofilm formation. In addition, it also showed biocompatibilities for example on cellular adhesion and proliferation. In 2013, Tsai et al. [84] produced a porous Zn-doped TiO_2 layer on Ti plate by cathodic arc deposition process. The results indicated that with 7.6 at. % of Zn, the composite $Ti(Zn)O_2$ coating not only provided antibacterial outcome, but also preserved the biocompatibility to bone cells. With different coating technologies by adopting a first plasma electrolytic oxidation (PEO) of surface modification and then a HiPIMS coating deposition. Recently, techniques have been rapidly developed to produce smart multifunctional nanomaterials by applying observations from nature materials which may be called "bio-inspired nanostructures". Sun et al. [102] fabricated fish-scale bio-inspired multifunctional ZnO nanostructures, and the nanostructured coatings showed tunable light refraction and reflection, modulated surface wettability and damage-tolerant mechanical properties for various applications, such as optical coatings, sensing or lens arrays for use in reflective displays, packing, advertising and solar energy harvesting; self-cleaning surfaces, including anti-smudge, anti-fouling and anti-fogging, and self-sterilizing surfaces; and mechanical/chemical barrier coatings. In 2020, Huang et al. [85] synthesized Ta(Zn)O films contained amorphous tantalum oxide and crystalline ZnO. The Ta(Zn)O films demonstrated a nice antibacterial effect on both Gram-positive and Gram-negative bacteria *S. aureus* and *A. actinomycetemcomitans*. Although Ta(Zn)O coating presented lower level of the cell viability in MG-63 bone cells, the performances of cell attachment and cell growth of human skin fibroblasts on Ta(Zn)O coating were good.

Vanadium (V) is a common metal element in titanium alloys. Especially for Ti-6Al-4V, it has been currently used as a material in 20–30% of medical instruments [103] as well as orthopedic and orthodontic implants. Although V has been proven to cause cytotoxic effects for tissues and aluminum is related to potential neurological diseases, the use of vanadium to produce additional oxide films, for example, Ti-V-O [86] as surface coatings still showed a good effect on inhibiting the growth of bacteria. It might be potential for the application of medical implants specifically requiring antibacterial requirements. In 2020, Suma et al. [104] synthesized nano-V_2O_5 with a plate like morphology and high surface area to have efficient antibacterial activity against *S. aureus* and *Escherichia coli*. They also used Raman spectroscopy to act as an excellent tool in studying the difference between intact and disrupted bacteria with reproducible molecular signatures used for analytical species identification based on the difference in outer membrane layers of bacterial species. ZnO nanoflowers decorated vanadium pentoxide (V_2O_5) nanowires heterojunction (ZVH) was fabricated using a facile water-bathing method by Sun et al. [105]. It demonstrated efficiently improved antibacterial activities against *S. aureus* than pure ZnO and pure V_2O_5 under light and dark conditions. Kayani et al. [106] also proved that the V doped ZnO coatings possessed antibacterial activity against *P. aeruginosa* and *S. aureus*.

The characteristics of various surface coatings is considered as an essential issue on the biomedical implants and devices due to the physical, chemical, biochemical, and biological interactions with the live bodies. It is a critical and huge issue to discuss the effects of the compositions, chemical bonds, microstructures, corrosion, and mechanical strength of biomaterials on the material–cell interactions. Once some parameter has been changed or regulated, the biocompatible effect of the material itself may be totally unexpected, such as inducing high level of wear particles or macrophages, inflammation, infection, cell necrosis, and the crash of live system. There is one thing for sure is the most important characteristic of biomaterials is definitely the biocompatibility. Ti, tantalum (Ta), zirconium (Zr), nitro (N), and carbon (C) have been addressed attentions as biomaterials applied to the orthopedic and dental implants or medical devices.

Ti-based materials have been used as biomedical implants in the orthopedics, dentistry, and vessel circulation for decades. Pure Ti and Ti alloys possess excellent mechanical strength and high corrosion resistance. The biocompatible performance of Ti is also superior. A variety of studies exhibit the performance of Ti-based substrate with different surface coatings or thin films to the live cells. In 2019, Ramos-Corella et al. demonstrate TiO_2 thin films with controlled crystalline phases deposited by radio frequency magnetron sputtering using different gas pressure, oxygen partial pressure, and magnetron power. And the thermal annealing was used to control the microstructure and crystalline phase of these thin films. These results indicate that plasma power is a critical parameter to control TiO_2 crystalline phase. MTT assay shows no significant difference in the cell proliferation of human pigment epithelium cells (ARPE-19) among TiO_2 crystalline phases, indicating that the biocompatibility might be mostly defined by the surface chemistry of the films [107]. In 2017, a temperature-controlled atomic layer deposition (ALD) method was used to provide nanostructured TiO_2 coatings on Ti substrates by Liu et al. In vitro studies show that the osteoblast adhesion and proliferation elevate when cells cultured onto the coatings. However, the performance of fibroblast adhesion and proliferation is suppressed. It also suggests that ALD-grown TiO_2-coated samples with an increase of nanoscale roughness and moderate surface hydrophilicity (surface energy) could contribute to increased protein adsorption selectively, as well as showed relatively promising antibacterial properties and desirable cellular functions [108]. In 2019, Saleem et al. report that the deposition of titania–alumina composite films by using pulsed DC magnetron sputtering system for various plasma excitation powers (100–200 W) to enhance the surface roughness and mechanical properties. The biocompatibility of deposited films was studied by culturing the MC3T3-E1 cells for 3 days. It shows a better performance of cell growth and proliferation (elongated morphology) on film prepared at 150 W [109].

Zirconium and its alloys have been used for centuries in a wide variety of ways. Recently, it suggests that zirconia (ZrO_2) performs favorably in both orthopedic and dental applications. The use of zirconia in the surface coating of biomedical implants has rapidly expanded over the past decade, driven by its superior mechanical properties, especially the highly resistant to corrosion. In 2011, a nanocomposite ZrCN/amorphous carbon (a-C) coatings with different carbon contents deposited on a biograde pure Ti implant material is produced by a cathodic-arc evaporation system with plasma enhanced duct equipment by Lai et al. The results suggests that the ZrCN/a-C coatings with carbon content higher than 12.7 at.% can improve antibacterial performance with excellent HGF cell compatibility as well [87]. An interesting work discusses about the morphological surface features, wear resistance, and in vitro-biocompatibility of ZrO_2 thin films deposited by the novel Pulsed Plasma Deposition (PPD) method by Bianchi et al. (2016). Cell adhesion, proliferation, and viability of the mesenchymal stem cells and osteoblast cells indicate a desirable biocompatibility of the nanostructured ZrO_2 thin films [89].

Surface modification of biomaterials can change the surface structure, morphology, and chemistry structure of titanium and its alloys, thereby improving the antibacterial performance or biocompatibility of the materials. The surface treatment of biomedical titanium alloy by MAO, also called PEO, can produce a rough porous oxide layer, which improves the hydrophilicity and increases the wear resistance and corrosion resistance of the titanium [110]. Owing to the high bonding strength and composition-controlled property, the MAO is an effective method to adjust the concentration of antibacterial elements, such as Ag, Zn, and Cu. In addition, the properties of the MAO-treated oxide coatings are mainly affected by the parameters including the applied high voltage, duty cycle, pulse frequency, and oxidization time. The MAO coating containing Zn, Ca, and P elements showed better apatite-inducing ability and antibacterial ability for *E. coli* and *S. aureus* at the same time [111]. In 2016, Tsai et al. also investigate the mechanical properties and biocompatibility of Ti-Zr alloy films deposited onto Ti substrates with MAO. The post-MAO-treated surface layers were characterized and the results show an increase of cell viability of MG-63 and SKF cells was found in MAO-treated Ti-Zr films with higher Zr content. Gene expression data by RT-PCR and agarose gel electrophoresis showed both two types of cells exhibited notable osteogenic gene expression, including Runx2, ALP, Dlx-5, OCN, BMP-2, and BMPR1A. It suggests that the MAO-treated Ti-Zr films may be potential and suitable for bone tissue- or soft tissue-derived cell growth and differentiation [88]. In 2020, Hu et al. [112] used ultrasonic MAO treatment to improve the antibacterial activity of Ti-Cu alloy, and the MAO-treated oxide coating of Ti-Cu does not cause cytotoxicity. The antibacterial ability was mainly controlled by CuO in contact sterilization mode. Rare earth elements, such as Yttrium (Y), doping had long been known as one of the most effective approach to promote photocatalytic performance. Zhang et al. [90] added rare earth elements Y as coating to titanium alloy (Ti-6Al-4V) via MAO. The fabricated Yttrium-doped TiO_2 (Y-TiO_2) coatings enhanced the adhesive, proliferative, spreading of osteoblastic precursor cells increasing bone morphogenetic protein and alkaline phosphatase expression. As for the biological coating applications, TiO_2 doped with Zn^{2+} and Y^{3+} via low cost sol-gel process exhibited better antibacterial effects than undoped TiO_2 [113]. Titanium can also be modified by the deposition of MAO-treated calcium phosphate coatings containing Ag, Sr, and Si elements. Sedelnikova et al. showed bacteriostatic effect against *S. aureus* using Ag-CP coatings. All the Ag containing coatings did not have a cytotoxic effect on fibroblasts [114]. Manganese (Mn), which is an essential trace element for human body, plays diverse roles in bone formation, metabolism of skeletal tissue along with other biological processes. Mn-incorporated TiO_2 coating can be fabricated through the facile one-step MAO treatment in the electrolyte containing Na2Mn-EDTA. Mn-incorporated TiO_2 coating exhibited an enhanced corrosion resistance, osteogenic activity and slightly improved antibacterial capability against *S. aureus* bacteria relative to that with the absence of Mn, which also exhibits negligible cytotoxicity [91].

Ta is a promising metal for biomedical implants or implant coating for orthopedic and dental applications because of its excellent corrosion resistance, fracture toughness, trabecular bone-like microstructure, and excellent biological properties. In 2014, tantalum carbide (TaC) and TaC/amorphous carbon (a-C) coatings with different carbon contents by using a twin-gun magnetron sputtering system are synthesized by Chang et al. [92]. To improve their biological properties, the carbon content in the deposited coatings was regulated by controlling the magnetron power ratio of the pure graphite and Ta cathodes. The results show that the TaC/a-C coating, which contained two metastable phases (TaC and a-C), was more biocompatible with MG-63 cells compared to the pure Ta coating [92]. In 2014, Chang et al. [93] produce a hydrophilic crystalline β-Ta_2O_5 coating deposited by using magnetron sputtering exhibiting great biocompatibility in human skin fibroblast cells (CCD-966SK).

Recently, Alias et al. [94] use SS 316L steel as substrates deposited with a nanocomposite layer of Ag and Ta_2O_5 by using PVD magnetron sputtering. The antibacterial performance on Ag/Ag-Ta_2O_5 at 400 °C indicated a significant zone of inhibition to *S. aureus* and *E. coli* when compared with SS 316L or Ag/Ag-Ta_2O_5 at 700 °C. The biocompatibility tests on Ag/Ag-Ta_2O_5 at 400 °C demonstrated an excellent in cell attachment, F-actin protein expression, and proliferation/viability of bone marrow derived mesenchymal stromal on day 14 when compared with uncoated or $Ag/Ag-Ta_2O_5$ at 700 °C. This study shows that the Ag segregation process, hydrophobicity, adhesion strength, crystallization, and hardness progressively improved after the annealing up to 400 °C [94]. Coating of amorphous calcium phosphate (ACP) on Ti implants is a promising technique for enhancing bone-forming ability. Ag and Ta can also be added into calcium phosphate coatings. In 2020, Wu et al. [115] fabricated Ag and Ta co-doped ACP coating films by RF magnetron sputtering. P, Ca, Ag, and Ta elements were homogeneously distributed. Antibacterial activity was obtained by the Ag ions release through the dissolution of Ag and Ta co-doped ACP coatings. Ag_2O nanoparticles were decorated on the edges of Ta_2O_5 nanotubes through a controlled PVD procedure by Sarraf et al. [116]. An effective antibacterial activity was obtained against *E. coli* via membrane disruption and intracellular leakage. Manipulation of the metallic implant surface by Ag_2O nanoparticles-decorated Ta_2O_5 nanotubes could provide a porous network for the proliferation and preservation of the differentiated human osteoblast cells with high viability and functionality.

An interesting material that may have potential to be applied to the surface coating of biomaterials for clinical use is DLC. It has excellent biocompatibility and is attracting attention as a means of surface modification to improve the antibacterial properties of biomaterials used in medical devices. DLC possesses many advantageous characteristics, such as a low friction coefficient, biocompatibility, and chemical stability. In 2019, Sakurai et al. [95] study the biological characteristics of DLC/SUS samples which were deposited on SUS substrates by the ionized vapor deposition method. The surface smoothness of the sample after immersion in an acidic solution and sterilization treatment prove that DLC film coating is a useful technique for improving the surfaces of medical devices. A fluorine-containing DLC (F-DLC) using an atmospheric pressure plasma surface treatment with O_2 gas can be effective as a surface treatment technology that can provide antibacterial properties [117]. Fluorinated DLC films can also be produced by PECVD by controlling the ratio of carbon tetrafluoride and methane [118]. F-DLC films possessed low stress and surface free energy. The antibacterial tests against *E. coli* showed the increase of DLC bactericidal activity as the increasing of F content. DLC films can be coated on textile material (cotton fibers) using a plasma-based ion implantation (PBII) [119]. The XPS results of the DLC coated on cotton showed that DLC coating formed a C–H bond on the cotton surface, and the prepared DLC coating is a promising method to restrain the increase of active bacteria (*S. aureus* and *K. pneumoniae*).

TFMGs have recently emerged as alternative advanced surfaces for many applications. Due to their microstructural feature and physico-chemical properties, TFMGs appear, for instance, as a rising candidate for coatings of implant and biomedical tools. Zr-based TFMGs appear as promising coatings for biomedical applications due to their microstructural and physicochemical properties. Ternary $Zr_{39}Cu_{39}Ag_{22}$ TFMG was elaborated by PVD magnetron sputtering using a multitarget reactor by Etiemble et al. [120]. It demonstrated that this thin film was amorphous, uniformly thick, and chemically homogenous. These TFMG films displayed high hardness and Young's modulus, as well as an interesting passive behavior and provided a strong biocide effect against *S. aureus*. In addition, utilizing TFMG is confirmed to minimize the adhesion of human cancer cells (breast cancer cell, colon cancer cell, and esophageal cancer cell), human and animal platelets, as reported by Chang et al. in 2018. TFMG and pure Ti are respectively grown on glass substrates to a thickness of 200 nm using magnetron sputtering. It shows that TFMG has the capability of reducing surface roughness of approximately 27% than that of Ti coating on glass samples. The concentrations of all major ions released from the TFMG were well below toxic levels. The application of TFMG to bare surfaces was shown to reduce the attachment area of human platelets by 77% and that of pig platelets by 63%. TFMG also reduced the attachment of cancer cells by up to ~87%. These findings demonstrate the considerable potential of TFMG coatings in the fabrication of medical instruments aimed at preventing the adhesion of platelet and cancer cells [96]. Jabed et al. [121] fabricated $Zr_{46}Ti_{40}Ag_{14}$ and $Zr_{46}Ti_{43}Al_{11}$ coating systems by magnetron sputtering in 2019. The electrochemical analyses of the ternary systems showed lower corrosion- and passive-current density when compared to 316L stainless steel and commercially pure titanium (cp-Ti). The $Zr_{46}Ti_{40}Ag_{14}$ exhibited significant antibacterial properties and reduced the growth of methicillin-resistant *S. aureus*. The combination of such desirable properties makes these TFMGs an excellent candidate material for a wide range of biomedical applications. Additionally, the bactericidal rate of copper (Cu) toward a variety of bacteria can reach more than 90% or even 100%, and it presents a long-term, broad-spectrum and safety during this bactericidal process [122]. In 2018, Cu and Ag were also added to form Zr-Cu-Ag TFMGs using magnetron sputtering by Nkou Bouala et al. [123]. It was clearly demonstrated that antibacterial activity is sensitive to Cu and Ag, while insensitive to Zr. A biocide activity against *E. coli* and *S. aureus* is obtained for an optimized 11 at. % of Ag content. In 2020, Ding et al. [124] used magnetron sputtering technology to prepare copper doped Ta_2O_5 multilayer composite coatings. The results showed that $Cu-Ta_2O_5$ multilayer coating obviously improves the hydrophobicity, corrosion resistance, and *S. aureus* antibacterial property of Ti6Al4V alloys.

9.4 CONCLUDING REMARKS

In this chapter, we try to summarize key recent developments in the production and use of nanostructured and composite coatings especially for their tribological and biomedical applications. At present, several PVD and CVD techniques are available for the deposition of nanostructured and nanocomposite coatings. Pre-treatment and post-treatment of the substrate materials are also combined as duplex treatment to provide proper coating assembly. Unlike conventional coatings, reported laboratory and field application test results show that nanostructured and nanocomposite coatings provide superior friction properties and wear resistance even under harsh tribological conditions, including high contact loads and elevated temperatures in corrosive environments. And under proper material selection and structure architecture, the nanostructured and nanocomposite coatings possess biocompatibilities and antibacterial properties with good mechanical performance at the same time. Overall, nanostructured and nanocomposite coatings hold great promise for applications involving various operating conditions in tribological and biomedical applications.

REFERENCES

[1] Montemor, M. F., Functional and smart coatings for corrosion protection: A review of recent advances. *Surface and Coatings Technology* 2014, 258, 17–37.

[2] Scharf, S.; Noeske, M.; Cavalcanti, W. L.; Schiffels, P., 4 - Multi-functional, self-healing coatings for corrosion protection: materials, design and processing. In *Handbook of Smart Coatings for Materials Protection*, Makhlouf, A. S. H., Ed. Cambridge, UK.: Woodhead Publishing, 2014, 75–104.

[3] Deyab, M. A., Anticorrosion properties of nanocomposites coatings: A critical review. *Journal of Molecular Liquids* 2020, 313, 113533.

[4] Calderon Velasco, S.; Cavaleiro, A.; Carvalho, S., Functional properties of ceramic-Ag nanocomposite coatings produced by magnetron sputtering. *Progress in Materials Science* 2016, 84, 158–191.

[5] Major, L.; Lackner, J. M.; Kot, M.; Major, B., Nanostructural aspects of the wear process of multilayer tribological coatings. *Tribology International* 2020, 151, 106411.

[6] Chang, Y.-Y.; Lai, C.-H.; Hsu, J.-T.; Tang, C.-H.; Liao, W.-C.; Huang, H.-L., Antibacterial properties and human gingival fibroblast cell compatibility of TiO_2/Ag compound coatings and ZnO films on titanium-based material. *Clinical Oral Investigations* 2012, 16, (1), 95–100.

[7] Chang, Y.-Y.; Huang, H.-L.; Chen, Y.-C.; Weng, J.-C.; Lai, C.-H. J. S.; Technology, C., Characterization and antibacterial performance of ZrNO–Ag coatings. *Surface and Coating Technology* 2013, 231, 224–228.

[8] Zhang, P.; Liu, Z.; Du, J.; Su, G.; Zhang, J.; Xu, C., On machinability and surface integrity in subsequent machining of additively-manufactured thick coatings: A review. *Journal of Manufacturing Processes* 2020, 53, 123–143.

[9] Mundotia, R.; Kothari, D. C.; Kale, A.; Mhatre, U.; Date, K.; Thorat, N.; Ghorude, T., Effect of ion bombardment and micro-blasting on the wear resistance properties of hard TiN coatings. *Materials Today: Proceedings* 2020, 26, 603–612.

[10] Voevodin, A. A.; Muratore, C.; Aouadi, S. M., Hard coatings with high temperature adaptive lubrication and contact thermal management: review. *Surface and Coatings Technology* 2014, 257, 247–265.

[11] Chang, Y.-Y.; Yang, Y.-J.; Weng, S.-Y., Effect of interlayer design on the mechanical properties of AlTiCrN and multilayered AlTiCrN/TiSiN hard coatings. *Surface and Coatings Technology* 2020, 389, 125637.

[12] Lin, Y.-W.; Chih, P.-C.; Huang, J.-H., Effect of Ti interlayer thickness on mechanical properties and wear resistance of TiZrN coatings on AISI D2 steel. *Surface and Coatings Technology* 2020, 394, 125690.

[13] Capote, G.; Lugo, D. C.; Gutiérrez, J. M.; Mastrapa, G. C.; Trava-Airoldi, V. J., Effect of amorphous silicon interlayer on the adherence of amorphous hydrogenated carbon coatings deposited on several metallic surfaces. *Surface and Coatings Technology* 2018, 344, 644–655.

[14] Ye, F.; Li, Y.; Sun, X.; Yang, Q.; Kim, C.-Y.; Odeshi, A. G., CVD diamond coating on WC-Co substrate with Al-based interlayer. *Surface and Coatings Technology* 2016, 308, 121–127.

[15] Cloutier, M.; Mantovani, D.; Rosei, F., Antibacterial Coatings: Challenges, Perspectives, and Opportunities. *Trends in Biotechnology* 2015, 33, (11), 637–652.

[16] Ahmed, W.; Zhai, Z.; Gao, C., Adaptive antibacterial biomaterial surfaces and their applications. *Materials Today Bio* 2019, 2, 100017.

[17] Wilson, C. J.; Clegg, R. E.; Leavesley, D. I.; Pearcy, M. J. J. T., Mediation of Biomaterial–Cell Interactions by adsorbed proteins: a review. *Tissue Engineering* 2005, 11, (1–2), 1–18.

[18] Campoccia, D.; Montanaro, L.; Arciola, C. R., A review of the biomaterials technologies for infection-resistant surfaces. *Biomaterials* 2013, 34, (34), 8533–8554.

[19] Chaloupka, K.; Malam, Y.; Seifalian, A. M., Nanosilver as a new generation of nanoproduct in biomedical applications. *Trends in Biotechnology* 2010, 28, (11), 580–588.

[20] Mijnendonckx, K.; Leys, N.; Mahillon, J.; Silver, S.; Van Houdt, R. J. B., Antimicrobial silver: Uses, toxicity and potential for resistance. *Biometals* 2013, 26, (4), 609–621.

[21] Beer, C.; Foldbjerg, R.; Hayashi, Y.; Sutherland, D. S.; Autrup, H., Toxicity of silver nanoparticles— Nanoparticle or silver ion? *Toxicology Letters* 2012, 208, (3), 286–292.

[22] Pang, S.; Gao, Y.; Wang, F.; Wang, Y.; Cao, M.; Zhang, W.; Liang, Y.; Song, M.; Jiang, G., Toxicity of silver nanoparticles on wound healing: A case study of zebrafish fin regeneration model. *Science of the Total Environment* 2020, 717, 137178.

[23] Huang, H.-L.; Tsai, M.-T.; Lin, Y.-J.; Chang, Y.-Y., Antibacterial and biological characteristics of tantalum oxide coated titanium pretreated by plasma electrolytic oxidation. *Thin Solid Films* 2019, 688, 137268.

[24] Mirhadi, S. M.; Hassanzadeh Nemati, N.; Tavangarian, F.; Daliri Joupari, M., Fabrication of hierarchical meso/macroporous TiO2 scaffolds by evaporation-induced self-assembly technique for bone tissue engineering applications. *Materials Characterization* 2018, 144, 35–41.

[25] Huang, H.-L.; Chang, Y.-Y.; Lai, M.-C.; Lin, C.-R.; Lai, C.-H.; Shieh, T.-M., Antibacterial TaN-Ag coatings on titanium dental implants. *Surface and Coatings Technology* 2010, 205, (5), 1636–1641.

[26] Dumée, L. F.; He, L.; Lin, B.; Ailloux, F.-M.; Lemoine, J.-B.; Velleman, L.; She, F.; Duke, M. C.; Orbell, J. D.; Erskine, G. J. J. Hodson, P. D. The fabrication and surface functionalization of porous metal frameworks–A review *Journal of Materials Chemistry A* 2013, 1, (48), 15185–15206.

[27] Tsai, M.-T.; Chang, Y.-Y.; Huang, H.-L.; Chen, Y.-C.; Wang, S.-P.; Lai, C.-H., Biological characteristics of human fetal skin fibroblasts and MG-63 human osteosarcoma cells on tantalum-doped carbon films. *Surface and Coatings Technology* 2014, 245, 16–21.

[28] Taheri, P.; Khajeh-Amiri, A., Antibacterial cotton fabrics via immobilizing silver phosphate nanoparticles onto the chitosan nanofiber coating. *International Journal of Biological Macromolecules* 2020, 158, 282–289.

[29] Sivaraj, D.; Vijayalakshmi, K., Enhanced antibacterial and corrosion resistance properties of Ag substituted hydroxyapatite/functionalized multiwall carbon nanotube nanocomposite coating on 316L stainless steel for biomedical application. *Ultrasonics Sonochemistry* 2019, 59, 104730.

[30] Gunputh, U. F.; Le, H.; Handy, R. D.; Tredwin, C., Anodised TiO2 nanotubes as a scaffold for antibacterial silver nanoparticles on titanium implants. *Materials Science and Engineering: C* 2018, 91, 638–644.

[31] Bag, R.; Panda, A.; Sahoo, A. K.; Kumar, R., Cutting tools characteristics and coating depositions for hard part turning of AISI 4340 martensitic steel: A review study. *Materials Today: Proceedings* 2020, 26, 2073–2078.

[32] Zhu, Y.; Qu, H.; Luo, M.; He, C.; Qu, J., Dry friction and wear properties of several hard coating combinations. *Wear* 2020, 456–457, 203352.

[33] Fox-Rabinovich, G. S.; Kovalev, A. I.; Aguirre, M. H.; Beake, B. D.; Yamamoto, K.; Veldhuis, S. C.; Endrino, J. L.; Wainstein, D. L.; Rashkovskiy, A. Y., Design and performance of AlTiN and TiAlCrN PVD coatings for machining of hard to cut materials. *Surface and Coatings Technology* 2009, 204, (4), 489–496.

[34] Chang, Y.-Y.; Wang, D.-Y.; Chang, C.-H.; Wu, W. J. S.; Technology, C., Tribological analysis of nano-composite diamond-like carbon films deposited by unbalanced magnetron sputtering. *Surface and Coatings Technology*. 2004, 184, (2–3), 349–355.

[35] Tyagi, A.; Walia, R. S.; Murtaza, Q.; Pandey, S. M.; Tyagi, P. K.; Bajaj, B., A critical review of diamond like carbon coating for wear resistance applications. *International Journal of Refractory Metals and Hard Materials* 2019, 78, 107–122.

[36] Grill, A., Review of the tribology of diamond-like carbon. *Wear* 1993, 168, (1), 143–153.

[37] Badaluddin, N. A.; Zamri, W. F. H. W.; Din, M. F. M.; Mohamed, I. F.; Ghani, J. A. J. I., Coatings of cutting tools and their contribution to improve mechanical properties: a brief review. *International Journal of Applied Engineering Research* 2018, 13, (14), 11653–11664.

[38] Donnet, C.; Erdemir, A., Historical developments and new trends in tribological and solid lubricant coatings. *Surface and Coatings Technology* 2004, 180–181, 76–84.

[39] Nemati, A.; Saghafi, M.; Khamseh, S.; Alibakhshi, E.; Zarrintaj, P.; Saeb, M. R., Magnetron-sputtered TixNy thin films applied on titanium-based alloys for biomedical applications: Composition-microstructure-property relationships. *Surface and Coatings Technology* 2018, 349, 251–259.

[40] Yeh-Liu, L.-K.; Hsu, S.-Y.; Chen, P.-Y.; Lee, J.-W.; Duh, J.-G., Improvement of CrMoN/SiNx coatings on mechanical and high temperature Tribological properties through biomimetic laminated structure design. *Surface and Coatings Technology* 2020, 393, 125754.

[41] Chang, Y.-Y.; Cai, M.-C., Mechanical property and tribological performance of AlTiSiN and AlTiBN hard coatings using ternary alloy targets. *Surface and Coatings Technology* 2019, 374, 1120–1127.

[42] Chu, J. P.; Jang, J. S. C.; Huang, J. C.; Chou, H. S.; Yang, Y.; Ye, J. C.; Wang, Y. C.; Lee, J. W.; Liu, F. X.; Liaw, P. K.; Chen, Y. C.; Lee, C. M.; Li, C. L.; Rullyani, C., Thin film metallic glasses: Unique properties and potential applications. *Thin Solid Films* 2012, 520, (16), 5097–5122.

[43] Chu, J. P.; Yu, C.-C.; Tanatsugu, Y.; Yasuzawa, M.; Shen, Y.-L., Non-stick syringe needles: Beneficial effects of thin film metallic glass coating. *Scientific Reports* 2016, 6, 31847.

[44] Lee, J.; Liou, M.-L.; Duh, J.-G., The development of a Zr-Cu-Al-Ag-N thin film metallic glass coating in pursuit of improved mechanical, corrosion, and antimicrobial property for bio-medical application. *Surface and Coatings Technology* 2017, 310, 214–222.

[45] Murty, B. S.; Yeh, J. W.; Ranganathan, S.; Bhattacharjee, P. P., 10 - High-entropy alloy coatings. In *High-Entropy Alloys* (Second Edition), Murty, B. S.; Yeh, J. W.; Ranganathan, S.; Bhattacharjee, P. P., Eds. Elsevier: 2019, 177–193.

[46] Tüten, N.; Canadinc, D.; Motallebzadeh, A.; Bal, B., Microstructure and tribological properties of TiTaHfNbZr high entropy alloy coatings deposited on Ti6Al4V substrates. *Intermetallics* 2019, 105, 99–106.

[47] Hahn, R.; Kirnbauer, A.; Bartosik, M.; Kolozsvári, S.; Mayrhofer, P. H., Toughness of Si alloyed high-entropy nitride coatings. *Materials Letters* 2019, 251, 238–240.

[48] Kirnbauer, A.; Spadt, C.; Koller, C. M.; Kolozsvári, S.; Mayrhofer, P. H., High-entropy oxide thin films based on Al–Cr–Nb–Ta–Ti. *Vacuum* 2019, 168, 108850.

[49] Bagdasaryan, A. A.; Pshyk, A. V.; Coy, L. E.; Kempiński, M.; Pogrebnjak, A. D.; Beresnev, V. M.; Jurga, S., Structural and mechanical characterization of (TiZrNbHfTa)N/WN multilayered nitride coatings. *Materials Letters* 2018, 229, 364–367.

[50] Chang, Y.-Y.; Cheng, C.-M.; Liou, Y.-Y.; Tillmann, W.; Hoffmann, F.; Sprute, T., High temperature wettability of multicomponent CrAlSiN and TiAlSiN coatings by molten glass. *Surface and Coatings Technology* 2013, 231, 24–28.

[51] Tillmann, W.; Fehr, A.; Stangier, D.; Dildrop, M., Influences of substrate pretreatments and Ti/Cr interlayers on the adhesion and hardness of CrAlSiN and TiAlSiN films deposited on Al2O3 and ZrO2-8Y2O3 thermal barrier coatings. *Results in Physics* 2019, 12, 2206–2212.

[52] Erdemir, A.; Martin, J. M., Superior wear resistance of diamond and DLC coatings. *Current Opinion in Solid State and Materials Science* 2018, 22, (6), 243–254.

[53] Drnovšek, A.; Rebelo de Figueiredo, M.; Vo, H.; Xia, A.; Vachhani, S. J.; Kolozsvári, S.; Hosemann, P.; Franz, R., Correlating high temperature mechanical and tribological properties of CrAlN and CrAlSiN hard coatings. *Surface and Coatings Technology* 2019, 372, 361–368.

[54] Zhang, S.; Wu, W.; Chen, W.; Yang, S., Structural optimisation and synthesis of multilayers and nano-composite AlCrTiSiN coatings for excellent machinability. *Surface and Coatings Technology* 2015, 277, 23–29.

[55] Vereschaka, A.; Gurin, V.; Oganyan, M.; Oganyan, G.; Bublikov, J.; Shein, A., Increase in tool life for end milling titanium alloys using tools with multilayer composite nanostructured modified coatings. *Procedia CIRP* 2019, 81, 1412–1416.

[56] Kumar, C. S.; Patel, S. K., Effect of duplex nanostructured TiAlSiN/TiSiN/TiAlN-TiAlN and TiAlN-TiAlSiN/TiSiN/TiAlN coatings on the hard turning performance of Al2O3-TiCN ceramic cutting tools. *Wear* 2019, 418–419, 226–240.

[57] Qiu, L.; Du, Y.; Wu, L.; Wang, S.; Zhu, J.; Cheng, W.; Tan, Z.; Yin, L.; Liu, Z.; Layyous, A., Microstructure, mechanical properties and cutting performances of TiSiCN super-hard nanocomposite coatings deposited using CVD method under the guidance of thermodynamic calculations. *Surface and Coatings Technology* 2019, 378, 124956.

[58] Singh, A.; Ghosh, S.; Aravindan, S., Flank wear and rake wear studies for arc enhanced HiPIMS coated AlTiN tools during high speed machining of nickel-based superalloy. *Surface and Coatings Technology* 2020, 381, 125190.

[59] Sanchette, F.; Ducros, C.; Schmitt, T.; Steyer, P.; Billard, A., Nanostructured hard coatings deposited by cathodic arc deposition: From concepts to applications. *Surface and Coatings Technology* 2011, 205, (23), 5444–5453.

[60] Aouadi, S. M.; Luster, B.; Kohli, P.; Muratore, C.; Voevodin, A. A., Progress in the development of adaptive nitride-based coatings for high temperature tribological applications. *Surface and Coatings Technology* 2009, 204, (6), 962–968.

[61] Aouadi, S. M.; Paudel, Y.; Simonson, W. J.; Ge, Q.; Kohli, P.; Muratore, C.; Voevodin, A. A., Tribological investigation of adaptive Mo2N/MoS2/Ag coatings with high sulfur content. *Surface and Coatings Technology* 2009, 203, (10), 1304–1309.

[62] Zha, X.; Chen, F.; Jiang, F.; Xu, X., Correlation of the fatigue impact resistance of bilayer and nanolayered PVD coatings with their cutting performance in machining Ti6Al4V. *Ceramics International* 2019, 45, (12), 14704–14717.

[63] Chang, Y.-Y.; Chiu, W.-T.; Hung, J.-P., Mechanical properties and high temperature oxidation of CrAlSiN/TiVN hard coatings synthesized by cathodic arc evaporation. *Surface and Coatings Technology* 2016, 303, 18–24.

[64] Xiao, B.; Liu, J.; Liu, F.; Zhong, X.; Xiao, X.; Zhang, T. F.; Wang, Q., Effects of microstructure evolution on the oxidation behavior and high-temperature tribological properties of AlCrN/TiAlSiN multilayer coatings. *Ceramics International* 2018, 44, (18), 23150–23161.

[65] Iram, S.; Wang, J.; Cai, F.; Zhang, J.; Ahmad, F.; Liang, J.; Zhang, S. J. S. E., Effect of bilayer number on mechanical and wear behaviours of the AlCrN/AlCrMoN coatings by AIP method. *Surface Engineering* 2020, 1–9.

[66] Bobzin, K.; Brögelmann, T.; Brugnara, R. H.; Arghavani, M.; Yang, T. S.; Chang, Y. Y.; Chang, S. Y., Investigation on plastic behavior of HPPMS CrN, AlN and CrN/AlN-multilayer coatings using finite element simulation and nanoindentation. *Surface and Coatings Technology* 2015, 284, 310–317.

[67] Esqué-de los Ojos, D.; Best, J. P.; Schwiedrzik, J.; Morstein, M.; Michler, J., A closed-form analytical approach for the simple prediction of hard-coating failure for tooling systems. *Surface and Coatings Technology* 2016, 308, 280–288.

[68] Campoccia, D.; Montanaro, L.; Speziale, P.; Arciola, C. R., Antibiotic-loaded biomaterials and the risks for the spread of antibiotic resistance following their prophylactic and therapeutic clinical use. *Biomaterials* 2010, 31, (25), 6363–6377.

[69] Pilliar, R., Dental implants: materials and design. *Journal (Canadian Dental Association)* 1990, 56, (9), 857–861.

[70] Norowski Jr, P. A.; Bumgardner, J. D., Biomaterial and antibiotic strategies for peri-implantitis: A review. *Journal of Biomedical Materials Research Part B: Applied Biomaterials: An Official Journal of The Society for Biomaterials, The Japanese Society for Biomaterials, and The Australian Society for Biomaterials and the Korean Society for Biomaterials* 2009, 88, (2), 530–543.

[71] Renvert, S.; Quirynen, M., Risk indicators for peri-implantitis. A narrative review. *Clinical Oral Implants Research* 2015, 26, 15–44.

[72] Cavaliere, E.; Benetti, G.; Banfi, F.; Gavioli, L., 11 - Antimicrobial nanostructured coating. In *Frontiers of Nanoscience*, Milani, P.; Sowwan, M., Eds. Amsterdam, The Netherlands: Elsevier: 2020; Vol. 15, pp 291–311.

[73] Li, M.; Liu, Y.; Wang, H.; Ye, Q.; Shen, H., Synthesis of TiO2 nanorings and nanorods on TCO substrate by potentiostatic anodization of titanium powder. *Crystal Research and Technology* 2011, 46, (4), 413–416.

[74] Guo, J.; Padilla, R. J.; Ambrose, W.; De Kok, I. J.; Cooper, L. F., The effect of hydrofluoric acid treatment of TiO2 grit blasted titanium implants on adherent osteoblast gene expression in vitro and in vivo. *Biomaterials* 2007, 28, (36), 5418–5425.

[75] Shuai, C.; Guo, W.; Wu, P.; Yang, W.; Hu, S.; Xia, Y.; Feng, P., A graphene oxide-Ag co-dispersing nano-system: Dual synergistic effects on antibacterial activities and mechanical properties of polymer scaffolds. *Chemical Engineering Journal* 2018, 347, 322–333.

[76] Ossowska, A.; Zieliński, A.; Olive, J.-M.; Wojtowicz, A.; Szweda, P. J. C., Influence of Two-Stage Anodization on Properties of the Oxide Coatings on the Ti–13Nb–13Zr Alloy. 2020, 10, (8), 707.

[77] Huang, H.-L.; Chang, Y.-Y.; Chen, H.-J.; Chou, Y.-K.; Lai, C.-H.; Chen, M. Y., Antibacterial properties and cytocompatibility of tantalum oxide coatings with different silver content. *Journal of Vacuum Science & Technology A: Vacuum, Surfaces, and Films* 2014, 32, (2), 02B117.

[78] Huang, H.-L.; Chang, Y.-Y.; Chen, Y.-C.; Lai, C.-H.; Chen, M. Y., Cytocompatibility and antibacterial properties of zirconia coatings with different silver contents on titanium. *Thin Solid Films* 2013, 549, 108–116.

[79] Brown, T.A.; Smith, D.G., The effects of silver nitrate on the growth and ultrastructure of the yeast cryptocaccus albidus. *Microbios Letter* 1976, 3, 155–162.

[80] Richards, R.; Odelola, H.; Anderson, B., Effect of silver on whole cells and spheroplasts of a silver resistant Pseudomonas aeruginosa. *Microbios* 1984, 39, 151–158.

[81] Yun'an Qing, L. C.; Li, R.; Liu, G.; Zhang, Y.; Tang, X.; Wang, J.; Liu, H.; Qin, Y. J. I., Potential antibacterial mechanism of silver nanoparticles and the optimization of orthopedic implants by advanced modification technologies. *International Journal of Nanomedicine* 2018, 13, 3311.

[82] Abbaszadegan, A.; Ghahramani, Y.; Gholami, A.; Hemmateenejad, B.; Dorostkar, S.; Nabavizadeh, M.; Sharghi, H. J. J., The effect of charge at the surface of silver nanoparticles on antimicrobial activity against gram-positive and gram-negative bacteria: a preliminary study. *Journal of Nanomaterials* 2015, 2015.

[83] Voicu, G.; Miu, D.; Ghitulica, C.-D.; Jinga, S.-I.; Nicoara, A.-I.; Busuioc, C.; Holban, A.-M., Co doped ZnO thin films deposited by spin coating as antibacterial coating for metallic implants. *Ceramics International* 2020, 46, (3), 3904–3911.

[84] Tsai, M.-T.; Chang, Y.-Y.; Huang, H.-L.; Hsu, J.-T.; Chen, Y.-C.; Wu, A. Y.-J., Characterization and antibacterial performance of bioactive Ti–Zn–O coatings deposited on titanium implants. *Thin Solid Films* 2013, 528, 143–150.

[85] Huang, H.-L.; Tsai, M.-T.; Chang, Y.-Y.; Lin, Y.-J.; Hsu, J.-T. J. M., Fabrication of a Novel Ta (Zn) O Thin Film on Titanium by Magnetron Sputtering and Plasma Electrolytic Oxidation for Cell Biocompatibilities and Antibacterial Applications. *Metals* 2020, 10, (5), 649.

[86] Chang, Y.-Y.; Zhang, J.-H.; Huang, H.-L., Effects of laser texture oxidation and high-temperature annealing of TiV alloy thin films on mechanical and antibacterial properties and cytotoxicity. *Materials* 2018, 11, (12), 2495.

[87] Lai, C.-H.; Chang, Y.-Y.; Huang, H.-L.; Kao, H.-Y., Characterization and antibacterial performance of ZrCN/amorphous carbon coatings deposited on titanium implants. *Thin Solid Films* 2011, 520, (5), 1525–1531.

[88] Tsai, M.-T.; Chang, Y.-Y.; Huang, H.-L.; Wu, Y.-H.; Shieh, T.-M., Micro-arc oxidation treatment enhanced the biological performance of human osteosarcoma cell line and human skin fibroblasts cultured on titanium–zirconium films. *Surface and Coatings Technology* 2016, 303, 268–276.

[89] Bianchi, M.; Gambardella, A.; Berni, M.; Panseri, S.; Montesi, M.; Lopomo, N.; Tampieri, A.; Marcacci, M.; Russo, A., Surface morphology, tribological properties and in vitro biocompatibility of nanostructured zirconia thin films. *Journal of Materials Science: Materials in Medicine* 2016, 27, (5), 96.

[90] Zhang, B.; Li, B.; Gao, S.; Li, Y.; Cao, R.; Cheng, J.; Li, R.; Wang, E.; Guo, Y.; Zhang, K.; Liang, J.; Liu, B., Y-doped TiO2 coating with superior bioactivity and antibacterial property prepared via plasma electrolytic oxidation. *Materials & Design* 2020, 192, 108758.

[91] Zhang, X.; Lv, Y.; Shan, F.; Wu, Y.; Lu, X.; Peng, Z.; Liu, B.; Yang, L.; Dong, Z., Microstructure, corrosion resistance, osteogenic activity and antibacterial capability of Mn-incorporated TiO2 coating. *Applied Surface Science* 2020, 531, 147399.

[92] Chang, Y. Y.; Huang, H. L.; Chen, Y. C.; Hsu, J. T.; Shieh, T. M.; Tsai, M. T., Biological characteristics of the MG-63 human osteosarcoma cells on composite tantalum carbide/amorphous carbon films. *PLoS One* 2014, 9, (4): e95590.

[93] Chang, Y.-Y.; Huang, H.-L.; Chen, H.-J.; Lai, C.-H.; Wen, C.-Y., Antibacterial properties and cytocompatibility of tantalum oxide coatings. *Surface and Coatings Technology* 2014, 259, 193–198.

[94] Alias, R.; Mahmoodian, R.; Genasan, K.; Vellasamy, K. M.; Abd Shukor, M. H.; Kamarul, T., Mechanical, antibacterial, and biocompatibility mechanism of PVD grown silver-tantalum-oxide-based nanostructured thin film on stainless steel 316L for surgical applications. *Materials Science & Engineering, C: Materials for Biological Applications* 2020, 107, 110304.

[95] Sakurai, K.; Hiratsuka, M.; Nakamori, H.; Namiki, K.; Hirakuri, K., Evaluation of sliding properties and durability of DLC coating for medical devices. *Diamond and Related Materials* 2019, 96, 97–103.

[96] Chang, C.-H.; Li, C.-L.; Yu, C.-C.; Chen, Y.-L.; Chyntara, S.; Chu, J. P.; Chen, M.-J.; Chang, S.-H., Beneficial effects of thin film metallic glass coating in reducing adhesion of platelet and cancer cells: Clinical testing. *Surface and Coatings Technology* 2018, 344, 312–321.

[97] Kronström, M.; Svenson, B.; Hellman, M.; Persson, G. R., Early implant failures in patients treated with Brånemark System titanium dental implants: a retrospective study. *The International Journal of Oral & Maxillofacial Implants* 2001, 16, (2), 201–207.

[98] Oprea, O.; Andronescu, E.; Ficai, D.; Ficai, A.; N Oktar, F.; Yetmez, M., ZnO applications and challenges. *Current Organic Chemistry* 2014, 18, (2), 192–203.

[99] Padmavathy, N.; Vijayaraghavan, R., Enhanced bioactivity of ZnO nanoparticles—an antimicrobial study. *Science and Technology of Advanced Materials* 2008, 9, (3), 035004.

[100] Hirota, K.; Sugimoto, M.; Kato, M.; Tsukagoshi, K.; Tanigawa, T.; Sugimoto, H., Preparation of zinc oxide ceramics with a sustainable antibacterial activity under dark conditions. *Ceramics International* 2010, 36, (2), 497–506.

[101] Li, M.; Zhu, L.; Lin, D., Toxicity of ZnO nanoparticles to *Escherichia coli*: mechanism and the influence of medium components. *Environmental Science & Technology* 2011, 45, (5), 1977–1983.

[102] Sun, Z.; Liao, T.; Li, W.; Dou, Y.; Liu, K.; Jiang, L.; Kim, S.-W.; Kim, J. H.; Dou, S. X., Fish-scale bio-inspired multifunctional ZnO nanostructures. *NPG Asia Materials* 2015, 7, (12), e232.

[103] Mureşan, L. M., Corrosion protective coatings for Ti and Ti alloys used for biomedical implants. In *Intelligent Coatings for Corrosion Control*, Elsevier: 2015; pp 585–602.

[104] Suma, P. R. P.; Nair, R. V.; Paul, W.; Jayasree, R. S., Vanadium pentoxide nanoplates: Synthesis, characterization and unveiling the intrinsic anti-bacterial activity. *Materials Letters* 2020, 269, 127673.

[105] Sun, H.; Yang, Z.; Pu, Y.; Dou, W.; Wang, C.; Wang, W.; Hao, X.; Chen, S.; Shao, Q.; Dong, M.; Wu, S.; Ding, T.; Guo, Z., Zinc oxide/vanadium pentoxide heterostructures with enhanced day-night antibacterial activities. *Journal of Colloid and Interface Science* 2019, 547, 40–49.

[106] Kayani, Z. N.; Bashir, H.; Riaz, S.; Naseem, S., Optical properties and antibacterial activity of V doped ZnO used in solar cells and biomedical applications. *Materials Research Bulletin* 2019, 115, 121–129.

[107] Ramos-Corella, K. J.; Sotelo-Lerma, M.; Gil-Salido, A. A.; Rubio-Pino, J. L.; Auciello, O.; Quevedo-Lopez, M. A., Controlling crystalline phase of TiO2 thin films to evaluate its biocompatibility. *Mater. Technology* 2019, 34, (8), 455–462.

[108] Liu, L. T.; Bhatia, R.; Webster, T. J., Atomic layer deposition of nano-TiO2 thin films with enhanced biocompatibility and antimicrobial activity for orthopedic implants. *International Journal of Nanomedicine* 2017, 12, 8711–8723.

[109] Saleem, S.; Ayub, R.; Ahmad, R.; Ikhlaq, U.; Arif, S.; Chu, P. K., Influence of plasma excitation power on mechanical property and biocompatibility of titania/alumina composite thin films for medical implant prepared by magnetron sputtering. *Materials Research Express* 2019, 6, 116418.

[110] Xue, W.; Wang, C.; Chen, R.; Deng, Z., Structure and properties characterization of ceramic coatings produced on Ti–6Al–4V alloy by microarc oxidation in aluminate solution. *Materials Letters* 2002, 52, (6), 435–441.

[111] Du, Q.; Wei, D.; Wang, Y.; Cheng, S.; Liu, S.; Zhou, Y.; Jia, D., The effect of applied voltages on the structure, apatite-inducing ability and antibacterial ability of micro arc oxidation coating formed on titanium surface. *Bioactive Materials* 2018, 3, (4), 426–433.

[112] Hu, J.; Li, H.; Wang, X.; Yang, L.; Chen, M.; Wang, R.; Qin, G.; Chen, D.-F.; Zhang, E., Effect of ultrasonic micro-arc oxidation on the antibacterial properties and cell biocompatibility of Ti-Cu alloy for biomedical application. *Materials Science and Engineering: C* 2020, 115, 110921.

[113] Wang, Y.; Yang, H.; Xue, X., Synergistic antibacterial activity of TiO2 co-doped with zinc and yttrium. *Vacuum* 2014, 107, 28–32.

[114] Sedelnikova, M. B.; Komarova, E. G.; Sharkeev, Y. P.; Ugodchikova, A. V.; Tolkacheva, T. V.; Rau, J. V.; Buyko, E. E.; Ivanov, V. V.; Sheikin, V. V., Modification of titanium surface via Ag-, Sr- and Si-containing micro-arc calcium phosphate coating. *Bioactive Materials* 2019, 4, 224–235.

[115] Wu, J.; Ueda, K.; Narushima, T., Fabrication of Ag and Ta co-doped amorphous calcium phosphate coating films by radiofrequency magnetron sputtering and their antibacterial activity. *Materials Science and Engineering: C* 2020, 109, 110599.

[116] Sarraf, M.; Dabbagh, A.; Abdul Razak, B.; Nasiri-Tabrizi, B.; Hosseini, H. R. M.; Saber-Samandari, S.; Abu Kasim, N. H.; Yean, L. K.; Sukiman, N. L., Silver oxide nanoparticles-decorated tantala nanotubes for enhanced antibacterial activity and osseointegration of Ti6Al4V. *Materials & Design* 2018, 154, 28–40.

[117] Onodera, S.; Fujii, S.; Moriguchi, H.; Tsujioka, M.; Hirakuri, K., Antibacterial property of F doped DLC film with plasma treatment. *Diamond and Related Materials* 2020, 107, 107835.

[118] Marciano, F. R.; Lima-Oliveira, D. A.; Da-Silva, N. S.; Corat, E. J.; Trava-Airoldi, V. J., Antibacterial activity of fluorinated diamond-like carbon films produced by PECVD. *Surface and Coatings Technology* 2010, 204, (18), 2986–2990.

[119] Kitahara, N.; Sato, T.; Isogawa, H.; Ohgoe, Y.; Masuko, S.; Shizuku, F.; Hirakuri, K. K., Antibacterial property of DLC film coated on textile material. *Diamond and Related Materials* 2010, 19, (7), 690–694.

[120] Etiemble, A.; Der Loughian, C.; Apreutesei, M.; Langlois, C.; Cardinal, S.; Pelletier, J. M.; Pierson, J. F.; Steyer, P., Innovative Zr-Cu-Ag thin film metallic glass deposed by magnetron PVD sputtering for antibacterial applications. *Journal of Alloys and Compounds* 2017, 707, 155–161.

[121] Jabed, A.; Khan, M. M.; Camiller, J.; Greenlee-Wacker, M.; Haider, W.; Shabib, I., Property optimization of Zr-Ti-X (X = Ag, Al) metallic glass via combinatorial development aimed at prospective biomedical application. *Surface and Coatings Technology* 2019, 372, 278–287.

[122] Zhang, E.; Li, F.; Wang, H.; Liu, J.; Wang, C.; Li, M.; Yang, K. J. M. S.; A new antibacterial titanium–copper sintered alloy: preparation and antibacterial property. *Materials Science and Engineering: C* 2013, 33, (7), 4280–4287.

[123] Nkou Bouala, G. I.; Etiemble, A.; Der Loughian, C.; Langlois, C.; Pierson, J. F.; Steyer, P., Silver influence on the antibacterial activity of multi-functional Zr-Cu based thin film metallic glasses. *Surface and Coatings Technology* 2018, 343, 108–114.

[124] Ding, Z.; Wang, Y.; Zhou, Q.; Ding, Z.; Liu, J.; He, Q.; Zhang, H. J. B., Microstructure, wettability, corrosion resistance and antibacterial property of Cu-MTa2O5 multilayer composite coatings with different Cu incorporation contents. *Biomolecules* 2020, 10, (1), 68.

10 The Wear Behavior and Mechanism of Graphene

Yuehua Huang
Southwest University, China

CONTENTS

10.1 INTRODUCTION

Friction and wear are important phenomena between solid contact interfaces. It has been reported that friction consumes nearly 1/3 of the world's disposable energy and wear caused by friction is the main reason for the failure of 80% mechanical parts [1, 2]. The wear issue not only exists in pumps, turbines, gears, engines, and other moving machine parts but also exists in magnetic storage systems such as the hard disk [3] and microelectromechanical systems (MEMS) [4, 5]. Moreover, as the specific surface area increases rapidly with the decreasing of the parts size, the body force acting as the driving force decreases faster than the surface force such as friction with the decreasing of the parts size, resulting in the wear issue more prominent and serious in micromachines. The wear of moving parts has become a significant factor affecting the function and service life of mechanical and microelectromechanical systems.

Various lubrication solutions to wear problems have been constantly searched and studied [6]. The gas lubrication only applies to a very limited range, which is often based on the hydrodynamics, physical adsorption or chemical reactions of specific gases on the surface of specific materials [7]. The most effective solution is to add a kind of liquid or solid lubricants to the sliding contact interface. Liquid film lubrication, which is based on hydrodynamic pressure effect, is not suitable for severe operation conditions such as low speeds, high loads, extreme temperature and vacuum. Moreover, liquid film lubrication cannot be used in micromechanical systems due to the surface tension and liquid viscosity problems. Solid film lubrication still works in the above cases, which makes solid film lubrication also widely used. In recent years, significant progress has been achieved on the basic and applied researches of solid lubricants [8, 9]. However, the traditional solid lubricants are sensitive to the working conditions such as the atmosphere, humidity and testing pressure to some extent. The effective thickness of solid lubricant and its consumption are large while the life time of lubrication is limited. The influence of solid lubricant film thickness is more prominent in precision machinery such as the MEMS. Due to the above shortages of traditional solid lubricants, to increase the lubrication life time and to reduce solid lubricant film thickness is one of the research objectives of tribology.

Compared with traditional solid lubricants, the emergence of two-dimensional materials like graphene provides a new choice for the ultra-thin solid lubricant film. The thickness of monolayer graphene is only 0.335nm [10], which is the thinnest material ever discovered. Due to the strong covalent bonding between graphene carbon atoms, its equivalent fracture strength, young's modulus and fatigue strength are as high as 130GPa, 1TPa [11] and 71GPa [12] respectively, making it one of the strongest materials in the world. What's more, monolayer graphene has lubrication effect comparable to that of conventional solid lubricant of bulk graphite [13]. With excellent mechanical properties and intrinsic lubrication effect, graphene is a promising candidate as a kind of ultra-thin solid lubricants.

Graphene has many potential applications. For wear problems in MEMS such as micro-static motors, micro-bearings and micro-resonators, graphene has become the main potential lubricant candidate [14]. For the wear problem in magnetic storage systems such as the hard disks, in which the spacing between the reader and writer is only a few nanometers, graphene with its ultra-thin thickness and excellent lubrication effect, has become a potential protecting material for the magnetic medium [15]. For electrical contact components such as silver and copper which require protecting coatings with lubrication effect and high conductivity, graphene has shown great potential with its excellent lubrication and electrical properties [16]. For the widely used stainless steel, monolayer or few layer graphene can significantly reduce friction and wear, which greatly reduce the effective thickness or amount of solid lubricant [17]. For light metals such as magnesium alloys and titanium alloys with strong application potential in the fields of aerospace and automobile industry, their poor wear resistance has become one of the main problems restricting their wide application. Hard coatings, which are often used to improve the wear resistance, are easy to generate extra stress between the substrate and the coating due to the mismatching of mechanical and thermal properties between the hard coating and protected substrate metals. With high flexibility and good lubrication property, graphene has shown the potential to improve the wear resistance of light metals.

10.2 THE WEAR BEHAVIOR OF GRAPHENE AT THE NANOSCALE

Since the radius of the atomic force microscope (AFM) tip is at nanometer scale, the contact area during the friction tests is at the nanoscale. As the friction test conducted on tribometers, the counterpart radius is often millimeter range and the contact area during the friction test is micron to millimeter. We classify the experiment and simulation with nanometer scale contact area as the nanoscale and those with micron to millimeter as the macroscale.

The frictional behavior and mechanism of graphene at the nanoscale have been extensively studied, but its wear behavior and mechanism are far less explored. Graphene has exhibited excellent friction-reducing property at the nanoscale and even monolayer graphene is comparable to the traditional solid lubricant of graphite in friction reduction [13]. As a kind of ultra-thin solid lubricant, in addition to the friction-reducing performance, the wear resistance is also important.

10.2.1 The Excellent Anti-Wear Property of Graphene at the Nanoscale

Sin et al. [18] conducted a nano-indenter experiment on mechanical exfoliated monolayer graphene on SiO_2 substrate, as shown in Figure 10.1(a). Under the condition that the indentation depth was 150 times of the monolayer graphene thickness (the indention depth was 50nm and the monolayer graphene was 0.335nm thick), monolayer graphene was not broken, where the average stress when graphene broke was 6.78GPa. The nano-indentation experiment demonstrates that monolayer graphene has the ability to withstand large deformation and large compressive stress without damage at the same time. In addition to the nano-indentation experiment in which only pressure was applied on graphene, Lin et al. [19] tested the wear resistance of multilayer graphene (6~7 layers) under the compressive stress and shear stress during sliding using an AFM, as shown in Figure 10.1(b). After 100 cycles under 4µN no detectable wear was found by AFM. When the normal load increased to 5µN after 100 cycle sliding test, there was a 0.47nm deep wear track over monolayer graphene thickness. Considering the AFM tip radius the pressure conducted on the graphene under 5µN was considerable. The sliding tests confirm that multilayer graphene is able to withstand reciprocating shear stress under high pressure. Moreover, monolayer graphene also exhibited excellent anti-wear property in nanoscale sliding tests. The friction coefficient of graphene was as low as 0.006 and remained during the 1000 cycles by 20nm radius AFM tip under 500nN, without detectable wear track after sliding test [20]. Qi et al. [21] found that monolayer graphene did not wear while sink-in appeared on the substrate SiO_2 surface after sliding test under 1726nN with 100nm radius AFM tip (the theoretical Hertzian stress was 5.39GPa) for 500 cycles, indicating that the yield strength of monolayer graphene was higher than that of SiO_2, as shown in Figure 10.1(c). When the normal load increased to 9150nN, monolayer graphene still survived with no detectable wear after more than 4000 cycles, exhibiting excellent anti-wear property at the nanoscale [21]. Klemenz et al. [22] studied the wear mechanism of monolayer graphene at the nanoscale with molecular dynamics simulation, as shown in Figure 10.1(d). In the simulation model, the substrate was platinum and the rigid counterpart slid on graphene under a certain load. The simulation results showed that with the increase in normal load, the process of deformation and destruction of monolayer graphene/platinum substrate was divided into three stages: Stage 1, under low load, the platinum substrate deformed elastically, graphene had no damage and the friction was low; Stage 2, as the load increased, the

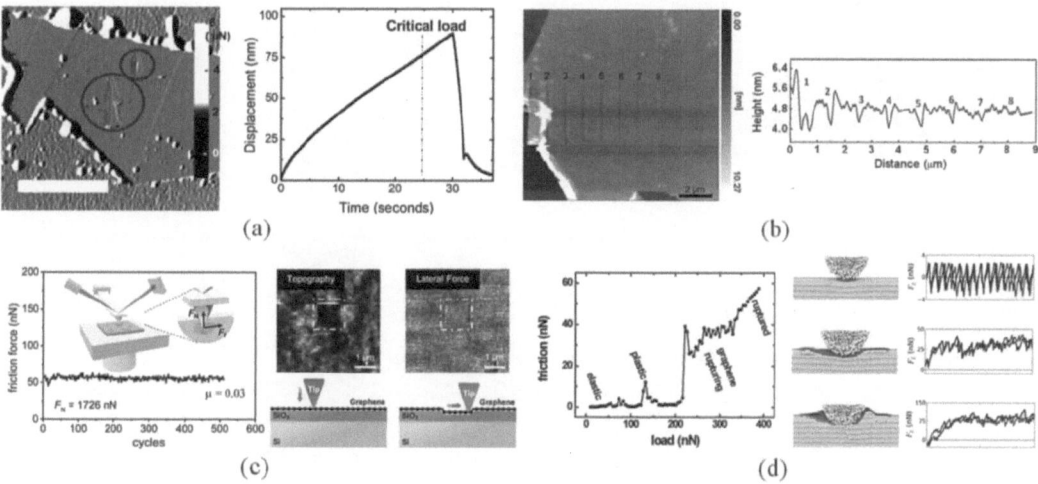

FIGURE 10.1 The excellent anti-wear property of graphene at nanoscale: (a) Scratch test on monolayer graphene by nano-indenter (Source: Shin, Y et al. 2011); (b) Sliding test on multilayer graphene by AFM (Source: Lin, L et al. 2011); (c) Sliding test on monolayer graphene by AFM (Source: Qi, Y et al. 2016); (d) Sliding test on graphene/Pt by AFM and the corresponding MD simulations (Source: Klemenz, A et al. 2014).

platinum substrate deformed plastically, graphene remained intact and the friction force increased; Stage 3, with further increase in normal load, graphene broke down and lost its protection effect, the platinum substrate deformed seriously and friction increased sharply. The experiment results were in good agreement with the simulation results. According to the simulation, the maximum in-plane tensile stress of monolayer graphene before breakage was 100GPa (35N/m), which is similar to the fracture strength (42N/m) of the suspended monolayer graphene measured by AFM [11]. It can be seen that the destruction of graphene covering the substrate is mainly determined by the intrinsic strength of graphene.

10.2.2 THE INFLUENCE OF GRAPHENE STRUCTURE ON ITS ANTI-WEAR PROPERTY

The above experiments and simulations demonstrate that mechanically exfoliated graphene (mechanically exfoliated graphene has few defects and is the closest to intrinsic graphene in performance) has excellent anti-wear property at the nanoscale. However, due to the limitation in size of mechanically exfoliated graphene, it's difficult to use in practical applications. The promising chemical vapor deposition (CVD) graphene often has a certain degree of defects, such as folds and breakage introduced during graphene synthesis and transfer. Since the thermal expansion coefficients of graphene and the growing base metals such as copper and nickel are different, the compressive stress in the cooling process leads to the graphene grown wrinkles [23]. The morphology of the growing substrate leads to larger wrinkles when graphene is transferred to other substrates [24]. Vasic et al. [25] studied the effect of CVD graphene wrinkles on its wear resistance at the nanoscale and they found that the breakage of graphene first occurred at graphene wrinkles, as shown in Figure 10.2(a). The possible associated mechanism was that graphene wrinkles increased the out-of-plane deformation of graphene, which led to the increase of contact area and interaction between graphene and AFM tip, and finally resulted in that graphene wrinkles were more prone to wear. Besides the influence of graphene wrinkles on its wear resistance, the effect of graphene step edges, which was also

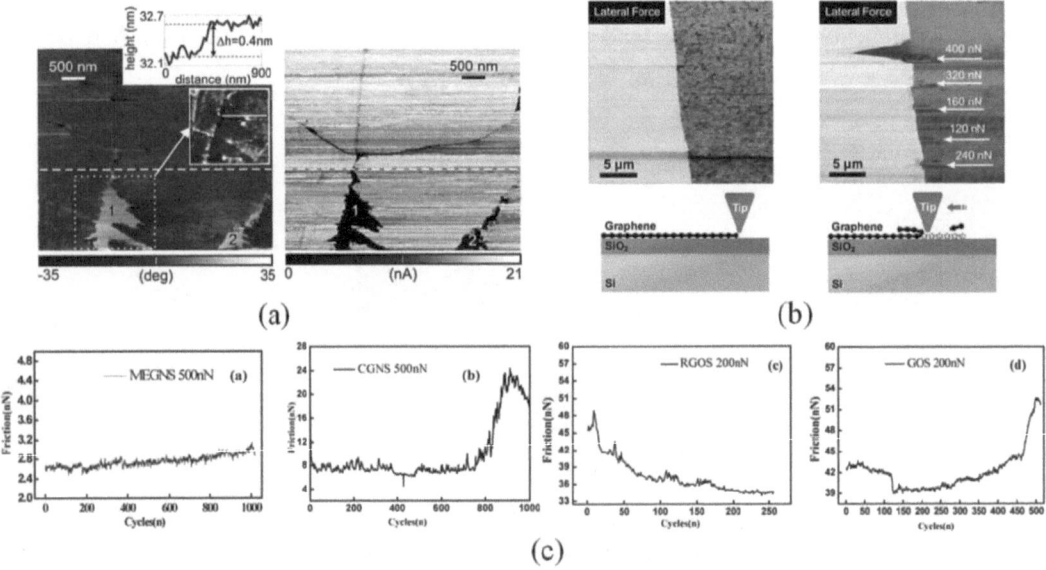

FIGURE 10.2 The influence of graphene structure on graphene anti-wear property at the nanoscale: (a) Sliding test on CVD graphene which contains wrinkles by AFM (Source: Vasic, B et al. 2016); (b) Sliding test on graphene edges by AFM (Source: Qi, Y et al. 2016); (c) Sliding test on mechanical exfoliated graphene nanosheet (MEGS), CVD graphene nanosheet (CGNS), reduced graphene oxide sheet (RGOS) and graphene oxide sheet (GOS) by AFM (Source: Peng Y et al. 2015).

hard to avoid in graphene practical application, on its anti-wear performance was also explored by researchers. The experiment by Qi et al. [21] showed that while graphene exhibited excellent anti-wear property in the interior region, it could be easily damaged at graphene step edge under a normal load of ~2 orders of magnitude smaller, as shown in Figure 10.2(b). The result of Vasic et al.'s study on graphene step edge [26] was consistent with that of Qi et al. The molecular dynamics simulation by Vasic et al. showed that with the AFM tip sliding across a graphene edge, the graphene step edge first underwent elastic deformation, then plastic deformation, accompanied by graphene wrinkling and partial stripping from the substrate. Finally, the graphene edge was completely stripped from the substrate and folded. Besides the above underlying mechanism, Qi et al. [21] found that the bonding between graphene and AFM tip might have considerable influence. Considering the bonding reaction between graphene and AFM tip, graphene was more likely to break due to the bonding adhesion between graphene carbon atom and the AFM tip and the dangling bond at graphene step edge was more likely to bond with the AFM tip [21]. The graphene edge also has significant influence on the friction performance [27–29] of graphene. The friction force during the step-up process of the graphene step is obviously higher, which is due to the synergy of the topography physical effect and the interfacial bonding chemical effect [29]. What's more, the environmental humidity influences the friction at the atomic steps but has no obvious influence on the friction on flat areas, which is due to the bonded water molecular at the step edges [28]. We think the higher friction force at the step edges will promote the wear of graphene initiated from the step edges. In addition to graphene wrinkles and step edges, the effect of graphene structure on its wear resistance at nanoscale was studied by Peng et al. [20]. As shown in Figure 10.2(c), they found that mechanically exfoliated graphene exhibited excellent anti-wear property and defects in CVD graphene decreased the anti-wear capability. Moreover, the graphene oxide (GO) and reduced graphene oxide (RGO) showed much worse wear resistance which may be due to the destroyed graphitic structure.

10.2.3 THE INFLUENCE OF THE SUBSTRATE AND THE GRAPHENE/SUBSTRATE INTERFACE ON GRAPHENE ANTI-WEAR PROPERTY

In addition to graphene itself, the substrate and graphene/substrate interface also have significant influence on graphene wear resistance at the nanoscale. Peng et al. [30] found that the friction of graphene on soft PDMS was significantly higher than that on SiO_2 substrate and they attributed the phenomenon to the larger contact area between AFM tip and graphene on softer substrate due to the substrate deformation. The first principle simulation results show that for a soft substrate, the substrate stiffness was the main factor determining the friction force [31]. Yao et al. [32] studied the wear resistance of monolayer CVD grown graphene transferred on PDMS, epoxy, SiO_2 and Al_2O_3 and conducted sliding test on graphene on these four different substrates with increased normal load. In the sliding tests the normal load increased as the cycle number increased at the sliding track and the normal load under which the friction force jumped abrupt was regarded as the critical normal load of graphene wear. They found that graphene was more prone to fracture on soft substrates such as PDMS and epoxy. According to a previous study on the wear mechanism of graphene in sliding test, the failure of 2D materials was primarily caused by the in-plane stretching [33]. The deformation of substrate with lower Young's modulus would lead to greater in-plane tensile stress in graphene assuming that the adhesion between graphene and different substrates were the same, which led to the failure of graphene. While the above researches on the effect of substrate did not consider the graphene/substrate interface, which is also a very important factor influencing graphene wear behavior. Zeng et al. [34] used plasma treatment of the SiO_2 substrate to increase the adhesion between graphene and substrate. They found that different interface adhesion strength significantly regulated the puckering effect and graphene friction properties in sliding test at the nanoscale [34]. Qi et al. [35] conducted sliding test on graphene on SiO_2 substrate with and without plasma treatment. They found that no matter in dry environment and humid environment the graphene wear critical normal load increased 2~3 times after the SiO_2 substrate was plasma treated, indicating that stronger graphene/substrate adhesion could increase anti-wear behavior of graphene. Zhao [36] et al. changed the adhesion between CVD graphene and the growing substrate copper by changing the substrate oxidation degree. They conducted sliding tests on graphene on copper with different oxidation degree and found that graphene was more likely to break from the wrinkles when the substrate was oxidized with higher degree. This phenomenon was attributed to that the substrate oxidation decreased the adhesion between graphene and the substrate, which was consistent with the result found by Qi et al. [35].

The Archard formula is the most common theoretical model for wear correlation, in which the wear amount is proportional to the sliding distance and the normal load l [37]. However, the predictions from the Archard law depend on empirical data, which cannot describe the wear problem at the nanoscale. As atomic thick material, the wear amount of graphene at specific parts is not gradual, but "0" or "1", namely, there is wear and there is no wear. The traditional wear model such as the Archard wear mode is obviously inapplicable for graphene. An atom-by-atom attrition model, in which the tip loses individual atoms during the sliding by a thermally activated process, is used to describe the wear issue at the nanoscale [38]. This atom-by-atom removal of silicon tip at the nanoscale is experimentally demonstrated by the Jacobs and Carpick [39]. While the unique two-dimensional structure of graphene with ultra-high covalent bonding is different with the previous studied materials and whether the atom-by-atom attrition model can be applied to graphene remains to be explored. Recently, some researchers have explored the graphene wear issue and help us to better understand the underlying mechanism. Qi et al. [21] had tried to explore graphene wear mechanism by molecular dynamics simulation and the failure mechanism of graphene in-plane in sliding tests is shown in Figure 10.3(a) and Figure 10.3(b). The results conducted by Qi et al. revealed that graphene in-plane should wear due to the in-plane tensile failure when the bonding between graphene and the counterpart was not considered and adhesive wear due to the bonding between graphene and the counterpart when considering the bonding between graphene and the counterpart. Due to the high tensile strength within graphene in-plane, the graphene in-plane could withstand high loads, which was consistent with the above sliding experiments results and the molecular dynamics simulation results of Klemenz et al. [22]. As the graphene step edge, graphene might wear due to the bonding between the counterpart and the graphene step edge or the peeling-induced rupture as shown in Figure 10.3(b). Huang et al. [33] tried to explore graphene wear mechanism in sliding test by finite element simulation and the schematic of the simulation model was shown in Figure 10.3(c), in which a rigid counterpart was pressed against graphene adhered to an elastoplastic substrate without considering the influence of sliding. To incorporate the influence of graphene/substrate interface, two scenarios, namely the tie (the graphene and the substrate were completely bonded) and the frictionless (the graphene was free to slide on the substrate) models were considered in the simulation. In Figure 10.3(d), the maximum compressive contact stress and the maximum in-plane stretching stress of graphene were monitored with the increasing of normal load when 5μm radius counterpart was chosen. As increasing in normal load, the graphene maximum contact stress saturated due to the yielding of substrate while the in-plane stretching stress inside graphene kept increasing and could easily reach the intrinsic strength of graphene. The finite element simulation result also suggested that failure of graphene was primarily caused by the in-plane stretching, which was consistent with the experiment and simulation results of Qi et al. [21] and Klemenz et al. [22].

In general, graphene exhibits excellent anti-wear performance at the nanoscale due to graphene ultra-high mechanical property and intrinsic lubrication effect. The failure of intrinsic graphene should be caused by the in-plane stretching or adhesion between graphene and the counterpart. Graphene anti-wear property at the nanoscale is influenced by many factors: graphene defects such as wrinkles, edges and structures will decrease its anti-wear property; the substrate stiffness and the graphene/substrate interface will influence graphene in-plane stress and result in different graphene wear behavior.

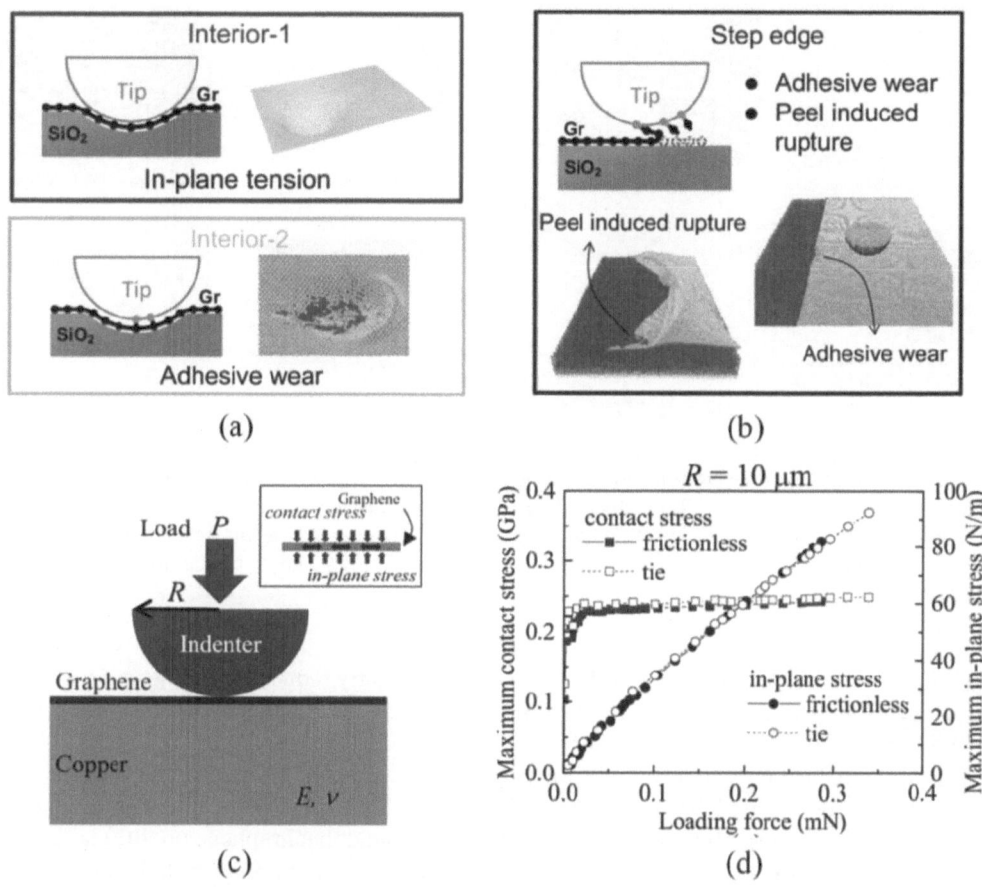

FIGURE 10.3 The wear mechanism of graphene: (a) The in-plane tension induced wear and adhesive induced wear for graphene in-plane revealed by molecular dynamics simulation (Source: Qi, Y et al. 2016); (b) The adhesive induced wear and peel induced rupture for graphene step revealed by molecular dynamics simulation (Source: Qi, Y et al. 2016); (c) The schematic of the finite element simulation model (Source: Huang Y et al. 2017); (d) Maximum contact stress and maximum in-plane 2D stress in graphene as a function of normal load when a radius of 10μm was used for the counterpart. The solid squares and dots are results using a frictionless interface and the open squares and circles are results using a tie interface (Source: Huang Y et al. 2017).

10.3 THE WEAR BEHAVIOR OF GRAPHENE AT THE MACROSCALE

10.3.1 GRAPHENE OFTEN WEARS AT THE MACROSCALE

Mechanically exfoliated graphene is the preferred choice for tribological research at the nanoscale due to its minimum defects. However, the size of mechanically exfoliated graphene is too small to be suitable for the study and application of tribology at the macroscale. CVD can be used to synthesize large area of graphene with high quality and has great application potential, which has attracted wide interest from researchers. Kim et al. [40] studied the tribological characteristic of CVD monolayer and multilayer graphene transferred on SiO₂/Si substrate. The experiment showed that graphene became worn even during the first sliding cycles under normal load ranging from 5mN to 70mN, as shown in Figure 10.4(a). Despite graphene was worn, it could effectively reduce the adhesion and friction of SiO₂/Si substrate. The multilayer graphene grown on nickel (reduce the COF from 0.7 to 0.12) showed a lower coefficient of friction than monolayer graphene grown on copper (reduce the COF from 0.7 to 0.22). The phenomenon was due to the transferred film on the counterpart

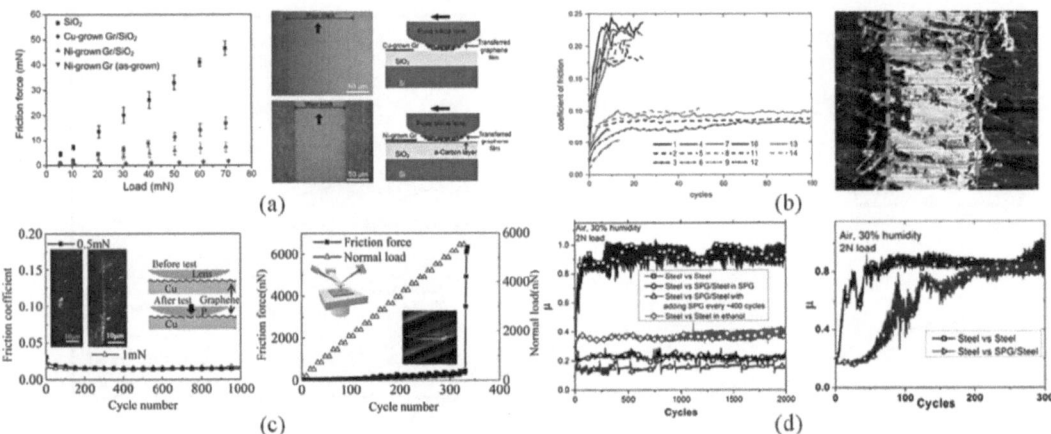

FIGURE 10.4 The graphene wear behavior at the macroscale: (a) Sliding test on CVD grown graphene at macroscale (Source: Kim K et al 2011); (b) Sliding test on epitaxial grown graphene on SiC under 0.1mN at macroscale (Source: Wählisch, F et al); (c) Sliding test on CVD graphene at macroscale and at nanoscale (Source: Huang Y et al. 2017); (d) Sliding test on solution processed graphene on steel at macroscale (Source: Berman D et al 2013).

after graphene wear has the lubrication effect, which significantly reduced the adhesion and friction of SiO₂/Si substrate. An amorphous carbon film formed between graphene and growing substrate nickel and transferred to the counterpart, which was regarded as the reason for the lower COF. Won et al. [41] explored the durability of multilayer graphene on copper at the macroscale and found that graphene completely lost the lubrication effect after a period of time. For the epitaxial graphene on SiC, the wear issue was also obvious. Marchetto et al. [42] found that graphene on SiC started to get worn even during the first cycles under 0.1mN normal load and the friction coefficient of graphene (0.05) was still significantly lower than that of graphite (0.05~0.1) and SiC substrate (0.2~0.3). This phenomenon was attributed to the graphitic interface layer terminating the SiC substrate [42, 43]. While due to the wear of graphene, this low-friction state was rather random and the friction force could be rather high in some cases, as shown in Figure 10.4(b) [43]. Those above experiment results demonstrated that the transferred film after graphene wear had the lubrication effect. Besides the effect of the transferred film, Huang et al. [33] demonstrated that partly worn graphene still had lubrication effect at the macroscale as shown in Figure 10.4(c). After sliding tests under 0.5mN and 1mN normal load for 1000 cycles, the SEM images of the sliding track demonstrated graphene partly wore during tests while in the whole sliding tests the friction coefficient remained below 0.02 at the macroscale. In contrast to the phenomenon at the macroscale, the friction force jumped abruptly when graphene was worn at nanoscale. The comparison suggested that the lubrication effect of worn graphene at the macroscale was due to the remained graphene at the friction contact interface. Huang et al. [33] further studied the wear evolution of monolayer graphene at the macroscale in detail and they found that there existed a critical normal load, below which no detectable wear of graphene was detected. This wearless regime had a 0.02 friction coefficient which was significantly lower than typical macroscale results. Beyond the critical load, graphene started to get worn and the friction increased gradually. The above experiment results show that graphene often wears at macroscale sliding tests and still has some lubrication effect to some extent due to the transferred film on the counterpart and the remaining graphene at the sliding interface. However, the lubrication effect with worn graphene didn't last long, as shown in Figure 10.4(d). The solution-processed graphene with its simple preparation and controllable deposition on arbitrary substrate shows high potential as ultra-thin solid lubricant at the macroscale. Berman et al. [17] found that multilayer solution-processed graphene could greatly reduce the friction coefficient of steel from

0.9 to 0.15 and decrease the wear rate by 3~4 orders of magnitude. However, this lubricating state could only be sustained by continuously replenishing graphene, which was due to the wear of graphene during the sliding test as confirmed by the Raman characterization. The lubrication failure caused by graphene wear was also demonstrated by Won et al. [41]. They focused on the durability of multilayer graphene on copper and found that the friction force would increase and completely lose the lubrication effect within 2000~6000 cycles, as shown in Table 10.1. The limited lubrication effect due to the wear of graphene can further be confirmed by more results by other researchers, as shown in Table 10.1.

Graphene exhibits excellent anti-wear property at the nanoscale while its anti-wear property at the macroscale is far from satisfactory. There're two possible reasons accounting for this phenomenon, as shown in Figure 10.5. First, the counterpart and the substrate both have asperities which are hard to completely avoid at the macroscale and the real local compressive contact stress can be much higher than the Hertzian stress due to these apices. This speculation was verified by Huang et al. work, in which graphene wear marks were inclined to appear at the apices and the ridges of the substrate at first [33]. Second, the contact interface at the macroscale is much larger than that at the nanoscale. Up till now, only the CVD graphene and solution-processed graphene film can reach centimeters wide and meet the dimensional requirement for application at the macroscale. While as shown in Figure 10.5(b), the wrinkles/grain boundaries of the CVD graphene and the steps of the solution-processed graphene, which is in combination of many graphene nanosheets, are hard to avoid in macroscale sliding tests. As shown in Figure 10.2(a) and Figure 10.2(b), graphene is much easier to wear initially from the graphene wrinkles and steps and this may result in the weak anti-wear property of graphene at the macroscale. In general, the unavoidable asperities on the counterpart/substrate and the graphene wrinkles/steps in the friction contact interface at the macroscale are the two main reasons for graphene unsatisfying anti-wear property at the macroscale.

10.3.2 The Monitoring of Graphene Wear at the Macroscale

Although graphene exhibits obvious lubrication effect at macroscale, which may mainly attribute to the remaining graphene and the transferred film in the sliding interface, the wear of graphene itself greatly limits the lubrication life of graphene and leads to the lubrication failure. To avoid lubrication failure which causes the damage of moving mechanical components, monitoring the wear of graphene is critical. Friction force has been widely used to monitor the wear state; however, it's not timely and accurate enough at the macroscale. The experiment by Huang et al. [59] demonstrated that the friction force remained barely changed when graphene was partly worn at the sliding track at the macroscale, as shown in Figure 10.4(c). This is due to the contact area at the macroscale sliding test is much larger than that at the nanoscale sliding test, in which the friction force will jump abruptly when graphene wears. Berman et al. [48] conducted sliding test on quartz with 1μm thick Au using a steel ball covered with 1μm thick titanium nitride layer and measured the friction force and the contact resistance simultaneously when graphene was applied on the substrate as solid lubricant, as shown in Figure 10.6(a). In Figure 10.6(b), as the cycle number increased, the friction force gradually increased from 0.15 to 0.3 in 3000 cycles and the topography image and the Raman mapping of the wear track demonstrated that graphene wore during the sliding tests. While in the whole process, the contact resistance remained almost unchanged. These experiment results showed that the friction force and contact resistance were unable to detect the initial wear of graphene in macroscale sliding tests. Huang et al. [59] conducted sliding test on monolayer graphene at the macroscale and measured the friction force and the graphene in-plane resistance simultaneously, as shown in Figure 10.6(c). Similarly, the friction force increased gradually during the whole sliding test while the graphene in-plane resistance jumped in the first four cycles and maintained almost unchanged till the end as shown in Figure 10.6(b). The SEM image of the sliding track after test demonstrated that graphene at the sliding track wore during the sliding test. The lubrication effect with graphene worn was attributed to the transferred film on the counterpart and led to the gradual

TABLE 10.1

The tribological behavior of graphene at the macroscale

Substrate	Graphene	Counterpart	Environment	Load/N	COF	Lubrication life/cycle	Literature	Interfacial energy
Cu	CVD monolayer	Glass	Air	~0.08	0.02	---	[33]	0.88 ± 0.018 Jm^{-2} [44]
	CVD multilayer	Steel	Air	0.02	0.18	2000~6000	[41]	
Ni	CVD multilayer	Steel	Air	1/2/3/5	0.02/0.04/0.06/0.08	3600~/3600~/3600~/3000	[45]	6.77 ± 0.056 Jm^{-2} [44]
		Ti-6AL-4V	0%/10%/32%/45%RH	1	0.6/0.15/0.15/0.1	1000~	[46]	
		Glass	Air	~0.07	0.03	---	[40]	
		H-DLC/ Steel/TiCN	Air	1	0.08/0.15/0.26	1500~	[47]	
Au	solution-processed multilayer	TiN	Air	1	0.15	2000~	[48]	$Au_{30meV/c}/Ag_{40meV/c}$ ($<Ni_{130meV/c}$ [49]
Ag	solution-processed multilayer	Ag/steel/W	Air	2	0.2/0.4/ 0.5	40000/2700/500	[16]	
Steel	CVD monolayer	Steel	H_2/N_2	1	0.25/0.25	6500/500	[50]	2.270 Jm^{-2} [51]
	solution-processed multilayer	Steel	H_2/N_2	1	0.25/0.25	47000/1200	[50]	
		GC15	N_2	1/2/3/5	0.20	2000~/600/200/20	[52]	
			N_2/30%/60%/90%RH	2	0.18	265/440/620/870	[53]	
Steel / bronze	solution-processed multilayer	Steel	Air	1	0.18/0.15	1000~/ 1000~	[54]	Steel>Cu [51]
SiO_2	CVD monolayer	Glass	Air	~0.005	0.22	---	[55]	$0.45\pm0.02Jm^{-2}$ [56]
	solution-processed multilayer	Si_3N_4/Graphene on Si_3N_4	Air	0.05	0.25/0.1	120~	[57]	
Si/Glass/AlTiC	CVD mono/bilayer	GC15	10%/30%/50%RH	0.5	0.15/0.20/0.25	1200~	[58]	
		Al_2O_3	50-60%RH	0.02	0.2/0.25/0.1~0.35	10000~/10000~/~10000	[15]	---

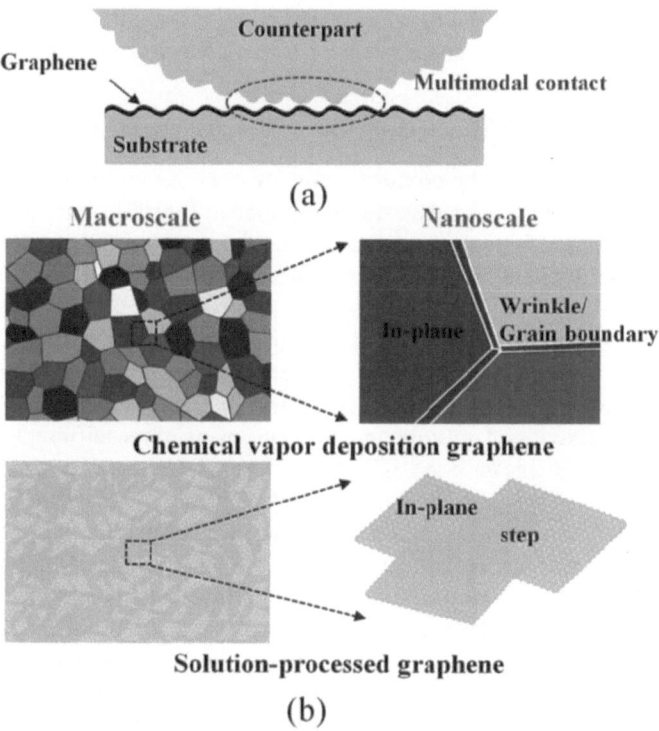

FIGURE 10.5 The possible reason for the weak anti-wear property of graphene at macroscale: (a) The multimodal contact between rough peaks at macroscale sliding tests; (b) The schematic showing the macroscale and nanoscale structure of chemical vapor deposition graphene and solution-processed graphene.

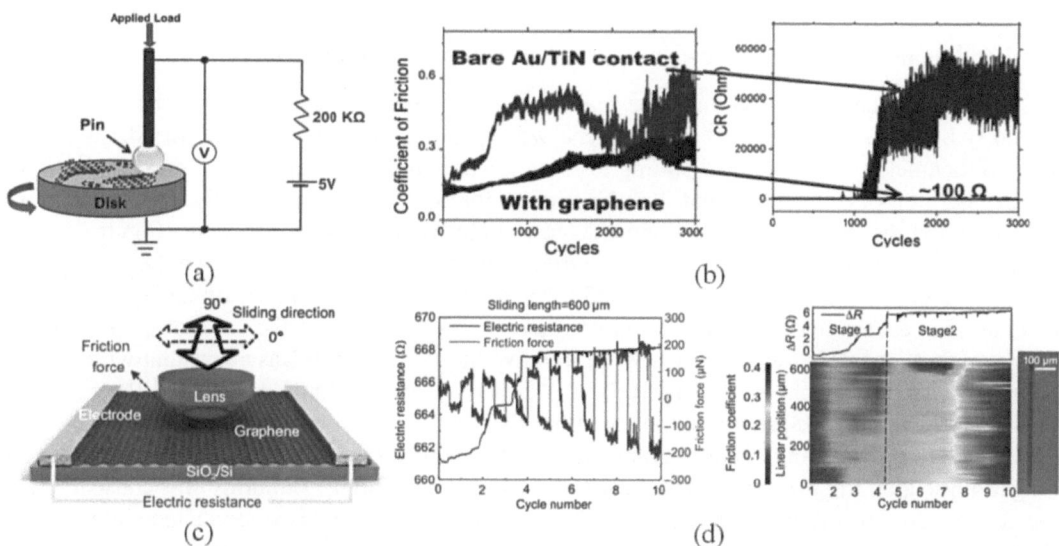

FIGURE. 10.6 The monitoring of graphene wear at macroscale: (a) The setup of measuring the friction and contact resistance (Source: Berman D et al. 2014); (b) The friction and contact resistance with and without graphene during the sliding tests (Source: Berman D et al. 2014); (c) The setup of measuring the friction and graphene in-plane resistance (Source: Huang Y et al. 2020); (d) The friction force and graphene in-plane resistance during the sliding tests (Source: Huang Y et al. 2020).

increase of friction force. Meanwhile the transferred film on the counterpart did not influence the graphene in-plane resistance, which could predict graphene wear prior to the friction force.

10.3.3 THE INFLUENCE OF THE COUNTERPART ON GRAPHENE WEAR BEHAVIOR

The wear behavior of graphene is the comprehensive result of many influence factors such as graphene property, the counterpart, the substrate, the graphene/substrate interface and the test conditions. Studying the influence factors of graphene wear behavior and the underlying mechanism is significant for graphene application as an ultra-thin solid lubricant and the guidance of graphene wear modulation.

Table 10.1 lists the typical results of the tribological behavior of graphene at the macroscale. The counterpart which directly contacts with graphene during the sliding test has great influence on graphene wear behavior at the macroscale. As shown in Table 10.1, Bhowmick et al. [47] tested the friction and wear behavior of multilayer graphene using three different counterparts, which were hydrogenated diamond-like carbon (H-DLC) coatings, TiCN coatings and uncoated steel balls. When using TiCN as the counterpart, the friction coefficient was 0.26 with severe wear while when using H-DLC as the counterpart, the friction coefficient was only 0.08 with no detectable transfer graphene film on the counterpart. The effect of H-DLC in the sliding test might be attributed to the two passivated surfaces, which were H-DLC terminated by H atoms and graphene by H and OH. Mao et al. [16] also studied the influence of the counterpart material on the wear behavior of multilayer graphene on Ag. The results showed that the lubrication life time was as long as 40000 cycles when Ag was used as the counterpart material while that with steel or W were only 2700 and 500 cycles. This phenomenon was attributed to the weakest interaction between Ag and graphene. Wu et al. [57] have used graphene coated Si_3N_4 ball as the counterpart, which exhibited lower friction. However, the graphene on the counterpart wore obviously which might be due to the weak adhesion between graphene and the Si_3N_4 ball. In general, the weak interaction between the counterpart and graphene may slow down the wear process of graphene and extend the lubrication life time of graphene at the macroscale.

10.3.4 THE INFLUENCE OF THE ENVIRONMENT ON GRAPHENE WEAR BEHAVIOR

Besides the influence of the counterpart materials, it has been demonstrated that the environment humidity has considerable effect on graphene wear behavior, as shown in Figure 10.7 and Table 10.1. Bhowmick et al. conducted sliding tests on CVD multilayer graphene on nickel at different relative humidity [46]. They found that the friction force would decrease as the relative humidity increased and the dry nitrogen environment led to high friction and severe wear (Figure 10.7(a)). Huang et al. [33] found that the humidity environment could obviously slow down the wear process of CVD monolayer graphene on copper and extended the life time of lubrication (Figure 10.7(b)). Li et al. [53] found that both the friction coefficient and wear rate of the solution-processed graphene on steel decreased with the increasing relative humidity, which was consistent with the phenomenon on CVD graphene (Figure 10.7(c)). The above experimental results and the molecular dynamics calculation [35] suggested that graphene's humidity sensitivity should be attributed to the passivation of the dangling bonds at graphene edge by chemisorbed H and OH ions from water. The longer lubrication life time or smaller wear rate of graphene at higher humidity was consistent with the atmosphere sensitivity of graphene as Berman et al. found [50]. They conducted sliding test on CVD monolayer graphene and solution-processed multilayer graphene in hydrogen and nitrogen atmosphere [50] and both types of graphene exhibited much longer lubrication life time in hydrogen than in nitrogen and they attributed the extraordinary wear performance to hydrogen passivation of the dangling bonds in the ruptured graphene [50]. Gao et al. [58] found that the response of solution-processed graphene tribological behavior to the environment humidity was also influenced by the oxygen-containing functional groups. Solution-processed graphene with less oxygen-containing

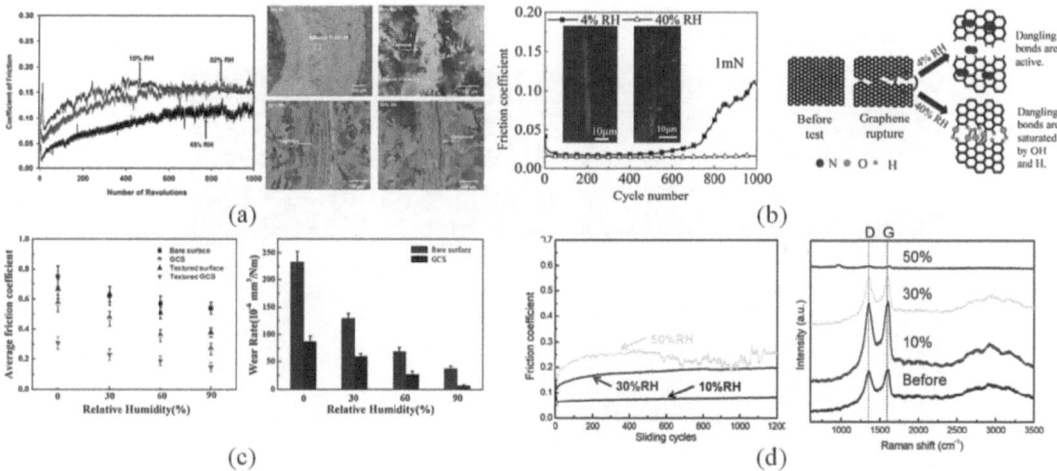

FIGURE 10.7 The influence of the environment humidity on graphene wear behavior at macroscale. (a) Sliding test on multilayer CVD grown graphene under N_2, 10%RH, 32%RH and 45%RH (Source: Bhowmick S et al. 2015); (b) Sliding test on monolayer CVD grown graphene under 4%RH and 40%RH (Source: Huang Y et al. 2017); (c) Sliding test on multilayer solution processed graphene under 0%RH, 30%RH, 60%RH and 90%RH (Source: Li Z et al. 2017); (d) Sliding test on multilayer solution processed graphene under 10%RH, 30%RH and 50%RH (Source: Gao X et al. 2018).

functional groups exhibited similar tribological behavior under varying relative humidity and that with more oxygen-containing functional groups exhibited higher friction and wear rate under higher relative humidity, which might be due to the fact that the water molecules form hydrogen bonds in the interlayer of graphene nanosheets and disrupt the ordered layered structure into amorphous carbon structure [58] (Figure 10.7(d)). Although the humidity has a great effect on graphene wear behavior, the strong adhesion between graphene and the substrate may suppress the humidity influence to some extent. Berman et al. [52] found in dry nitrogen the solution-processed graphene on steel could significantly reduce the friction from 1 to 0.15 and this low-friction state could last more than 2000 cycles, reducing the wear volume by 2 orders of magnitude compared with bare steel. The first principles calculations showed that graphene binds strongly to native iron surfaces highly reducing their surface energy, which might account for the excellent friction-reducing and anti-wear performance of graphene on steel even in dry N_2 [51]. This mechanism was further confirmed by the experiment result by Qi. et al. [35], in which the critical normal load of mechanically exfoliated graphene on plasma-treated SiO_2 substrate was not influenced by the relative humidity. In general, the environment humidity plays an important role on graphene wear behavior at the macroscale while other influence factors such as the graphene structure and the graphene/substrate adhesion should also be considered in combination.

The wear of graphene can be effectively regulated by controlling the environmental conditions. Besides the influence of the environmental humidity and atmosphere on graphene wear behavior, Huang et al. [55] have studied the effect of airborne contaminants on CVD graphene macroscale wear behavior, as shown in Figure 10.8. AFM was used to characterize the new graphene and aged graphene in contact mode in Figure 10.8(a). From both the topographic map and lateral force map, no obvious difference could be found from the first and second scanning at the same location, while there're obvious strips or peaks image on aged graphene, whose intensity decreased at the second scanning. This was due to the mechanical cleaning by the AFM tip [60, 61], which demonstrated the existence of contaminants on the aged graphene. The contaminants on aged graphene were further confirmed by the AFM characterization of the wear track and the XPS spectrum. The aged CVD graphene exhibited better anti-wear properties than the as-prepared samples at the macroscale, as

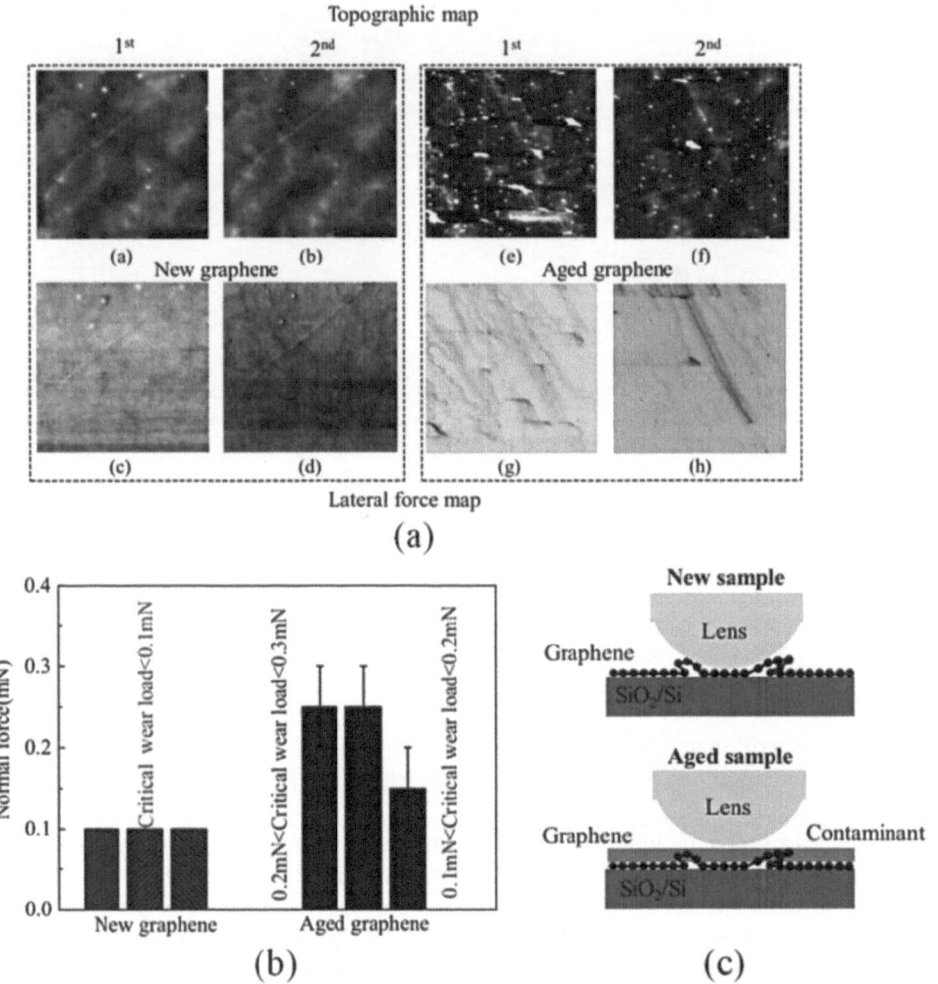

FIGURE. 10.8 The influence of the airborne contaminants on graphene wear behavior at the macroscale (Source: Huang Y et al. 2020). (a) The contaminants on aged graphene demonstrated by AFM; (b) The critical normal load of new graphene and added graphene; (c) Schematic showing the wear mechanism of new and aged graphene.

shown in Figure 10.8(b). As for CVD graphene, the graphene wrinkles were induced in the synthesis [62] and transfer process [24] and were unavoidable in the macroscale sliding length range, which could be seen in the AFM topographic and lateral force map in Figure 10.8(a). The difference on graphene anti-wear property after the airborne contaminants were introduced was attributed to the separation of the counterpart from the graphene wrinkles by the contaminants, from where graphene wear was initiated.

10.3.5 THE INFLUENCE OF THE NORMAL LOAD ON GRAPHENE WEAR BEHAVIOR

Besides the environment, the sliding parameters such as the normal load also plays a role in graphene macroscale wear behavior. As shown in Table 10.1, Berman et al. conducted sliding tests on multilayer solution-processed graphene on steel under 1N, 2N, 3N and 5N normal load in N_2 atmosphere [52]. From the lubrication life we can see that the lubrication effect of the solution-processed graphene was more pronounced at lower loads and lost quickly at higher loads due to the fact that graphene was worn out much more quickly under higher load. This load dependence of graphene wear behavior was also verified on CVD multilayer graphene on Ni, whose friction coefficient increased with the increasing normal load as shown in Table 10.1. The lubrication life time was 3000 cycles under 5N normal load and that under lower load were longer than 3600 cycles [45]. Huang et al. also found that graphene wear was highly related to the normal load and they tried to explore the underlying mechanism by finite element analysis [33]. The simulation result showed that with increasing normal load, the substrate's plastic deformation and the in-plane stress of graphene increased with the increasing normal load, leading to the breakage of graphene initially from the rough peaks eventually and the peeling off and tribochemical reaction of graphene subsequently. Under higher load, graphene is more likely to fracture and the subsequent wear of graphene is accelerated.

10.3.6 THE INFLUENCE OF THE GRAPHENE/SUBSTRATE INTERFACE AND THE SUBSTRATE ON GRAPHENE WEAR BEHAVIOR

The strong adhesion between graphene and the substrate is supposed to suppress the wear of graphene according to the nanoscale research. This method is also applied to modulate graphene wear behavior at the macroscale. Shaojun Qi et al. modified steel by a silane layer (3-Aminopropyl) triethoxysilane (APTES) prior to the assembly of solution-processed graphene and the lubrication life time increased 25 times than the graphene coating without the silane layer [63]. Based on the APTES layer, a layer of dopamine was further added before graphene coating, which exhibited better anti-wear properties than those with only APTES transition layer and this improvement might also be attributed to the increased adhesion between graphene and the substrate [64].

As the influence of substrate on graphene wear behavior, the adhesion between graphene and the substrate also affects graphene wear behavior. From Table 10.1, we can see that the graphene on Ni exhibited smaller or similar COF and longer lubrication life time than that on Cu under much higher load. Kim et al. found that detectable wear of monolayer graphene on Cu was observed under lower load than that on Ni [40]. Besides the difference in the mechanical property, the difference in the adhesion energy of graphene on two substrates was widely believed to be the main cause of this phenomenon. The ultra-thin thickness of graphene made it difficult to measure the adhesion energy of graphene to the substrate. Meanwhile both the results measured by the nano-indenter using the scratch mode [65] and that measured by the blister test [44] showed that the adhesion energy of as-grown graphene on Ni was much higher than that on Cu, which was consistent with the theoretical calculations which showed that graphene weakly adsorbed on Cu, Ag and Au while interacted much stronger including chemisorption on Ni [49]. The first principle calculations showed that graphene bonded strongly to native iron surfaces highly reducing their surface energy, which might account for the excellent friction-reducing and anti-wear performance of graphene on steel [51]. The Raman spectra of the counterpart after sliding test demonstrated that the transferred layer on the counterpart was amorphous carbon when steel was used as the substrate while the transferred layer on the counterpart was still graphene with obvious 2D and G peak when bronze was used as the substrate, indicating that graphene reacted much stronger with steel than with bronze [54]. While due to the limited testing cycles, the tribological behavior of graphene on steel and bronze had no obvious difference under the same conditions as shown in Table 10.1 [54]. What's more, the Ag/graphene/Ag contact exhibited excellent lubrication life under 2N in ambient atmosphere and even more durable under 1N, which maintained lower and stable friction more than 150,000 cycles [16], while the adhesion strength between graphene and Ag was rather weak. The Raman and XPS spectra of the wear track confirmed that there was no reaction between graphene and Ag. In macroscale sliding tests, the tribo-chemistry behavior of graphene might have a great influence whose effect suppressed the effect of graphene/substrate interface. The substrate roughness which influences the adhesion between graphene film and the substrate also played a role on graphene wear behavior. Graphene on the rougher substrate AlTiC that exhibited worse wear resistance than that on smoother Si substrate

[15]. Due to the combined effect of the graphene/substrate interfacial strength, graphene tribo-chemistry behavior and the substrate surface roughness, the influence of the substrate on graphene wear behavior is rather complicated and should be considered in combination with the above factor.

10.4 GRAPHENE-BASED SOLID LUBRICANT AT THE MACROSCALE

Besides graphene exhibited lubrication effect at the macroscale, the combination of graphene and other materials showed even much better friction and wear-reducing effects under specific conditions. The combination of graphene sheets and nanodiamond particles exhibited ultralow friction coefficient of 0.004 and very low wear at the macroscale under 1 N load in dry N_2 with diamond-like carbon (DLC) as the counterpart [66]. The confirmation of graphene nanoscrolls formation by TEM and the molecule dynamics simulation results suggested that this macroscale superlubricity might be due to the fact that graphene sheets wrapped around nanodiamonds to form nanoscrolls, which reduced the real contact area during the sliding test with DLC counterpart and resulted in the ultralow friction and wear. However, there were some constraints to the realization of the macroscale superlu-bricity which was based on the graphene/nanodiamond nanoscrolls. Firstly, it could only achieve in dry N_2 while failed at humidity atmosphere, in which the graphene sheets were more likely to adhere to the counterpart than form the nanoscrolls with nanodiamonds under the effect of water molecules. Secondly, it was unable to realize the macroscale superlubricity when the DLC counterpart was replaced by other materials such as polytetrafluoroethylene polymer, alumina and silicon nitride in the same conditions. It was speculated that the adhesion between graphene and these materials was larger than that between DLC, which resulted that the nanoscrolls structure could not form during the sliding. What's more, this macroscale superlubricity was also load dependent, which preferred the low load and lost under high load. Due to the above restrictions, the practicability and applicability of this nanoscrolls structure based on graphene still needed to be further improved. This macroscale superlubricity was also realized by the combination of MoS_2 nanosheets and nanodiamond particles [67]. Similarly, the CeO_2 nanoparticles were utilized in combination with solution-processed graphene, whose lubrication life time was more than seven times than the film without CeO_2, which was attributed to the fact that the CeO_2 nanoparticles reduced the real contact area during the sliding test and provided facile shearing between the mating surfaces [68]. The combination of graphene and other materials may further improve the lubrication effect of solid lubricant.

10.5 WEAR CHARACTERISTICS OF OTHER 2D MATERIALS AT THE MACROSCALE

Compared to the study on graphene frictional behavior, much less researches have been focused on graphene wear behavior. Although the wear of graphene seems hard to avoid at macroscale, graphene exhibits excellent lubrication effect no matter at nanoscale or macroscale. Other 2D materials such as MoS_2 exhibit fundamental properties and application potential complementary to graphene, while they are much less explored. The AFM experiments demonstrated the high mechanical strength [69] and excellent friction-reducing effect [13, 70] of monolayer MoS_2 at the nanoscale. As shown in Figure 10.9, Huang et al. [71] executed sliding test on monolayer MoS_2 with a 5 mm radius counterpart and found that there existed a critical normal load, below which there was no detect-able wear of monolayer MoS_2 at the macroscale. Under the critical load, monolayer MoS_2 exhibited excellent friction-reducing effect which decreased the friction from 0.7 (on bare SiO_2) to 0.02, and this low-friction state remained stable in the whole sliding test (1000 cycles) without wear of mono-layer MoS_2. What's more, the critical normal load of monolayer MoS_2 was greatly affected by the residual strain in MoS_2, which should be induced in the CVD synthesis and varies in different MoS_2 samples [72]. The experiment and the finite-element simulation results revealed that MoS_2 fractured primarily due to in-plane stretching and the tensile residual strain would add to the in-plane strain in MoS_2, leading to the degradation of MoS_2 antiwear property.

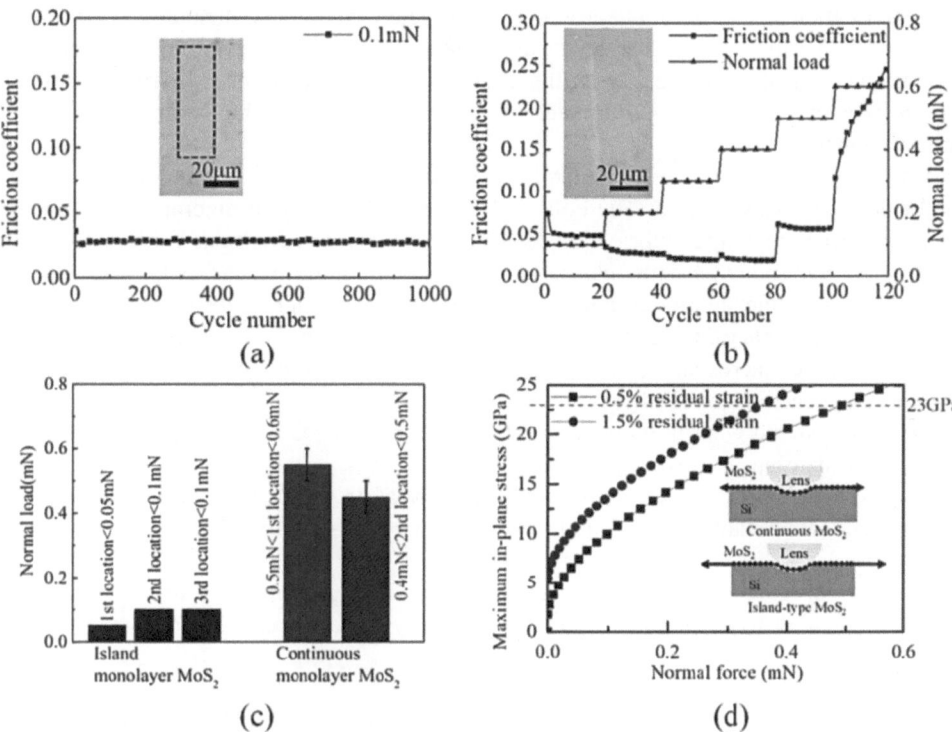

FIGURE 10.9 The wear behavior of monolayer MoS$_2$ at the macroscale: (a) The sliding test on MoS$_2$ under 0.1mN; (b) The sliding test on MoS$_2$ with normal load increasing; (c) The critical normal load of MoS$_2$ with different residual strain; (d) Maximum in-plane 2D stress in MoS$_2$ with 0.5% and 1.5% residual strain as a function of the normal force (Source: Huang Y et al. 2018).

10.6 SUMMARY AND PROSPECTS

With the rapid development of the society and technology, searching for ultra-thin solid lubricants is a goal pursued by researchers in the field of tribology, and graphene has the potential to become a kind of ultra-thin solid lubricants. Currently, most researchers focus on graphene friction behavior at the nanoscale and researches on the wear behavior of graphene are still relatively lacking. In this book, we provide an overview of the recent studies on graphene wear behavior and the underling mechanism in terms of the nanoscale and macroscale. These studies have confirmed the following important findings.

- The intrinsic graphene exhibited excellent anti-wear property, which was mainly attributed to graphene ultra-high mechanical strength brought by the strong covalent bonding between graphene carbon atoms.
- The anti-wear property of graphene at the nanoscale would significantly decrease at graphene wrinkles, step edges and structure defects. Increasing the adhesion between graphene and the substrate could increase graphene anti-wear property to some extent.
- The failure of graphene should be attributed to that the graphene in-plane stress exceeded its fracture strength. While the bonding between graphene and the counterpart, the peeling from the graphene step edges might bring the failure of graphene forward.
- The anti-wear property of graphene at the macroscale was far from satisfactory, which might be due to the asperities-induced high local contact stress, and the unavoidable graphene wrinkles and edges, from where the failure of graphene initiated.

- The lubrication effects of graphene at the macroscale were mainly attributed to the transferred film and the partly remaining graphene at the friction interface. To avoid the lubrication failure due to the wear of graphene, measuring graphene in-plane resistance was demonstrated to be a sensitive and simple method to detect the wear of graphene, which was comparable to the conventional friction force.
- It was desired that graphene exhibited long lubrication life time at the macroscale, which was influenced by many factors: The counterpart material with weak interaction with graphene was preferred for longer lubrication life; The humidity environment could slow down the graphene wear and extend the lubrication life, and the humidity sensitivity of graphene wear behavior was influenced by graphene structure and the graphene/substrate adhesion; The air-borne contaminants on CVD graphene were confirmed to increase the anti-wear property of graphene at the macroscale to some extent; Increasing the graphene/substrate adhesion could extend graphene lubrication life.
- Lower friction and wear at the macroscale could be achieved when graphene was combined with other materials to form rolling friction.
- Monolayer CVD MoS_2 exhibited excellent lubrication effect under wearless state at the macroscale, whose anti-wear property was influenced by its in-plane residual tensile stress.

Although the tribological properties of graphene have been preliminarily understood, the wear behavior and mechanism of graphene are still not fully understood. The following aspects are worthy of attention:

- Although graphene exhibited excellent anti-wear performance at the nanoscale, it often wears at the macroscale and leads to lubrication failure. How to improve the anti-wear property or the lubrication life of graphene at the macroscale is a subject to be explored further.
- Graphene exhibited different wear behavior on different substrates, which were influenced by the substrate stiffness, graphene/substrate adhesion and tribo-chemistry. Figure out those factors on graphene wear behavior is significant to regulate the wear behavior of graphene.

REFERENCES

[1] P. Dašić, F. Franek, E. Assenova, M. Radovanović, International standardization and organizations in the field of tribology, *Industrial Lubrication and Tribology*, 2003, 55(6): 287–291

[2] K. Holmberg, P. Andersson, A. Erdemir, Global energy consumption due to friction in passenger cars, *Tribology International*, 2012, 47: 221–234

[3] R. Wood, M. Williams, A. Kavcic, J. Miles, The feasibility of magnetic recording at 10 terabits per square inch on conventional media, *IEEE Transactions on Magnetics*, 2009, 45(2): 917–923

[4] M.P. De Boer, T.M. Mayer, Tribology of MEMS, *Mrs Bulletin*, 2001, 26(4): 302–304

[5] W. Wang, Y. Wang, H. Bao, B. Xiong, M. Bao, Friction and wear properties in MEMS, *Sensors and Actuators A: Physical*, 2002, 97: 486–491

[6] Dowson D. *History of tribology*. Seconded. London: Professional Engineering Publishing, 1998.

[7] D.B. Asay, M.T. Dugger, S.H. Kim, In-situ vapor-phase lubrication of MEMS, *Tribology Letters*, 2008, 29(1): 67–74

[8] C. Donnet, A. Erdemir, Historical developments and new trends in tribological and solid lubricant coatings, *Surface and coatings technology*, 2004, 180: 76–84

[9] C. Donnet, A. Erdemir, Solid lubricant coatings: recent developments and future trends, *Tribology Letters*, 2004, 17(3): 389–397

[10] J.C. Meyer, A.K. Geim, M.I. Katsnelson, K.S. Novoselov, T.J. Booth, S. Roth, The structure of suspended graphene sheets, *Nature*, 2007, 446(7131): 60–63

[11] C. Lee, X. Wei, J.W. Kysar, J. Hone, Measurement of the elastic properties and intrinsic strength of monolayer graphene, *Science*, 2008, 321(5887): 385–388

[12] T. Cui, S. Mukherjee, P.M. Sudeep, G. Colas, F. Najafi, J. Tam, P.M. Ajayan, C.V. Singh, Y. Sun, T. Filleter, Fatigue of graphene, *Nature Materials*, 2020: 1–7

[13] C. Lee, Q. Li, W. Kalb, X.-Z. Liu, H. Berger, R.W. Carpick, J. Hone, Frictional characteristics of atomically thin sheets, *Science*, 2010, 328(5974): 76–80

[14] S. Zhang, T. Ma, A. Erdemir, Q. Li, Tribology of two-dimensional materials: From mechanisms to modulating strategies, *Materials Today*, 2019, 26: 67–86

[15] N. Dwivedi, T.K. Patra, J.-B. Lee, R.J. Yeo, S. Srinivasan, T. Dutta, K. Sasikumar, C. Dhand, S. Tripathy, M.S. Saifullah, Slippery and Wear Resistant Surfaces Enabled by Interface Engineered Graphene, *Nano Letters*, 2020, 20(2): 905–917

[16] F. Mao, U. Wiklund, A.M. Andersson, U. Jansson, Graphene as a lubricant on Ag for electrical contact applications, *Journal of materials science*, 2015, 50(19): 6518–6525

[17] D. Berman, A. Erdemir, A.V. Sumant, Few layer graphene to reduce wear and friction on sliding steel surfaces, *Carbon*, 2013, 54: 454–459

[18] Y.J. Shin, R. Stromberg, R. Nay, H. Huang, A.T. Wee, H. Yang, C.S. Bhatia, Frictional characteristics of exfoliated and epitaxial graphene, *Carbon*, 2011, 49(12): 4070–4073

[19] L.-Y. Lin, D.-E. Kim, W.-K. Kim, S.-C. Jun, Friction and wear characteristics of multi-layer graphene films investigated by atomic force microscopy, *Surface and Coatings Technology*, 2011, 205(20): 4864–4869

[20] Y. Peng, Z. Wang, K. Zou, Friction and wear properties of different types of graphene nanosheets as effective solid lubricants, *Langmuir*, 2015, 31(28): 7782–7791

[21] Y. Qi, J. Liu, J. Zhang, Y. Dong, Q. Li, Wear resistance limited by step edge failure: the rise and fall of graphene as an atomically thin lubricating material, *ACS applied materials & interfaces*, 2016, 9(1): 1099–1106

[22] A. Klemenz, L. Pastewka, S.G. Balakrishna, A. Caron, R. Bennewitz, M. Moseler, Atomic scale mechanisms of friction reduction and wear protection by graphene, *Nano letters*, 2014, 14(12): 7145–7152

[23] A.N. Obraztsov, E.A. Obraztsova, A.V. Tyurnina, A.A. Zolotukhin, Chemical vapor deposition of thin graphite films of nanometer thickness, *Carbon*, 2007, 45(10): 2017–2021

[24] N. Liu, Z. Pan, L. Fu, C. Zhang, B. Dai, Z. Liu, The origin of wrinkles on transferred graphene, *Nano Research*, 2011, 4(10): 996–1004

[25] B. Vasić, A. Zurutuza, R. Gajić, Spatial variation of wear and electrical properties across wrinkles in chemical vapour deposition graphene, *Carbon*, 2016, 102: 304–310

[26] B. Vasić, A. Matković, R. Gajić, I. Stanković, Wear properties of graphene edges probed by atomic force microscopy based lateral manipulation, *Carbon*, 2016, 107: 723–732

[27] H. Holscher, D. Ebeling, U.D. Schwarz, Friction at atomic-scale surface steps: experiment and theory, *Phys Rev Lett*, 2008, 101(24): 246105

[28] P. Egberts, Z. Ye, X.Z. Liu, Y. Dong, A. Martini, R.W. Carpick, Environmental dependence of atomic-scale friction at graphite surface steps, *Physical Review B*, 2013, 88(3): 035409

[29] A.K. Zhe Chen, Ashlie Martini, Seong H. Kim, Chemical and physical origins of friction on surfaces with atomic steps, *Science advances*, 2019, 5(8):eaaw0513

[30] Y. Peng, X. Zeng, L. Liu, X. Cao, K. Zou, R. Chen, Nanotribological characterization of graphene on soft elastic substrate, *Carbon*, 2017, 124: 541–546

[31] H. Zhang, Z. Guo, H. Gao, T. Chang, Stiffness-dependent interlayer friction of graphene, *Carbon*, 2015, 94: 60–66

[32] Q. Yao, Y. Qi, J. Zhang, S. Zhang, P. Zhao, H. Wang, X.-Q. Feng, Q. Li, Impacts of the substrate stiffness on the anti-wear performance of graphene, *AIP Advances*, 2019, 9(7): 075317

[33] Y. Huang, Q. Yao, Y. Qi, Y. Cheng, H. Wang, Q. Li, Y. Meng, Wear evolution of monolayer graphene at the macroscale, *Carbon*, 2017, 115: 600–607

[34] X. Zeng, Y. Peng, H. Lang, A novel approach to decrease friction of graphene, *Carbon*, 2017, 118: 233–240

[35] Y. Qi, J. Liu, Y. Dong, X.-Q. Feng, Q. Li, Impacts of environments on nanoscale wear behavior of graphene: Edge passivation vs. substrate pinning, *Carbon*, 2018, 139: 59–66

[36] S. Zhao, Z. Zhang, Z. Wu, K. Liu, Q. Zheng, M. Ma, The Impacts of Adhesion on the Wear Property of Graphene, *Advanced Materials Interfaces*, 2019, 6(18): 1900721

[37] J. Archard, Contact and rubbing of flat surfaces, *Journal of applied physics*, 1953, 24(8): 981–988

[38] B. Gotsmann, M.A. Lantz, Atomistic wear in a single asperity sliding contact, *Phys Rev Lett*, 2008, 101(12): 125501

[39] T.D. Jacobs, R.W. Carpick, Nanoscale wear as a stress-assisted chemical reaction, *Nature nanotechnology*, 2013, 8(2): 108–112

[40] K.-S. Kim, H.-J. Lee, C. Lee, S.-K. Lee, H. Jang, J.-H. Ahn, J.-H. Kim, H.-J. Lee, Chemical vapor deposition-grown graphene: the thinnest solid lubricant, *ACS nano*, 2011, 5(6): 5107–5114

[41] M.-S. Won, O.V. Penkov, D.-E. Kim, Durability and degradation mechanism of graphene coatings deposited on Cu substrates under dry contact sliding, *Carbon*, 2013, 54: 472–481

[42] D. Marchetto, C. Held, F. Hausen, F. Wählisch, M. Dienwiebel, R. Bennewitz, Friction and wear on single-layer epitaxial graphene in multi-asperity contacts, *Tribology Letters*, 2012, 48(1): 77–82

[43] F. Wählisch, J. Hoth, C. Held, T. Seyller, R. Bennewitz, Friction and atomic-layer-scale wear of graphitic lubricants on SiC (0001) in dry sliding, *Wear*, 2013, 300(1): 78–81

[44] W. Chang, S. Rajan, B. Peng, C. Ren, M. Sutton, C. Li, Adhesion energy of as-grown graphene on nickel substrates via StereoDIC based blister experiments, *Carbon*, 2019, 153: 699–706

[45] Y. Sun, K. Kandan, S. Shivareddy, F. Farukh, R. Bailey, Effect of sliding conditions on the macroscale lubricity of multilayer graphene coatings grown on nickel by CVD, *Surface and Coatings Technology*, 2019, 358: 247–255

[46] S. Bhowmick, A. Banerji, A.T. Alpas, Role of humidity in reducing sliding friction of multilayered graphene, *Carbon*, 2015, 87: 374–384

[47] S. Bhowmick, A. Banerji, A.T. Alpas, Friction reduction mechanisms in multilayer graphene sliding against hydrogenated diamond-like carbon, *Carbon*, 2016, 109: 795–804

[48] D. Berman, A. Erdemir, A.V. Sumant, Graphene as a protective coating and superior lubricant for electrical contacts, *Applied Physics Letters*, 2014, 105(23): 231907

[49] P. Khomyakov, G. Giovannetti, P. Rusu, G.V. Brocks, J. Van Den Brink, P.J. Kelly, First-principles study of the interaction and charge transfer between graphene and metals, *Physical Review B*, 2009, 79(19): 195425

[50] D. Berman, S.A. Deshmukh, S.K. Sankaranarayanan, A. Erdemir, A.V. Sumant, Extraordinary macroscale wear resistance of one atom thick graphene layer, *Advanced Functional Materials*, 2014, 24(42): 6640–6646

[51] P. Restuccia, M.C. Righi, Tribochemistry of graphene on iron and its possible role in lubrication of steel, *Carbon*, 2016, 106: 118–124

[52] D. Berman, A. Erdemir, A.V. Sumant, Reduced wear and friction enabled by graphene layers on sliding steel surfaces in dry nitrogen, *Carbon*, 2013, 59: 167–175

[53] Z.-Y. Li, W.-J. Yang, Y.-P. Wu, S.-B. Wu, Z.-B. Cai, Role of humidity in reducing the friction of graphene layers on textured surfaces, *Applied Surface Science*, 2017, 403: 362–370

[54] D. Marchetto, P. Restuccia, A. Ballestrazzi, M.C. Righi, A. Rota, S. Valeri, Surface passivation by graphene in the lubrication of iron: A comparison with bronze, *Carbon*, 2017, 116: 375–380

[55] Y. Huang, Q. Li, J. Zhang, Y. Qi, H. Wang, P. Zhao, Y. Meng, Effect of airborne contaminants on the macroscopic anti-wear performance of chemical vapor deposition graphene, *Surface and Coatings Technology*, 2020, 383: 125276

[56] S.P. Koenig, N.G. Boddeti, M.L. Dunn, J.S. Bunch, Ultrastrong adhesion of graphene membranes, *Nature Nanotechnology*, 2011, 6(9): 543–546

[57] P. Wu, X. Li, C. Zhang, X. Chen, S. Lin, H. Sun, C.-T. Lin, H. Zhu, J. Luo, Self-assembled graphene film as low friction solid lubricant in macroscale contact, *ACS applied materials & interfaces*, 2017, 9(25): 21554–21562

[58] X. Gao, L. Chen, L. Ji, X. Liu, H. Li, H. Zhou, J. Chen, Humidity-sensitive macroscopic lubrication behavior of an as-sprayed graphene oxide coating, *Carbon*, 2018, 140: 124–130

[59] Y. Huang, Q. Li, J. Zhang, H. Wang, P. Zhao, Y. Meng, Electric resistance as a sensitive measure for detecting graphene wear during macroscale tribological tests, *Science China Technological Sciences*, 2021, 64(1):179–186.

[60] A.M. Goossens, V.E. Calado, A. Barreiro, K. Watanabe, T. Taniguchi, L.M.K. Vandersypen, Mechanical cleaning of graphene, *Applied Physics Letters*, 2012, 100(7): 073110

[61] N. Lindvall, A. Kalabukhov, A. Yurgens, Cleaning graphene using atomic force microscope, *Journal of Applied Physics*, 2012, 111(6): 064904

[62] X. Li, W. Cai, J. An, S. Kim, J. Nah, D. Yang, R. Piner, A. Velamakanni, I. Jung, E. Tutuc, Large-area synthesis of high-quality and uniform graphene films on copper foils, *Science*, 2009, 324(5932): 1312–1314

[63] S. Qi, X. Li, H. Dong, Improving the macro-scale tribology of monolayer graphene oxide coating on stainless steel by a silane bonding layer, *Materials Letters*, 2017, 209: 15–18

[64] C. Wang, G. Zhang, Z. Li, Y. Xu, X. Zeng, S. Zhao, J. Deng, H. Hu, Y. Zhang, T. Ren, Microtribological properties of Ti6Al4V alloy treated with self-assembled dopamine and graphene oxide coatings, *Tribology International*, 2019, 137: 46–58

[65] S. Das, D. Lahiri, D.-Y. Lee, A. Agarwal, W. Choi, Measurements of the adhesion energy of graphene to metallic substrates, *Carbon*, 2013, 59: 121–129

[66] D. Berman, S.A. Deshmukh, S.K. Sankaranarayanan, A. Erdemir, A.V. Sumant, Macroscale superlubricity enabled by graphene nanoscroll formation, *Science*, 2015, 348(6239): 1118–1122

[67] D. Berman, B. Narayanan, M.J. Cherukara, S. Sankaranarayanan, A. Erdemir, A. Zinovev, A.V. Sumant, Operando tribochemical formation of onion-like-carbon leads to macroscale superlubricity, *Nat Commun*, 2018, 9(1): 1164

[68] G. Bai, J. Wang, Z. Yang, H. Wang, Z. Wang, S. Yang, Self-assembly of ceria/graphene oxide composite films with ultra-long antiwear lifetime under a high applied load, *Carbon*, 2015, 84: 197–206

[69] S. Bertolazzi, J. Brivio, A. Kis, Stretching and breaking of ultrathin MoS2, *ACS nano*, 2011, 5(12): 9703–9709

[70] J. Quereda, A. Castellanos-Gomez, N. Agraït, G. Rubio-Bollinger, Single-layer MoS2 roughness and sliding friction quenching by interaction with atomically flat substrates, *Applied Physics Letters*, 2014, 105(5): 053111

[71] Y. Huang, Q. Yao, Z. Lu, L. Jiao, S. Zhang, Q. Li, Y. Meng, Antiwear Performance of Monolayer MoS2 Modulated by Residual Straining, *ACS Applied Nano Materials*, 2018, 1(12): 7092–7097

[72] S. Kataria, S. Wagner, T. Cusati, A. Fortunelli, G. Iannaccone, H. Pandey, G. Fiori, M.C. Lemme, Growth-Induced Strain in Chemical Vapor Deposited Monolayer MoS2: Experimental and Theoretical Investigation, *Advanced Materials Interfaces*, 2017, 4(17): 1700031

Index